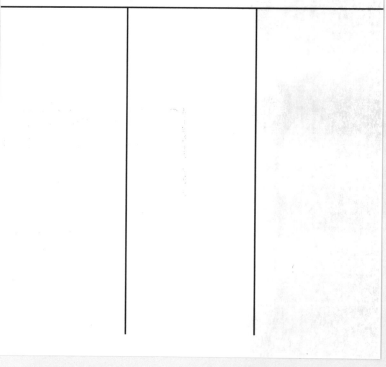

Spectral Methods and Their Applications

Spectral Methods and Their Applications

Guo Ben-Yu

Shanghai University
P R China

Published by

World Scientific Publishing Co. Pte. Ltd.
P O Box 128, Farrer Road, Singapore 912805
USA office: Suite 1B, 1060 Main Street, River Edge, NJ 07661
UK office: 57 Shelton Street, Covent Garden, London WC2H 9HE

Library of Congress Cataloging-in-Publication Data
Guo, Ben-yu, 1942–
Spectral methods and their applications / Guo Ben-yu.
p. cm.
Includes bibliographical references.
ISBN 9810233337 (alk. paper)
1. Spectral theory (Mathematics) I. Title.
QA320.K84 1998
515'.7222--dc21 97-44642
CIP

British Library Cataloguing-in-Publication Data
A catalogue record for this book is available from the British Library.

This book is printed on acid-free paper.

Printed in Singapore by Uto-Print

PREFACE

In the past two decades, spectral methods have developed rapidly. They have become important tools for numerical solutions of partial differential equations, and have been widely applied to numerical simulations in various fields.

The purpose of this book is to present the basic algorithms, the main theoretical results and some applications of spectral methods. Particular attention is paid to the applications to nonlinear problems.

The outline of this book is as follows. Chapter 1 is a colloquial introduction to spectral methods. In Chapter 2, we discuss various orthogonal approximations in Sobolev spaces, used in spectral methods. We also discuss the filterings and recovering the spectral accuracy. Chapter 3 is a survey of the theory of stability and convergence. Chapter 4 consists of two parts. In the first part, we present some basic spectral methods with their applications to nonlinear problems. In the second part, we consider the spectral penalty methods, the spectral viscosity methods and the spectral approximations of isolated solutions. Chapter 5 is devoted to the spectral approximations of multi-dimensional and high order problems. The spectral domain decomposition methods and the spectral multigrid methods are also introduced. We consider the mixed spectral methods for semi-periodic problems in Chapter 6, and some combined spectral methods with their applications in Chapter 7. The final chapter focuses on the spectral methods on the spherical surface.

I am grateful to Prof. Wong, R. S. C. and Dr. Dai, H. H.. They organized the Workshop at City University of Hong Kong, which motivated me to write this book. I thank Dr. Caldwell, J. for his help in polishing up the manuscript. I also thank Misters Chan, H. C. and Cha, K. H. who typed the manuscript.

Guo Ben-yu
Shanghai University
City University of Hong Kong
December 1996

CONTENTS

Spectral Methods and Their Applications

CHAPTER 1

INTRODUCTION

Spectral methods have developed rapidly in the past two decades. They have been applied successfully to numerical simulations in many fields, such as heat conduction, fluid dynamics, quantum mechanics and so on. Nowadays, they are some of the powerful tools as well as finite difference methods and finite element methods, for numerical solutions of partial differential equations.

The main feature of the spectral methods is to take various orthogonal systems of infinitely differentiable global functions as trial functions. Different trial functions lead to different spectral approximations. For instance, trigonometric polynomials for periodic problems, Legendre polynomials and Chebyshev polynomials for non-periodic problems, Laguerre polynomials for problems on the half line, and Hermite polynomials for problems on the whole line.

The fascinating merit of spectral methods is the high accuracy, the so-called convergence of "infinite order". For example, we consider numerical solution of the Laplace equation on a cube, and divide the domain into N^3 uniform subcubes. If we use linear finite element methods, then the error in the L^2-norm between the genuine solution and the numerical one is of order N^{-2}, no matter how smooth the genuine solution is. If we use the standard "nine points" difference scheme with the uniform mesh size $\frac{1}{N}$, then the error is of the same order. However, the smoother the genuine solution, the higher the convergence rate of the numerical solution by spectral methods. In particular, when the genuine solution is infinitely differentiable, the numerical one converges faster than $N^{-\alpha}, \alpha$ being any positive constant.

The basic idea of spectral methods stems from Fourier analysis. In 1820, Navier considered a thin plate problem governed by the equation

$$\frac{\partial^4 U}{\partial x^4} + 2\frac{\partial^4 U}{\partial^2 x \partial^2 y} + \frac{\partial^4 U}{\partial y^4} = f, \qquad 0 < x,\, y < \pi, \tag{1.1}$$

with the boundary conditions

$$U = \frac{\partial^2 U}{\partial x^2} = 0, \quad x = 0, \pi, \tag{1.2}$$

$$U = \frac{\partial^2 U}{\partial y^2} = 0, \quad y = 0, \pi. \tag{1.3}$$

Suppose that $f(x,y)$ can be expanded as

$$f(x,y) = \sum_{l=1}^{\infty}\sum_{m=1}^{\infty} f_{l,m} \sin lx \sin my.$$

By the orthogonality of trigonometric functions,

$$f_{l,m} = \frac{4}{\pi^2}\int_0^{\pi}\int_0^{\pi} f(x,y) \sin lx \sin my \, dx dy.$$

We look for the solution as

$$U(x,y) = \sum_{l=1}^{\infty}\sum_{m=1}^{\infty} U_{l,m} \sin lx \sin my. \tag{1.4}$$

Obviously it satisfies the boundary conditions (1.2) and (1.3). By substituting (1.4) into (1.1) and using the orthogonality again, we deduce that

$$U_{l,m} = \frac{f_{l,m}}{l^4 + 2l^2 m^2 + m^4}$$

and so

$$U(x,y) = \sum_{l=1}^{\infty}\sum_{m=1}^{\infty} \frac{f_{l,m}}{l^4 + 2l^2 m^2 + m^4} \sin lx \sin my. \tag{1.5}$$

Finally we truncate this series to obtain the approximate solution

$$u_N(x,y) = \sum_{l=1}^{N}\sum_{m=1}^{N} U_{l,m} \sin lx \sin my$$

which is a solution of the Fourier spectral method for problem (1.1)–(1.3).

Spectral methods can also be applied to evolutionary problems. Consider the heat conduction problem

$$\begin{cases} \dfrac{\partial U}{\partial t} = \dfrac{\partial^2 U}{\partial x^2}, & 0 < x < \pi,\, t > 0, \\ U(0,t) = U(\pi,t) = 0, & t > 0, \\ U(x,0) = U_0(x), & 0 \le x \le \pi \end{cases} \tag{1.6}$$

where $U_0(x)$ is a given continuous function and vanishes at $x = 0, \pi$. Let

$$U(x,t) = \sum_{l=1}^{\infty} U_l(t) \sin lx \tag{1.7}$$

and substitute this expansion into (1.6), we find that

$$\begin{cases} \dfrac{dU_l}{dt} = -l^2 U_l, \\ U_l(0) = U_{0,l}, \quad l = 1, 2, \dots, \end{cases} \tag{1.8}$$

where $U_{0,l}$ is the Fourier coefficient of the initial state $U_0(x)$,

$$U_{0,l} = \frac{2}{\pi} \int_0^{\pi} U_0(x) \sin lx \, dx. \tag{1.9}$$

By (1.8),

$$U_l(t) = U_{0,l} e^{-l^2 t}.$$

We now truncate the series (1.7) and denote the corresponding one by $u_N(x,t)$,

$$u_N(x,t) = \sum_{l=1}^{N} U_l(t) \sin lx.$$

This is an exceedingly good approximation to (1.7) for any $t > 0$. In fact,

$$\begin{aligned} |U(x,t) - u_N(x,t)| &= \Big| \sum_{l=N+1}^{\infty} U_{0,l} e^{-l^2 t} \sin lx \Big| \\ &\le 2 \max_{0 \le x \le \pi} |U_0(x)| \int_N^{\infty} e^{-ty^2} \, dy. \end{aligned}$$

Thus the error goes to zero very rapidly as $N \to \infty$.

There are several kinds of spectral methods, called spectral methods, pseudospectral methods and Tau methods, respectively. All of them can be derived from the method of weighted residuals. Let Ω be a spatial domain with the boundary $\partial\Omega$, and $\bar{\Omega} = \Omega \cup \partial\Omega$. We consider here an initial-boundary value problem as follows

$$\begin{cases} LU = f, & x \in \Omega, \, t > 0, \\ BU = 0, & x \in \partial\Omega, \, t > 0, \\ U(x,0) = U_0(x), & x \in \bar{\Omega} \end{cases} \tag{1.10}$$

where L is a differential operator and B is a linear boundary operator, $f(x,t)$ and $U_0(x)$ are given functions. Assume that some conditions are fulfilled for ensuring the existence, uniqueness and regularity of solution of (1.10). The method of weighted residuals is to find an approximate solution of the form

$$u_N(x,t) = u_B(x,t) + \sum_{l=0}^{N} u_{N,l}(t)\phi_l(x) \tag{1.11}$$

where the trial functions $\phi_l(x), 0 \le l \le N$ are linearly independent, $u_B(x,t)$ is chosen in such a way that $u_N(x,t)$ fits the boundary conditions. The unknown coefficients $u_{N,l}(t), 0 \le l \le N$, are determined by the equation

$$\begin{cases} \int_\Omega (Lu_N(x,t) - f(x,t))\,\psi_j(x)\,dx = 0, & 0 \le j \le N,\ t > 0, \\ \int_\Omega (u_N(x,0) - U_0(x))\,\psi_j(x)\,dx = 0, & 0 \le j \le N \end{cases} \tag{1.12}$$

where the weight functions $\psi_j(x), 0 \le j \le N$, are linearly independent.

We first derive the spectral methods. In this case, each $\phi_l(x)$ satisfies the boundary conditions and so $u_B(x,t) \equiv 0$. In addition, $\psi_j(x) = \phi_j(x)$. For simplicity, we introduce the inner product and the norm of the space $L^2(\Omega)$, i.e., for any $v, w \in L^2(\Omega)$,

$$(v,w) = \int_\Omega v(x)w(x)\,dx, \qquad \|v\| = (v,v)^{\frac{1}{2}}.$$

Accordingly, (1.12) stands for

$$\begin{cases} (Lu_N(t), \phi_l) = (f(t), \phi_l), & 0 \le l \le N, t > 0, \\ (u_N(0), \phi_l) = (U_0, \phi_l), & 0 \le l \le N. \end{cases} \tag{1.13}$$

It is more convenient to describe scheme (1.12) by a projection operator P_N. To do this, we set a finite dimensional space V_N as

$$V_N = \text{span}\,\{\phi_l \mid 0 \le l \le N\}$$

which is called the trial function space. For any $v \in L^2(\Omega)$, the L^2-projection $P_N v \in V_N$, satisfies

$$(v - P_N v, \phi_l) = 0, \quad 0 \le l \le N.$$

Since $\phi_l(x), 0 \le l \le N$ are linearly independent, $P_N v$ is determined uniquely. By using such notations, scheme (1.13) is equivalent to

$$\begin{cases} P_N L u_N(x,t) = P_N f(x,t), & t > 0, \\ u_N(x,0) = P_N U_0(x). \end{cases} \tag{1.14}$$

We next deduce the pseudospectral methods. In this case, $\phi_l(x)$ are the same as in spectral methods and $u_B(x,t) \equiv 0$. We choose suitable collocation points $x^{(j)}, 0 \le j \le N$, such that

$$\begin{vmatrix} \phi_0(x^{(0)}) & \cdots & \phi_0(x^{(N)}) \\ \cdots & \cdots & \cdots \\ \phi_N(x^{(0)}) & \cdots & \phi_N(x^{(N)}) \end{vmatrix} \neq 0. \tag{1.15}$$

The weight functions are

$$\psi_j(x) = \delta(x - x^{(j)}), \quad 0 \le j \le N$$

where $\delta(x)$ is the Dirac delta function. Then (1.12) stands for

$$\begin{cases} Lu_N(x^{(j)},t) = f(x^{(j)},t), & 0 \le j \le N,\ t > 0, \\ u_N(x^{(j)},0) = U_0(x^{(j)}), & 0 \le j \le N. \end{cases} \tag{1.16}$$

It is also more convenient to express scheme (1.16) by an interpolation operator I_N. For any $v \in C(\bar{\Omega})$, the interpolant $I_N v \in V_N$, satisfies

$$I_N v(x^{(j)}) = v(x^{(j)}), \quad 0 \le j \le N.$$

By (1.15), the interpolant $I_N v$ is determined uniquely. Thus (1.16) is equivalent to

$$\begin{cases} I_N L u_N(x,t) = I_N f(x,t), & t > 0, \\ u_N(x,0) = I_N U_0(x). \end{cases} \tag{1.17}$$

There is another way to express pseudospectral methods. For instance, if $x^{(j)}$ are distributed uniformly and if we define the discrete inner product as

$$(v,w)_N = \frac{1}{N+1} \sum_{j=0}^{N} v(x^{(j)}) w(x^{(j)}),$$

then (1.16) and (1.17) are equivalent to

$$\begin{cases} (Lu_N(t), \phi_l)_N = (f(t), \phi_l)_N\,, & 0 \le l \le N,\ t > 0, \\ (u_N(0), \phi_l)_N = (U_0, \phi_l)_N\,, & 0 \le l \le N. \end{cases} \tag{1.18}$$

We now turn to the Tau methods. Suppose that $\phi_l(x), 0 \le l \le N$ are orthogonal in $L^2(\Omega)$, but do not fulfill the boundary conditions. The component $u_B(x,t)$ in (1.11) is of the form

$$u_B(x,t) = \sum_{l=N+1}^{N+k} u_{N,l}(t)\phi_l(x)$$

where k is the number of independent boundary conditions. We take $\psi_j(x) = \phi_j(x), 0 \le l \le N$. Then (1.12) is read as

$$\begin{cases} (Lu_N(t), \phi_l) = (f(t), \phi_l), & 0 \le l \le N, t > 0, \\ (u_N(0), \phi_l) = (U_0, \phi_l), & 0 \le l \le N \end{cases} \tag{1.19}$$

with k additional equations given by the boundary conditions. The Tau methods can also be described by a projection operator, namely

$$\begin{cases} P_N L u_N(x,t) = P_N f(x,t), & t > 0, \\ u_N(x,0) = P_N U_0(x) \end{cases} \tag{1.20}$$

with k additional equations on $\partial\Omega$.

In practice, mostly used methods for problems with non-homogeneous boundary conditions are some combinations of pseudospectral methods with Tau methods. It means that partial differential equations are approximated by collocation approach in which $\phi_l(x), 0 \le l \le N$, do not satisfy the boundary conditions, and the boundary conditions are treated as in Tau methods.

When we solve initial-boundary value problems numerically, we need to discretize the derivatives of unknown functions with respect to time t. Denote by $[T]$ the integer part of any fixed positive constant T. Let τ be the mesh size for the variable t and

$$R_\tau(T) = \left\{ t = k\tau \;\middle|\; k = 1, 2, \dots, \left[\frac{T}{\tau}\right] \right\}, \quad \bar{R}_\tau(T) = R_\tau(T) \cup \{0\}. \tag{1.21}$$

We use the following notations

$$\begin{aligned} \hat{v}(x,t) &= \frac{1}{2}v(x,t+\tau) + \frac{1}{2}v(x,t-\tau), \\ D_\tau v(x,t) &= \frac{1}{\tau}\Big(v(x,t+\tau) - v(x,t)\Big), \\ \bar{D}_\tau v(x,t) &= \frac{1}{\tau}\Big(v(x,t) - v(x,t-\tau)\Big), \\ \hat{D}_\tau v(x,t) &= \frac{1}{2\tau}\Big(v(x,t+\tau) - v(x,t-\tau)\Big), \\ \hat{D}_{\tau\tau} v(x,t) &= \frac{1}{\tau^2}\Big(v(x,t+\tau) - 2v(x,t) + v(x,t-\tau)\Big). \end{aligned}$$

$D_\tau v$ and $\bar{D}_\tau v$ approximate $\dfrac{\partial v}{\partial t}$ at time t with the truncation error of order $\mathcal{O}(\tau)$, provided that $\left|\dfrac{\partial^2 v}{\partial t^2}\right|$ is bounded. $\hat{D}_\tau v$ approximates $\dfrac{\partial v}{\partial t}$ with the error $\mathcal{O}(\tau^2)$, when

$\left|\frac{\partial^3 v}{\partial t^3}\right|$ is bounded. Similarly, $\hat{D}_{\tau\tau}v$ approximates $\frac{\partial^2 v}{\partial t^2}$ with the error $\mathcal{O}(\tau^2)$, as long as that $\left|\frac{\partial^4 v}{\partial t^4}\right|$ is bounded. But $D_\tau v$ approximates $\frac{\partial v}{\partial t}$ at $t+\frac{\tau}{2}$ with the truncation error $\mathcal{O}(\tau^2)$. It can be verified that

$$2\Big(D_\tau v(t), v(t)\Big) = D_\tau \|v(t)\|^2 - \tau\|D_\tau v(t)\|^2, \tag{1.22}$$

$$2\left(\hat{D}_\tau v(t), \hat{v}(t)\right) = \hat{D}_\tau \|v(t)\|^2, \tag{1.23}$$

$$2\left(\hat{D}_{\tau\tau} v(t), \hat{D}_\tau v(t)\right) = D_\tau \|\bar{D}_\tau v(t)\|^2. \tag{1.24}$$

As an example, we consider the problem

$$\begin{cases} \dfrac{\partial U}{\partial t} = \dfrac{\partial^2 U}{\partial x^2} + f, & 0 < x < \pi,\ 0 < t \le T, \\ U(0,t) = U(\pi,t) = 0, & 0 < t \le T, \\ U(x,0) = U_0(x), & 0 \le x \le \pi \end{cases} \tag{1.25}$$

where $f(x,t)$ and $U_0(x)$ are given continuous functions, and $U_0(x)$ vanishes at $x = 0, \pi$. We take $\phi_l(x) = \sin lx$ in (1.13), and let

$$u_N(x,t) = \sum_{l=1}^{N} u_{N,l}(t) \sin lx.$$

A spectral scheme of Crank-Nicolson type for (1.25) is

$$\begin{cases} \Big(D_\tau u_N(t), \phi_l\Big) = \Big(\dfrac{1}{2}\dfrac{\partial^2 u_N}{\partial x^2}(t+\tau) + \dfrac{1}{2}\dfrac{\partial^2 u_N}{\partial x^2}(t) \\ \qquad\qquad +f(x, t+\dfrac{\tau}{2}), \phi_l\Big), & 0 \le l \le N,\ t \in \bar{R}_\tau(T), \\ \Big(u_N(0), \phi_l\Big) = \Big(U_0, \phi_l\Big), & 0 \le l \le N. \end{cases} \tag{1.26}$$

Further, by the orthogonality of trigonometric functions, we obtain from (1.26) that

$$\begin{cases} D_\tau u_{N,l}(t) = -\frac{l^2}{2} u_{N,l}(t+\tau) - \frac{l^2}{2} u_{N,l}(t) + f_l\left(t + \frac{\tau}{2}\right), & 0 \le l \le N, t \in \bar{R}_\tau(T), \\ u_{N,l}(0) = U_{0,l}, & 0 \le l \le N \end{cases}$$

where $U_{0,l}$ is the same as in (1.9), and

$$f_l\left(t + \frac{\tau}{2}\right) = \frac{2}{\pi}\int_0^\pi f\left(x, t + \frac{\tau}{2}\right) \sin lx \, dx.$$

Finally

$$u_{N,l}(t+\tau) = \left(1+\frac{\tau l^2}{2}\right)^{-1}\left(\left(1-\frac{\tau l^2}{2}\right)u_{N,l}(t)+\tau f_l\left(t+\frac{\tau}{2}\right)\right), \quad t \in \bar{R}_\tau(T).$$

Therefore we can evaluate the coefficients $u_{N,l}(t), 0 \le l \le N$, $t \in \bar{R}_\tau(T)$ explicitly. This is indeed another advantage of spectral methods. Conversely, if a finite difference scheme of Crank-Nicolson type is used for (1.25), then we have to solve a linear algebraic system for the values of unknown function at each time step. It is also true for any finite element scheme of Crank-Nicolson type. On the other hand, the matrices occurring in finite element methods may weaken the stability of computation. Sometimes, the "lumping" technique is used to overcome this trouble, but it usually lowers the accuracy.

Even though the spectral methods have many advantages, they were not used widely for a long time. The main reason is the expensive cost of computational time. However, the discovery of the Fast Fourier Transformation (**FFT**), see Cooley and Tukey (1965), removed this obstacle. Let N be the number of terms in one-dimensional Fourier expansion. Then **FFT** reduces the total number of operations from $\mathcal{O}(N^2)$ to $\mathcal{O}(N\log_2 N)$. Especially, **FFT** saves a lot of work for multi-dimensional problems, since trial functions in several dimensions are the products of several trial functions in one dimension. For example, we consider the summation

$$\begin{aligned}&\sum_{l=0}^{N}\sum_{m=0}^{N}\sum_{n=0}^{N} a_{l,m,n}\exp\left\{2\pi i\left(\frac{lp}{N+1}+\frac{mq}{N+1}+\frac{nr}{N+1}\right)\right\}\\ &= \sum_{l=0}^{N} e^{\frac{2\pi lpi}{N+1}}\left(\sum_{m=0}^{N} e^{\frac{2\pi mqi}{N+1}}\left(\sum_{n=0}^{N} a_{l,m,n}e^{\frac{2\pi nri}{N+1}}\right)\right).\end{aligned}$$

If we use the usual method to evaluate the left-hand side of the above formula, then the total number of operations is $\mathcal{O}(N^6)$. In contrast, it is reduced to $\mathcal{O}(3N^2\log_2 N)$, if the right-hand side is calculated by **FFT**. Since Chebyshev polynomials can be changed into trigonometric polynomials by a transformation of independent variables, **FFT** is also available for Chebyshev spectral approximations. Furthermore, the coming of the Fast Legendre Transformation (**FLT**), see Alpert and Rokhlin (1991), and the possible parallelization of **FFT** and **FLT** make the spectral methods more efficient.

The first serious application of the spectral methods to partial differential equations was due to Silberman (1954). However, they have become practical only after Orszag (1969, 1970), and Eliasen, Machenhauer and Rasmussen (1970) developed transform technique for evaluating convolution sums arising from quadratic non-linearities. The collocation approach was first used by Slater (1934) and Kantorovich (1934) in specific applications, and the foundation of orthogonal collocation was laid by Lanczos (1938). But the earliest application of the pseudospectral methods to partial differential equations was made by Kreiss and Oliger (1972), and Orszag (1972). Lanczos (1938) also developed the Tau methods. Gottlieb and Orszag (1977) summarized the state of art in the theory and application of spectral methods. They provided numerical analysis for linear problems. Hald (1981), Maday and Quarteroni (1981, 1982a, 1982b), and Bernardi, Canuto and Maday (1986) considered nonlinear problems. While Guo (1981a, 1985) and Kuo (1983) developed spectral methods for nonlinear problems with numerical analysis independently. There have been numerous introductory and review articles, e.g., Mercier (1981), Hussaini, Salas and Zang (1985), Gottlieb (1985), and Quarteroni (1991). Some new developments in this field are contained in the books by Canuto, Hussaini, Quarteroni and Zang (1988), by Bernardi and Maday (1992a), and by Guo (1996). On the other hand, applications of spectral methods in meteorology have been covered in the review by Jarraud and Baede (1985), and in the book by Haltiner and Williams (1980).

CHAPTER 2

ORTHOGONAL APPROXIMATIONS IN SOBOLEV SPACES

The remarkable convergences of spectral methods owe much to the rapid convergences of expansions in series of orthogonal systems of smooth functions. We present a summary of the relevant theory in this chapter. Fourier approximation, Legendre approximation, Chebyshev approximation, Laguerre approximation and Hermite approximation are considered. The error estimations of different projections in Sobolev spaces are discussed. We also introduce the filterings which are used to improve the stability of pseudospectral methods for nonlinear partial differential equations. In order to recover the spectral accuracy of orthogonal expansions for discontinuous functions, some essentially nonoscillatory approximations and reconstruction methods are used.

2.1. Preliminaries

Let $\mathbb{R}^n$ be the n-dimensional Euclidean space and $x = (x_1, x_2, \dots, x_n) \in \mathbb{R}^n$. Denote by Ω a bounded open domain in $\mathbb{R}^n$ with the boundary $\partial\Omega$. If for any $x \in \partial\Omega$, there exists a system of orthogonal coordinates $y = (y_1, y_2, \dots, y_n)$, a hypercube $U_x = \sqcap_{q=1}^{n} [-a_q, a_q]$ and a Lipschitz continuous mapping Φ_x from $\sqcap_{q=1}^{n-1} [-a_q, a_q]$ into $[-\frac{1}{2}a_n, \frac{1}{2}a_n]$ such that

$$
\begin{aligned}
\Omega \cap U_x &= \{y \in U_x, y_n > \Phi_x(y_1, \dots, y_{n-1})\}, \\
\partial\Omega \cap U_x &= \{y \in U_x, y_n = \Phi_x(y_1, \dots, y_{n-1})\},
\end{aligned}
$$

then we say that $\partial\Omega$ is Lipschitz continuous. Further, let m be a non-negative integer. A relatively open part $\Gamma \in \partial\Omega$ is said to be of class $C^{m,1}$, if for any $x \in \Gamma$, we can choose such a mapping Φ_x that its derivatives up to the order m are Lipschitz continuous. In this book, we assume that $\partial\Omega$ is at least Lipschitz continuous, and $\bar{\Omega} = \Omega \cup \partial\Omega$.

Let $\mathcal{D}(\Omega)$ be the space of infinitely differentiable functions with compact supports in Ω, and $C^\infty\left(\bar{\Omega}\right)$ be the space of infinitely differentiable functions on $\bar{\Omega}$. The dual space $\mathcal{D}'(\Omega)$ of $\mathcal{D}(\Omega)$ is the space of distributions on Ω. By the theory of distributions (see Schwarz (1966)), for any $v \in \mathcal{D}'(\Omega)$, the associated partial derivatives of any order exist. The notation ∂_q stands for $\dfrac{\partial}{\partial x_q}$, and $\nabla = (\partial_1, \partial_2, \ldots, \partial_n)$. Let $k_q, 1 \le q \le n$ be non-negative integers. For any n-tuple $k = (k_1, k_2, \ldots, k_n)\,, |k| = k_1 + k_2 + \cdots + k_n$. The symbol $\partial_x^k = \partial_1^{k_1}\partial_2^{k_2}\cdots\partial_n^{k_n}$. When $n = 1, \partial_x = \partial_1$. In addition, $\partial_t = \dfrac{\partial}{\partial t}$, etc..

We now introduce the Sobolev spaces. For any real number $p, 1 \le p \le \infty$, let

$$L^p(\Omega) = \left\{v \mid v \text{ is measurable and } \|v\|_{L^p(\Omega)} < \infty\right\}$$

with the norm

$$\begin{aligned} \|v\|_{L^p(\Omega)} &= \left(\int_\Omega |v(x)|^p\,dx\right)^{\frac{1}{p}}, & 1 \le p < \infty, \\ \|v\|_{L^p(\Omega)} &= \operatorname*{ess\,sup}_{x\in\Omega}|v(x)|, & p = \infty. \end{aligned}$$

$L^p(\Omega)$ is a Banach space. In particular, $L^2(\Omega)$ is a Hilbert space equipped with the inner product

$$(v, w)_{L^2(\Omega)} = \int_\Omega v(x)\bar{w}(x)\,dx, \qquad \forall\ v, w \in L^2(\Omega)$$

where $\bar{w}$ is the conjugate of w, if w is a complex-valued function. Since $L^2(\Omega)$ contains $\mathcal{D}(\Omega)$ and $C^\infty\left(\bar{\Omega}\right)$ as dense subspaces, the duality pairing between $\mathcal{D}'(\Omega)$ and $\mathcal{D}(\Omega)$ is an extension of the inner product in $L^2(\Omega)$.

Next, let m be any non-negative integer, and set

$$W^{m,p}(\Omega) = \left\{v \;\middle|\; \partial_x^k v \in L^p(\Omega),\ \ |k| \le m\right\}.$$

It is provided with the following semi-norm and norm,

$$|v|_{W^{m,p}(\Omega)} = \left(\sum_{|k|=m} \|\partial^k v\|^p_{L^p(\Omega)}\right)^{\frac{1}{p}}, \tag{2.1}$$

$$\|v\|_{W^{m,p}(\Omega)} = \left(\sum_{l=0}^{m} |v|^p_{W^{l,p}(\Omega)}\right)^{\frac{1}{p}}. \tag{2.2}$$

$W^{m,p}(\Omega)$ is a Banach space. In particular, $W^{m,2}(\Omega) = H^m(\Omega)$. It is a Hilbert space for the associated inner product

$$(v,w)_{H^m(\Omega)} = \sum_{|k|\leq m} \left(\partial_x^k v, \partial_x^k w\right)_{L^2(\Omega)}, \quad \forall\, v,w \in H^m(\Omega).$$

In order to define the space $W^{r,p}(\Omega)$ for any non-negative number r, we need the theory of space interpolation. Firstly, let B be a Banach space with the norm $\|\cdot\|_B$. For any real numbers a and b, define the Banach space

$$L^p(a,b;B) = \{f \mid f:(a,b)\to B \text{ is strongly measurable}, \ \ \|f(\cdot)\|_B \in L^p(a,b)\}$$

with the norm

$$\|f\|_{L^p(a,b;B)} = \left(\int_a^b \|f(t)\|_B^p\,dt\right)^{\frac{1}{p}}, \quad 1\leq p<\infty,$$

$$\|f\|_{L^p(a,b;B)} = \operatorname*{ess\,sup}_{t\in(a,b)} \|f(t)\|_B, \quad p=\infty$$

where the integral on the interval (a,b) is the Bochner integral. Now let B_1 and B_2 be two Banach spaces. B_1 is continuously imbedded and dense in B_2. For any real numbers ν and $p\geq 1$, we denote by $W(\nu,p;B_1,B_2)$ the space containing all functions f such that

$$t^\nu f \in L^p(0,\infty;B_1), \qquad t^\nu \partial_t f \in L^p(0,\infty;B_2).$$

This is also a Banach space with the norm

$$\|f\|_{W(\nu,p;B_1,B_2)} = \max\left(\|t^\nu f\|_{L^p(0,\infty;B_1)},\ \|t^\nu \partial_t f\|_{L^p(0,\infty;B_2)}\right).$$

Assume $\theta = \frac{1}{p}+\nu<1$. Denote by $T(\nu,p;B_1,B_2)$ the set of initial values of functions in $W(\nu,p;B_1,B_2)$. It is also a Banach space with the norm

$$\|v\|_{T(\nu,p;B_1,B_2)} = \inf_{\substack{f\in W(\nu,p;B_1,B_2)\\ f(0)=v}} \|f\|_{W(\nu,p;B_1,B_2)}.$$

If $0<\theta<1$, then for any $v\in T(\nu,p;B_1,B_2)$,

$$\|v\|_{T(\nu,p;B_1,B_2)} = \inf_{\substack{f\in W(\nu,p;B_1,B_2)\\ f(0)=v}} \|t^\nu f\|^{1-\theta}_{L^p(0,\infty;B_1)} \|t^\nu\partial_t f\|^{\theta}_{L^p(0,\infty;B_2)}.$$

If in addition $v \in B_1$, then $v \in T(\nu, p; B_1, B_2)$ and

$$\|v\|_{T(\nu,p;B_1,B_2)} \leq c_0 \|v\|_{B_1}^{1-\theta} \|v\|_{B_2}^{\theta}.$$

Hereafter c_l denotes some positive constant independent of any function v. On the other hand, let $\tilde{B}_1$ be continuously imbedded and dense in $\tilde{B}_2$ as well. If L is a linear operator from B_1 into $\tilde{B}_1$, and for any $v \in B_1$,

$$\|Lv\|_{\tilde{B}_l} \leq c_l \|v\|_{B_l}, \quad l = 1, 2,$$

then there exists a unique continuous extension, denoted by L, such that for any $v \in B_2$,

$$\|Lv\|_{\tilde{B}_2} \leq \max(c_1, c_2) \|v\|_{B_2}.$$

If in addition $0 < \theta < 1$, then for any $v \in T(\nu, p; B_1, B_2)$, we have that $Lv \in T(\nu, p; \tilde{B}_1, \tilde{B}_2)$ and

$$\|Lv\|_{T(\nu,p;\tilde{B}_1,\tilde{B}_2)} \leq c_1^{1-\theta} c_2^{\theta} \|v\|_{T(\nu,p;B_1,B_2)}.$$

As an example, if $1 \leq p \leq q < \infty$ and $0 < \theta < 1$, then $T\big(\nu, p; L^q(\Omega), L^p(\Omega)\big)$ is continuously imbedded in $L^\gamma(\Omega)$, $\dfrac{1}{\gamma} = \dfrac{1-\theta}{q} + \dfrac{\theta}{p}$.

Now let $1 < p < \infty$. For any integer $r = m \geq 0$, $W^{r,p}(\Omega) = W^{m,p}(\Omega)$. If $r = m + \sigma$ and $0 < \sigma < 1$, then define the Sobolev space $W^{r,p}(\Omega)$ as

$$W^{r,p}(\Omega) = \left\{ v \;\middle|\; v \in W^{m,p}(\Omega), \partial_x^k v \in T\Big(1 - \sigma - \frac{1}{p}, p, W^{1,p}(\Omega), L^p(\Omega)\Big), |k| = m \right\}.$$

It is also a Banach space with the norm

$$\|v\|_{W^{r,p}(\Omega)} = \left(\|v\|_{W^{m,p}(\Omega)}^p + \sum_{|k|=m} \|\partial_x^k v\|_{T\left(1-\sigma-\frac{1}{p}, p, W^{1,p}(\Omega), L^p(\Omega)\right)}^p \right)^{\frac{1}{p}}. \tag{2.3}$$

This definition can be generalized to the case $p = 1$ or $p = \infty$ (see Adams (1975)). An equivalent norm is as follows,

$$\|v\|_{W^{r,p}(\Omega)} = \left(\|v\|_{W^{m,p}(\Omega)}^p + \sum_{|k|=m} \int_\Omega \int_\Omega \frac{\left|\partial_x^k v(x) - \partial_y^k v(y)\right|^p}{|x-y|^{n+p\sigma}} \, dx \, dy \right)^{\frac{1}{p}}, \quad 1 \leq p < \infty,$$

$$\|v\|_{W^{r,\infty}(\Omega)} = \max \left(\|v\|_{W^{m,\infty}(\Omega)}, \max_{|k|=m} \operatorname*{ess\,sup}_{\substack{x,y\in\Omega \\ x \neq y}} \frac{\left|\partial_x^k v(x) - \partial_y^k v(y)\right|}{|x-y|^\sigma} \right).$$

For $r \geq 0$, let $W_0^{r,p}(\Omega)$ be the closure of $\mathcal{D}(\Omega)$ in $W^{r,p}(\Omega)$. Then for $r < 0$ and $1 < p < \infty$, define the Sobolev space $W^{r,p}(\Omega)$ by

$$W^{r,p}(\Omega) = \left[W_0^{-r,q}(\Omega)\right]', \quad \frac{1}{p} + \frac{1}{q} = 1.$$

In particular, $W^{r,2}(\Omega) = H^r(\Omega)$. For $r > 0$, it is a Hilbert space for the related inner product. For $r < 0$,

$$\|v\|_{H^r(\Omega)} = \sup_{\substack{g \in H_0^{-r}(\Omega) \\ g \neq 0}} \frac{\langle v, g \rangle}{|g|_{H^{-r}(\Omega)}}. \tag{2.4}$$

It is more convenient to use another notation for the interpolation between two separable Hilbert spaces X_1 and X_2, X_1 being continuously imbedded and dense in X_2. For $0 < \theta < 1$, let

$$[X_1, X_2]_\theta = T\left(\theta - \frac{1}{2}, 2, X_1, X_2\right).$$

In this case, we have

$$\|v\|_{[X_1,X_2]_\theta} \leq \|v\|_{X_1}^{1-\theta} \|v\|_{X_2}^{\theta}. \tag{2.5}$$

Moreover if $0 \leq \lambda < \delta \leq 1$, then

$$\left[[X_1, X_2]_\lambda, [X_1, X_2]_\delta\right]_\theta = [X_1, X_2]_{(1-\theta)\lambda + \theta\delta}.$$

In particular, for $0 \leq \mu \leq r$,

$$[H^r(\Omega), H^\mu(\Omega)]_\theta = H^{(1-\theta)r + \theta\mu}(\Omega). \tag{2.6}$$

When $\partial\Omega$ is of class $C^{m-1,1}$ for certain positive integer m, a general interpolation exists for three spaces. It states that if $\lambda - \frac{1}{2}$ is not an integer, $\lambda \geq 0, 0 < \theta < 1$ and $0 \leq \mu \leq r \leq m$, then

$$\left[H^r(\Omega) \cap H_0^\lambda(\Omega), H^\mu(\Omega) \cap H_0^\lambda(\Omega)\right]_\theta = H^{(1-\theta)r+\theta\mu}(\Omega) \cap H_0^\lambda(\Omega).$$

This property also holds when Ω is a hypercube without limitation on r and μ.

Now let $\Omega = (0, 2\pi)^n$ and $\mathcal{D}_p(\Omega)$ be the space of infinitely differentiable functions with the period 2π for all variables. Similarly we can define $L_p^q(\Omega), H_p^r(\Omega)$ and $W_p^{r,q}(\Omega)$. Moreover let

$$A(v) = \int_\Omega v(x)\, dx, \qquad \forall v \in L^1(\Omega)$$

and

$$L_0^2(\Omega) = \left\{ v \;\middle|\; v \in L^2(\Omega), A(v) = 0 \right\}.$$

Then for any $v \in H_0^1(\Omega)$ or $v \in H_p^1(\Omega) \cap L_0^2(\Omega)$, there exists the Poincaré inequality

$$\|v\|_{L^2(\Omega)} \leq c \|\nabla v\|_{L^2(\Omega)}. \tag{2.7}$$

Hereafter c denotes a generic positive constant depending only on the geometry of Ω. Thus for any non-negative integer m, the semi-norm $|\cdot|_{H^m(\Omega)}$ is a norm of $H_0^m(\Omega)$ or $H_p^m(\Omega) \cap L_0^2(\Omega)$, equivalent to the norm $\|\cdot\|_{H^m(\Omega)}$.

We give a simple version of the Sobolev imbedding theorem. It states that $W^{m,p}(\Omega)$ is continuously imbedded in $L^q(\Omega)$ for $\frac{1}{q} = \frac{1}{p} - \frac{m}{n} > 0$, in $C\left(\bar{\Omega}\right)$ for $\frac{1}{p} - \frac{m}{n} < 0$, and in $L^q(\Omega)$ for $m = \frac{n}{p}$ and any $q, 1 \leq q < \infty$.

We now turn to the trace theory. Let $\nu = (\nu_1, \ldots, \nu_n)$ be the unit outward normal vector to $\partial\Omega$, and $\partial_\nu = \sum\limits_{q=1}^{n} \nu_q \partial_q$. If $\partial\Omega$ is of class $C^{m-1,1}$, then for a relatively open part Γ of $\partial\Omega$, the mapping T_m^Γ:

$$v \to T_m^\Gamma v = \left(v|_\Gamma, \partial_\nu v|_\Gamma, \ldots, \partial_\nu^{m-1}|_\Gamma \right)$$

defined on $C^\infty\left(\bar{\Omega}\right)$, can be extended by a density argument to any space $H^r(\Omega), r > m - \frac{1}{2}$. If $\partial\Omega$ is a union of finite number of $\bar{\Gamma}_j$, and Γ_j are relatively open parts of class $C^{m-1,1}$, then for any $r < m + \frac{1}{2}$ such that $r - \frac{1}{2}$ is not an integer, we have

$$H_0^r(\Omega) = \left\{ v \;\middle|\; v \in H^r(\Omega), T_{[r+\frac{1}{2}]}^{\Gamma_j} v = 0 \text{ for all } j \right\}.$$

On the other hand, let $H^{r-\frac{1}{2}}(\Gamma)$ denote the image of $H^r(\Omega)$ by the trace mapping T_1^Γ with $r > \frac{1}{2}$. It is shown that

$$\|\phi\|_{H^{r-\frac{1}{2}}(\Gamma)} = \inf_{\substack{v \in H^r(\Omega) \\ v|_\Gamma = \phi}} \|v\|_{H^r(\Omega)}. \tag{2.8}$$

The details of the above results can be found in Lions and Magenes (1972) and Adams (1975).

Finally we give the following Lax-Milgram Lemma (1954).

Lemma 2.1. *Let B be a reflexive Banach space and $a(\cdot,\cdot)$ be a continuous bilinear elliptic form on $B \times B$, i.e., there exists a positive constant α such that*

$$a(w,w) \geq \alpha \|w\|_B^2, \qquad \forall\, w \in B. \tag{2.9}$$

Then for any $g \in B'$, there exists $v \in B$ uniquely such that

$$a(v,w) = < g, w >, \qquad \forall\, w \in B. \tag{2.10}$$

For simplicity, the inner products $(\cdot,\cdot)_{L^2(\Omega)}, (\cdot,\cdot)_{H^r(\Omega)}$ and the norms $\|\cdot\|_{L^2(\Omega)}, \|\cdot\|_{H^r(\Omega)}, \|\cdot\|_{L^p(\Omega)}, \|\cdot\|_{W^{m,p}(\Omega)}$ will be denoted by $(\cdot,\cdot), (\cdot,\cdot)_r, \|\cdot\|, \|\cdot\|_r, \|\cdot\|_{L^p}, \|\cdot\|_{W^{m,p}}$ respectively. Similarly the spaces $C\left(a,b;H^r(\Omega)\right), C\left(a,b;W^{m,p}(\Omega)\right), H^\gamma\left(a,b;H^r(\Omega)\right)$ and $H^\gamma\left(a,b;W^{m,p}(\Omega)\right)$ will be denoted by $C\left(a,b;H^r\right), C\left(a,b;W^{m,p}\right), H^\gamma\left(a,b;H^r\right)$ and $H^\gamma\left(a,b;W^{m,p}\right)$, etc..

2.2. Fourier Approximations

We consider the Fourier approximations in this section. Let $n=1$ and $\Lambda = (0,2\pi)$. The set of functions e^{ilx}, $l = 0, \pm 1, \ldots$, is an orthogonal system in $L^2(\Lambda)$. The Fourier transformation of a function $v \in L^2(\Lambda)$ is

$$Sv = \sum_{l=-\infty}^{\infty} \hat{v}_l e^{ilx} \tag{2.11}$$

where $\hat{v}_l$ is the Fourier coefficient,

$$\hat{v}_l = \frac{1}{2\pi}\int_\Lambda v(x) e^{-ilx}\, dx, \qquad l = 0, \pm 1, \ldots.$$

According to Riesz Theorem, Sv converges to v in $L^2(\Lambda)$, and the Parseval equality holds, namely

$$\|v\|^2 = 2\pi \sum_{l=-\infty}^{\infty} |\hat{v}_l|^2 . \tag{2.12}$$

Conversely for any complex sequence $\{a_l\}$ such that $\sum\limits_{l=-\infty}^{\infty} |a_l|^2 < \infty$, there exists a unique function in $L^2(\Lambda)$ such that its Fourier coefficient $\hat{v}_l = a_l$ for all l. Thus we can write

$$v = \sum_{l=-\infty}^{\infty} \hat{v}_l e^{ilx}.$$

A general result is the Young-Hausdroff inequality, see Hardy, Littlewood and Pólya (1952). It states that for $2 \le p < \infty, \dfrac{1}{p} + \dfrac{1}{q} = 1$,

$$\left(\frac{1}{2\pi}\int_\Lambda |v(x)|^p\,dx\right)^{\frac{1}{p}} \le \left(\sum_{l=-\infty}^{\infty} |\hat{v}_l|^q\right)^{\frac{1}{q}}, \tag{2.13}$$

$$\left(\sum_{l=-\infty}^{\infty} |\hat{v}_l|^p\right)^{\frac{1}{p}} \le \left(\frac{1}{2\pi}\int_\Lambda |v(x)|^q\,dx\right)^{\frac{1}{q}}. \tag{2.14}$$

The convergence in $L^2(\Lambda)$ does not imply the pointwise convergence of Sv to v at all points of $\bar{\Lambda}$. However, Carleson (1966) proved that Sv converges to v for all x outside a set of zero measures in $\bar{\Lambda}$. Some other results are as follows,

(i) if v is continuous, periodic and of bounded variation on $\bar{\Lambda}$, then Sv is uniformly convergent to v on $\bar{\Lambda}$;

(ii) if v is of bounded variation on $\bar{\Lambda}$, then Sv is convergent-pointwise to $\frac{1}{2}v(x^+) + \frac{1}{2}v(x^-)$ for any $x \in \bar{\Lambda}$;

(iii) if v is continuous and periodic, Sv does not necessarily converge at every point $x \in \bar{\Lambda}$.

For discussion of differentiation, let m be a positive integer and $\tilde{H}_p^m(\Lambda)$ be the subspace of $H^m(\Lambda)$ which consists of all functions whose first $m-1$ derivatives are periodic. In this space, it is permissible to differentiate termwise the Fourier series m times, in the sense of square mean. Indeed, $\partial_x e^{ilx} = ile^{ilx}$ and so

$$2\pi\widehat{(\partial_x v)}_l = \left(\partial_x v, e^{ilx}\right) = -\left(v, \partial_x e^{ilx}\right) = il\left(v, e^{ilx}\right) = 2\pi il\hat{v}_l.$$

Therefore, for any $v \in \tilde{H}_p^1(\Lambda)$,

$$\partial_x v(x) = \sum_{l=-\infty}^{\infty} il\hat{v}_l e^{ilx}.$$

By repeating the above procedure, we obtain

$$\partial_x^m v(x) = \sum_{l=-\infty}^{\infty} (il)^m \hat{v}_l e^{ilx}.$$

Consequently, the norm $\|v\|_m$ in $\tilde{H}_p^m(\Lambda)$ is equivalent to

$$\left(\sum_{l=-\infty}^{\infty}(1+l^2)^m|\hat{v}_l|^2\right)^{\frac{1}{2}}.$$

For $r=m+\sigma$ and $0<\sigma<1$, by the space interpolation (2.6),

$$H_p^r(\Lambda)=\left[H_p^{m+1}(\Lambda),H_p^m(\Lambda)\right]_{1-\sigma}$$

and by (2.5), for $v\in H_p^r(\Lambda)$,

$$\|v\|_r\le\|v\|_{m+1}^{\sigma}\|v\|_m^{1-\sigma}. \tag{2.15}$$

For $r<0$, let $H_p^r(\Lambda)=\left(H_p^{-r}(\Lambda)\right)'$. Therefore, for any $r, H_p^r(\Lambda)$ has the equivalent norm

$$\left(\sum_{l=-\infty}^{\infty}\left(1+l^2\right)^r|\hat{v}_l|^2\right)^{\frac{1}{2}}. \tag{2.16}$$

In particular for $r\ge 0$, the space $H_p^r(\Omega)$ has the equivalent semi-norm

$$\left(\sum_{l=-\infty}^{\infty}l^{2r}|\hat{v}_l|^2\right)^{\frac{1}{2}}.$$

Now let N be any positive integer and $\tilde{V}_N$ be the set of all trigonometric polynomials of degree at most N, i.e.,

$$\tilde{V}_N=\operatorname{span}\left\{e^{ilx}\mid |l|\le N\right\}. \tag{2.17}$$

V_N denotes the subset of $\tilde{V}_N$ consisting of all real-valued functions. In numerical analysis of spectral methods, we need some kind of inverse inequalities. The first one is due to Nikolskii (1951), stated below.

Theorem 2.1. *If* $1\le p\le q\le\infty$, *then for any* $\phi\in\tilde{V}_N$,

$$\|\phi\|_{L^q}\le\left(\frac{Np_0+1}{2\pi}\right)^{\frac{1}{p}-\frac{1}{q}}\|\phi\|_{L^p}$$

where p_0 *is the least even number greater than or equal to* p.

Proof. For simplicity, let $\phi \in V_N$ and $D_N(t)$ be the Dirichlet kernel,

$$D_N(t) = \frac{\sin\frac{2N+1}{2}t}{2\sin\frac{t}{2}} = \frac{1}{2} + \sum_{l=1}^{N} \cos lt.$$

Then

$$\phi(x) = \frac{1}{\pi}\int_\Lambda \phi(t) D_N(x-t)\,dt \tag{2.18}$$

and thus by the Cauchy inequality,

$$\|\phi\|_{L^\infty} \le \frac{1}{\pi}\|\phi\|\|D_N\|.$$

Since $\|D_N\| = \sqrt{\pi\left(N+\frac{1}{2}\right)}$, we get

$$\|\phi\|_{L^\infty} \le \left(\frac{2N+1}{2\pi}\right)^{\frac{1}{2}} \|\phi\|.$$

Furthermore, $\phi^{\frac{p_0}{2}} \in V_{\frac{p_0 N}{2}}$ and so

$$\|\phi^{\frac{p_0}{2}}\|_{L^\infty} \le \left(\frac{Np_0+1}{2\pi}\right)^{\frac{1}{2}} \|\phi^{\frac{p_0}{2}}\| \le \left(\frac{Np_0+1}{2\pi}\right)^{\frac{1}{2}} \|\phi\|_{L^\infty}^{\frac{p_0-p}{2}} \|\phi\|_{L^p}^{\frac{p}{2}}.$$

Hence

$$\|\phi\|_{L^\infty} \le \left(\frac{Np_0+1}{2\pi}\right)^{\frac{1}{p}} \|\phi\|_{L^p}.$$

Since $p \le q$, we get

$$\|\phi\|_{L^q}^q \le \|\phi\|_{L^\infty}^{q-p}\|\phi\|_{L^p}^p \le \left(\frac{Np_0+1}{2\pi}\right)^{\frac{q-p}{p}} \|\phi\|_{L^p}^q.$$

■

Another kind of inverse inequalities is called the Bernstein inequalities. Some of them can be found in Butzer and Nessel (1971).

Theorem 2.2. *Let m be a non-negative integer and $1 \le p \le \infty$. Then for any $\phi \in \tilde{V}_N$,*

$$\|\partial_x^m \phi\|_{L^p} \le (2N)^m \|\phi\|_{L^p}.$$

Also for any $r \ge 0$,

$$\|\phi\|_r \le cN^r \|\phi\|.$$

Proof. Let $D_N(t)$ be the Dirichlet kernel. By differentiating (2.18),

$$\partial_x \phi(x) = \frac{1}{\pi} \int_\Lambda \phi(x+t) \left(\sum_{l=1}^{N} l \sin lt \right) dt.$$

Since $\phi \in \tilde{V}_N$, we may add the sum $\sum\limits_{l=1}^{N-1} l \sin(2N-l)t$ to the term in brackets, which does not change the value of the integral. Therefore

$$\partial_x \phi(x) = \frac{2N}{\pi} \int_\Lambda \phi(x+t) \sin Nt F_{N-1}(t)\, dt$$

where $F_{N-1}(t)$ is the Fejér kernel, i.e.,

$$F_{N-1}(t) = \frac{1}{N} \sum_{l=0}^{N-1} D_l(t) = \frac{1}{2} + \sum_{l=1}^{N-1} \left(1 - \frac{l}{N}\right) \cos lt.$$

On use of the Hölder inequality,

$$\|\partial_x \phi\|_{L^p} \le \frac{2N}{\pi} \int_\Lambda \|\phi(\cdot + t)\|_{L^p} F_{N-1}(t)\, dt \le 2N \|\phi\|_{L^p}.$$

Then an iteration leads to the first conclusion. By (2.15) and the previous result for $p = 2$, we obtain that for $r = m + \sigma, 0 < \sigma < 1$,

$$\|\phi\|_r \le \|\phi\|_{m+1}^{\sigma} \|\phi\|_m^{1-\sigma} \le cN^r \|\phi\|.$$

■

The L^2-orthogonal projection $P_N : L^2(\Lambda) \to \tilde{V}_N$ is such a mapping that for any $v \in L^2(\Lambda)$,

$$(v - P_N v, \phi) = 0, \qquad \forall\, \phi \in \tilde{V}_N.$$

Indeed,

$$P_N v = \sum_{|l| \le N} \hat{v}_l e^{ilx}.$$

Moreover (2.12) implies that

$$\|v - P_N v\| = \left(2\pi \sum_{|l| > N} |\hat{v}_l|^2 \right)^{\frac{1}{2}}.$$

If v is suitably smooth, then

$$\|v - P_N v\|_{L^\infty} \le \sum_{|l|>N} |\hat{v}_l| .$$

Thus the error between v and $P_N v$ depends on how fast the Fourier coefficients decay to zero. This in turn depends on the regularity and the periodicity of v. In fact, for any $v \in C^1\left(\bar{\Lambda}\right)$ and $l \neq 0$,

$$2\pi \hat{v}_l = -\frac{1}{il}\left(v\left(2\pi^-\right) - v\left(0^+\right)\right) + \frac{1}{il}\int_\Lambda \partial_x v(x) e^{-ilx}\, dx$$

and so $\hat{v}_l = \mathcal{O}\left(\frac{1}{l}\right)$. Because the last integral is the Fourier coefficient of $\partial_x v$, we conclude that if $v \in C^2\left(\bar{\Lambda}\right)$ and $v\left(2\pi^-\right) = v\left(0^+\right)$, then $\hat{v}_l = \mathcal{O}\left(\frac{1}{l^2}\right)$. The repetition of the above procedure tells us that if $v \in C^m\left(\bar{\Lambda}\right)$ and $\partial_x^k v$ are periodic for all $k \le m-2$, then $\hat{v}_l = \mathcal{O}\left(l^{-m}\right)$. In the early work of Kreiss and Oliger (1972), Guo (1981, 1985) and Kuo (1983), the error estimates between solutions of differential equations and their spectral approximations were carried out based on the above facts. But recent work is associated with global approximations, started by Pasciak (1980).

Theorem 2.3. *If $r \ge 0$ and $\mu \le r$, then for any $v \in H_p^r(\Lambda)$,*

$$\|v - P_N v\|_\mu \le c N^{\mu - r} |v|_r .$$

Proof. For $\mu \ge 0$, we have from (2.16) that

$$\begin{aligned} \|v - P_N v\|_\mu^2 &\le c \sum_{|l|>N} \left(1 + l^2\right)^\mu |\hat{v}_l|^2 \le c N^{2\mu - 2r} \sum_{|l|>N} l^{2r - 2\mu} \left(1 + l^2\right)^\mu |\hat{v}_l|^2 \\ &\le c N^{2\mu - 2r} \sum_{|l|>N} \left(1 + l^2\right)^r |\hat{v}_l|^2 \le c N^{2\mu - 2r} |v|_r^2 . \end{aligned}$$

For $\mu < 0$, we have that for any $g \in H_p^{-\mu}(\Lambda)$,

$$|(v - P_N v, g)| = |(v - P_N v, g - P_N g)| \le \|v - P_N v\| \|g - P_N g\|$$

and so

$$\|v - P_N v\|_\mu = \sup_{\substack{g \in H_p^{-\mu}(\Lambda) \\ g \neq 0}} \frac{(v - P_N v, g)}{\|g\|_{-\mu}} \le c N^{\mu - r} |v|_r .$$

∎

We have seen that $P_N\partial_x v = \partial_x P_N v$. Hence P_N is also the best approximation of v in $\tilde{V}_N$ for the norm $\|\cdot\|_m, m$ being any non-negative integer. However it is not so for the L^p-norm, $1 \le p \le \infty, p \ne 2$. But we still have some results. For instance,

$$\inf_{\phi\in\tilde{V}_N} \|v-\phi\|_{L^p} \le cN^{-m}\|\partial_x^m v\|_{L^p}, \qquad 1 \le p \le \infty.$$

In particular, $c = \frac{\pi}{2}$ for $p = \infty$. Also, we have

$$\|v - P_N v\|_{L^p} \le c(1+\sigma(p)\ln N) \inf_{\phi\in\tilde{V}_N} \|v-\phi\|_{L^p}$$

with $\sigma(p) = 0$ for $1 < p < \infty$ and $\sigma(p) = 1$ for $p = 1, \infty$. Anyhow $P_N v$ approximates v in the L^p-norms with the same order as the best approximation, and $\|v - P_N v\|_{L^p} \to 0$ as $N \to \infty$.

In many applications, the numerical algorithms based on Fourier transformation cannot be implemented precisely. So we prefer to use discrete Fourier transformation. Let Λ_N be the set of interpolation points,

$$\Lambda_N = \left\{ x^{(j)} \;\middle|\; x^{(j)} = \frac{2\pi j}{2N+1}, \quad j = 0, 1, \ldots, 2N \right\}. \tag{2.19}$$

The discrete Fourier transformation of function $v \in C\left(\bar{\Lambda}\right)$ is

$$I_N v(x) = \sum_{|l|\le N} \tilde{v}_l e^{ilx} \tag{2.20}$$

such that

$$I_N v\left(x^{(j)}\right) = v\left(x^{(j)}\right), \quad 0 \le j \le 2N.$$

Due to the orthogonality, i.e.,

$$\sum_{j=0}^{2N} e^{ipx^{(j)}} = \begin{cases} 2N+1, & \text{if } p = (2N+1)q,\ q \text{ is an integer}, \\ 0, & \text{otherwise}, \end{cases}$$

we have

$$\tilde{v}_l = \frac{1}{2N+1} \sum_{j=0}^{2N} v\left(x^{(j)}\right) e^{-ilx^{(j)}}.$$

The coefficient $\tilde{v}_l$ can also be expressed in terms of the coefficient $\hat{v}_l$. In fact,

$$\tilde{v}_l = \frac{1}{2N+1} \sum_{j=0}^{2N} \left(\sum_{p=-\infty}^{\infty} \hat{v}_p e^{i(p-l)x^{(j)}} \right)$$

$$= \frac{1}{2N+1} \sum_{p=-\infty}^{\infty} \left(\sum_{j=0}^{2N} \hat{v}_p e^{i(p-l)x^{(j)}} \right)$$

$$= \sum_{p=-\infty}^{\infty} \hat{v}_{l+p(2N+1)}. \tag{2.21}$$

The interpolation I_N can be regarded as an orthogonal projection upon $\tilde{V}_N$ with respect to the discrete inner product. To do this, we define the discrete inner product and the norm as

$$(v,w)_N = \frac{2\pi}{2N+1} \sum_{j=0}^{2N} v\left(x^{(j)}\right) \bar{w}\left(x^{(j)}\right), \quad \|v\|_N = (v,v)_N^{\frac{1}{2}}. \tag{2.22}$$

Then

$$(v,w)_N = (v,\phi), \quad \forall v,\phi \in \tilde{V}_N$$

and by (2.20),

$$(I_N v, \phi)_N = (v,\phi)_N, \qquad \forall v \in C\left(\bar{\Lambda}\right), \phi \in \tilde{V}_N. \tag{2.23}$$

This shows that $I_N v$ is the orthogonal projection of v for the inner product $(\cdot,\cdot)_N$.

As in the case of Fourier transformation, the following results hold,

(i) if v is continuous, periodic and of bounded variation on $\bar{\Lambda}, I_N v$ tends to v uniformly on $\bar{\Lambda}$;

(ii) if v is of bounded variation on $\bar{\Lambda}, I_N v$ is uniformly bounded on $\bar{\Lambda}$ and converges to v at every continuity point for v;

(iii) for any integer $l \neq 0$, and any positive N such that $N > |l|$, let $\tilde{v}_l = \tilde{v}_l^{(N)}$ be the l-th Fourier coefficient of $I_N v$. If $v \in C_p^\infty\left(\bar{\Lambda}\right)$, then $\left|\tilde{v}_l^{(N)}\right|$ decays faster than cl^{-m}, m being any positive integer, uniformly in N. If v satisfies the hypotheses for which $\hat{v}_l = \mathcal{O}\left(l^{-m}\right)$, then $\tilde{v}_l^{(N)}$ possesses the same asymptotic behavior.

We now deal with the global error between v and $I_N v$.

Theorem 2.4. *If $r > \frac{1}{2}$ and $0 \leq \mu \leq r$, then for any $v \in H_p^r(\Lambda)$,*

$$\|v - I_N v\|_\mu \leq cN^{\mu-r}|v|_r.$$

Proof. Let $E_N v = P_N v - I_N v$. By (2.21),

$$
\begin{aligned}
\|E_N v\|_\mu^2 &\le c \sum_{|l|\le N} \left(1+l^2\right)^\mu |\hat{v}_l - \tilde{v}_l|^2 \\
&\le cN^{2\mu} \sum_{|l|\le N} \left| \sum_{p\neq 0} \hat{v}_{l+p(2N+1)} \right|^2 \\
&\le cN^{2\mu} \sum_{|l|\le N} \left(\sum_{p\neq 0} |l+p(2N+1)|^{-2r} \right) \left(\sum_{p\neq 0} |l+p(2N+1)|^{2r} \left|\hat{v}_{l+p(2N+1)}\right|^2 \right) \\
&\le cN^{2\mu} \sum_{p\neq 0} \left(N^{-2r}|p|^{-2r} \sum_{|q|>N} |q|^{2r} |\hat{v}_q|^2 \right) \\
&\le cN^{2\mu-2r} \sum_{p\neq 0} |p|^{-2r} |v|_r^2 \le cN^{2\mu-2r}|v|_r^2
\end{aligned}
$$

which together with Theorem 2.3 imply the desired result. ∎

The following result provides the estimate in L^∞-norm,

$$\|v - I_N v\|_{L^\infty} \le c(\log N) N^{-r} |v|_{W^{r,\infty}}.$$

Kenneth (1978) studied the estimates in C^m-norms.

Finally $\partial_x I_N v \neq I_N \partial_x v$ unlike P_N. But for any $v \in H_p^r(\Lambda)$ with $r > \frac{1}{2}$,

$$\|I_N \partial_x v - \partial_x I_N v\| \le cN^{1-r}\|v\|_r.$$

2.3. Orthogonal Systems of Polynomials in a Finite Interval

Let $\omega(x)$ be a non-negative, continuous and integrable real-valued function in the interval $\Lambda = (-1,1)$. The associated weighted space of real-valued functions is

$$L_\omega^p(\Lambda) = \left\{ v \;\middle|\; \|v\|_{L_\omega^p(\Lambda)} < \infty \right\}$$

equipped with the norm

$$\|v\|_{L_\omega^p(\Lambda)} = \left(\int_\Lambda |v(x)|^p \omega(x)\, dx \right)^{\frac{1}{p}}, \quad 1 \le p < \infty.$$

If $p=2$, then $L_\omega^2(\Lambda)$ is a Hilbert space for the weighted inner product

$$(v,w)_{L_\omega^2(\Lambda)} = \int_\Lambda v(x) w(x) \omega(x)\, dx, \quad \forall v, w \in L_\omega^2(\Lambda).$$

For any non-negative integer m, the Sobolev space $W_\omega^{m,p}(\Lambda)$ is defined as

$$W_\omega^{m,p}(\Lambda) = \left\{ v \;\middle|\; \partial_x^k v \in L_\omega^p(\Lambda), k \le m \right\}.$$

For any real $r \ge 0, W_\omega^{r,p}(\Lambda)$ is defined by the space interpolation with the corresponding norm. $W_{0,\omega}^{r,p}(\Lambda)$ is the closure of $\mathcal{D}(\Lambda)$ in $W_\omega^{r,p}(\Omega)$. For $r < 0, W_\omega^{r,p}(\Omega) = \left(W_{0,\omega}^{-r,q}(\Omega)\right)', \frac{1}{p} + \frac{1}{q} = 1$. In particular, $W_\omega^{r,2}(\Lambda) = H_\omega^r(\Lambda)$. It is a Hilbert space for the related inner product when $r \ge 0$. While for $r < 0$,

$$\|v\|_{H_\omega^r(\Lambda)} = \sup_{\substack{g \in H_{0,\omega}^{-r}(\Lambda) \\ g \ne 0}} \frac{\langle v, g \rangle}{|g|_{H_\omega^{-r}(\Lambda)}}.$$

For simplicity, we denote $(\cdot,\cdot)_{L_\omega^2(\Lambda)}, \|\cdot\|_{L_\omega^2(\Lambda)}, \|\cdot\|_{L_\omega^p(\Lambda)}, \|\cdot\|_{H_\omega^r(\Lambda)}$ and $\|\cdot\|_{W_\omega^{r,p}(\Lambda)}$ by $(\cdot,\cdot)_\omega$, $\|\cdot\|_\omega, \|\cdot\|_{L_\omega^p}, \|\cdot\|_{r,\omega}$ and $\|\cdot\|_{W_\omega^{r,p}}$, respectively. Also, $C\left(a, b; W_\omega^{r,p}(\Lambda)\right)$ and $H^\gamma\left(a, b; W_\omega^{r,p}(\Lambda)\right)$ are denoted by $C(a, b; W_\omega^{r,p})$ and $H^\gamma(a, b; W_\omega^{r,p})$ with the norms $\|\cdot\|_{C(a,b;W_\omega^{r,p})}$ and $\|\cdot\|_{H^\gamma\left(a,b;W_\omega^{r,p}\right)}$, respectively, etc..

Let $\phi_l(x), l = 0, 1, \ldots,$ be an orthogonal system of algebraic polynomials with respect to the weighted inner product $(\cdot,\cdot)_\omega$. By Weierstrass Theorem, such a system is complete in $L_\omega^2(\Lambda)$. The formal series of a function $v \in L_\omega^2(\Lambda)$ in terms of $\{\phi_l\}$ is

$$Sv = \sum_{l=0}^{\infty} \hat{v}_l \phi_l(x) \tag{2.24}$$

with

$$\hat{v}_l = \frac{1}{\|\phi_l\|_\omega^2} \int_\Lambda v(x)\phi_l(x)\omega(x)\, dx.$$

By the completeness of the system $\{\phi_l\}$, Sv is convergent to v in $L_\omega^2(\Lambda)$. The rate of convergence depends on the choice of the weight function $\omega(x)$. The different choices correspond to different spectral methods.

Denote by $\mathbb{P}_N$ the set of all algebraic polynomials of degree at most N in Λ. In this set, several inverse inequalities are valid. We first consider the relation between $\|\phi\|_{L_\omega^p}$ and $\|\phi\|_{L_\omega^q}, 1 \le p \le q \le \infty$. Let L be a linear operator defined on $\mathbb{P}_N$. L is said to be of (p, q) type, if there exists a positive constant d depending only on p, q and N such that $\|L\phi\|_{L_\omega^q} \le d\|\phi\|_{L_\omega^p}$ for any $\phi \in \mathbb{P}_N$. According to Riesz-Thorin Theorem, we know that if L is of both (p_1, q_1) type and (p_2, q_2) type for

$1 \le p_1, p_2 < \infty, 1 \le q_1, q_2 \le \infty$, then for

$$p = \frac{p_1 p_2}{(1-\theta)p_2 + \theta p_1}, \quad q = \frac{q_1 q_2}{(1-\theta)q_2 + \theta q_1}, \quad 0 \le \theta \le 1, \tag{2.25}$$

the operator L is also of (p,q) type. If in addition $\|L\phi\|_{L^{q_l}_\omega} \le d_l \|\phi\|_{L^{p_l}_\omega}, l = 1, 2$, then

$$\|L\phi\|_{L^q_\omega} \le c(p_1, p_2) d_1^{1-\theta} d_2^{\theta} \|\phi\|_{L^p_\omega} \tag{2.26}$$

where $c(p_1, p_2)$ is a positive constant depending only on p_1 and p_2.

Theorem 2.5. *If $\phi_l \in L^p_\omega(\Lambda)$ for all l, and for certain positive constant c_0 and real number δ,*

$$\|\phi_0\|_{L^\infty} \le c_0, \qquad \|\phi_l\|_{L^\infty} \le c_0 l^\delta \|\phi_l\|_\omega, \quad l \ge 1, \tag{2.27}$$

then for any $\phi \in \mathbb{P}_N$ and all $1 \le p \le q \le \infty$,

$$\|\phi\|_{L^q_\omega} \le c\sigma^{\frac{1}{p}-\frac{1}{q}}(N) \|\phi\|_{L^p_\omega}$$

where $\sigma(N) = N^{2\delta+1}$ for $\delta > -\frac{1}{2}, \sigma(N) = \ln N$ for $\delta = -\frac{1}{2}$, and $\sigma(N) = 1$ for $\delta < -\frac{1}{2}$.

Proof. Let us consider, for example, the case $\delta > -\frac{1}{2}$. By (2.24) and (2.27),

$$\|\phi\|_{L^\infty} \le \|\phi\|_{L^1_\omega} \sum_{l=0}^{N} \frac{\|\phi_l\|^2_{L^\infty}}{\|\phi_l\|^2_\omega} \le cN^{2\delta+1} \|\phi\|_{L^1_\omega}. \tag{2.28}$$

Thus the identity operator L is of $(1,\infty)$ type with the constant $d = cN^{2\delta+1}$. Clearly for any $r = \frac{pq-q+p}{p}, 1 \le p \le q \le \infty, L$ is of (r,r) type with $d = 1$. For $1 \le p < \infty$, we set $p_1 = 1, q_1 = \infty, p_2 = q_2 = r$ and $\theta = \frac{r}{q}$ in (2.25). Then (2.26) implies the desired result. On the other hand, we know from (2.28) that L is also of (1,1) type and (∞,∞) type. The proof is complete. ∎

We now turn to another kind of inverse inequalities. The first result in the following theorem is due to Markov (1898), e.g., see Timan (1963).

Theorem 2.6. *For any $\phi \in \mathbb{P}_N$,*

$$|\partial_x \phi(x)| \le N \min\left(\frac{1}{\sqrt{1-x^2}}, N\right) \|\phi\|_{L^\infty}.$$

If, in addition,

$$\|\partial_x \phi\|_\omega \le cN^2 \|\phi\|_\omega, \tag{2.29}$$

then for all $2 \le p \le \infty$,

$$\|\partial_x \phi\|_{L^p_\omega} \le cN^2 \|\phi\|_{L^p_\omega}.$$

Proof. Set

$$x = \cos y, \qquad v(x) = v(\cos y). \tag{2.30}$$

From the proof of Theorem 2.2,

$$|\partial_x \phi(\cos y) \sin y| \le N \max_{0 \le y \le 2\pi} |\phi(\cos y)|$$

and so

$$|\partial_x \phi(x)| \le \frac{N}{\sqrt{1-x^2}} \|\phi\|_{L^\infty}. \tag{2.31}$$

Next let ϕ_N be an algebraic polynomial of degree N, and $x^{(j)}, 0 \le j \le N-1$ be distinct points in $\bar{\Lambda}$. Set

$$q_{N-1,j}(x) = \frac{q_N(x)}{\partial_x q_N\left(x^{(j)}\right)\left(x - x^{(j)}\right)}, \qquad q_N(x) = \sqcap_{j=0}^{N-1} \left(x - x^{(j)}\right).$$

Then we have the Lagrange interpolation as

$$\partial_x \phi_N(x) = \sum_{j=0}^{N-1} \partial_x \phi_N\left(x^{(j)}\right) q_{N-1,j}(x).$$

By taking

$$x^{(j)} = \cos \frac{\pi(2j+1)}{2N}, \quad 0 \le j \le N-1,$$

we obtain that

$$\begin{aligned} q_N(x) &= \frac{1}{2^{N-1}} \cos(N \arccos x), \\ \partial_x q_N\left(x^{(j)}\right) &= \frac{(-1)^j N}{2^{N-1} \sin \frac{\pi(2j+1)}{2N}}. \end{aligned}$$

Consequently,

$$\partial_x \phi_N(x) = \frac{\cos(N \arccos x)}{N} \sum_{j=0}^{N-1} (-1)^j \partial_x \phi_N\left(x^{(j)}\right) \frac{\sqrt{1 - \left(x^{(j)}\right)^2}}{x - x^{(j)}}.$$

Hence by virtue of (2.31),

$$|\partial_x \phi_N(x)| \le |\cos(N \arccos x)| \|\phi_N\|_{L^\infty} \sum_{j=0}^{N-1} \frac{1}{|x - x^{(j)}|}.$$

If $x^{(0)} < |x| \le 1$, then all the differences $x - x^{(j)}$ have the same sign and so

$$\begin{aligned} |\partial_x \phi_N(x)| &\le \|\phi_N\|_{L^\infty} \left| \sum_{j=0}^{N-1} \frac{\cos(N \arccos x)}{x - x^{(j)}} \right| \\ &\le \|\phi_N\|_{L^\infty} \left| \frac{N \sin(N \arccos x)}{\sqrt{1-x^2}} \right| \le N^2 \|\phi_N\|_{L^\infty}. \end{aligned}$$

If $x^{(0)} \ge x$, then the same conclusion follows immediately from (2.31). If (2.29) holds, then the second conclusion comes from the space interpolation and the previous result as shown in Quarteroni (1984). ■

The L^2_ω-orthogonal projection $P_N : L^2_\omega \to \mathbb{P}_N$ is such a mapping that for any $v \in L^2_\omega(\Lambda)$,

$$(v - P_N v, \phi)_\omega = 0, \qquad \forall\, \phi \in \mathbb{P}_N. \tag{2.32}$$

Indeed,

$$P_N v(x) = \sum_{l=0}^{N} \hat{v}_l \phi_l(x).$$

There are some close relations between orthogonal systems and Gauss-type integrations. We have the following results.

(i) **Gauss integration.** Let $x^{(0)}, \ldots, x^{(N)}$ be the roots of the $(N+1)$-th orthogonal polynomial $\phi_{N+1}(x)$, and $\omega^{(0)}, \ldots, \omega^{(N)}$ be the solution of the linear system

$$\sum_{j=0}^{N} \left(x^{(j)}\right)^m \omega^{(j)} = \int_\Lambda x^m \omega(x)\, dx, \quad 0 \le m \le N. \tag{2.33}$$

Then $\omega^{(j)} > 0$ for $0 \le j \le N$, and for any $\phi \in \mathbb{P}_{2N+1}$,

$$\sum_{j=0}^{N} \phi\left(x^{(j)}\right) \omega^{(j)} = \int_\Lambda \phi(x) \omega(x)\, dx. \tag{2.34}$$

(ii) **Gauss-Radau integration.** Let $x^{(0)}, \ldots, x^{(N)}$ be the $N+1$ roots of the polynomial $\psi(x) = \phi_{N+1}(x) - \phi_{N+1}(-1)\phi_N^{-1}(-1)\phi_N(x)$, and $\omega^{(0)}, \ldots, \omega^{(N)}$ be the solution of (2.33). Then (2.34) holds for any $\phi \in \mathbb{P}_{2N}$.

(iii) **Gauss-Lobatto integration.** Let $x^{(0)},\dots,x^{(N)}$ be the $N+1$ roots of the polynomial $(1-x^2)\psi(x)$ such that $\psi(x)=\phi_{N+1}(x)+a\phi_N(x)+b\phi_{N-1}(x)$ and $\psi(-1)=\psi(1)=0$, and $\omega^{(0)},\dots,\omega^{(N)}$ be the solution of (2.33). Then (2.34) is valid for any $\phi\in\mathbb{P}_{2N-1}$.

The points $x^{(j)}, 0\le j\le N$ are called Gauss-type interpolation points. Clearly $x^{(0)}=-1$ in Gauss-Radau integration and Gauss-Lobatto integration, and $x^{(N)}=1$ in Gauss-Lobatto integration. The numbers $\omega^{(j)}, 0\le j\le N$ are called Gauss-type weights.

In pseudospectral methods, the fundamental representations of a smooth function v are in terms of its values at Gauss-type intepolation points. The interpolant is given by

$$I_N v(x)=\sum_{l=0}^{N}\tilde{v}_l\phi_l(x) \tag{2.35}$$

such that

$$I_N v\left(x^{(j)}\right)=v\left(x^{(j)}\right),\quad 0\le j\le N.$$

We introduce the discrete inner product and the discrete norm as

$$(v,w)_{N,\omega}=\sum_{j=0}^{N}v\left(x^{(j)}\right)w\left(x^{(j)}\right)\omega^{(j)},\quad \|v\|_{N,\omega}=(v,v)_{N,\omega}^{\frac{1}{2}}. \tag{2.36}$$

The Gauss-type integrations imply that

$$(v,w)_{N,\omega}=(v,w)_\omega,\quad \forall\, vw\in\mathbb{P}_{2N+\lambda} \tag{2.37}$$

where $\lambda=1,0,-1$ for the Gauss interpolation, the Gauss-Radau interpolation and the Gauss-Lobatto integration respectively. Obviously by (2.35),

$$(I_N v,w)_{N,\omega}=(v,w)_{N,\omega}\quad \forall\, v,w\in C\left(\bar{\Lambda}\right). \tag{2.38}$$

Thus $I_N v$ is the orthogonal projection of v upon $\mathbb{P}_N$ with respect to the inner product given in (2.36). Furthermore, the orthogonality of $\{\phi_l(x)\}$ together with (2.37) give

$$(\phi_l,\phi_m)_{N,\omega}=\gamma_l\delta_{l,m},\quad \gamma_l=\sum_{j=0}^{N}\phi_l^2\left(x^{(j)}\right)\omega^{(j)},\quad 0\le l,m\le N$$

where $\delta_{l,m}$ is the Kronecker function. Therefore by virtue of (2.38),

$$(v,\phi_l)_{N,\omega} = (I_N v,\phi_l)_{N,\omega} = \sum_{p=0}^{N} \tilde{v}_p(\phi_p,\phi_l)_{N,\omega} = \gamma_l \tilde{v}_l, \quad 0 \le l \le N.$$

The coefficient $\tilde{v}_l$ can be expressed in terms of the coefficients $\{\hat{v}_l\}$, i.e.,

$$\tilde{v}_l = \hat{v}_l + \frac{1}{\gamma_l}\sum_{p>N}(\phi_p,\phi_l)_{N,\omega}\hat{v}_p. \tag{2.39}$$

There are some relations between $\|\cdot\|_\omega$ and $\|\cdot\|_{N,\omega}$, and between $(\cdot,\cdot)_\omega$ and $(\cdot,\cdot)_{N,\omega}$. In fact, for the Gauss integration and the Gauss-Radau integration,

$$\|\phi\|_{N,\omega} = \|\phi\|_\omega, \quad \forall\, \phi \in \mathbb{P}_N. \tag{2.40}$$

Moreover, (2.37) and (2.38) imply that for any $v, w \in C\left(\bar{\Lambda}\right)$,

$$(v,w)_\omega - (v,w)_{N,\omega} = (v,w)_\omega - (I_N v, w)_{N,\omega}$$

and so for the Gauss integration and the Gauss-Radau integration,

$$|(v,\phi)_\omega - (v,\phi)_{N,\omega}| \le \|v - I_N v\|_\omega \|\phi\|_\omega, \quad \forall\, \phi \in \mathbb{P}_N. \tag{2.41}$$

For the Gauss-Lobatto integration, $\|\phi\|_{N,\omega} \ne \|\phi\|_\omega$ usually. But for mostly used orthogonal systems in Λ, they are equivalent, namely, for certain positive constants c_1 and c_2,

$$c_1\|\phi\|_\omega \le \|\phi\|_{N,\omega} \le c_2\|\phi\|_\omega.$$

In this case, we derive from (2.37) that for any $\phi \in \mathbb{P}_N$,

$$\begin{aligned} |(v,\phi)_\omega - (v,\phi)_{N,\omega}| &\le |(v,\phi)_\omega - (P_{N-1}v,\phi)_\omega| + |(P_{N-1}v,\phi)_\omega - (I_N v,\phi)_{N,\omega}| \\ &\le c(\|v - P_{N-1}v\|_\omega + \|P_{N-1}v - I_N v\|_{N,\omega})\,\|\phi\|_\omega \\ &\le c(2\|v - P_{N-1}v\|_\omega + \|v - I_N v\|_\omega)\,\|\phi\|_\omega. \end{aligned} \tag{2.42}$$

This formula plays an important role in numerical analysis of pseudospectral methods.

2.4. Legendre Approximations

Let $\Lambda = (-1,1)$ and $\omega(x) \equiv 1$ in this section. The Legendre polynomial of degree l is

$$L_l(x) = \frac{(-1)^l}{2^l l!}\partial_x^l \left(1-x^2\right)^l. \tag{2.43}$$

It is the l-th eigenfunction of the singular Sturm-Liouville problem

$$\partial_x\left(\left(1-x^2\right)\partial_x v\right) + \lambda v = 0, \quad x \in \Lambda, \tag{2.44}$$

related to the l-th eigenvalue $\lambda_l = l(l+1)$. Clearly $L_0(x) = 1, L_1(x) = x$, and they satisfy the recurrence relations

$$L_{l+1}(x) = \frac{2l+1}{l+1}xL_l(x) - \frac{l}{l+1}L_{l-1}(x), \quad l \geq 1, \tag{2.45}$$

$$(2l+1)L_l(x) = \partial_x L_{l+1}(x) - \partial_x L_{l-1}(x), \quad l \geq 1. \tag{2.46}$$

It can be checked that $L_l(1) = 1, L_l(-1) = (-1)^l, \partial_x L_l(1) = \frac{1}{2}l(l+1)$ and $\partial_x L_l(-1) = (-1)^{l+1}\frac{1}{2}l(l+1)$. Moreover,

$$|L_l(x)| \leq 1, \quad |\partial_x L_l(x)| \leq \frac{1}{2}l(l+1), \quad x \in \Lambda. \tag{2.47}$$

The set of Legendre polynomials is the L^2-orthogonal system in Λ, i.e.,

$$\int_\Lambda L_l(x)L_m(x)\,dx = \left(l+\frac{1}{2}\right)^{-1}\delta_{l,m}. \tag{2.48}$$

By integrating by parts, we deduce that

$$\int_\Lambda \left(\partial_x L_l(x)\right)^2 dx = l(l+1). \tag{2.49}$$

The Legendre expansion of a function $v \in L^2(\Lambda)$ is

$$v(x) = \sum_{l=0}^{\infty} \hat{v}_l L_l(x) \tag{2.50}$$

with

$$\hat{v}_l = \left(l+\frac{1}{2}\right)\int_\Lambda v(x)L_l(x)\,dx.$$

We now consider the differentiation. Let $\hat{v}_l^{(1)}$ be the coefficients of Legendre expansion of $\partial_x v, l = 0, 1, \ldots$. By (2.46),

$$\begin{aligned}
\partial_x v(x) &= \sum_{l=0}^{\infty} \frac{\hat{v}_l^{(1)}}{2l+1} \partial_x L_{l+1}(x) - \sum_{l=0}^{\infty} \frac{\hat{v}_l^{(1)}}{2l+1} \partial_x L_{l-1}(x) \\
&= \sum_{l=1}^{\infty} \frac{\hat{v}_{l-1}^{(1)}}{2l-1} \partial_x L_l(x) - \sum_{l=-1}^{\infty} \frac{\hat{v}_{l+1}^{(1)}}{2l+3} \partial_x L_l(x) \\
&= \sum_{l=1}^{\infty} \left(\frac{\hat{v}_{l-1}^{(1)}}{2l-1} - \frac{\hat{v}_{l+1}^{(1)}}{2l+3} \right) \partial_x L_l(x).
\end{aligned}$$

On the other hand,

$$\partial_x v(x) = \sum_{l=1}^{\infty} \hat{v}_l \partial_x L_l(x).$$

Hence

$$\hat{v}_l^{(1)} = (2l+1) \sum_{\substack{p=l+1 \\ p+l \text{ odd}}}^{\infty} \hat{v}_p. \tag{2.51}$$

This formula is valid for any $v \in H^1(\Lambda)$.

We turn to the inverse inequalities in the space $\mathbb{P}_N$.

Theorem 2.7. *For any $\phi \in \mathbb{P}_N$ and $1 \le p \le q \le \infty$,*

$$\|\phi\|_{L^q} \le \left((p+1)N^2\right)^{\frac{1}{p}-\frac{1}{q}} \|\phi\|_{L^p}.$$

Proof. Assume that $|\phi|$ reaches the maximum value at $x^* \in \bar{\Lambda}$. Thus

$$|\phi(x) - \phi(x^*)| \le |x - x^*| \|\partial_x \phi\|_{L^\infty},$$

and by Theorem 2.6,

$$|\phi(x^*)| - |\phi(x)| \le N^2 |x - x^*| |\phi(x^*)|.$$

We suppose that $x^* \le 0$ for definiteness. Then

$$|\phi(x^*)|^p \int_{x^*}^{x^* + \frac{1}{N^2}} \left(1 - N^2 |x - x^*|\right)^p dx \le \int_\Lambda |\phi(x)|^p dx$$

i.e.,

$$\|\phi\|_{L^\infty} \le (p+1)^{\frac{1}{p}} N^{\frac{2}{p}} \|\phi\|_{L^p}.$$

Further, we have

$$\|\phi\|_{L^q}^q = \|\phi\|_{L^\infty}^{q-p}\|\phi\|_{L^p}^p \le \left((p+1)N^2\right)^{\frac{q-p}{p}} \|\phi\|_{L^p}^q$$

which leads to the conclusion. ∎

We can derive the above theorem from Theorem 2.5 apart from the constant. Indeed by (2.47) and (2.48), we have $\delta = \frac{1}{2}$ in (2.27).

Theorem 2.8. *Let m be a non-negative integer and $2 \le p \le \infty$. Then for any $v \in \mathbb{P}_N$,*

$$\|\partial_x^m \phi\|_{L^p} \le cN^{2m}\|\phi\|_{L^p}.$$

Also for any $r \ge 0$,

$$\|\phi\|_r \le cN^{2r}\|\phi\|.$$

Proof. Let $\hat{\phi}_l$ and $\hat{\phi}_l^{(1)}$ be the coefficients of Legendre expansions of ϕ and $\partial_x\phi$. By (2.51),

$$\begin{aligned}\|\partial_x\phi\|^2 &= \sum_{l=0}^{N} \frac{2}{2l+1}\left(\hat{\phi}_l^{(1)}\right)^2 \\ &= 4\sum_{l=0}^{N}\left(l+\frac{1}{2}\right)\left(\sum_{\substack{p=l+1\\ p+l \text{ odd}}}^{N} \hat{\phi}_p\right)^2 \le N^2\left(N+\frac{1}{2}\right)^2 \sum_{l=0}^{N}\frac{2}{2l+1}\hat{\phi}_l^2.\end{aligned}$$

Let $a_N = 1 + \frac{1}{2N} \le \frac{3}{2}$. Then

$$\|\partial_x\phi\| \le a_N N^2\|\phi\|$$

from which, Theorem 2.6 and a repetition, the first conclusion follows. We now prove the second one. If $r = m+\sigma$ and m is a non-negative integer, $0 < \sigma < 1$, then by (2.15),

$$\|\partial_x\phi\|_r \le \|\partial_x\phi\|_{m+1}^{\sigma}\|\partial_x\phi\|_m^{1-\sigma} \le cN^{2r}\|\phi\|.$$

∎

The L^2-orthogonal projection $P_N : L^2(\Lambda) \to \mathbb{P}_N$ is such a mapping that for any $v \in L^2(\Lambda)$,

$$(v - P_N v, \phi) = 0, \qquad \forall\, \phi \in \mathbb{P}_N,$$

or equivalently,

$$P_N v(x) = \sum_{l=0}^{N} \hat{v}_l L_l(x).$$

We now estimate the difference between v and $P_N v$.

Lemma 2.2. *For any $v \in H^r(\Lambda)$ and $r \geq 0$,*

$$\|v - P_N v\| \leq cN^{-r}\|v\|_r.$$

Proof. Assume at first that $r = 2m, m$ being an integer. We have

$$\|v - P_N v\|^2 = \sum_{l=N+1}^{\infty} \hat{v}_l^2 \|L_l\|^2.$$

Let $Av = -\partial_x\left((1-x^2)\,\partial_x v\right)$. By (2.44),

$$\begin{aligned} \hat{v}_l &= \frac{1}{\|L_l\|^2}\int_\Lambda v(x)L_l(x)\,dx = \frac{1}{l(l+1)\|L_l\|^2}\int_\Lambda v(x)AL_l(x)\,dx \\ &= \frac{1}{l(l+1)\|L_l\|^2}\int_\Lambda Av(x)L_l(x)\,dx. \end{aligned}$$

Iterating this procedure yields

$$\hat{v}_l = \frac{1}{l^m(l+1)^m\|L_l\|^2}\int_\Lambda A^m v(x)L_l(x)\,dx$$

and thus

$$\|v - P_N v\|^2 \leq N^{-4m}\sum_{l=0}^{\infty}\|L_l\|^2\left(\frac{\int_\Lambda A^m v(x)L_l(x)\,dx}{\|L_l\|^2}\right)^2 \leq N^{-4m}\|A^m v\|^2$$

which is the desired result for $r = 2m$. For any $r \geq 0$, we deduce the result from the space interpolation as before. ∎

Usually $\partial_x P_N v(x) \neq P_N \partial_x v(x)$. But we have the following result.

Lemma 2.3. *For any $v \in H^r(\Lambda)$ and $0 \leq \mu \leq r-1$,*

$$\|P_N\partial_x v - \partial_x P_N v\|_\mu \leq cN^{2\mu - r + \frac{3}{2}}\|v\|_r.$$

Proof. According to the space interpolation, it suffices to consider the case $r \geq 1$, $\mu = 0$. Assume first that $\partial_x v$ is continuous and so (2.51) is valid. Also let

$$\partial_x P_N v(x) = \sum_{l=0}^{N-1} \hat{w}_l L_l(x), \quad \hat{w}_l = (2l+1) \sum_{\substack{p=l+1 \\ p+l \text{ odd}}}^{N} \hat{v}_p.$$

Then by (2.51),

$$\frac{1}{2l+1}\left(\hat{v}_l^{(1)} - \hat{w}_l\right) = \begin{cases} \sum\limits_{\substack{p=N+2 \\ p+N \text{ even}}}^{\infty} \hat{v}_p = \dfrac{\hat{v}_{N+1}^{(1)}}{2N+3}, & \text{if } l+N \text{ is odd,} \\ \sum\limits_{\substack{p=N+1 \\ p+N \text{ odd}}}^{\infty} \hat{v}_p = \dfrac{\hat{v}_N^{(1)}}{2N+1}, & \text{if } l+N \text{ is even.} \end{cases}$$

Accordingly,

$$P_N \partial_x v(x) - \partial_x P_N v(x) = \begin{cases} \dfrac{1}{2N+1}\hat{v}_N^{(1)}\phi_N^0(x) + \dfrac{1}{2N+3}\hat{v}_{N+1}^{(1)}\phi_N^1(x), & \text{if } N \text{ is even,} \\ \dfrac{1}{2N+3}\hat{v}_{N+1}^{(1)}\phi_N^0(x) + \dfrac{1}{2N+1}\hat{v}_N^{(1)}\phi_N^1(x), & \text{if } N \text{ is odd} \end{cases} \tag{2.52}$$

where

$$\phi_N^0(x) = \sum_{\substack{0 \leq p \leq N \\ p \text{ even}}} (2p+1)L_p(x), \quad \phi_N^1(x) = \sum_{\substack{1 \leq p \leq N \\ p \text{ odd}}} (2p+1)L_p(x).$$

If $v \in H^1(\Lambda)$, it can be approximated by a sequence of infinitely differentiable functions for which (2.52) holds. Then we can pass to the limit. Next by Lemma 2.2,

$$|\hat{v}_N^{(1)}| \leq \left(N + \frac{1}{2}\right)^{\frac{1}{2}} \|\partial_x v - P_{N-1}\partial_x v\| \leq cN^{\frac{3}{2}-r}\|v\|_r.$$

A similar estimate exists for $\left|\hat{v}_{N+1}^{(1)}\right|$. On the other hand, (2.48) leads to

$$\|\phi_N^0\|^2 = \sum_{\substack{0 \leq p \leq N \\ p \text{ even}}} 2(2p+1) \leq cN^2$$

and a similar estimate is valid for $\|\phi_N^1\|^2$. Finally the orthogonality of ϕ_N^0 and ϕ_N^1 implies

$$\|P_N \partial_x v - \partial_x P_N v\| \leq cN^{\frac{3}{2}-r}\|v\|_r$$

which completes the proof. ∎

Theorem 2.9. *For any* $v \in H^r(\Lambda), r \geq 0$ *and* $\mu \leq r$,

$$\|v - P_N v\|_\mu \leq cN^{\sigma(\mu,r)}\|v\|_r$$

where

$$\sigma(\mu,r) = \begin{cases} 2\mu - r - \frac{1}{2}, & \text{for } \mu \geq 1, \\ \frac{3}{2}\mu - r, & \text{for } 0 \leq \mu \leq 1, \\ \mu - r, & \text{for } \mu < 0. \end{cases} \tag{2.53}$$

Proof. Lemma 2.2 implies the desired result for $\mu = 0$. Now let $\mu > 0$. The space interpolation allows us to consider positive integer μ only. We now use the induction. Assume that the result is true for $\mu - 1$. Then

$$\|\partial_x v - P_N \partial_x v\|_{\mu-1} \leq cN^{\sigma(\mu-1,r-1)}\|v\|_r$$

from which and Lemma 2.3, it follows that

$$\begin{aligned} \|v - P_N v\|_\mu &\leq \|\partial_x v - P_N \partial_x v\|_{\mu-1} + \|P_N \partial_x v - \partial_x P_N v\|_{\mu-1} + \|v - P_N v\| \\ &\leq cN^{\sigma(\mu-1,r-1)}\|v\|_r + cN^{\sigma(\mu,r)}\|v\|_r. \end{aligned}$$

Since $\sigma(\mu-1, r-1) \leq \sigma(\mu,r)$, the conclusion for $\mu > 0$ follows. Finally a duality argument leads to the conclusion for $\mu < 0$. ■

The result of Theorem 2.9 is provided by Canuto and Quarteroni (1982a), except the case $\mu < 0$. Other estimates are as follows,

$$\begin{aligned} \|v - P_N v\|_{L^\infty} &\leq cN^{\frac{1}{2}-m} V\left(\partial_x^m v\right), \\ \|v - P_N v\|_{L^\infty} &\leq cN^{\frac{3}{4}-r}\|v\|_r \end{aligned}$$

where m is any positive integer and $r \geq \frac{3}{4}, V(v)$ denotes the total variation of v. The best approximation error in L^p-norm, $2 < p \leq \infty$ decays as the truncation error in L^2-norm, i.e.,

$$\inf_{\phi \in \mathbb{P}_N} \|v - \phi\|_{L^p} \leq cN^{-r}\|v\|_{W^{r,p}}.$$

The convergence rate in Theorem 2.9 is not optimal for $\mu > 0$. Let m be non-negative integer. Since $H^m(\Lambda)$ is a Hilbert space, the best approximation to v should be the orthogonal projection of v upon $\mathbb{P}_N$ in the following inner product

$$(v,w)_m = \sum_{k=0}^{m} \left(\partial_x^k v, \partial_x^k w\right), \qquad \forall\, v,w \in H^m(\Lambda).$$

The corresponding orthogonal projection satisfies

$$(v - P_N^m v, \phi)_m = 0, \qquad \forall\, \phi \in \mathbb{P}_N \tag{2.54}$$

and thus

$$\|v - P_N^m v\|_m = \inf_{\phi \in \mathbb{P}_N} \|v - \phi\|_m. \tag{2.55}$$

Furthermore, when the spectral methods are used for partial differential equations with homogeneous boundary conditions, other kinds of projections are needed for obtaining the optimal error estimates. Let

$$\mathbb{P}_N^{m,0} = \left\{\phi \in \mathbb{P}_N \mid \partial_x^k \phi(-1) = \partial_x^k \phi(1) = 0, 0 \le k \le m-1\right\} \tag{2.56}$$

and $\mathbb{P}_N^0 = \mathbb{P}_N^{1,0}$ for simplicity. Set

$$a_m(v,w) = (\partial_x^m v, \partial_x^m w)\,, \quad \forall\, v, w \in H_0^m(\Lambda)$$

and $a(v,w) = a_1(v,w)$. The commonly used inner product in $H_0^m(\Lambda)$ is $a_m(v,w)$. The H_0^m-orthogonal projection $P_N^{m,0} : H_0^m(\Lambda) \to \mathbb{P}_N^{m,0}$ is such a mapping that for any $v \in H_0^m(\Lambda)$,

$$a_m\left(v - P_N^{m,0} v, \phi\right) = 0, \qquad \forall\, \phi \in \mathbb{P}_N^{m,0}. \tag{2.57}$$

We now estimate the difference between v and $P_N^m v$, and the difference between v and $P_N^{m,0} v$.

Theorem 2.10. *Let m be a positive integer. Then for any $v \in H^r(\Lambda)$ and $0 \le \mu \le m \le r$,*

$$\|v - P_N^m v\|_\mu \le c N^{\mu - r} \|v\|_r.$$

Proof. We first assume that $\mu = m$ and let

$$W = \left\{w \in H^m(\Lambda) \mid A\left(\partial_x^k v\right) = 0, \quad 0 \le k \le m-1\right\}.$$

Because for any $w \in W$ and $k \le m-1$, $\partial_x^k v$ vanishes at certain point, so the Poincaré inequality (2.7) is valid. Hence W is a Hilbert space for the inner product $a_m(\cdot,\cdot)$, and the norm $\|\cdot\|_W = |\cdot|_m$ is equivalent to the norm $\|\cdot\|_m$. For each $v \in H^m(\Lambda)$, let

$$v^*(x) = q(x) + \int_{-1}^{x} P_{N-1}^{m-1} \partial_y v(y)\, dy.$$

The polynomial $q(x)$ of degree $m-1$, is chosen in such a way that $A\left(\partial_x^k v^* - \partial_x^k v\right) = 0$ for $0 \le k \le m-1$. We now use the induction. By Theorem 2.9, the conclusion is true for $m = 0$. Now suppose that it is valid for $m-1$. Then

$$\|v - v^*\|_W = |v - v^*|_m = \left|\partial_x v - P_{N-1}^{m-1}\partial_x v\right|_{m-1} \le cN^{m-r}\|v\|_r$$

and thus by (2.55),

$$\|v - P_N^m v\|_m \le \|v - v^*\|_m \le c\|v - v^*\|_W \le cN^{m-r}\|v\|_r.$$

This completes the induction and implies the conclusion for $\mu = m$.

Next we consider the case $\mu = 0$. Let $g \in L^2(\Lambda)$ and consider the auxiliary problem

$$(w, z)_m = (g, z), \qquad \forall\, z \in H^m(\Lambda). \tag{2.58}$$

We know from Lemma 2.1 that (2.58) possesses a unique solution. Moreover, the regularity tells us that $\|w\|_{2m} \le c\|g\|$. Putting $z = v - P_N^m v$ in (2.58), we get

$$(v - P_N^m v, g) = (v - P_N^m v, w)_m = (v - P_N^m v, w - P_N^m w)_m\,.$$

Consequently,

$$|(v - P_N^m, g)| \le cN^{-m}\|w\|_{2m}\|v - P_N^m v\|_m \le cN^{-m}\|g\|\|v - P_N^m v\|_m.$$

Then by a duality argument,

$$\|v - P_N^m v\| = \sup_{\substack{g\in L^2(\Lambda)\\ g\ne 0}} \frac{(v - P_N^m v, g)}{\|g\|} \le cN^{-m}\|v - P_N^m v\|_m \le cN^{-r}\|v\|_r.$$

The result for $0 \le \mu \le m$ follows from the space interpolation. ∎

Theorem 2.11. *Let m be a positive integer. Then for any $v \in H^r(\Lambda)\cap H_0^m(\Lambda)$ and $0 \le \mu \le m \le r$,*

$$\|v - P_N^{m,0} v\|_\mu \le cN^{\mu-r}\|v\|_r.$$

Proof. We first assert that for any $v \in H_0^m(\Lambda)$,

$$P_N^{m,0} v(x) = \int_{-1}^{x} P_{N-1}^{m-1,0}\partial_y v(y)\,dt, \quad m \ge 1. \tag{2.59}$$

Indeed the function on the right-hand side of (2.59) is in $\mathbb{P}_N$, vanishes at $x = -1$, and its derivatives up to the order of $m-1$ vanish at $x = -1, 1$. So it suffices to verify that it also vanishes at $x = 1$. Let $\xi = (1-x^2)^{m-1}$. By the definition of $P_N^{m,0}$,

$$\int_\Lambda \partial_x^{m-1}\left(P_{N-1}^{m-1,0}\partial_x v(x)\right)\partial_x^{m-1}\xi(x)\,dx = \int_\Lambda \partial_x^{m-1}\left(\partial_x v(x)\right)\partial_x^{m-1}\xi(x)\,dx.$$

Integrating $m-1$ times by parts, we deduce that

$$\begin{aligned}(2(m-1))!\int_\Lambda P_{N-1}^{m-1,0}\partial_x v(x)\,dx &= (2(m-1))!\int_\Lambda \partial_x v\,dx \\ &= (2(m-1))!(v(1)-v(-1)) = 0\end{aligned}$$

which ends the proof of (2.59). Furthermore,

$$\begin{aligned}|v - P_N^{m,0}v|_m &= |\partial_x v - P_{N-1}^{m-1,0}\partial_x v|_{m-1} = \cdots = \|\partial_x^m v - P_{N-m}\left(\partial_x^m v\right)\| \\ &\le c(N-m)^{m-r}\|\partial_x^m v\|_{r-m} \le cN^{m-r}\|v\|_r\end{aligned}$$

whence the conclusion is true for $\mu = m$. For $\mu = 0$, we consider the problem

$$a_m(w,z) = (g,z), \qquad \forall\, z \in H_0^m(\Lambda).$$

It has a unique solution and $\|w\|_{2m} \le c\|g\|$. Following the same line as in the proof of Theorem 2.10, we complete the derivation for $\mu = 0$. The general result is a consequence of the space interpolation. ∎

The proof of Theorem 2.11 is given by Bernardi and Maday (1994).

We now turn to the discrete Legendre approximation. There are three kinds of Gauss-type interpolations.

(i) **Legendre-Gauss interpolation**. In this case, $x^{(j)}$ are the $N+1$ roots of $L_{N+1}(x)$, and

$$\omega^{(j)} = \frac{2}{\left(1-\left(x^{(j)}\right)^2\right)\left(\partial_x L_{N+1}\left(x^{(j)}\right)\right)^2}, \qquad 0 \le j \le N.$$

(ii) **Legendre-Gauss-Radau interpolation**. In this case, $x^{(j)}$ are the $N+1$ roots of $L_N(x) + L_{N+1}(x)$, and

$$\omega^{(0)} = \frac{2}{(N+1)^2}, \qquad \omega^{(j)} = \frac{1}{(N+1)^2}\frac{1-x^{(j)}}{\left(L_N\left(x^{(j)}\right)\right)^2}, \qquad 1 \le j \le N.$$

(iii) **Legendre-Gauss-Lobatto interpolation.** In this case, $x^{(0)} = -1, x^{(N)} = 1, x^{(j)}, 1 \leq j \leq N-1$ are the roots of $\partial_x L_N(x)$, and

$$\omega^{(j)} = \frac{2}{N(N+1)} \frac{1}{\left(L_N\left(x^{(j)}\right)\right)^2}, \qquad 0 \leq j \leq N.$$

The Legendre interpolant of a function $v \in C\left(\bar{\Lambda}\right)$ is

$$I_N v(x) = \sum_{l=0}^{N} \tilde{v}_l L_l(x)$$

where

$$\tilde{v}_l = \frac{1}{\gamma_l} \sum_{l=0}^{N} v\left(x^{(j)}\right) L_l\left(x^{(j)}\right) \omega^{(j)},$$

$\gamma_l = \left(l + \frac{1}{2}\right)^{-1}$ for $l < N, \gamma_N = \left(N + \frac{1}{2}\right)^{-1}$ for the Legendre-Gauss interpolation and the Legendre-Gauss-Radau interpolation, and $\gamma_N = \dfrac{2}{N}$ for the Legendre-Gauss-Lobatto interpolation. Consequently in the third case,

$$\|\phi\| \leq \|\phi\|_{N,\omega} \leq \sqrt{2 + \frac{1}{N}} \|\phi\|, \quad \forall\, \phi \in \mathbb{P}_N. \tag{2.60}$$

Usually $\partial_x I_N v \neq I_N \partial_x v$. In order to estimate the error between v and $I_N v$, we need the following lemma.

Lemma 2.4. *Let $0 < s < 1$. There exists a projection $\Pi_N^s : H^\mu(\Lambda) \rightarrow \mathbb{P}_N$ such that for any $v \in H^r(\Lambda)$ and $0 \leq \mu \leq s \leq r$,*

$$\|v - \Pi_N^s v\|_\mu \leq c N^{\mu - r} \|v\|_r$$

where Π_N^s is the orthogonal projection with respect to a suitable inner product, whose associated norm is equivalent to the standard norm $\|\cdot\|_s$.

Proof. By an affine mapping, we transform Λ into $\Lambda^* = (0, \pi)$. For any $v \in H^s(\Lambda^*)$, denote by Mv the function obtained by the even reflection with respect to the origin and so $Mv \in H_p^s(-\pi, \pi)$. This space is equipped with the inner product

$$(v, w)_{H^s(-\pi,\pi)} = \sum_{l=-\infty}^{\infty} \left(1 + l^2\right)^s \hat{v}_l \bar{\hat{w}}_l$$

where $\hat{v}_l$ and $\hat{w}_l$ are the coefficients in Fourier expansions of v and w. So we define the following inner product in $H^s(\Lambda^*)$,

$$(v, w)_{H^s(\Lambda^*)} = \frac{1}{2}(Mv, Mw)_{H^s(-\pi,\pi)}$$

and denote by Π_N^s the orthogonal projection upon $\mathbb{P}_N$ with respect to this inner product. It is easily checked that $(v, v)_{H^s(\Lambda^*)}^{\frac{1}{2}}$ is a norm equivalent to $\|\cdot\|_s$. In order to prove this lemma, we first consider the case $\mu = s$. We have

$$\|v - \Pi_N^s v\|_s \leq c\|v\|_s, \quad \forall v \in H^s(\Lambda^*). \tag{2.61}$$

Further let $\Pi_N^1 : H^1(0,\pi) \to \mathbb{P}_N$ be the H^1-orthogonal projection. Then for any $v \in H^r(\Lambda^*)$ and $r \geq 1$,

$$\|v - \Pi_N^s v\|_s \leq c\|v - \Pi_N^1 v\|_s \leq cN^{s-r}\|v\|_r. \tag{2.62}$$

When $s \leq r \leq 1$, we use interpolation taking into account (2.61) and (2.62).

We next consider the case $\mu = 0$, arguing by a standard duality argument. Denote by $J_s : (H^s(\Lambda^*))' \to H^s(\Lambda^*)$ the Riesz isomorphism. It means that for any $w \in H^s(\Lambda^*)$,

$$(J_s g, w)_{H^s(\Lambda^*)} = \langle g, w\rangle_{\mathcal{L}((H^s(\Lambda^*))', H^s(\Lambda^*))}.$$

We assert that J_s maps $L^2(\Lambda^*)$ into $H^{2s}(\Lambda^*)$ with continuity. To prove it, let $\hat{w}_l, \hat{g}_l$ and $\hat{f}_l$ be the Fourier coefficients of Mw, Mg and $MJ_s g$ respectively. Then the above expression is equivalent to

$$\sum_{l=-\infty}^{\infty}\left(1+l^2\right)^s \hat{f}_l\hat{w}_l = \sum_{l=-\infty}^{\infty}\hat{g}_l\hat{w}_l, \quad \forall\, v \in H^s(\Lambda^*).$$

Notice that $\hat{w}_l = \hat{w}_{-l}, \hat{g}_l = \hat{g}_{-l}$ and $\hat{f}_l = \hat{f}_{-l}$. Taking $w = \cos lx$, we get

$$\left(1+l^2\right)^s \hat{f}_l = \hat{g}_l$$

whence $MJ_s g \in H^{2s}(-\pi,\pi)$ with

$$(MJ_s g, MJ_s g)_{H^{2s}(-\pi,\pi)} = (Mg, Mg)_{L^2(-\pi,\pi)}.$$

It means that J_s maps $L^2(\Lambda^*)$ into $H^{2s}(\Lambda^*)$ with the continuity. By this fact, (2.62) and the property of Riesz isomorphism described in the above, we obtain that

$$\begin{aligned}\|v-\Pi_N^s v\| &= \sup_{\substack{g\in L^2(\Lambda^*)\\ \|g\|=1}} (v-\Pi_N^s v, g) = \sup_{\substack{g\in L^2(\Lambda^*)\\ \|g\|=1}} (v-\Pi_N^s v, J_s g)_{H^s(\Lambda^*)} \\ &= \sup_{\substack{g\in L^2(\Lambda^*)\\ \|g\|=1}} (v-\Pi_N^s v, J_s g - \Pi_N^s J_s g)_{H^s(\Lambda^*)} \\ &\le cN^{s-r}\|v\|_r\cdot N^{-s}\sup_{\substack{g\in L^2(\Lambda^*)\\ \|g\|=1}} \|J_s g\|_{2s} \le cN^{-r}\|v\|_r.\end{aligned}$$

We deduce the result with $0<\mu<s$ by the space interpolation. ■

Theorem 2.12. *For any $v\in H^r(\Lambda), r>\frac{1}{2}$ and $0\le\mu\le r$,*

$$\|v-I_N v\|_\mu \le cN^{2\mu-r+\frac{1}{2}}\|v\|_r.$$

Proof. We first consider the case $\mu=0$. For any $\varepsilon>0$,

$$v-I_N v = \left(v-\Pi_N^{\frac{1}{2}+\varepsilon}v\right) - I_N\left(v-\Pi_N^{\frac{1}{2}+\varepsilon}v\right).$$

By (2.60),

$$\|I_N\left(v-\Pi_N^{\frac{1}{2}+\varepsilon}v\right)\|^2 \le \left\|I_N\left(v-\Pi_N^{\frac{1}{2}+\varepsilon}v\right)\right\|_N^2 = \left\|v-\Pi_N^{\frac{1}{2}+\varepsilon}v\right\|_N^2 \le c\left\|v-\Pi_N^{\frac{1}{2}+\varepsilon}v\right\|_{L^\infty}^2.$$

Let us denote by $B^s_{p,q}(\Lambda)$ the Besov space of order s and index p,q, see Bergh and Löfström (1976). We know that $B^{\frac{1}{2}}_{2,1}(\Lambda)\subseteq L^\infty(\Lambda)$ with continuous injection, and for any $\varepsilon>0, B^{\frac{1}{2}}_{2,1} = \left(H^{\frac{1}{2}+\varepsilon}(\Lambda), H^{\frac{1}{2}-\varepsilon}(\Lambda)\right)_{\frac{1}{2},1}$. Thus

$$\left\|v-\Pi_N^{\frac{1}{2}+\varepsilon}v\right\|_{L^\infty}^2 \le c\left\|v-\Pi_N^{\frac{1}{2}+\varepsilon}v\right\|_{B^{\frac{1}{2}}_{2,1}(\Lambda)}^2 \le c\left\|v-\Pi_N^{\frac{1}{2}+\varepsilon}v\right\|_{H^{\frac{1}{2}+\varepsilon}(\Lambda)}\left\|v-\Pi_N^{\frac{1}{2}+\varepsilon}v\right\|_{H^{\frac{1}{2}-\varepsilon}(\Lambda)}.$$

Using Lemma 2.4 and letting $\varepsilon\to 0$, we get the desired result . For $\mu>0$, we have

$$\begin{aligned}\|v-I_N v\|_\mu &\le \|v-P_N v\|_\mu + cN^{2\mu}\|P_N v - I_N v\| \\ &\le cN^{\sigma(\mu,r)}\|v\|_r + cN^{2\mu}(\|v-P_N v\| + \|v-I_N v\|) \\ &\le cN^{\sigma(\mu,r)}\|v\|_r + cN^{2\mu-r+\frac{1}{2}}\|v\|_r.\end{aligned}$$

■

Canuto and Quarteroni (1982a) first proved Theorem 2.12. It could be improved in some special cases. Bernardi and Maday (1992a,1992b) showed that for the Legendre-Gauss interpolation and any $v \in H^r(\Lambda), r > \frac{1}{2}$,

$$\|v - I_N v\| \leq cN^{-r}\|v\|_r,$$

and for the Legendre-Gauss-Lobatto interpolation and any $v \in H^r(\Lambda), 0 \leq \mu \leq 1, \mu < 2r - 1$,

$$\|v - I_N v\|_\mu \leq cN^{\mu-r}\|v\|_r. \tag{2.63}$$

They also showed that for the Legendre-Gauss-Lobatto interpolation and any positive integer m,

$$|v - I_N v|_1 \leq cN^{1-m}\|v\|_m.$$

We also refer to Maday (1990).

In the end of this section, it is noted that Theorem 2.12 and (2.42) imply that for any $v \in H^r(\Lambda), r > \dfrac{1}{2}$ and $\phi \in \mathbb{P}_N$,

$$|(v, \phi) - (v, \phi)_N| \leq cN^{\lambda - r}\|v\|_r\|\phi\| \tag{2.64}$$

where $\lambda = \frac{1}{2}$ in general. But $\lambda = 0$ for the Legendre-Gauss interpolation and the Legendre-Gauss-Lobatto interpolation.

2.5. Chebyshev Approximations

Let $\Lambda = (-1, 1)$ and $\omega(x) = (1 - x^2)^{-\frac{1}{2}}$ in this section. The Chebyshev polynomial of the first kind of degree l is

$$T_l(x) = \cos(l \arccos x). \tag{2.65}$$

It is the l-th eigenfunction of the singular Sturm-Liouville problem

$$\partial_x\left(\left(1 - x^2\right)^{\frac{1}{2}}\partial_x v\right) + \lambda\left(1 - x^2\right)^{-\frac{1}{2}} v = 0, \quad x \in \Lambda, \tag{2.66}$$

related to the l-th eigenvalue $\lambda_l = l^2$. Obviously $T_0(x) = 1, T_1(x) = x$, and they satisfy the recurrence relations

$$T_{l+1}(x) = 2xT_l(x) - T_{l-1}(x), \quad l \geq 1, \tag{2.67}$$

$$2T_l(x) = \frac{1}{l+1}\partial_x T_{l+1}(x) - \frac{1}{l-1}\partial_x T_{l-1}(x), \quad l \geq 1. \tag{2.68}$$

It can be verified that $T_l(1) = 1, T_l(-1) = (-1)^l, \partial_x T_l(1) = l^2$ and $\partial_x T_l(-1) = (-1)^{l+1} l^2$. Moreover

$$|T_l(x)| \leq 1, \quad |\partial_x T_l(x)| \leq l^2, \quad x \in \Lambda. \tag{2.69}$$

The set of Chebyshev polynomials is the L^2_ω-orthogonal system in Λ, i.e.,

$$\int_\Lambda T_l(x) T_m(x) \omega(x)\, dx = \frac{\pi}{2} c_l \delta_{l,m} \tag{2.70}$$

with $c_0 = 2$ and $c_l = 1$ for $l \geq 1$. The Chebyshev expansion of a function $v \in L^2_\omega(\Lambda)$ is

$$v(x) = \sum_{l=0}^{\infty} \hat{v}_l T_l(x) \tag{2.71}$$

with

$$\hat{v}_l = \frac{2}{\pi c_l} \int_\Lambda v(x) T_l(x) \omega(x)\, dx.$$

We consider the differentiation. Let $\hat{v}_l^{(1)}$ be the coefficients of Chebyshev expansion of $\partial_x v, l = 0, 1, \dots$. By (2.68),

$$2l\hat{v}_l = c_{l-1}\hat{v}_{l-1}^{(1)} - \hat{v}_{l+1}^{(1)}$$

and so

$$\hat{v}_l^{(1)} = \frac{2}{c_l} \sum_{\substack{p=l+1 \\ p+l \text{ odd}}}^{\infty} p\hat{v}_p. \tag{2.72}$$

This formula is true for any $v \in H^1_\omega(\Lambda)$.

We now present the inverse inequalities.

Theorem 2.13. *For any $\phi \in \mathbb{P}_N$ and $1 \leq p \leq q \leq \infty$,*

$$\|\phi\|_{L^q_\omega} \leq \left(\frac{Np_0 + 1}{\pi}\right)^{\frac{1}{p} - \frac{1}{q}} \|\phi\|_{L^p_\omega}$$

where p_0 is the same as in Theorem 2.1.

Proof. Let $\Lambda^* = (0, 2\pi)$ and make use of the transformation

$$x = \cos y, \qquad v^*(y) = v(\cos y). \tag{2.73}$$

Since $\dfrac{dy}{dx} = -\omega(x)$, we have

$$\|\phi\|_{L^p_\omega} = \left(\frac{1}{2}\right)^{\frac{1}{p}} \|\phi^*\|_{L^p(\Lambda^*)}, \quad 1 \le p \le \infty. \tag{2.74}$$

Then the desired result comes from Theorem 2.1. ∎

Theorem 2.14. *Let m be a non-negative integer and $2 \le p \le \infty$. Then for any $\phi \in \mathbb{P}_N$,*

$$\|\partial_x^m \phi\|_{L^p_\omega} \le cN^{2m} \|\phi\|_{L^p_\omega}.$$

Also for any $r \ge 0$,

$$\|\phi\|_{r,\omega} \le cN^{2r} \|\phi\|_\omega.$$

Proof. Let $\hat{\phi}_l$ and $\hat{\phi}_l^{(1)}$ be the coefficients of Chebyshev expansions of ϕ and $\partial_x \phi$ respectively. By (2.72),

$$\left(c_l \hat{\phi}_l^{(1)}\right)^2 \le 4 \left(\sum_{\substack{p=l+1 \\ p+l \text{ odd}}}^{N} p^2 \right) \left(\sum_{\substack{p=l+1 \\ p+l \text{ odd}}}^{N} \hat{\phi}_l^2 \right) \le \frac{2}{3} N(N+1)(2N+1) \sum_{l=0}^{N} \hat{\phi}_l^2.$$

Accordingly,

$$\|\partial_x \phi\|_\omega^2 = \frac{\pi}{2} \sum_{l=0}^{N-1} c_l \left(\hat{\phi}_l^{(1)}\right)^2 \le \frac{\pi}{3} N(N+1)(2N+1) \sum_{l=0}^{N-1} \frac{1}{c_l} \sum_{l=0}^{N} \hat{\phi}_l^2 \le 4N^4 \|\phi\|_\omega^2$$

from which, Theorem 2.6 and a repetition, the first conclusion follows. We now prove the second one. If $r = m + \sigma$ and m is a non-negative integer, $0 < \sigma < 1$, then by an inequality parallel to (2.15),

$$\|\phi\|_{r,\omega} \le \|\phi\|_{m+1,\omega}^{\sigma} \|\phi\|_{m,\omega}^{1-\sigma} \le cN^{2r} \|\phi\|_\omega.$$

∎

The L^2_ω-orthogonal projection $P_N : L^2_\omega(\Lambda) \to \mathbb{P}_N$ is such a mapping that for any $v \in L^2_\omega(\Lambda)$,

$$(v - P_N v, \phi)_\omega = 0, \qquad \forall \phi \in \mathbb{P}_N,$$

or equivalently,

$$P_N v(x) = \sum_{l=0}^{N} \hat{v}_l T_l(x).$$

We now deal with the error between v and $P_N v$.

Lemma 2.5. *For any* $v \in H^r_\omega(\Lambda)$ *and* $r \geq 0$,

$$\|v - P_N v\|_\omega \leq cN^{-r}\|v\|_{r,\omega}.$$

Proof. By the space interpolation, it is enough to prove it for non-negative integer r. So by transformation (2.73), $v^* \in H^r_p(\Lambda^*)$. Since $\left|\frac{dx}{dy}\right| = |\sin y| \leq 1$, we get

$$\|v^*\|_{H^r(\Lambda^*)} \leq c\|v\|_{r,\omega}.$$

Let P_N^* be the symmetric truncation of the Fourier transformation up to degree N as in Section 2.2. Then for any $v \in L^2_\omega(\Lambda)$,

$$(P_N v)^* = P_N^* v^*. \tag{2.75}$$

Therefore by Theorem 2.3 and (2.74),

$$\|v - P_N v\|_\omega = \frac{1}{\sqrt{2}}\|v^* - P_N^* v^*\|_{L^2(\Lambda^*)} \leq cN^{-r}\|\partial_y^r v^*\|_{L^2(\Lambda^*)} \leq cN^{-r}\|v\|_{r,\omega}.$$

■

Generally $\partial_x P_N v(x) \neq P_N \partial_x v(x)$. But we have the following result.

Lemma 2.6. *For any* $v \in H^r_\omega(\Lambda)$ *and* $0 \leq \mu \leq r-1$,

$$\|P_N \partial_x v - \partial_x P_N v\|_{\mu,\omega} \leq cN^{2\mu - r + \frac{3}{2}}\|v\|_{r,\omega}.$$

Proof. It suffices to prove it for $r \geq 1$ and $\mu = 0$. As in the proof of Lemma 2.3, we can assume that v is a smooth function. Let

$$\partial_x P_N v(x) = \sum_{l=0}^{N-1} \hat{w}_l T_l(x), \qquad c_l \hat{w}_l = 2 \sum_{\substack{p=l+1 \\ p+l \text{ odd}}}^{N} p\hat{v}_p.$$

By virtue of (2.72),

$$c_l\left(\hat{v}_l^{(1)} - \hat{w}_l\right) = \begin{cases} 2 \displaystyle\sum_{\substack{p=N+2 \\ p+N \text{ even}}}^{\infty} p\hat{v}_p = \hat{v}_{N+1}^{(1)}, & \text{if } l+N \text{ is odd}, \\ 2 \displaystyle\sum_{\substack{p=N+1 \\ p+N \text{ odd}}}^{\infty} p\hat{v}_p = \hat{v}_N^{(1)}, & \text{if } l+N \text{ is even}. \end{cases}$$

So

$$P_N \partial_x v(x) - \partial_x P_N v(x) = \begin{cases} \hat{v}_N^{(1)} \phi_N^0(x) + \hat{v}_{N+1}^{(1)} \phi_N^1(x), & \text{if } N \text{ is even,} \\ \hat{v}_{N+1}^{(1)} \phi_N^0(x) + \hat{v}_N^{(1)} \phi_N^1(x), & \text{if } N \text{ is odd} \end{cases}$$

where

$$\phi_N^0(x) = \sum_{\substack{0 \le p \le N \\ p \text{ even}}} \frac{1}{c_p} T_p(x), \quad \phi_N^1(x) = \sum_{\substack{1 \le p \le N \\ p \text{ odd}}} T_p(x).$$

By Lemma 2.5, we have

$$\left|\hat{v}_N^{(1)}\right| \le c\|\partial_x v - P_{N-1}\partial_x v\|_\omega \le cN^{1-r}\|v\|_{r,\omega}$$

and a similar result for $\left|\hat{v}_{N+1}^{(1)}\right|$. On the other hand, we know from (2.70) that $\|\phi_N^0\|_\omega \simeq N^{\frac{1}{2}}$ and $\|\phi_N^1\|_\omega \simeq N^{\frac{1}{2}}$. Thus, noting that ϕ_N^0 and ϕ_N^1 are orthogonal, we obtain

$$\|P_N \partial_x v - \partial_x P_N v\|_\omega \le cN^{\frac{3}{2}-r}\|v\|_{r,\omega}.$$

■

Theorem 2.15. *For any $v \in H_\omega^r(\Lambda), r \ge 0$ and $\mu \le r$,*

$$\|v - P_N v\|_{\mu,\omega} \le cN^{\sigma(\mu,r)}\|v\|_{r,\omega}$$

where $\sigma(\mu, r)$ is given by (2.53).

Proof. The conclusion for $\mu = 0$ is true by Lemma 2.5. Now let $\mu > 0$. The space interpolation allows us to consider positive integer μ only. We shall use induction. Assume that the result is valid for $\mu - 1$. Then by Lemma 2.6,

$$\begin{aligned} \|v - P_N v\|_{\mu,\omega} &\le \|\partial_x v - P_N \partial_x v\|_{\mu-1,\omega} + \|P_N \partial_x v - \partial_x P_N v\|_{\mu-1,\omega} + \|v - P_N v\|_\omega \\ &\le cN^{\sigma(\mu-1,r-1)}\|v\|_{r,\omega} + cN^{\sigma(\mu,r)}\|v\|_{r,\omega} \le cN^{\sigma(\mu,r)}\|v\|_{r,\omega}. \end{aligned}$$

Finally a duality argument completes the proof for $\mu < 0$. ■

The proof of the above theorem is due to Canuto and Quarteroni (1982a), except the case $\mu < 0$. Another estimation is that for any $v \in W^{r,p}(\Lambda)$ and $1 \le p \le \infty$,

$$\|v - P_N v\|_{L_\omega^p} \le c\sigma_N(p)N^{-r}\|v\|_{W_\omega^{r,p}}$$

where $\sigma_N(p) = 1$ for $1 < p < \infty$ and $\sigma_N(1) = \sigma_N(\infty) = 1 + \ln N$.

In order to define the best approximation in $H_\omega^m(\Lambda)$, we introduce the inner product

$$(v, w)_{m,\omega} = \sum_{k=0}^{m} \left(\partial_x^k v, \partial_x^k w\right)_\omega, \quad \forall\, v, w \in H_\omega^m(\Lambda). \tag{2.76}$$

The H_ω^m-orthogonal projection $P_N^m : H_\omega^m(\Lambda) \to \mathbb{P}_N$ is such a mapping that for any $v \in H_\omega^m(\Lambda)$,

$$(v - P_N^m v, \phi)_{m,\omega} = 0, \quad \forall \phi \in \mathbb{P}_N.$$

Consequently,

$$\|v - P_N^m v\|_{m,\omega} = \inf_{\phi \in \mathbb{P}_N} \|v - \phi\|_{m,\omega}. \tag{2.77}$$

In the numerical analysis of Chebyshev spectral methods applied to partial differential equations with homogeneous boundary conditions, we need other kinds of projections for the derivations of optimal error estimates. To do this, let

$$\tilde{a}_{m,\omega}(v, w) = (\partial_x^m v, \partial_x^m w)_\omega, \tag{2.78}$$

$$a_{m,\omega}(v, w) = (\partial_x^m v, \partial_x^m (w\omega)). \tag{2.79}$$

In particular, $\tilde{a}_\omega(v, w) = \tilde{a}_{1,\omega}(v, w)$ and $a_\omega(v, w) = a_{1,\omega}(v, w)$. We first analyze some properties of $a_{m,\omega}(v, w)$. For the sake of simplicity, we consider the case $m = 1$ only.

Lemma 2.7. *For any $v \in H_{0,\omega}^1(\Lambda)$,*

$$\|v\omega^2\|_\omega \le |v|_{1,\omega}.$$

Proof. Let $g(t) = \frac{1}{t}\int_0^t f(s)\, ds$. From Theorem 4.1 in Chapter 3 of Lions (1967),

$$\int_0^\infty |g(t)|^2 t^{-\frac{1}{2}}\, dt \le \frac{16}{9}\int_0^\infty |f(t)|^2 t^{-\frac{1}{2}}\, dt.$$

Set $t = x + 1$. Then

$$\int_{-1}^\infty |g(x+1)|^2 (x+1)^{-\frac{1}{2}}\, dx \le \frac{16}{9}\int_{-1}^\infty |f(x+1)|^2 (x+1)^{-\frac{1}{2}}\, dx.$$

Now take $f(x+1) = \partial_x v(x)$ for $-1 \le x \le 0$ and $f(x+1) = 0$ for $x > 0$. Then $g(x+1) = \frac{1}{x+1} v(x)$ and so

$$\int_{-1}^0 |v(x)|^2 \omega^5(x)\, dx \le \frac{16}{9}\sqrt{2}\int_{-1}^0 |\partial_x v(x)|^2 \omega(x)\, dx.$$

We can get a similar result on the interval [0,1]. The combination of them reads

$$\|v\omega^2\|_\omega^2 \leq \frac{16}{9}\sqrt{2}|v|_{1,\omega}^2$$

which implies that $v^2(x)\omega^3(x)$ tends to zero as x goes to ± 1. Furthermore, by integrating by parts,

$$\int_\Lambda \partial_x v \partial_x(v\omega)\,dx = \int_\Lambda |\partial_x v|^2\omega\,dx - \frac{1}{2}\int_\Lambda \left(2x^2+1\right) v^2\omega^5\,dx, \tag{2.80}$$

$$\int_\Lambda \partial_x v \partial_x(v\omega)\,dx = \int_\Lambda |\partial_x(v\omega)|^2\omega^{-1}\,dx + \frac{1}{2}\int_\Lambda v^2\omega^5\,dx. \tag{2.81}$$

Subtracting (2.81) from (2.80) yields

$$\int_\Lambda |\partial_x v|^2\omega\,dx - \int_\Lambda \left(x^2+1\right) v^2\omega^5\,dx = \int_\Lambda |\partial_x(v\omega)|^2\omega^{-1}\,dx \geq 0$$

and the desired result follows. ∎

Lemma 2.8. *For any* $v, w \in H^1_{0,\omega}(\Lambda)$,

$$\begin{aligned} a_\omega(v,v) &\geq \frac{1}{4}\|v\|_{1,\omega}^2, \\ |a_\omega(v,w)| &\leq 2|v|_{1,\omega}|w|_{1,\omega}. \end{aligned}$$

Proof. We get from (2.80) and (2.81) that

$$a_\omega(v,v) = \frac{1}{4}\|\partial_x v\|_\omega^2 + \frac{1}{4}\int_\Lambda v^2(x)\omega^3(x)\,dx + \frac{3}{4}\int_\Lambda |\partial_x(v(x)\omega(x))|^2\omega^{-1}(x)\,dx$$

which leads to the first conclusion. Next by Lemma 2.7, for any $z \in L^2_\omega(\Lambda)$ and $w \in H^1_{0,\omega}(\Lambda)$,

$$\begin{aligned} |(z, \partial_x(w\omega))| &\leq |(z, \partial_x w)_\omega| + \left|\left(z, xw\omega^2\right)_\omega\right| \\ &\leq \|z\|_\omega |w|_{1,\omega} + \|z\|_\omega \|w\omega^2\|_\omega \leq 2\|z\|_\omega |w|_{1,\omega}. \end{aligned}$$

Putting $z = \partial_x v$, we complete the proof. ∎

The $H^1_{0,\omega}$-orthogonal projection $\tilde{P}_N^{1,0} : H^1_{0,\omega}(\Lambda) \to \mathbb{P}_N^0$ is such a mapping that for any $v \in H^1_{0,\omega}(\Lambda)$,

$$\tilde{a}_\omega\left(v - \tilde{P}_N^{1,0}v, \phi\right) = 0, \quad \forall\, \phi \in \mathbb{P}_N^0.$$

The other $H^1_{0,\omega}$-orthogonal projection $P_N^{1,0} : H^1_{0,\omega}(\Lambda) \to \mathbb{P}_N^0$ is such a mapping that for any $v \in H^1_{0,\omega}(\Lambda)$,

$$a_\omega\left(v - P_N^{1,0}v, \phi\right) = 0, \quad \forall\, \phi \in \mathbb{P}_N^0. \tag{2.82}$$

We now estimate the difference between v, $P_N^1 v$, $\tilde{P}_N^{1,0} v$ and $P_N^{1,0} v$.

Theorem 2.16. *For any $v \in H^r_\omega(\Lambda)$ and $0 \le \mu \le 1 \le r$,*

$$\|v - P_N^1 v\|_{\mu,\omega} \le cN^{\mu-r}\|v\|_{r,\omega}.$$

Proof. We first set $\mu = 1$. Let

$$\begin{aligned} A_\omega(w) &= \int_\Lambda w(x)\omega(x)\,dx, \\ W_\omega &= \left\{w \in H^1_\omega(\Lambda) \mid A_\omega(w) = 0\right\}. \end{aligned}$$

W_ω is a Hilbert space for the inner product $\tilde{a}_\omega(\cdot,\cdot)$. In fact, the Poincaré inequality (2.7) is valid for any $w \in W_\omega$ and the norm $\|\cdot\|_{W_\omega} = \tilde{a}_\omega^{\frac{1}{2}}(\cdot,\cdot)$ is a norm equivalent to $\|\cdot\|_{1,\omega}$. For each $v \in H^1_\omega(\Lambda)$, put

$$v^*(x) = q + \int_{-1}^{x} P_{N-1}\partial_y v(y)\,dy.$$

The constant q is chosen in such a way that $A_\omega(v - v^*) = 0$. Thus by Theorem 2.15,

$$\|v - v^*\|_{W_\omega} = \|\partial_x v - P_{N-1}\partial_x v\|_\omega \le cN^{1-r}\|v\|_{r,\omega}$$

from which and (2.77), we complete the proof for $\mu = 1$.

We next deal with the case $\mu = 0$. Let $g \in L^2_\omega(\Lambda)$ and consider the problem

$$(w, z)_{1,\omega} = (g, z)_\omega, \qquad \forall\, z \in H^1_\omega(\Lambda). \tag{2.83}$$

Since the left-hand side is the inner product of $H^1_\omega(\Lambda)$, the existence and uniqueness of w is ensured by Riesz representation form. Taking $z = w$ in (2.83), we get that $\|w\|_{1,\omega} \le c\|g\|_\omega$. Now, let w vary in $\mathcal{D}(\Lambda)$ and then in the sense of distributions,

$$-\,\partial_x\left(\partial_x w(x)\omega(x)\right) = \left(g(x) - w(x)\right)\omega(x). \tag{2.84}$$

Since

$$\begin{aligned} |\partial_x w(x)\omega(x) - \partial_x w(x^*)\omega(x^*)| &= \left|\int_{x^*}^{x} (g(y) - w(y))\,\omega(y)\,dy\right| \\ &\le \|g - w\|_\omega \,|\arccos x - \arccos x^*|^{\frac{1}{2}}\,, \end{aligned}$$

$\partial_x w(x)\omega(x)$ is meaningful at $x = \pm 1$. Multiplying (2.84) by any $z \in H^1_\omega(\Lambda)$ and integrating by parts, we obtain that

$$\begin{aligned} &\partial_x w(1)z(1)\omega(1) - \partial_x w(-1)z(-1)\omega(-1) = \\ &\qquad \int_\Lambda \partial_x w(x)\partial_x z(x)\omega(x) - \int_\Lambda (g(x) - w(x))\,z(x)\omega(x)\,dx, \quad \forall\, z \in H^1_\omega(\Lambda). \end{aligned}$$

Thus by (2.83), $\partial_x w(1)\omega(1) = \partial_x w(-1)\omega(-1) = 0$. Furthermore by (2.84), $-\partial_x^2 w(x) = g(x) - w(x) + \partial_x w(x)\partial_x\omega(x)\omega^{-1}(x)$. Therefore it remains to prove that $\partial_x w\partial_x\omega\omega^{-1} \in L^2_\omega(\Lambda)$. Because of $\partial_x\omega(x)\omega^{-1}(x) = x\omega^2(x)$, we deduce from (2.84) that

$$\begin{aligned} \int_{-1}^{0} \left(\partial_x w(x)\partial_x\omega(x)\right)^2 \omega^{-1}(x)\,dx &\le \int_{-1}^{0} \left(\partial_x w(x)\right)^2 \omega^5(x)\,dx \\ &\le \int_{-1}^{0} \left[\omega^2(x) \int_{-1}^{x} (w(y) - g(y))\omega(y)\,dy\right]^2 \omega^{-1}(x)\,dx \\ &\le c\int_{-1}^{0} \left[\frac{1}{x+1} \int_{-1}^{x} (w(y) - g(y))\omega(y)\,dy\right]^2 \sqrt{x+1}\,dx. \end{aligned}$$

Using Hardy inequality (see Hardy, Littlewood and Pólya (1952)), we obtain

$$\int_{-1}^{0} \left(\partial_x w(x)\right)^2 \omega^5(x)\,dx \le c\int_{-1}^{0} (w(x) - g(x))^2\omega(x)\,dx.$$

A similar result holds on the interval (0,1). So

$$\|\partial_x^2 w\|_\omega \le c\left(\|w\|_\omega + \|g\|_\omega\right).$$

By taking $z = v - P^1_N v$ in (2.83), it follows that

$$\begin{aligned} \left|(v - P^1_N v, g)_\omega\right| &= \left|(w, v - P^1_N)_{1,\omega}\right| = \left|(w - P^1_N w, v - P^1_N v)_{1,\omega}\right| \\ &\le cN^{-1}\|w\|_{2,\omega}\|v - P^1_N v\|_{1,\omega} \le cN^{-1}\|g\|_\omega\|v - P^1_N v\|_{1,\omega}. \end{aligned}$$

Then by the previous result,

$$\|v - P^1_N v\|_\omega \le \sup_{\substack{g \in L^2_\omega(\Lambda) \\ g \ne 0}} \frac{(v - P^1_N v, g)_\omega}{\|g\|_\omega} \le cN^{-1}\|v - P^1_N v\|_{1,\omega} \le cN^{-r}\|v\|_{r,\omega}.$$

We can derive the result for $0 < \mu < 1$ by the space interpolation and complete the proof. ∎

Theorem 2.17. *For any* $v \in H^r_\omega(\Lambda) \cap H^1_{0,\omega}(\Lambda)$ *and* $r \geq 1$,

$$\|v - \tilde{P}^{1,0}_N v\|_{1,\omega} \leq cN^{1-r}\|v\|_{r,\omega}.$$

Proof. Let

$$\begin{aligned} v^*(x) &= \int_{-1}^{x} P_{N-1}\partial_y v(y)\,dy, \\ Mv(x) &= \int_{-1}^{x} \left(P_{N-1}\partial_y v(y)\,dy - \frac{1}{2}v^*(1)\right)\,dx. \end{aligned}$$

Then $Mv \in \mathbb{P}^0_N$ and

$$\|\partial_x v - \partial_x(Mv)\|_\omega \leq \|\partial_x v - P_{N-1}\partial_x v\|_\omega + \frac{1}{2}\left(\int_\Lambda \omega(x)\,dx\right)^{\frac{1}{2}} |v^*(1)|\,.$$

On the other hand,

$$\begin{aligned} |v^*(1)| = |v(1) - v^*(1)| &= \left|\int_\Lambda (\partial_x v(x) - P_{N-1}\partial_x v(x))\,dx\right| \\ &\leq \left(\int_\Lambda \omega^{-1}(x)\,dx\right)^{\frac{1}{2}} \|\partial_x v - P_{N-1}\partial_x v\|_\omega \end{aligned}$$

and so

$$\|\partial_x v - \partial_x(Mv)\|_\omega \leq cN^{1-r}\|v\|_{r,\omega}.$$

Finally the conclusion comes, since $\tilde{P}^{1,0}_N v$ is the best approximation to v in the norm associated to the inner product (2.78). ■

Theorem 2.18. *For any* $v \in H^r_\omega(\Lambda) \cap H^1_{0,\omega}(\Lambda)$ *and* $0 \leq \mu \leq 1 \leq r$,

$$\|v - P^{1,0}_N v\|_{\mu,\omega} \leq cN^{\mu-r}\|v\|_{r,\omega}.$$

Proof. By Lemma 2.8, for any $\phi \in \mathbb{P}^0_N$,

$$\begin{aligned} \left\|v - P^{1,0}_N v\right\|^2_{1,\omega} &\leq ca_\omega\left(v - P^{1,0}_N v, v - P^{1,0}_N v\right) = ca_\omega\left(v - P^{1,0}_N v, v - \phi\right) \\ &\leq c\|v - P^{1,0}_N v\|_{1,\omega}\|v - \phi\|_{1,\omega}. \end{aligned}$$

Therefore

$$\|v - P^{1,0}_N v\|_{1,\omega} \leq c \inf_{\phi \in \mathbb{P}^0_N} \|v - \phi\|_{1,\omega}$$

and the conclusion for $\mu = 1$ follows from Theorem 2.17.

For the case $\mu = 0$, let us consider the auxiliary problem

$$a_\omega(z, w) = (g, z)_\omega, \quad \forall\, z \in H^1_{0,\omega}(\Lambda). \tag{2.85}$$

Since the transpose form $a_\omega(w, z)$ satisfies the condition (2.9), this problem has a unique solution. Further, we can verify that $\|w\|_{2,\omega} \le c\|g\|_\omega$. By virtue of (2.82), (2.85), Lemma 2.8 and the result for $\mu = 1$, we know that

$$\begin{aligned}
\left|\left(v - P_N^{1,0}v, g\right)_\omega\right| &= \left|a_\omega\left(v - P_N^{1,0}v, w\right)\right| = \left|a_\omega\left(v - P_N^{1,0}v, w - P_N^{1,0}w\right)\right| \\
&\le c\|v - P_N^{1,0}v\|_{1,\omega}\|w - P_N^{1,0}w\|_{1,\omega} \\
&\le cN^{-r}\|w\|_{2,\omega}\|v\|_{r,\omega} \le cN^{-r}\|g\|_\omega\|v\|_{r,\omega}
\end{aligned}$$

whence

$$\|v - P_N^{1,0}v\|_\omega = \sup_{\substack{g \in L^2_\omega(\Lambda) \\ g \ne 0}} \frac{\left(v - P_N^{1,0}v, g\right)_\omega}{\|g\|_\omega} \le cN^{-r}\|v\|_{r,\omega}.$$

The result for $0 < \mu < 1$ follows from the space interpolation. ∎

The first proof of the above three theorems can be found in Maday and Quarteroni (1981).

For the discrete Chebyshev approximations, there are also three kinds of Gauss-type interpolations.

(i) **Chebyshev-Gauss interpolation**. In this case, $x^{(j)} = \cos\frac{\pi(2j+1)}{2N+1}$ and $\omega^{(j)} = \frac{\pi}{N+1}$ for $0 \le j \le N$.

(ii) **Chebyshev-Gauss-Radau interpolation**. In this case, $x^{(j)} = \cos\frac{2\pi j}{2N+1}$ for $0 \le j \le N$, and $\omega^{(0)} = \frac{\pi}{2N+1}, \omega^{(j)} = \frac{\pi}{N+1}$ for $1 \le j \le N$.

(iii) **Chebyshev-Gauss-Lobatto interpolation**. In this case, $x^{(j)} = \cos\frac{\pi j}{N}$ for $0 \le j \le N, \omega^{(0)} = \omega^{(N)} = \frac{\pi}{2N}$ and $\omega^{(j)} = \frac{\pi}{N}$ for $1 \le j \le N-1$.

The Chebyshev interpolant of a function $v \in C(\bar{\Lambda})$ is

$$I_N v = \sum_{l=0}^{N} \tilde{v}_l T_l(x)$$

where

$$\tilde{v}_l = \frac{1}{\gamma_l}\sum_{j=0}^{N} v\left(x^{(j)}\right) T_l\left(x^{(j)}\right)\omega^{(j)}$$

$\gamma_l = \frac{\pi}{2}c_l$ for $l < N$; $\gamma_N = \frac{\pi}{2}$ for the Chebyshev-Gauss interpolation and the Chebyshev-Gauss-Radau interpolation, and $\gamma_N = \pi$ for the Chebyshev-Gauss-Lobatto interpolation. As a consequence, for Chebyshev-Gauss-Lobatto interpolation and any $\phi \in \mathbb{P}_N$,

$$\|\phi\|_\omega \le \|\phi\|_{N,\omega} \le \sqrt{2}\|\phi\|_\omega. \tag{2.86}$$

In general, $\partial_x I_N v \neq I_N \partial_x v$. But we have the following result.

Theorem 2.19. *For any* $v \in H^r_\omega(\Lambda), r > \frac{1}{2}$ *and* $0 \le \mu \le r$,

$$\|v - I_N v\|_{\mu,\omega} \le cN^{2\mu - r}\|v\|_{r,\omega}.$$

Proof. We use the transformation (2.73) and denote by I_N^* the corresponding interpolation associated with the interpolation points in Λ^*. It is easy to see that $(I_N v)^* = I_N^* v^*$ and thus

$$\|v - I_N v\|_\omega = \frac{1}{\sqrt{2}}\|v^* - I_N^* v\|_{L^2(\Lambda^*)}.$$

Hence the result for $\mu = 0$ comes from Theorem 2.4 immediately. For $\mu > 0$, we have from Theorem 2.14 that

$$\|v - I_N v\|_{\mu,\omega} \le \|v - P_N v\|_{\mu,\omega} + cN^{2\mu}\|P_N v - I_N v\|_\omega.$$

Then we reach the aim, using Theorem 2.15 and the previous result. ∎

Theorem 2.19 is cited from Canuto and Quarteroni (1982a). Another estimate is that for $r > \frac{1}{2}$,

$$\|v - I_N v\|_{L^\infty} \le cN^{\frac{1}{2} - r}\|v\|_{r,\omega}.$$

We also refer to Bernardi and Maday (1989), and Maday (1990). Finally, by Theorem 2.19 and (2.42), for any $v \in H^r_\omega(\Lambda), r > \dfrac{1}{2}$ and $\phi \in \mathbb{P}_N$,

$$|(v,\phi)_\omega - (v,\phi)_{N,\omega}| \le cN^{-r}\|v\|_{r,\omega}\|\phi\|_\omega. \tag{2.87}$$

In the end of this section, we state a close relation between Legendre transformation and Chebyshev transformation. Let $\Gamma(y)$ be the Gamma function and $\psi(y) = \Gamma\left(y+\frac{1}{2}\right)\Gamma^{-1}(y+1)$. Denote by A_N and B_N a pair of $(N+1)\times(N+1)$ matrices with the elements $A_{N,j,k}$ and $B_{N,j,k}$,

$$A_{N,j,k} = \begin{cases} \frac{1}{\pi}\psi^2\left(\frac{k}{2}\right), & \text{if } 0 = j \le k < N+1 \text{ and } k \text{ is even}, \\ \frac{2}{\pi}\psi\left(\frac{k-j}{2}\right)\psi\left(\frac{k+j}{2}\right), & \text{if } 0 < j \le k < N+1 \text{ and } j+k \text{ is even}, \\ 0, & \text{otherwise}, \end{cases}$$

$$B_{N,j,k} = \begin{cases} 1, & \text{if } j = k = 0, \\ \frac{\sqrt{\pi}}{2\psi(j)}, & \text{if } 0 < j = k < N+1, \\ \frac{-k\left(j+\frac{1}{2}\right)}{(k+j+1)(k-j)}\psi\left(\frac{k-j-2}{2}\right)\psi\left(\frac{k+j-1}{2}\right), & \text{if } 0 \le j < k < N+1 \text{ and } j+k \text{ is even}, \\ 0, & \text{otherwise}. \end{cases}$$

Now, suppose that $v(x)$ has a finite Legendre expansion of the form

$$v(\cos y) = \sum_{l=0}^{N} a_l L_l(\cos y). \tag{2.88}$$

Then it also has a finite Chebyshev expansion of the form

$$v(\cos y) = \sum_{l=0}^{N} b_l T_l(\cos y) \tag{2.89}$$

where $\boldsymbol{a} = (a_0, \ldots, a_N)$ and $\boldsymbol{b} = (b_0, \ldots, b_N)$ are related by the equation

$$\boldsymbol{b} = A_N \boldsymbol{a}.$$

Conversely, if v is a function given by (2.89), then it may be expressed in the form of (2.88), where

$$\boldsymbol{a} = B_N \boldsymbol{b}.$$

Based on the above fact, Alpert and Rokhlin (1991) developed the Fast Legendre Transformation (**FLT**). For a Legendre expansion of degree N, **FLT** produces its values at $N+1$ Chebyshev points in Λ with a cost proportional to $(N+1)\ln(N+1)$. Similarly, **FLT** produces the Legendre expansion of degree N from the values of v at $N+1$ Chebyshev points. The cost of this algorithm is roughly three times that of **FFT** of length $N+1$, provided that the calculations are performed to single precision accuracy. In double precision, the ratio is approximately 5.5. **FLT** saves a lot of work in Legendre spectral methods and Legendre pseudospectral methods, and makes them more efficient and applicable to numerical solutions of partial differential equations.

2.6. Some Orthogonal Approximations in Infinite Intervals

A number of physical problems are set in unbounded domains. The first remark with spectral methods on these problems, is that polynomials are not integrable on unbounded domains. So the idea is to work with weighted Sobolev spaces where the weight must be of exponential type in order to avoid restrictions on the degree of the polynomials. We first consider the case $\Lambda = (0, \infty)$ and $\omega(x) = e^{-x}$. Define the spaces $L^p_\omega(\Lambda), p \geq 1, W^{r,p}_\omega(\Lambda), H^r_\omega(\Lambda), C\left(a, b; W^{r,p}_\omega(\Lambda)\right), H^\gamma\left(a, b; W^{r,p}_\omega(\Lambda)\right)$ and their semi-norms and norms in the same way as in Section 2.3. Also we set the inner product $(\cdot, \cdot)_\omega$ of $L^2_\omega(\Lambda)$ following the same lines as in that section. Clearly for any bounded interval $\Lambda^* \subset \Lambda$, the restrictions to Λ^* of all functions in $H^r_\omega(\Lambda)$ belong to $H^r_\omega(\Lambda^*)$. We quote some essential properties of $H^r_\omega(\Lambda)$. They state that

(i) for any $r \geq 0$, the space $C^\infty\left(\bar{\Lambda}\right)$ is dense in $H^r_\omega(\Lambda)$;

(ii) for any $r \geq 0$, the mapping: $v(x) \to e^{-\frac{x}{2}} v(x)$ is an isomorphism from $H^r_\omega(\Lambda)$ onto $H^r(\Lambda)$;

(iii) for any $0 \leq \mu < r$ and $0 < \theta < 1$,

$$[H^r_\omega(\Lambda), H^\mu_\omega(\Lambda)]_\theta = H^{(1-\theta)r+\theta\mu}_\omega(\Lambda);$$

(iv) for any $r > \frac{1}{2}$, the space $H^r_\omega(\Lambda)$ is contained in $C\left(\bar{\Lambda}\right)$, and for any $v \in H^r_\omega(\Lambda)$,

$$\sup_{x \geq 0} \left| e^{-\frac{x}{2}} v(x) \right| \leq c \|v\|_{r,\omega}.$$

For technical reasons, we also need another Sobolev space associated with a positive integer α,

$$H^r_\omega(\Lambda, \alpha) = \left\{ v \in H^r_\omega(\Lambda) \mid x^{\frac{\alpha}{2}} v \in H^r_\omega(\Lambda) \right\}.$$

This space is provided with the associated natural norm $\|v\|_{r,\omega,\alpha} = \|v(1+x)^{\frac{\alpha}{2}}\|_{r,\omega}$.

The Laguerre polynomial of degree l is

$$\mathcal{L}_l(x) = \frac{1}{l!} e^x \partial_x^l \left(x^l e^{-x} \right). \tag{2.90}$$

It is the l-th eigenfunction of the singular Sturm-Liouville problem

$$\partial_x \left(x e^{-x} \partial_x v(x) \right) + \lambda e^{-x} v(x) = 0,$$

related to the l-th eigenvalue $\lambda_l = l$. Clearly $\mathcal{L}_0(x) = 1, \mathcal{L}_1(x) = 1 - x$, and they satisfy the recurrence relations

$$(l+1)\mathcal{L}_{l+1}(x) = (2l+1-x)\mathcal{L}_l(x) - (l-1)\mathcal{L}_{l-1}(x), \quad l \geq 1, \tag{2.91}$$

$$\mathcal{L}_l(x) = \partial_x \mathcal{L}_l(x) - \partial_x \mathcal{L}_{l+1}(x), \quad l \geq 0. \tag{2.92}$$

It can be checked that $\mathcal{L}_l(0) = 1, \partial_x \mathcal{L}_l(0) = l$ for $l \geq 1$, and

$$|\mathcal{L}_l(x)| \leq e^{\frac{x}{2}}, \qquad x \in \Lambda. \tag{2.93}$$

The set of Laguerre polynomials is the L^2_ω-orthogonal system on the half line Λ, i.e.,

$$\int_\Lambda \mathcal{L}_l(x) \mathcal{L}_m(x) \omega(x)\, dx = \delta_{l,m}.$$

By integrating by parts, we deduce that

$$\int_\Lambda \partial_x \mathcal{L}_l(x) \partial_x \mathcal{L}_m(x) x \omega(x)\, dx = l \delta_{l,m}. \tag{2.94}$$

The Laguerre expansion of a function $v \in L^2_\omega(\Lambda)$ is

$$v(x) = \sum_{l=0}^{\infty} \hat{v}_l \mathcal{L}_l(x) \tag{2.95}$$

with

$$\hat{v}_l = \int_\Lambda v(x) \mathcal{L}_l(x) \omega(x)\, dx.$$

Now let $\mathbb{P}_N$ be the space of restrictions to Λ of polynomials of degree at most N. Several inverse inequalities exist in this space. To do this, we need some preparations. Let $\Lambda^* = (a, b), -\infty \leq a < b \leq \infty$, and $L^p_\chi(\Lambda^*)$ be the space of functions, p-th power integrable with the weight $\chi(x) \geq 0$. We define the norm $\|\cdot\|_{L^p_\chi}$ in the usual way. Let

$\psi_l(x), l = 0, 1, \ldots,$ be an orthogonal system such that $\psi_l \in L^p_\chi(\Lambda^*)$ for all $1 \le p \le \infty$, and suppose that there exist positive constant c_0 and some real δ such that

$$\|\psi_0\|_{L^\infty} \le c_0, \quad \|\psi_l\|_{L^\infty} \le c_0 l^\delta \|\psi_l\|_{L^2_\chi}, \quad l > 0. \tag{2.96}$$

For any $w \in L^p_\chi(\Lambda^*)$, its formal expansion in terms of $\psi_l(x)$ is

$$w(x) = \sum_{l=0}^{\infty} \hat{w}_l \psi_l(x)$$

with

$$\hat{w}_l = \frac{1}{\|\psi_l\|^2_{L^2_\chi}} \int_{\Lambda^*} w(x)\psi_l(x)\chi(x)\,dx.$$

Consider the Cesàro mean of order k, k being some non-negative integer. Set $A^k_N = \frac{(N+k)!}{N!k!}$. Then the Cesàro mean of the function w is of the form

$$C(w, N, k) = \sum_{l=0}^{N} \frac{A^k_{N-l}}{A^k_N} \hat{w}_l \psi_l(x).$$

The system $\{\psi_l\}$ is said to be regular, if for some k^*, there exists a constant $c_1 > 0$ such that for any $w \in L^p_\chi$ and $1 \le p \le \infty$,

$$\|C(w, N, k^*)\|_{L^p_\chi} \le c_1 \|w\|_{L^p_\chi}.$$

Denote by Q_N the span $\{\psi_l(x) \mid 0 \le l \le N\}$. Nessel and Wilmes (1976) proved the following result.

Lemma 2.9. *Let $\{\psi_l\}$ be regular and (2.96) hold. Then for any $\psi \in Q_N$ and $1 \le p \le q \le \infty$,*

$$\|\psi\|_{L^q_\chi} \le c\sigma^{\frac{1}{p}-\frac{1}{q}}(N)\|\psi\|_{L^p_\chi}$$

where $\sigma(N)$ is the same as in Theorem 2.5.

Proof. For example, we consider the case $\delta > -\frac{1}{2}$. By (2.96),

$$\|\psi\|_{L^\infty} \le \|\psi\|_{L^1_\chi} \sum_{l=0}^{N} \frac{\|\psi_l\|^2_{L^\infty}}{\|\psi_l\|^2_{L^2_\chi}} \le cN^{2\delta+1}\|\psi\|_{L^1_\chi}. \tag{2.97}$$

Let $\lambda(x)$ be a smooth function on Λ^* such that $\lambda(x) = 1$ for $0 \le x \le 1$ and $\lambda(x) = 0$ for $x \ge 2$. Then the mean

$$L_N w(x) = \sum_{l=0}^{\infty} \lambda\left(\frac{l}{N}\right) \hat{w}_l \psi_l(x)$$

possesses the following properties:

(i) $L_N w \in Q_{2N}$ for all $w \in L^p_\chi(\Lambda^*)$;

(ii) $L_N \psi = \psi$ for all $\psi \in Q_N$;

(iii) for any $w \in L^p_\chi(\Lambda^*)$ and $1 \le p \le \infty$,

$$\|L_N w\|_{L^p_\chi} \le c_1 \|w\|_{L^p_\chi} \int_0^2 \left| y^{k^*} \partial_y^{k^*+1} \lambda(y) \right| dy;$$

(iv) for any $w \in L^1_\chi(\Lambda^*)$,

$$\|L_N w\|_{L^\infty} \le c N^{2\delta+1} \|w\|_{L^1_\chi}.$$

Indeed, the results (i) and (ii) are obvious. The result (iii) follows by a multiplier criterion as given in Butzer, Nessel and Trebels (1972). The result (iv) is an immediate consequence of (i),(iii) and (2.97). Thus L_N is of types $(1,1), (1,\infty)$ and (∞,∞). So the classical Riesz-Thorin convexity theorem implies that L_N is of type (p,q) for all $1 \le p \le q \le \infty$. In view of (ii), we get the desired conclusion. ∎

Theorem 2.20. *For any $\phi \in \mathbb{P}_N$ and $1 \le p \le q \le \infty$,*

$$\|\phi\|_{L^q_\omega} \le c N^{\frac{1}{p}-\frac{1}{q}} \|\phi\|_{L^p_\omega}.$$

Proof. Let $a = 0, b = \infty$ and $\chi(x) \equiv 1$. Consider the functions

$$\psi_l(x) = \frac{1}{l!} e^{\frac{x}{2}} \partial_x^l \left(x^l e^{-x} \right), \quad l = 0, 1, \ldots.$$

By (2.93) and the orthogonality of $\mathcal{L}_l(x)$, we know that (2.96) is valid and $\delta = 0$. Since $\{\psi_l(x)\}$ is regular, Lemma 2.9 leads to the conclusion of this theorem. ∎

Theorem 2.21. *For any $\phi \in \mathbb{P}_N$,*

$$\|\partial_x \phi\|_\omega \le c N \|\phi\|_\omega.$$

Proof. By (2.92),

$$\partial_x \phi(x) = \sum_{l=0}^{N} \hat{\phi}_l \partial_x \mathcal{L}_l(x) = -\sum_{l=1}^{N} \left(\hat{\phi}_l \sum_{p=0}^{l-1} \mathcal{L}_p(x) \right) = -\sum_{p=0}^{N-1} \mathcal{L}_p(x) \left(\sum_{l=p+1}^{N} \hat{\phi}_l \right)$$

and so

$$\|\partial_x \phi\|_\omega^2 = \sum_{p=0}^{N-1} \left(\sum_{l=p+1}^{N} \hat{\phi}_l \right)^2 \le \sum_{l=0}^{N} \hat{\phi}_l^2 \sum_{p=0}^{N-1} (N-p).$$

This gives the result. ■

The L_ω^2-orthogonal projection $P_N : L_\omega^2(\Lambda) \to \mathbb{P}_N$ is such a mapping that for any $v \in L_\omega^2(\Lambda)$,

$$(v - P_N v, \phi)_\omega = 0, \qquad \forall\, \phi \in \mathbb{P}_N,$$

or equivalently,

$$P_N v(x) = \sum_{l=0}^{N} \hat{v}_l \mathcal{L}_l(x).$$

Lemma 2.10. *Let $r \ge 0$ and α be the largest integer for which $\alpha < r+1$. Then for any $v \in H_\omega^r(\Lambda, \alpha)$,*

$$\|v - P_N v\|_\omega \le c N^{-\frac{r}{2}} \|v\|_{r,\omega,\alpha}.$$

Proof. We have

$$\|v - P_N v\|_\omega^2 = \sum_{l=N+1}^{\infty} \hat{v}_l^2.$$

According to the Sturm-Liouville equation, we set

$$Av(x) = -e^x \partial_x \left(x e^{-x} \partial_x v(x) \right) = -x \partial_x^2 v(x) + (x-1) \partial_x v(x).$$

It is a continuous mapping from $H_\omega^{\gamma+2}(\Lambda, \beta+2)$ into $H_\omega^\gamma(\Lambda, \beta)$ where $\gamma, \beta \ge 0$ and γ is an integer. When $r = \alpha$ is an even integer, we have from integration by parts that

$$\hat{v}_l = \int_\Lambda v(x) \mathcal{L}_l(x) e^{-x}\, dx = l^{-\frac{\alpha}{2}} \int_\Lambda A^{\frac{\alpha}{2}} v(x) \mathcal{L}_l(x) e^{-x}\, dx \tag{2.98}$$

whence

$$\begin{aligned} \|v - P_N v\|_\omega^2 &\le N^{-\alpha} \sum_{l=N+1}^{\infty} \left(\int_\Lambda A^{\frac{\alpha}{2}} v(x) \mathcal{L}_l(x) e^{-x}\, dx \right)^2 \\ &\le N^{-\alpha} \|A^{\frac{\alpha}{2}} v\|_\omega^2 \le N^{-\alpha} \|v\|_{\alpha,\omega,\alpha}. \end{aligned}$$

When $r = \alpha$ is an odd integer, we have from integration by parts again that

$$\begin{aligned}\hat{v}_l &= l^{-\frac{\alpha-1}{2}} \int_\Lambda A^{\frac{\alpha-1}{2}} v(x)\mathcal{L}_l(x)e^{-x}\,dx \\ &= l^{-\frac{\alpha+1}{2}} \int_\Lambda \partial_x \left(A^{\frac{\alpha-1}{2}} v(x)\right) \partial_x\mathcal{L}_l(x)xe^{-x}\,dx.\end{aligned} \tag{2.99}$$

By (2.94),

$$\begin{aligned}\|v - P_N v\|_\omega^2 &\le N^{-\alpha} \sum_{l=N+1}^{\infty} \left(\int_\Lambda \partial_x \left(A^{\frac{\alpha-1}{2}} v(x)\right) \partial_x\mathcal{L}_l(x)xe^{-x}\,dx\right)^2 \\ &\quad \left(\int_\Lambda (\partial_x\mathcal{L}_l(x))^2\, xe^{-x}\,dx\right)^{-1} \\ &\le N^{-\alpha}\|\partial_x \left(A^{\frac{\alpha-1}{2}} v\right) x^{\frac{1}{2}}\|_\omega^2 \le N^{-\alpha}\|v\|_{\alpha,\omega,\alpha}^2.\end{aligned}$$

The general result follows by an interpolation argument between the spaces $H_\omega^{\alpha+1}(\Lambda, \alpha+1)$ and $H_\omega^\alpha(\Lambda, \alpha+1)$. ■

Theorem 2.22. *For any $v \in H_\omega^r(\Lambda, \alpha)$ and $0 \le \mu \le r$,*

$$\|v - P_N v\|_{\mu,\omega} \le cN^{\mu-\frac{r}{2}}\|v\|_{r,\omega,\alpha}$$

where α is the largest integer such that $\alpha < r + 1$.

Proof. Firstly, let μ be an integer. We prove the result by induction. Clearly it is true for $\mu = 0$ by Lemma 2.10. Assume that it holds for $\mu - 1$. We have

$$\|v - P_N v\|_{\mu,\omega} \le \|v - P_N v\|_\omega + \|\partial_x v - P_N \partial_x v\|_{\mu-1,\omega} + \|P_N \partial_x v - \partial_x P_N v\|_{\mu-1,\omega}. \tag{2.100}$$

We observe from (2.92) that

$$\begin{aligned}P_N \partial_x v(x) &= -\sum_{l=0}^{N} \mathcal{L}_l(x) \left(\sum_{p=l+1}^{\infty} \hat{v}_p\right), \\ \partial_x P_N v(x) &= -\sum_{l=0}^{N-1} \mathcal{L}_l(x) \left(\sum_{p=l+1}^{N} \hat{v}_p\right).\end{aligned}$$

Thus

$$P_N \partial_x v(x) - \partial_x P_N v(x) = -\left(\sum_{l=0}^{N} \mathcal{L}_l(x)\right) \left(\sum_{l=N+1}^{\infty} \hat{v}_l\right) = \partial_x \mathcal{L}_{N+1}(x) \sum_{l=N+1}^{\infty} \hat{v}_l.$$

Using Theorem 2.21 and (2.94), we obtain that

$$\|\partial_x \mathcal{L}_{N+1}\|_{\mu-1,\omega} \leq cN^{\mu-1}\|\partial_x \mathcal{L}_{N+1}\|_{\omega} \leq cN^{\mu-\frac{1}{2}}.$$

If r is an integer, we have from (2.98) and (2.99) that

$$\|P_N\partial_x v - \partial_x P_N v\|_{\mu-1,\omega} \leq cN^{\mu-\frac{r}{2}}\|v\|_{r,\omega,\alpha}.$$

If r is not an integer, then we still get the above result by the space interpolation. Combining this line with (2.100), we complete the induction.

When μ is not an integer, the result is proven by an interpolation argument between the two inequalities

$$\begin{aligned}
\|v - P_N v\|_{[\mu]+1,\omega} &\leq cN^{[\mu]+1-\frac{[r]+1}{2}}\|v\|_{[r+1],\omega,\alpha+1}, \\
\|v - P_N v\|_{[\mu],\omega} &\leq cN^{[\mu]-\frac{r}{2}}\|v\|_{r,\omega,\alpha+1}.
\end{aligned}$$

■

Theorem 2.22 was given by Maday, Pernaud-Thomas and Vandeven (1985). Sometimes, we take $\Lambda = (0,\infty)$ and $\omega(x) = e^x$. The choice between the two types depends on the behavior at infinity which is enforced on the solution of the exact problem.

In practice, we evaluate $\mathcal{L}_l(x)$ by various asymptotic approximations, see Bateman (1953). For instance, for $0 \leq x < \infty$ and large l,

$$\mathcal{L}_l(x) \sim \frac{1}{\sqrt{\pi}}\, e^{\frac{x}{2}} x^{-\frac{1}{4}} l^{-\frac{1}{4}} \cos\left(2\sqrt{lx} - \frac{\pi}{4}\right).$$

We now consider the case $\Lambda = (-\infty,\infty)$ and $\omega(x) = e^{-x^2}$. We define $L^p_\omega(\Lambda), p \geq 1, H^r_\omega(\Lambda)$ and their norms in the same way as in Section 2.3. The Hermite polynomial of degree l is

$$H_l(x) = (-1)^l e^{x^2}\partial_x^l\left(e^{-x^2}\right). \tag{2.101}$$

It is the l-th eigenfunction of the singular Sturm-Liouville problem

$$\partial_x\left(e^{-x^2}\partial_x v(x)\right) + \lambda e^{-x^2}v(x) = 0,$$

related to the l-th eigenvalue $\lambda_l = 2l$. Clearly $H_0(x) = 1, H_1(x) = 2x$, and they satisfy the recurrence relations

$$H_{l+1}(x) - 2xH_l(x) + 2lH_{l-1}(x) = 0, \quad l \geq 1, \tag{2.102}$$

$$\partial_x H_l(x) = 2lH_{l-1}(x), \quad l \geq 1. \tag{2.103}$$

It can be checked that

$$|H_l(x)| \le c_2 2^{\frac{1}{2l}} (l!)^{\frac{1}{2}} e^{\frac{x^2}{2}}, \quad x \in \Lambda, \tag{2.104}$$

where $c_2 \sim 1.086435$. The set of Hermite polynoimals is the L^2_ω-orthogonal system on the whole line Λ, i.e.,

$$\int_\Lambda H_l(x) H_m(x) \omega(x)\, dx = 2^l l! \sqrt{\pi} \delta_{l,m} \tag{2.105}$$

from which and (2.103),

$$\int_\Lambda \partial_x H_l(x) \partial_x H_m(x) \omega(x)\, dx = 2^{l+1} l l! \sqrt{\pi} \delta_{l,m}. \tag{2.106}$$

The Hermite expansion of a function $v \in L^2_\omega(\Lambda)$ is

$$v(x) = \sum_{l=0}^{\infty} \hat{v}_l H_l(x)$$

with

$$\hat{v}_l = \frac{1}{2^l l! \sqrt{\pi}} \int_\Lambda v(x) H_l(x) \omega(x)\, dx.$$

Now let $\mathbb{P}_N$ be the space of polynomials of degree at most N. Several inverse inequalities exist in this space.

Theorem 2.23. *For any $\phi \in \mathbb{P}_N$ and $1 \le p \le q \le \infty$,*

$$\|\phi\|_{L^q_\omega} \le c N^{\frac{5}{6}\left(\frac{1}{p} - \frac{1}{q}\right)} \|\phi\|_{L^p_\omega}.$$

Proof. Let $a = -\infty, b = \infty$ and $\chi(x) = 1$ in Lemma 2.9. Consider the Hermite functions

$$\psi_l(x) = (-1)^l \left(2^l l! \sqrt{\pi}\right)^{-\frac{1}{2}} e^{\frac{x^2}{2}} \partial_x^l \left(e^{-x^2}\right).$$

By Hille and Phillips (1957), (2.96) is valid and $\delta = -\frac{1}{12}$. Since the system $\{\psi_l\}$ is regular, e.g., $k^* = 1$, Lemma 2.9 implies the conclusion of this theorem. ∎

Theorem 2.24. *For any $\phi \in \mathbb{P}_N$,*

$$\|\partial_x \phi\|_\omega \le c\sqrt{N} \|\phi\|_\omega.$$

Proof. By (2.103),

$$\partial_x \phi(x) = 2\sum_{l=0}^{N} l\hat{\phi}_l H_{l-1}(x)$$

and so

$$\|\partial_x \phi\|_\omega^2 \leq 4\sum_{l=0}^{N-1} 2^l (l+1)^2 l! \sqrt{\pi}\hat{\phi}_{l+1}^2 \leq cN\|\phi\|_\omega^2.$$

■

The L^2_ω-orthogonal projection $P_N : L^2_\omega(\Lambda) \to \mathbb{P}_N$ is such a mapping that for any $v \in L^2_\omega(\Lambda)$,

$$(v - P_N v, \phi)_\omega = 0, \qquad \forall\, \phi \in \mathbb{P}_N,$$

or equivalently,

$$P_N v(x) = \sum_{l=0}^{N} \hat{v}_l H_l(x).$$

Lemma 2.11. *For any $v \in H^r_\omega(\Lambda)$ and $r \geq 0$,*

$$\|v - P_N v\|_\omega \leq cN^{-\frac{r}{2}}\|v\|_{r,\omega}.$$

Proof. We have

$$\|v - P_N v\|_\omega^2 \leq \sum_{l=N+1}^{\infty} 2^l l! \sqrt{\pi}\hat{v}_l^2.$$

According to the corresponding Sturm-Liouville equation, we set

$$Av(x) = -e^{x^2}\partial_x\left(e^{-x^2}\partial_x v(x)\right) = -\partial_x^2 v(x) + 2x\partial_x v(x).$$

It is a continuous mapping from $H^{\gamma+2}_\omega(\Lambda)$ into $H^\gamma_\omega(\Lambda)$ where γ is any non-negative integer. When $r = \alpha$ is an even integer, we have from the integration by parts that

$$\int_\Lambda v(x)H_l(x)e^{-x^2}\,dx = (2l)^{-\frac{\alpha}{2}}\int_\Lambda A^{\frac{\alpha}{2}}v(x)H_l(x)e^{-x^2}\,dx.$$

Thus

$$\|v - P_N v\|_\omega^2 \leq cN^{-\alpha}\|A^{\frac{\alpha}{2}}v\|_\omega^2 \leq cN^{-\alpha}\|v\|_{r,\omega}^2.$$

When $r = \alpha$ is an odd integer, we have

$$\begin{aligned}\int_\Lambda v(x)H_l(x)e^{-x^2}\,dx &= (2l)^{-\frac{\alpha-1}{2}}\int_\Lambda A^{\frac{\alpha-1}{2}}v(x)H_l(x)e^{-x^2}\,dx \\ &= (2l)^{-\frac{\alpha+1}{2}}\int_\Lambda \partial_x\left(A^{\frac{\alpha-1}{2}}v(x)\right)\partial_x H_l(x)e^{-x^2}\,dx.\end{aligned}$$

Thus by (2.105) and (2.106),

$$\begin{aligned}\|v - P_N v\|_\omega^2 &\leq \sum_{l=N+1}^{\infty} \frac{1}{2^{l+\alpha+1} l^{\alpha+1} l! \sqrt{\pi}} \left(\int_\Lambda \partial_x \left(A^{\frac{\alpha-1}{2}} v(x) \right) \partial_x H_l(x) e^{-x^2} \, dx \right)^2 \\ &\leq (2N)^{-\alpha} \|\partial_x \left(A^{\frac{\alpha-1}{2}} v \right) \|_\omega^2 \leq (2N)^{-\alpha} \|v\|_{r,\omega}^2.\end{aligned}$$

■

Theorem 2.25. *For any $v \in H_\omega^r(\Lambda)$ and $0 \leq \mu \leq r$,*

$$\|v - P_N v\|_{\mu,\omega} \leq c N^{\frac{\mu}{2} - \frac{r}{2}} \|v\|_{r,\omega}.$$

Proof. It suffices to consider any integer μ. Clearly, it is true for $\mu = 0$ by Lemma 2.11. Assume that it is true for $\mu - 1$. We have

$$\|v - P_N v\|_{\mu,\omega} \leq \|v - P_N v\|_\omega + \|\partial_x v - P_N \partial_x v\|_{\mu-1,\omega} + \|P_N \partial_x v - \partial_x P_N v\|_{\mu-1,\omega}.$$

By (2.103),

$$P_N \partial_x v(x) - \partial_x P_N v(x) = 2(N+1) \hat{v}_{N+1} H_N(x).$$

Using Theorem 2.24 and (2.105),

$$\|H_N\|_{\mu-1,\omega}^2 \leq c 2^N N^{\mu-1} N!.$$

Moreover

$$|\hat{v}_{N+1}|^2 \leq c 2^{-N-1} N^{-r} ((N+1)!)^{-1} \|v\|_{r,\omega}^2.$$

Therefore

$$\|P_N \partial_x v - \partial_x P_N v\|_{\mu-1,\omega}^2 \leq c N^{\mu-r} \|v\|_{r,\omega}^2$$

from which the induction is complete. ■

Sometimes, we take $\Lambda = (-\infty, \infty)$ and $\omega(x) = e^{x^2}$. In practice, we calculate $H_l(x)$ by various asymptotic approximations, see Bateman (1953). For example, for $-\infty < x < \infty$ and large l,

$$H_l(x) \sim e^{\frac{x^2}{2}} \frac{l!}{\left(\frac{l}{2}\right)!} \cos\left(\sqrt{2l+1}\, x - \frac{1}{2} l\pi \right).$$

For the applications of Hermite approximation, we refer to Funaro and Kavian (1990).

So far, we have considered the spectral approximations based on algebraic polynomials. However, we can also use other kinds of orthogonal systems. Christov (1982), Boyd (1987) and Weideman (1992) used rational functions. For instance, let $\Lambda = (-\infty, \infty), \omega(x) = (x^2+1)^{-1}$ and

$$R_l(x) = \cos(l \operatorname{arccot} x), \quad l = 0, 1, \ldots.$$

The set of $\{R_l(x)\}$ is an orthogonal system, i.e.,

$$\int_\Lambda R_l(x) R_m(x) \omega(x)\, dx = \frac{1}{2} c_l \pi \delta_{l,m}$$

where $c_0 = 2$ and $c_l = 1$ for $l \geq 1$. The expansion of a function $v \in L^2_\omega(\Lambda)$ is

$$v(x) = \sum_{l=0}^{\infty} \hat{v}_l(x) R_l(x)$$

with

$$\hat{v}_l(x) = \frac{2}{\pi c_l} \int_\Lambda v(x) R_l(x) \omega(x)\, dx.$$

Let

$$\mathrm{I\!R}_N = \operatorname{span}\{R_l(x) \mid 0 \leq l \leq N\}.$$

The L^2_ω-orthogonal projection $P_N : L^2_\omega(\Lambda) \to \mathrm{I\!R}_N$ is such a mapping that for any $v \in L^2_\omega(\Lambda)$,

$$(v - P_N v, \phi)_\omega = 0, \qquad \forall\, \phi \in \mathrm{I\!R}_N,$$

or equivalently,

$$P_N v(x) = \sum_{l=0}^{N} \hat{v}_l(x) R_l(x).$$

We can also estimate the error $\|v - P_N v\|_\omega$.

2.7. Filterings and Recovering the Spectral Accuracy

Spectral methods have the convergence rate of "infinite order", also called the spectral accuracy, and so often provide good numerical solutions of partial differential equations. But this merit may be destroyed by two facts. The first is the instability of nonlinear computation. The second is the lower accuracy caused by the discontinuity of data. In order to remedy these deficiencies, various techniques have been developed, such as filterings, non-oscillatory polynomial interpolations and reconstructions of orthogonal approximations.

We first discuss the improvement of stability. As we know, pseudospectral approximations are preferable in practice, since they are only needed to evaluate unknown functions at interpolation points and it is much easier to deal with nonlinear terms. But they are usually less stable than the corresponding spectral approximations, due to aliasing interaction in calculations of nonlinear terms and derivatives, such as the term $v\partial_x v$ in the Burgers' equation. Kreiss and Oliger (1979) first proposed the filtering operator R_N for Fourier approximations. The simplest one is based on Cesàro mean. Let $\Lambda = (0, 2\pi), \phi \in V_N$ and $\hat{\phi}_l$ be the Fourier coefficients of ϕ, $\hat{\phi}_{-l} = \bar{\hat{\phi}}_l$. Then the filtering is given by

$$R_N\phi(x) = \sum_{|l|\le N} \left(1 - \left|\frac{l}{N}\right|\right) \hat{\phi}_l e^{ilx}.$$

If $\phi(x)$ has the error $\tilde{\phi}(x)$, then $R_N I_N \left(\tilde{\phi}(x)\partial_x\tilde{\phi}(x)\right)$ weakens the effect of high frequency components of $\tilde{\phi}(x)$ and thus improves the stability. On the other hand, if $v \in C\left(\bar{\Lambda}\right)$, then $R_N P_N v(x)$ converges to $v(x)$ uniformly in $\bar{\Lambda}$, while it is not so for standard Fourier approximation. The drawback of this approach is that the convergence rate of $R_N P_N v(x)$ in L^∞-norm is limited to $\mathcal{O}\left(\frac{1}{N}\right)$, no matter how smooth the function $v(x)$ is. Kuo (1983) proposed another filtering operator $R_N(\alpha, \beta)$ based on a generalization of Bochner summation, i.e.,

$$R_N(\alpha,\beta)\phi(x) = \sum_{|l|\le N} \left(1 - \left|\frac{l}{N}\right|^\alpha\right)^\beta \hat{\phi}_l e^{ilx}, \quad \alpha, \beta \ge 1. \tag{2.107}$$

This approach is analyzed in Ma and Guo (1986), Guo and Cao (1988), and Guo and Ma (1991). The key point is to improve the stability and keep the same accuracy as the standard pseudospectral methods.

Theorem 2.26. *For any* $\phi \in V_N$ *and* $0 \le r - \mu \le \alpha$,

$$\|R_N(\alpha,\beta)\phi - \phi\|_\mu \le c\beta N^{\mu - r}\|\phi\|_r.$$

If in addition $\mu \geq 0$, then

$$\|R_N(\alpha,\beta)\phi - \phi\|_\mu \leq c\beta N^{\mu-r}|\phi|_r.$$

Proof. Let

$$\psi(x) = 1-(1-x)^\beta, \quad 0 \leq x \leq 1.$$

Then $\psi(0)=0$ and $0 \leq \partial_x\psi(x) = \beta(1-x)^{\beta-1} \leq \beta$ for $0 \leq x \leq 1$. Therefore

$$\begin{aligned}
\|R_N(\alpha,\beta)\phi - \phi\|_\mu^2 &\leq c \sum_{0<|l|\leq N} \left(1+l^2\right)^\mu \psi^2\left(\left|\frac{l}{N}\right|^\alpha\right)\left|\hat{\phi}_l\right|^2 \\
&\leq c\beta^2 \sum_{0<|l|\leq N} \left(1+l^2\right)^\mu \left|\frac{l}{N}\right|^{2\alpha}\left|\hat{\phi}_l\right|^2 \\
&\leq c\beta^2 N^{2(\mu-r)} \sum_{0<|l|\leq N} \left(1+l^2\right)^\mu \left(1+l^2\right)^{r-\mu}\left|\hat{\phi}_l\right|^2 = c\beta^2 N^{2(\mu-r)}\|\phi\|_r^2.
\end{aligned}$$

If $\mu \geq 0$, then $|\phi|_\mu$ is well-defined. Since $l \neq 0$, the second result follows. ∎

Theorem 2.3 and Theorem 2.26 imply that for $v \in H_p^r(\Lambda)$ and $0 \leq r-\mu \leq \alpha$,

$$\|R_N(\alpha,\beta)P_Nv - v\|_\mu \leq c(\beta+1)N^{\mu-r}\|v\|_r. \tag{2.108}$$

If in addition $r > \frac{1}{2}$, then by Theorem 2.4,

$$\|R_N(\alpha,\beta)I_Nv - v\|_\mu \leq c(\beta+1)N^{\mu-r}\|v\|_r. \tag{2.109}$$

For fixed value of α, the accuracy of filtered Fourier approximation is still limited. However we can take $\alpha = \alpha(N)$ and $\alpha(N) \to \infty$ as $N \to \infty$. In this case, for any $0 \leq \mu \leq r$ and suitably large N,

$$\begin{aligned}
\|R_N\left(\alpha(N),\beta\right)I_Nv - v\|_\mu^2 &\leq c\beta^2 \sum_{0<|l|\leq N} \left(1+l^2\right)^\mu \left|\frac{l}{N}\right|^{2(r-\mu)}\left|\frac{l}{N}\right|^{2(\alpha(N)-r+\mu)}|\hat{v}_l|^2 \\
&\quad + \|I_Nv - v\|_\mu^2 \\
&\leq c(\beta^2+1)N^{2\mu-2r}\|v\|_r^2.
\end{aligned}$$

The L^∞-error estimates between $R_N(\alpha,\beta)P_Nv(x)$ and $v(x)$ also exist. Let

$$\omega(v;\rho) = \max_{|x-x'|\leq\rho} |v(x)-v(x')|$$

and $\omega_m(v;\rho) = \omega(\partial_x^m v;\rho)$ for positive integer m. For $\beta = 1$, Cheng and Chen (1956) showed that if $\alpha > m+1$ or $\alpha = m+1, m$ being even, then for any $v \in C_p^m\left(\bar{\Lambda}\right)$,

$$R_N(\alpha,1)P_N v(x) - v(x) = \mathcal{O}\left(\frac{1}{N^m}\omega_m\left(v;\frac{1}{N}\right)\right),$$

and if $\alpha = m+1, m$ being odd, then for any $v \in C_p^m\left(\bar{\Lambda}\right)$,

$$R_N(\alpha,1)P_N v(x) - v(x) = \mathcal{O}\left(\frac{\ln N}{N^m}\omega_m\left(v;\frac{1}{N}\right)\right).$$

A general filtering for $\phi \in V_N$ is of the form

$$R_N\phi(x) = \sum_{|l|\le N} \sigma_{N,l}\hat{\phi}_l e^{ilx} \tag{2.110}$$

where $\sigma_{N,l} = \sigma_{N,-l}$ are real numbers, which decay smoothly from one to zero when $|l|$ goes from zero to N. Let $v * w$ be the convolution between v and w, defined as

$$v * w(x) = \frac{1}{2\pi}\int_\Lambda v(y)w(x-y)\,dy. \tag{2.111}$$

Then (2.110) is equivalent to

$$R_N\phi(x) \equiv \phi * K_N(x) \tag{2.112}$$

with

$$K_N(x) = \sum_{|l|\le N} \sigma_{N,l} e^{ilx}.$$

The coefficients $\sigma_{N,l}$ could be produced by a function $\sigma(x), 0 \le x \le 1$ such that $\sigma(0) = 1, \sigma(1) = 0$ and $\partial_x^k\sigma(0) = \partial_x^k\sigma(1) = 0$ for $1 \le k \le m$. Set $\sigma_{N,l} = \sigma\left(\left|\frac{l}{N}\right|\right)$ in (2.110). Then for smooth function $v(x)$, $R_N P_N v(x)$ converges to $v(x)$ pointwise in the order of $\mathcal{O}\left(\frac{1}{N^{m+1}}\right)$, see Vandeven (1991). In actual computations, we could take $\sigma(x) = e^{-\alpha x^{2\gamma}}$ where α is chosen so that $\sigma_{N,N}$ is the machine zero and 2γ is called the order of the exponential filtering. Majda, McDonough and Osher (1978) discussed the effects of discontinuous data on the stabilities and the convergences of numerical schemes for hyperbolic equations, and used the filtering (2.110) of exponential type. Harvey (1981) also proposed a technique to control the oscillations of expansions of functions.

Mulholland and Sloan (1991) discussed the filtering (2.107) with $\beta = 1$ and (2.110). They also used the filterings for the discrete Fourier transformations of derivatives of functions $\phi \in V_N$. The corresponding filtering operator R_N is defined by

$$R_N \partial_x^m \phi(x) = \sum_{|l| \leq N} (il)^m \lambda_l \tilde{\phi}_l e^{ilx} \tag{2.113}$$

where λ_l are non-negative real numbers, $\lambda_0 = 1, \lambda_l = \lambda_{-l}$ and λ_l decreases as l increases. Several kinds of λ_l are taken, such as $\lambda_l = e^{-\alpha l^2}, \lambda_l = \left(\sin \frac{2\pi l}{2N+1}\right)\left(\frac{2\pi l}{2N+1}\right)^{-1}$ and $\lambda_l = \frac{1}{2}\left(1 + \cos\left(\frac{2\pi l}{2N+1}\right)\right)$. In particular, let $\{a_p(N)\}$ be a series, $a_0(N) = 1$ and $a_p(N)$ decays as N and p increase. Set

$$\lambda_l = \sum_{p=0}^{\infty} a_p(N)(il)^{2p}.$$

Then for $v \in C_p\left(\bar{\Lambda}\right)$ and all interpolation points $x^{(j)}$,

$$R_N I_N \partial_x^m v\left(x^{(j)}\right) = \partial_x^m v\left(x^{(j)}\right) + \sum_{p=1}^{\infty} a_p(N) \partial_x^{m+2p} v\left(x^{(j)}\right).$$

If we use it for the problem

$$\begin{cases} \partial_t U + LU = 0, & x \in \Lambda,\ 0 < t \leq T, \\ U(x+2\pi, t) = U(x,t), & 0 \leq t \leq T, \end{cases}$$

where $LU = \sum_{q=1}^{d} b_q \partial_x^q U$, then it is equivalent to solving an artificial viscosity problem, i.e.,

$$\partial_t U + LU + L_\lambda U = 0$$

where

$$L_\lambda = \sum_{q=1}^{d} b_q \left(\sum_{p=1}^{\infty} a_p \partial_x^{q+2p}\right).$$

In the early work of Gottlieb and Turkel (1980), the filterings are coupled with the mesh size τ of time t for revolutionary problems. Let $\boldsymbol{U} = \left\{U^{(1)}, U^{(2)}, \ldots, U^{(m)}\right\}$ and A be an $m \times m$ matrix. Consider the problem

$$\begin{cases} \partial_t \boldsymbol{U} = A \partial_x \boldsymbol{U}, & x \in \Lambda,\ 0 < t \leq T, \\ \boldsymbol{U}(x,t) = \boldsymbol{U}(x+2\pi, t), & 0 \leq t \leq T, \\ \boldsymbol{U}(x,0) = \boldsymbol{U}_0(x), & x \in \bar{\Lambda}. \end{cases}$$

The corresponding approximate problem is

$$U(x,t+\tau)=U(x,t-\tau)+2\tau A\partial_x U(x,t).$$

Further, we use discrete Fourier approximation in the space and denote by $u_N(x,t)$ the numerical solution. The last term in the above formula is evaluated as

$$2\sum_{|l|\le N} il\tau\left(\widetilde{Au_N(t)}\right)_l e^{ilx}.$$

Gottlieb and Turkel (1980) replaced the above term by

$$2\sum_{|l|\le N} i\lambda_l\left(\widetilde{Au_N(t)}\right)_l e^{\lambda lx}$$

where

$$\lambda_l=\frac{\sin(la\tau)}{a},\quad a=\alpha\rho(A),\quad \alpha\ge 1,$$

and $\rho(A)$ is the spectral radius of matrix A. For hyperbolic systems, there exists a matrix B such that $BAB^{-1}=D, D=[d_1,d_2,\ldots,d_m]$. Let $\sin(l\tau D)=[\sin l\tau d_1,\ldots,\sin l\tau d_m]$. We can take $\lambda_l=B^{-1}\sin(l\tau D)B$. An alternative is to replace $l\tau A$ by a Padé approximation. In this case, the term $l\tau\left(\widetilde{Au_N(t)}\right)_l$ is replaced by θ_l, and $(I+\alpha^2l^2\tau^2A^2)\,\theta_l=l\tau\widetilde{A(u_N)}_l$. If $A=I$, we can take $\lambda_l=\frac{1}{i\alpha}\left(e^{\alpha l\tau i}-1\right)$.

We next consider the recovering spectral accuracy. The usual Fourier expansion of a function with discontinuity converges slowly. For example, consider a sawtooth-like function

$$f(x,x',A)=\begin{cases}-Ax, & \text{for } x\le x',\\ 2\pi A-Ax, & \text{for } x>x'\end{cases}\tag{2.114}$$

where x' is the location of the discontinuity and $A=[f]_{x'}=\frac{f(x'^{+},x',A)-f(x'^{-},x',A)}{2\pi}$ is the jump of $f(x,x',A)$ across x'. Let $v_N(x)=P_Nv(x)$. Then

$$f_N(x,x',A)=\sum_{|l|\le N}\hat{f}_l(x',A)\,e^{ilx}\tag{2.115}$$

with $\hat{f}_0(x',A)=A\pi-Ax'$, and

$$\hat{f}_l(x',A)=\frac{Ae^{-ilx'}}{il},\qquad |l|\ge 1.$$

We can see that $\hat{f}_l(x', A)$ only decays like $\mathcal{O}\left(\frac{1}{l}\right)$ as $l \to \infty$. As a result, the convergence of (2.115) will only be of the first order, and moreover, the Gibbs oscillations near x' will be of the order of $\mathcal{O}(1)$. In order to get rid of them, several techniques have been proposed.

Cai, Gottlieb and Shu (1989) provided a method based on the property of sawtooth-like function (2.114). Assume that $v(x)$ is a periodic piecewise C^∞ function with one discontinuity at x' and $A = [v]_{x'}$. Let x^* and A^* be the approximations to x' and A respectively. Define

$$v_N^*(x) = \sum_{|l|\le N} \hat{v}_l e^{ilx} + \sum_{|l|>N} \frac{A^*}{il} e^{il(x-x^*)}. \tag{2.116}$$

Since the last sum is actually $f(x, x^*, A^*) - f_N(x, x^*, A^*)$, we have

$$v_N^*(x) = \sum_{|l|\le N} \left(\hat{v}_l - \hat{f}_l(x^*, A^*)\right) e^{ilx} + f(x, x^*, A^*).$$

(2.116) yields an essentially nonoscillatory approximation provided that the approximations for the location x' and the jump A are quite accurate. Indeed, it is shown that

$$\begin{aligned} V(v_N^*) &\le V(v) + c\frac{\ln N}{N} + cN\ln N\,|x' - x^*| + c\ln N\,|A - A^*|, \\ \left\|v - v_N^*\right\|_{L^1} &\le c\frac{\ln N}{N^2} + c\ln N\,|x' - x^*| + c\,|A - A^*| \end{aligned}$$

where $V(v)$ is the total variation of v. Moreover, both $|x' - x^*|$ and $|A - A^*|$ are of the order of $\mathcal{O}\left(\frac{1}{N^2}\right)$, if x^* and A^* are determined by the following system

$$\begin{cases} \frac{A^*}{i(N+1)} e^{-i(N+1)x^*} = \hat{v}_{N+1}, \\ \frac{A^*}{i(N+2)} e^{-i(N+2)x^*} = \hat{v}_{N+2}. \end{cases}$$

A more efficient approach is related to the idea of Harten, Engquist, Osher and Chakravarthy (1987). Let $x^{(j)}$ be the same as in (2.19) and $v(x)$ have only one discontinuity at x'. Define a polynomial interpolant of degree m for $v(x)$ at point $x^{(j)}$, i.e.,

$$\begin{aligned} &Q_m\left(x^{(j)}; v\right) = v\left(x^{(j)}\right), && 0 \le j \le 2N, \\ &Q_m(x; v) = q_{m,j+\frac{1}{2}}(x; v), && \text{for } x^{(j)} \le x \le x^{(j+1)}, \end{aligned}$$

$q_{m,j+\frac{1}{2}}(x;v)$ interpolates $v(x)$ at $(m+1)$ successive points $x^{(\lambda)}, \lambda_m(j) \le \lambda \le \lambda_m(j)+m$. The stencil of these $(m+1)$ points will be chosen according to the smoothness of the data $v\left(x^{(\lambda)}\right)$ around $x^{(j)}$. A recursive algorithm to define $\lambda_m(j)$ starts with

$$\lambda_1(j) = j. \tag{2.117}$$

It means that $q_{1,j+\frac{1}{2}}(x)$ is the first degree polynomial which interpolates $v(x)$ at $x^{(j)}$ and $x^{(j+1)}$. Assume that $q_{p,j+\frac{1}{2}}(x)$ is the p-th degree polynomial which interpolates $v(x)$ at the points

$$x^{(\lambda_p(j))}, \dots, x^{(\lambda_p(j)+p)}. \tag{2.118}$$

Then we need one additional point for determining $q_{p+1,j+\frac{1}{2}}(x)$. That point may be the nearest one to the left of stencil (2.118), i.e., $x^{(\lambda_p(j)-1)}$, or the nearest one to the right of stencil (2.118), i.e., $x^{(\lambda_p(j)+p+1)}$. The choice will be based on the absolute values of the corresponding $(p+1)$-th order divided differences, namely

$$\lambda_{p+1}(j) = \begin{cases} \lambda_p(j)-1, & \text{if } v\left[x^{(\lambda_p(j)-1)}, \dots, x^{(\lambda_p(j)+p)}\right] < v\left[x^{(\lambda_p(j))}, \dots, x^{(\lambda_p(j)+p+1)}\right], \\ \lambda_p(j), & \text{otherwise.} \end{cases} \tag{2.119}$$

The piecewise polynomial $Q_m(x;v)$ gives uniform nonoscillatory approximation to $v(x)$ up to the discontinuity. In fact, it can be shown that for $0 \le k \le m$,

$$\partial_x^k Q_m(x;v) = \partial_x^k v(x) + \mathcal{O}\left(N^{k-m-1}\right), \tag{2.120}$$

except for the cell containing x'. Using sub-cell resolutions in Harten (1989), this is also correct in the shocked cell.

Cai and Shu (1991) provided a uniform approximation by combining the idea in the previous paragraph and the filtering (2.110). Assume that the discontinuity has been detected within an interval $[x_l', x_r']$. Denote all the points inside the interval $[x_l', x_r']$ as $x^{(j_l)}, \dots, x^{(j_r)}$. Then we define a piecewise m-th order polynomial $\psi(x)$ which interpolates the function $v(x)$ at the points $x^{(j)}, j_l \le j \le j_r$,

$$\psi(x) = \begin{cases} q_{m,j+\frac{1}{2}}(x), & \text{if } x \in \left[x^{(j)}, x^{(j+1)}\right] \cap [x_l', x_r'] \text{ for some } j, \\ p_l(x), & \text{if } x \in [0, x_l'], \\ p_r(x), & \text{if } x \in [x_r', 2\pi], \end{cases}$$

where $q_{m,j+\frac{1}{2}}(x)$ is defined by (2.117)–(2.119), and $p_l(x)$ and $p_r(x)$ are both m'-th order polynomials on the interval $[0, x'_l]$ and $[x'_r, 2\pi]$ respectively, $m' = 2m+1$, and for $0 \le k \le m$,

$$\begin{cases} \partial_x^k p_l(x'_l) = \partial_x^k q_{m,j_l-\frac{1}{2}}(x'_l), \\ \partial_x^k p_r(x'_r) = \partial_x^k q_{m,j_r+\frac{1}{2}}(x'_r), \\ \partial_x^k p_l(x'_r - 2\pi) = \partial_x^k q_{m,j_r+\frac{1}{2}}(x'_r), \\ \partial_x^k p_r(x'_l + 2\pi) = \partial_x^k q_{m,j_l-\frac{1}{2}}(x'_l). \end{cases} \tag{2.121}$$

This condition ensures that $\psi(x)$ is at least globally C^m continuous. There are $2m+2 = m'+1$ constraints on the m'-th polynomials $p_l(x)$ and $p_r(x)$ respectively. Therefore, they are uniquely defined. By (2.120),

$$\psi\left(x^{(j)}\right) = v\left(x^{(j)}\right), \quad j_l \le j \le j_r, \tag{2.122}$$

$$\psi(x) - v(x) = \mathcal{O}\left(N^{-m-1}\right), \quad x \in [x'_l, x'_r]. \tag{2.123}$$

Now we consider the difference $w(x) = v(x) - \psi(x)$. $w(x)$ is a C^m function in $\bar{\Lambda}$ except at x', and $[w]_{x'} = \mathcal{O}(N^{-m-1})$. Moreover $w\left(x^{(j)}\right) = 0, j_l \le j \le j_r$. Let $\sigma_{N,l} = \exp\left(-\alpha\left|\frac{l}{N}\right|^{2\gamma}\right)$ and

$$w_{N,\sigma}(x) = \sum_{|l|\le N} \sigma_{N,l}\tilde{w}_l e^{ilx}$$

where

$$\begin{aligned} \tilde{w}_l &= \frac{1}{2N+1}\sum_{j=0}^{2N}\left(v\left(x^{(j)}\right) - \psi\left(x^{(j)}\right)\right)e^{-ilx^{(j)}} \\ &= \frac{1}{2N+1}\left(\sum_{j<j_l}\left(v\left(x^{(j)}\right) - p_l\left(x^{(j)}\right)\right)e^{-ilx^{(j)}} + \sum_{j>j_r}\left(v\left(x^{(j)}\right) - p_r\left(x^{(j)}\right)\right)e^{-ilx^{(j)}}\right). \end{aligned}$$

Finally we define the uniform approximation by

$$P_N^* v(x) = \psi(x) + w_{N,\sigma}(x). \tag{2.124}$$

The derivatives of $v(x)$ will be approximated by those of $P_N v(x)$, i.e.,

$$\partial_x^k v(x) \sim \partial_x^k P_N^* v(x), \quad k > 0.$$

It is proved that

$$\|v - P_N^* v\| = \mathcal{O}\left(N^{-m-1}\right)$$

which establishes the uniform $(m+1)$-th order convergence of the approximation $P_N^* v$ to v in L^2-norm. In the regions outside $[x_l', x_r']$, (2.124) has the convergence rate of "infinite order".

Another uniform approximation is based on the one-sided filtering. Suppose that the function $v(x)$ is analytic over $[0,2\pi)$, but $v(0) \neq v(2\pi)$. Vandeven (1991) considered the two-sided filtering (2.110) with

$$\sigma_{N,l} = 1 - \frac{(2p-1)!}{((p-1)!)^2} \int_0^{\frac{l}{N}} |y(1-y)|^{p-1}\, dy, \quad p = N^{\frac{\varepsilon}{4}}, \quad 0 < \varepsilon < 1, \tag{2.125}$$

and proved that

$$\max_{\delta \le x \le 2\pi - \delta} |v(x) - R_N P_N v(x)| \le \frac{N^\lambda}{\left(cN^{\frac{\varepsilon}{2}}\right)^{N^{\frac{\varepsilon}{4}}}}$$

where $\delta = N^{\varepsilon - 1}$, and λ is a certain constant independent of N. Cai, Gottlieb and Shu (1992) defined the one-sided filtering as

$$R_N^* \phi(x) = \sum_{|l| \le N} \sigma_{N,l}^* \hat{\phi}_l e^{ilx}, \quad \forall\, \phi \in V_N \tag{2.126}$$

where

$$\sigma_{N,l}^* = \sigma_{N,l} \sum_{p=1}^{m} (-1)^{p+1} \frac{m!}{p!(m-p)!} \exp\left(iplN^{\varepsilon-1}\right), \quad m = N^{\frac{\varepsilon}{4}},$$

and $\sigma_{N,l}$ is given by (2.125). They proved that if $v(x)$ is periodic, analytic in $[0,2\pi)$, but $v(0) \neq v(2\pi)$, then for any $0 < \varepsilon < \frac{4}{7}$,

$$\max_{0 \le x \le 2\pi - \delta} |v(x) - R_N^* P_N v(x)| \le \frac{N^\lambda}{\left(cN^{\frac{\varepsilon}{2}}\right)^{N^{\frac{\varepsilon}{4}}}}$$

where $\delta = 2N^{\frac{5}{4}\varepsilon - 1}$. The above filtering is naturally labelled "right-sided" filtering since it can recover the spectral accuracy up to the discontinuity $x = 0$ from the right. The "left-sided" filtering can be constructed similarly. But a very large N is needed for the reasonable size of m in (2.126). Cai, Gottlieb and Shu (1992) used a least square procedure to create $\sigma_{N,l}^*$ for $8 \le N \le 32$. The numerical experiments showed the advantages of this approach.

Recently Gottlieb, Shu, Solomonoff and Vandeven (1992) developed a new method for recovering the spectral accuracy. The main idea of this method is to reconstruct the Fourier expansion by using Gegenbauer polynomials. The Gegenbauer polynomial of degree l with parameter $\lambda \geq 0$ is given by

$$C_{l,\lambda}(x) = G_{l,\lambda}\left(1-x^2\right)^{\frac{1}{2}-\lambda}\partial_x^l\left(\left(1-x^2\right)^{l+\lambda-\frac{1}{2}}\right) \tag{2.127}$$

where

$$G_{l,\lambda} = \frac{(-1)^l\Gamma\left(\lambda+\frac{1}{2}\right)\Gamma(l+2\lambda)}{2^l l!\Gamma(2\lambda)\Gamma\left(l+\lambda+\frac{1}{2}\right)}, \quad \text{for } \lambda > 0,$$
$$G_{l,0} = \frac{(-1)^l\sqrt{\pi}}{2^{l-1}l\Gamma\left(l+\frac{1}{2}\right)}, \quad \text{for } l \geq 1$$

and

$$G_{0,0} = 1.$$

By this standardization, $C_{0,0}(x) = 1$ and for $l \geq 0$,

$$C_{l,0}(x) = \frac{2}{l}T_l(x), \qquad C_{l,\frac{1}{2}}(x) = L_l(x). \tag{2.128}$$

The Gegenbauer polynomials satisfy the recurrence relations

$$\partial_x\left(\left(1-x^2\right)^{\lambda-\frac{1}{2}}C_{l,\lambda}(x)\right) = \frac{G_{l,\lambda}}{G_{l+1,\lambda-1}}\left(1-x^2\right)^{\lambda-\frac{3}{2}}C_{l+1,\lambda-1}(x) \tag{2.129}$$

and

$$2(l+\lambda)C_{l,\lambda}(x) = \partial_x C_{l+1,\lambda}(x) - \partial_x C_{l-1,\lambda}(x). \tag{2.130}$$

It is shown that $C_{l,\lambda}(1) = \frac{\Gamma(l+2\lambda)}{l!\Gamma(2\lambda)}$ and $|C_{l,\lambda}(x)| \leq C_{l,\lambda}(1)$ for $|x| \leq 1$. Let $\Lambda = (-1,1)$ and $\omega_\lambda(x) = (1-x^2)^{\lambda-\frac{1}{2}}$. The set of Gegenbauer polynomials is the $L^2_{\omega_\lambda}$-orthogonal system in Λ, i.e.,

$$\int_\Lambda C_{l,\lambda}(x)C_{m,\lambda}(x)\omega_\lambda(x)\,dx = h_{l,\lambda}\delta_{l,m} \tag{2.131}$$

where for $\lambda > 0$,

$$h_{l,\lambda} = \frac{\sqrt{\pi}\,C_{l,\lambda}(1)\Gamma\left(\lambda+\frac{1}{2}\right)}{(l+\lambda)\Gamma(\lambda)}.$$

There exists a constant c_0 independent of l and λ such that

$$\frac{\lambda^{\frac{1}{2}}C_{l,\lambda}(1)}{c_0(l+\lambda)} \leq h_{l,\lambda} \leq \frac{c_0\lambda^{\frac{1}{2}}C_{l,\lambda}(1)}{l+\lambda}.$$

The Gegenbauer expansion of a function $v \in L^2_{\omega_\lambda}(\Lambda)$ is

$$v(x) = \sum_{l=0}^{\infty} \hat{v}_{l,\lambda} C_{l,\lambda}(x) \tag{2.132}$$

with the coefficients

$$\hat{v}_{l,\lambda} = \frac{1}{h_{l,\lambda}} \int_\Lambda v(x) C_{l,\lambda}(x) \omega_\lambda(x)\, dx, \quad l = 0, 1, \ldots .$$

The corresponding orthogonal projection is

$$v_{N,\lambda}(x) = \sum_{l=0}^{N} \hat{v}_{l,\lambda} C_{l,\lambda}(x). \tag{2.133}$$

Now, we consider an analytic, but non-periodic function $v(x)$ in $\bar{\Lambda}$. If it is extended periodically with the period 2, then $v(x)$ has a discontinuity at $x = \pm 1$. The Fourier coefficients of $v(x)$ are defined as

$$\hat{v}_l = \frac{1}{2} \int_\Lambda v(x) e^{-il\pi x}\, dx.$$

The traditional truncated Fourier sum is

$$v_N(x) = \sum_{|l| \le N} \hat{v}_l e^{il\pi x}.$$

Let $J_\lambda(x)$ be the Bessel function and $\hat{w}_{l,\lambda}$ be the coefficients of Gegenbauer expansion of $v_N(x)$. Then they can be expressed as

$$\hat{w}_{l,\lambda} = \delta_{l,0} \hat{v}_0 + i^l (l + \lambda) \Gamma(\lambda) \sum_{0 < |p| \le N} J_{l+\lambda}(p\pi) \left(\frac{2}{p\pi} \right)^\lambda \hat{v}_p.$$

The reconstructed expansion of $v(x)$ is

$$v_N^*(x) = \sum_{l=0}^{M} \hat{w}_{l,\lambda} C_{l,\lambda}(x). \tag{2.134}$$

Assume that there is a positive constant $c(\rho)$ depending only on $\rho \ge 1$ such that

$$\max_{|x| \le 1} \left| \partial_x^k v(x) \right| \le c(\rho) \frac{k!}{\rho^k}. \tag{2.135}$$

This is a standard assumption for analytic functions. The quantity ρ is the distance from $\bar{\Lambda}$ to the nearest singularity of $v(x)$ in the complex plane. Gottlieb, Shu,

Solomonoff and Vandeven (1992) proved that if (2.135) holds, $\lambda = M = \beta N$ and $\beta < \frac{2}{27}\pi e$, then

$$\max_{|x|\le 1} |v(x) - v_N^*(x)| \le b_1 N^2 q_1^N + b_2 q_2^N$$

where b_1 and b_2 are certain positive constants related to $\max\limits_{0\le l\le\infty} |\hat{v}_l|$, and

$$q_1 = \left(\frac{27\beta}{2\pi e}\right)^\beta < 1, \qquad q_2 = \left(\frac{27}{32\rho}\right)^\beta < 1.$$

The meanings of q_1 and q_2 are explained in their paper. (2.134) gives a uniform approximation to $v(x)$, including the discontinuity, and recovers the spectral accuracy. We also refer to work in Gottlieb and Shu (1994).

All of the methods in the previous paragraphs are also available for Legendre approximation and Chebyshev approximation. For instance, let $v \in \mathbb{P}_N$ and $\hat{v}_l$ be the Chebyshev coefficients of $v(x), 0 \le l \le N$. Guo and Li (1993) offered a filtering. Ma and Guo (1994) considered $R_N(\alpha,\beta)$ and $R_N^0(\alpha,\beta)$ for $\phi \in \mathbb{P}_N$, defined by

$$R_N(\alpha,\beta)\phi(x) = \sum_{l=0}^{N} \left(1 - \left|\frac{l}{N}\right|^\alpha\right)^\beta \hat{\phi}_l T_l(x), \qquad \alpha,\beta \ge 1, \tag{2.136}$$

$$\begin{aligned} R_N^0(\alpha,\beta)\phi(x) = & \sum_{l=0}^{N-2} \left(1 - \left|\frac{l}{N}\right|^\alpha\right)^\beta \hat{\phi}_l T_l(x) \\ & + a_{N-1}T_{N-1}(x) + a_N T_N(x), \quad \alpha,\beta \ge 1, \end{aligned} \tag{2.137}$$

where the constants a_{N-1} and a_N are determined by

$$R_N^0(\alpha,\beta)\phi(-1) = \phi(-1), \qquad R_N^0(\alpha,\beta)\phi(1) = \phi(1).$$

Let $\omega(x) = (1-x^2)^{-\frac{1}{2}}$. It can be checked that

$$\left(R_N(\alpha,\beta)\phi,\psi\right)_\omega = \left(\phi, R_N(\alpha,\beta)\psi\right)_\omega, \quad \forall\, \phi,\psi \in \mathbb{P}_N, \tag{2.138}$$

$$\left(R_N(\alpha,\beta)\phi,\psi\right)_\omega = \left(\phi, R_N^0(\alpha,\beta)\psi\right)_\omega, \quad \forall\, \phi \in \mathbb{P}_{N-2},\ \psi \in \mathbb{P}_N. \tag{2.139}$$

Theorem 2.27. *For any* $\phi \in \mathbb{P}_N$ *and* $0 \le r \le \alpha$,

$$\left\| R_N(\alpha,\beta)\phi - \phi \right\|_\omega \le c\beta N^{-r} \|\phi\|_{r,\omega}.$$

Proof. We have

$$\begin{aligned}
\|R_N(\alpha,\beta)\phi-\phi\|_\omega^2 &\le \frac{1}{2}\pi\beta^2\sum_{l=1}^N\left|\frac{l}{N}\right|^{2\alpha}\hat{\phi}_l^2\le c\beta^2N^{-2r}\sum_{l=1}^N l^{2r}\hat{\phi}_l^2\\
&\le c\beta^2N^{-2r}\sum_{l=0}^N l^{2r}\hat{\phi}_l^2\le c\beta^2N^{-2r}\int_{-\pi}^{\pi}\left|\partial_y^r\left(\sum_{|l|\le N}\hat{\phi}_{|l|}e^{ily}\right)\right|^2dy\\
&\le c\beta^2N^{-2r}\int_{-\pi}^{\pi}\left|\partial_y^r\left(\sum_{l=0}^N\hat{\phi}_lT_l(\cos y)\right)\right|^2dy\le c\beta^2N^{-2r}\|v\|_{r,\omega}^2.
\end{aligned}$$

■

Theorem 2.28. *For any* $\phi\in\mathbb{P}_N$,

$$|R_N(\alpha,\beta)\phi|_{1,\omega}\le cb_{\alpha,\beta}N^2\|\phi\|_\omega$$

with

$$b_{\alpha,\beta}=\left(\frac{\Gamma(2\beta+1)\Gamma\left(\frac{3}{\alpha}\right)}{\alpha\Gamma\left(2\beta+1+\frac{3}{\alpha}\right)}\right)^{\frac{1}{2}}.$$

If in addition $\phi\in\mathbb{P}_N^0$, *then*

$$\left|R_N^0(\alpha,\beta)\phi\right|_{1,\omega}\le cd_{\alpha,\beta}N^2\|\phi\|_\omega$$

with

$$d_{\alpha,\beta}=\left(\frac{\Gamma(2\beta+2)\Gamma\left(\frac{1}{\alpha}\right)}{\alpha\Gamma\left(2\beta+1+\frac{1}{\alpha}\right)}\right)^{\frac{1}{2}}.$$

Proof. Let

$$\partial_xR_N(\alpha,\beta)\phi(x)=\sum_{l=0}^{N-1}a_lT_l(x).$$

Then

$$a_l=\frac{2}{c_l}\sum_{\substack{p=l+1\\p+l\text{ odd}}}p\left(1-\left|\frac{p}{N}\right|^\alpha\right)^\beta\hat{\phi}_p$$

and thus

$$\begin{aligned}
|R_N(\alpha,\beta)\phi|_{1,\omega}^2 &= \frac{\pi}{2}\sum_{l=0}^{N-1}c_la_l^2\le cN\left(\sum_{l=1}^N l\left(1-\left|\frac{l}{N}\right|^\alpha\right)^\beta|\hat{\phi}_l|\right)^2\\
&\le cN\sum_{l=1}^N l^2\left(1-\left|\frac{l}{N}\right|^\alpha\right)^{2\beta}\sum_{l=1}^N\hat{\phi}_l^2\le cN^3\sum_{l=1}^N\left|\frac{l}{N}\right|^2\left(1-\left|\frac{l}{N}\right|^\alpha\right)^{2\beta}\sum_{l=1}^N\hat{\phi}_l^2\\
&\le cN^4\|\phi\|_\omega^2\int_0^1y^2(1-y^\alpha)^{2\beta}dy=cN^4b_{\alpha,\beta}^2\|\phi\|_\omega^2.
\end{aligned}$$

The second result can be proved similarly. ■

A general form of filtering for Chebyshev approximation is

$$R_N\phi(x) = \sum_{l=0}^{N} \sigma_{N,l}\hat{\phi}_l T_l(x), \quad \forall\, \phi \in \mathbb{P}_N$$

where $0 \leq \sigma_{N,l} \leq 1, \sigma_{N,l} \sim 1$ for $l \ll N$ and $\sigma_{N,l} \ll 1$ for $l \sim N$. Weideman and Trefethen (1988) used this approach for evaluating eigenvalues of matrix related to the second order differentiations of Chebyshev expansions, and took

$$\sigma_{N,l} = \begin{cases} 1, & 0 \leq l \leq l_0, \\ \exp\left(-\alpha\left(\frac{l-l_0}{N-l_0}\right)^4\right), & \alpha > 0,\ l_0 < l \leq N. \end{cases}$$

In the early work of Fulton and Taylor (1984), the filtering is coupled with the mesh size τ for the variable t, for the revolutionary problem

$$\partial_t U + \partial_x U = 0.$$

Let $\left(\widehat{u_N(t)}\right)_l^{(1)}$ be the coefficients of the first order derivative of the corresponding numerical solution by Chebyshev approximation. Then the filtered ones are

$$\left(\widehat{u_N(t)}\right)_l^{(1)*} = \frac{2}{c_l} \sum_{\substack{p=l+1 \\ p+l \text{ odd}}}^{N} p\lambda\left(\alpha p^2 \tau\right)\left(\widehat{u_N(t)}\right)_p$$

where

$$\lambda(x) = \frac{8\sin x - \sin(2x)}{6x}$$

or

$$\lambda(x) = \frac{1}{6x}\left[11 - 18e^{-x} + 9e^{-2x} - 2e^{-3x}\right].$$

Various methods for recovering the spectral accuracy can also be generalized to Legendre approximation and Chebyshev approximation. For example, Gottlieb and Shu (1995a) developed reconstruction methods for these two approximations. Let $v(x)$ be a function in $L^1(\Lambda)$, which is analytic in a subinterval $[a,b] \subset \bar{\Lambda}$ and satisfies a condition like (2.135). Its Gegenbauer projection $v_{N,\mu}(x)$ is given by (2.133). Set $\varepsilon = \frac{1}{2}(b-a), \delta = \frac{1}{2}(b+a)$ and

$$\hat{w}_{l,\lambda} = \frac{1}{h_{l,\lambda}} \int_\Lambda v_{N,\mu}(\varepsilon x + \delta) C_{l,\lambda}(x)\omega_\lambda(x)\, dx. \tag{2.140}$$

If $\mu \geq 0$ and $\lambda = M = \beta\varepsilon N$ with $\beta < \frac{2}{27}$, then

$$\max_{|y|\leq 1}\left|v(\varepsilon y+\delta) - \sum_{l=0}^{M} \hat{w}_{l,\lambda} C_{l,\lambda}(y)\right| \leq b\left(q_1^{\varepsilon N} + q_2^{\varepsilon N}\right)$$

where

$$q_1 = \left(\frac{27\beta}{2}\right)^{\beta} < 1, \qquad q_2 = \left(\frac{27\varepsilon}{32\rho}\right)^{\beta} < 1,$$

and b grows at most as $N^{1+2\mu}$. The proof is given in Gottlieb and Shu (1995a). (2.140) gives a reconstructed uniform approximation for Gegenbauer expansion. By (2.128), it covers Legendre approximation and Chebyshev approximation. In Gottlieb and Shu (1996), a sharper result exists for Legendre approximation. On the other hand, we can also recover the exponential accuracy from the values of piecewise analytic functions at interpolation points, see Gottlieb and Shu (1995b).

CHAPTER 3

STABILITY AND CONVERGENCE

The theory of stability and convergence plays an important role in numerical solutions of differential equations. Courant, Friedrichs and Lewy (1928) first considered the convergence, and Von Neumann and Goldstine (1947) started the study of stability for linear problems. Kantorovich (1948) developed a general framework for approximations of linear operator equations in Banach space, and Lax and Richtmyer (1956) dealt with linear initial-value problems. For nonlinear problems, Guo (1965), Stetter (1966) and Keller (1975) provided several kinds of stability, which also imply the convergence sometimes. This chapter is a survey of those results.

3.1. Stability and Convergence for Linear Problems

We consider a general framework. Let B_1 and B_2 be two Banach spaces with the norms $\|\cdot\|_{B_1}$ and $\|\cdot\|_{B_2}$ respectively. For any $v \in B_q$ and $R > 0$, the balls are defined as

$$S_q(v,R) = \left\{w \in B_q \mid \|w - v\|_{B_q} < R\right\}, \quad q = 1, 2.$$

Let L be an operator from B_1 to B_2, and f be a given element in B_2. Consider the operator equation

$$LU = f. \tag{3.1}$$

In order to solve (3.1) approximately, we first approximate the spaces B_1, B_2 and L, f. Let N be any positive integer, and $B_{q,N}$ be finite dimensional Banach spaces with the norms $\|\cdot\|_{B_{q,N}}, q = 1, 2$. For any $v \in B_{q,N}$ and $R > 0$, the balls are defined by

$$S_{q,N}(v,R)=\left\{w\in B_{q,N}\ \mid\ \|w-v\|_{B_{q,N}}<R\right\},\quad q=1,2.$$

Assume that $\mathrm{Dim}(B_{1,N})=\mathrm{Dim}(B_{2,N})$, and $B_{q,N}$ approximates $B_q, q=1,2$. Let $p_{q,N}$ be continuous operators from B_q to $B_{q,N}$ such that

$$\lim_{N\to\infty}\|p_{1,N}v\|_{B_{1,N}}=\|v\|_{B_1},\quad \forall\, v\in B_1,\tag{3.2}$$

$$\lim_{N\to\infty}\|p_{2,N}v\|_{B_{2,N}}=\|v\|_{B_2},\quad \forall\, v\in B_2.\tag{3.3}$$

Let q_N be a continuous operator from $B_{2,N}$ to B_2, satisfying

$$p_{2,N}q_Nv=v,\quad \forall\, v\in B_{2,N}.\tag{3.4}$$

We denote by B_3 another Banach space with the norm $\|\cdot\|_{B_3}$. σ_N is a continuous operator from $B_{1,N}$ to B_3, while ω is a continuous operator from B_1 to B_3. Suppose that there exist positive constant c_0 and positive integer N_0 such that for all $N\geq N_0$,

$$\|\sigma_N\|=\sup_{v\in B_{1,N}}\frac{\|\sigma_Nv\|_{B_3}}{\|v\|_{B_{1,N}}}\leq c_0,\tag{3.5}$$

and for any $v\in B_1$,

$$\lim_{N\to\infty}\|\sigma_Np_{1,N}v-\omega v\|_{B_3}=0.\tag{3.6}$$

In this section, we assume that all operators $L, p_{1,N}, p_{2,N}, q_N, \sigma_N$ and ω are linear.

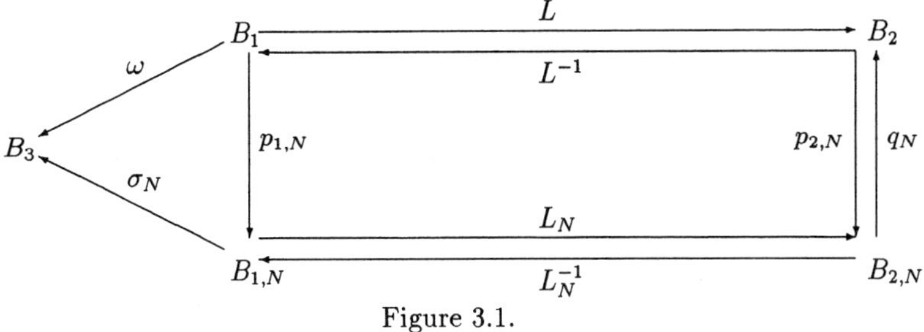

Figure 3.1.

Let L_N be an invertible operator from $B_{1,N}$ to $B_{2,N}$. We approximate (3.1) by

$$L_Nu_N=p_{2,N}f.\tag{3.7}$$

The approximation error is defined as

$$R_N(v) = Lv - q_N L_N p_{1,N} v, \quad \forall\, v \in B_1. \tag{3.8}$$

Let D be a subset of solutions of (3.1), and F be the set of all f related to $U \in D$. Assume that F is dense in B_2. The approximation (3.7) is said to be consistent with (3.1), if for any $v \in D$,

$$\lim_{N\to\infty} \|R_N(v)\|_{B_2} = 0. \tag{3.9}$$

If for all U, the solutions of (3.1) and u_N, the corresponding solutions of (3.7), there holds

$$\lim_{N\to\infty} \|\sigma_N u_N - \omega U\|_{B_3} = 0,$$

then we say that (3.7) is convergent.

Usually the data f possesses certain error, denoted by $\tilde{f}$. It induces the error of u_N, denoted by $\tilde{u}_N$. If there exists a positive constant c_1 such that for any $\tilde{f}$ and $N \geq N_0$,

$$\|\sigma_N \tilde{u}_N\|_{B_3} \leq c_1 \|\tilde{f}\|_{B_2}, \tag{3.10}$$

then we say that (3.7) is stable. Since L_N and σ_N are linear, it is equivalent to

$$\|\sigma_N u_N\|_{B_3} \leq c_1 \|f\|_{B_2}. \tag{3.11}$$

Since $u_N = L_N^{-1} p_{2,N} f$, it is also equivalent to

$$\left\|\sigma_N L_N^{-1} p_{2,N}\right\| = \sup_{\substack{v\in B_2\\ v\neq 0}} \frac{\|\sigma_N L_N^{-1} p_{2,N} v\|_{B_3}}{\|v\|_{B_2}} \leq c_1. \tag{3.12}$$

Theorem 3.1. *If L^{-1} exists, and (3.7) is consistent with (3.1), then the stability is equivalent to the convergence.*

Proof. If (3.7) is convergent, then for all f,

$$\lim_{N\to\infty} \|\sigma_N L_N^{-1} p_{2,N} f - \omega L^{-1} f\|_{B_3} = 0.$$

Thus $\|\sigma_N L_N^{-1} p_{2,N} f\|_{B_3}$ is bounded uniformly for N. By the Uniform Boundedness Theorem, (3.12) is valid.

Now assume that (3.7) is stable. For any $U \in D$ and the corresponding approximate solution u_N,

$$\|\sigma_N u_N - \omega U\|_{B_3} \le \|\sigma_N p_{1,N} U - \omega U\|_{B_3} + \|\sigma_N u_N - \sigma_N p_{1,N} U\|_{B_3}. \tag{3.13}$$

The first term on the right-hand side tends to zero by (3.6). Since $L_N p_{1,N} U \in B_{2,N}$, (3.4) implies

$$\begin{aligned}\|\sigma_N u_N - \sigma_N p_{1,N} U\|_{B_3} &= \|\sigma_N L_N^{-1} p_{2,N} f - \sigma_N L_N^{-1} p_{2,N} q_N L_N p_{1,N} U\|_{B_3} \\ &= \|\sigma_N L_N^{-1} p_{2,N} \left(LU - q_N L_N p_{1,N} U\right)\|_{B_3} \\ &\le \|\sigma_N L_N^{-1} p_{2,N}\| \|R_N(U)\|_{B_2}. \end{aligned} \tag{3.14}$$

By (3.9) and (3.12), the above right-hand side tends to zero as $N \to \infty$. Hence for any $U \in D$,

$$\lim_{N\to\infty} \|\sigma_N u_N - \omega U\|_{B_3} = 0. \tag{3.15}$$

If U is a solution of (3.1), but $U \bar{\in} D$, then there exists a sequence $\{U^{(l)}\} \in D$, related to the data $\{f^{(l)}\}$ such that

$$\lim_{l\to\infty} \|f - f^{(l)}\|_{B_2} = 0. \tag{3.16}$$

Let $u_N^{(l)}$ be the corresponding approximate solutions of (3.7) with the data $f^{(l)}$. Then we have

$$\begin{aligned}\|\sigma_N u_N - \omega U\|_{B_3} &\le \|\sigma_N u_N - \sigma_N u_N^{(l)}\|_{B_3} + \|\sigma_N u_N^{(l)} - \omega U^{(l)}\|_{B_3} + \|\omega U^{(l)} - \omega U\|_{B_3} \\ &\le \|\sigma_N L_N^{-1} p_{2,N}\| \|f - f^{(l)}\|_{B_2} + \|\sigma_N u_N^{(l)} - \omega U^{(l)}\|_{B_3} \\ &\quad + \|\omega\| \|L^{-1}\| \|f - f^{(l)}\|_{B_2}. \end{aligned} \tag{3.17}$$

By (3.12), (3.15) and (3.16), the above right-hand side goes to zero as $N, l \to \infty$. This completes the proof. ■

Theorem 3.1 can be found in Guo (1988a). We now turn to the convergence rate.

Theorem 3.2. *If L^{-1} exists, and (3.7) is stable and consistent with (3.1), then*

$$\|\sigma_N u_N - \omega U\|_{B_3} \le c\left(\|\sigma_N p_{1,N} U - \omega U\|_{B_3} + \|R_N(U)\|_{B_2}\right).$$

Proof. The conclusion comes immediately from (3.13), (3.14) and (3.17). ∎

If ω is an isomorphism between B_1 and B_2, then the above estimate also holds for $\|\omega^{-1}\sigma_N u_N - U\|_{B_1}$.

If $B_3 = B_q = B_{q,N}, q = 1,2$, then $p_{1,N}, p_{2,N}, q_N, \sigma_N$ and ω are the identity operators. If in addition $D = B_1$, then Theorems 3.1 and 3.2 turn to be the Basic Convergence Theorem of Kantorovich (1948). Besides, N could be replaced by any real parameter. It is also noted that in spectral methods, $B_{q,N}$ are certain subspaces of B_q, and $p_{q,N}$ are some projections from B_q to $B_{q,N}, q = 1,2$, usually. In this case, we can take $B_3 = B_1$ and $\|\cdot\|_{B_{q,N}} = \|\cdot\|_{B_q}, q = 1,2$. Therefore q_N, σ_N and ω are the identity operators, and so

$$\|u_N - U\|_{B_1} \le c\left(\|p_{1,N}U - U\|_{B_1} + \|R_N(U)\|_{B_2}\right).$$

3.2. Generalized Stability for Nonlinear Problems

In this section, we deal with nonlinear problems. We still consider problem (3.1), but L may be nonlinear. The notations $B_q, B_{q,N}$ and $p_{q,N}$ have the same meanings as before. They also fulfill (3.2) and (3.3), but $p_{1,N}$ and $p_{2,N}$ could be nonlinear. Let L_N and f_N approximate L and f respectively. Consider the approximate problem

$$L_N u_N = f_N. \tag{3.18}$$

The approximation error is defined as

$$R_N(v) = p_{2,N}Lv - L_N p_{1,N} v + f_N - p_{2,N} f.$$

If U is a solution of (3.1), then $R_N(U) = f_N - L_N p_{1,N} U$. Let D be the same as before. The approximation (3.18) is consistent with (3.1), if for any $v \in D$ and $f \in F$,

$$\lim_{N\to\infty} \|R_N(v)\|_{B_{2,N}} = 0.$$

If for all U, the solutions of (3.1) and u_N, the corresponding solutions of (3.18),

$$\lim_{N\to\infty} \|u_N - p_{1,N}U\|_{B_{1,N}} = 0,$$

then we say that (3.18) is convergent.

How to define the stability is a problem. The simplest way is to follow (3.10). It means that there exists a positive constant c_0 such that for any $v, w \in B_{1,N}$ and $N \geq N_0$,

$$\|v - w\|_{B_{1,N}} \leq c_0 \|L_N v - L_N w\|_{B_{2,N}}. \tag{3.19}$$

Such kind of stability is called linear stability. It is an alternative expression to the stability in the previous section. But nonlinear problems usually do not possess this property. Indeed, this stability is very strong so that it is essentially suitable for linear problems and some specific nonlinear problems, such as sine-Gordon equation. Firstly, we see that the constant c_0 is uniform for all u_N, whereas the corresponding constant depends on the exact solution of (3.18) for most nonlinear problems. Next, (3.19) is valid for any $\|L_N v - L_N w\|_{B_{2,N}}$. However, the feature of nonlinear problems is that only when it lies in certain region, the difference between v and w is controlled. But $\|v - w\|_{B_{1,N}}$ may grow rapidly, once $\|L_N v - L_N w\|_{B_{2,N}}$ exceeds some critical bound. Thus (3.19) cannot describe the essence of nonlinear problems properly. Finally, (3.19) holds for all $v, w \in B_{1,N}$. Clearly this is not possible for nonlinear problems with several solutions. These facts motivated us to look for new definitions of stability. Strang (1965) served the weak stability for linear problems, which is also used in spectral methods, see Gottlieb and Orsag (1977). Stetter (1966, 1973) developed the nonlinear stability applied to initial-value problems of nonlinear ordinary differential equations, while Guo (1965, 1982) provided several kinds of generalized stability used for numerical solutions of nonlinear partial differential equations. All of these works are suitable only for problems with unique solutions. Keller (1975) generalized the stability in a different way and proposed the local stability, which has been applied successfully to boundary-value problems of nonlinear ordinary differential equations. In 1985, Guo unified the above work, see Guo (1988a, 1989). We shall present this general result below.

We first introduce some terminologies. The operator L is said to be stable in the ball $S_1(v, R)$, if there are positive constants R and c_R such that

$$\|w - z\|_{B_1} \leq c_R \|Lw - Lz\|_{B_2}, \quad \forall\, w, z \in S_1(v, R). \tag{3.20}$$

If U is a solution of (3.1) and there exists a positive constant R such that L is stable in $S_1(U,R)$, then we say that U is a stable solution. Obviously such a solution is unique in $S_1(U,R)$. Assume that the data f in (3.1) possesses the error $\tilde{f}$ which induces the error of solution, denoted by $\tilde{U}$. If the operator L is linear and stable in the ball $S_1(U,R)$, then $\|\tilde{U}\|_{B_1} \leq c_R\|\tilde{f}\|_{B_2}$. Indeed, let $\|\tilde{U}\|_{B_1} \neq 0$ and $\tilde{U}' = \dfrac{R\tilde{U}}{\|\tilde{U}\|_{B_1}}$. Then $\|\tilde{U}'\|_{B_1} \leq R$ and

$$L\left(U+\tilde{U}'\right) = f + \frac{R\tilde{f}}{\|\tilde{U}\|_{B_1}}.$$

By (3.20),

$$\|\tilde{U}'\|_{B_1} \leq \frac{c_R R\|\tilde{f}\|_{B_2}}{\|\tilde{U}\|_{B_1}}$$

and the desired result follows. Another kind of solution is called isolated solution. If there exists a linear continuous operator $L'(v)$ such that

$$L(v+w) - Lv = L'(v)w + E(v,w)$$

and

$$\lim_{\|w\|_{B_1}\to 0} \frac{\|E(v,w)\|_{B_2}}{\|w\|_{B_1}} = 0,$$

then we say that L is Frèchet differentiable. $L'(v)$ is called the Frèchet derivative of L at the point v. Further, if $L'(v)w = 0$ implies $w = 0$, then we say that $L'(v)$ is nonsingular at v. Moreover, if U is a solution of (3.1) and $L'(v)$ is nonsingular, then we say that U is an isolated solution of (3.1). There is a close relationship between stable solutions and isolated solutions. Keller (1975) proved that

(i) if U is a stable solution of (3.1) and $L'(U)$ exists, then U is an isolated solution;

(ii) if U is an isolated solution of (3.1) and $L'(U)$ is Lipschitz continuous in $S_1(U,R)$, then U is a stable solution.

Now, we give the definition of the stability for the approximation (3.18). If there are positive constant $c_0(v,N)$ and non-negative constant $c_1(v,N)$ such that for all $N \geq N_0$ and $w \in S_{1,N}(v,R)$, the inequality

$$\|L_N v - L_N w\|_{B_{2,N}} \leq c_0(v,N) \tag{3.21}$$

implies

$$\|v - w\|_{B_{1,N}} \leq c_1(v, N)\|L_N v - L_N w\|_{B_{2,N}}, \tag{3.22}$$

then we say that (3.18) is of generalized weak stability for v in the ball $S_{1,N}(v, R)$. Especially, if for all $N \geq N_0$ and $w^{(q)} \in S_{1,N}(v, R)$, the inequalities

$$\|L_N w^{(q)} - L_N v\|_{B_{2,N}} \leq \bar{c}_0(v, N), \quad q = 1, 2 \tag{3.23}$$

imply

$$\|w^{(1)} - w^{(2)}\|_{B_{1,N}} \leq \bar{c}_1(v, N)\|L_N w^{(1)} - L_N w^{(2)}\|_{B_{2,N}}, \tag{3.24}$$

then we say that (3.18) is of generalized uniform weak stability for v in the ball $S_{1,N}(v, R)$.

We have the following result.

Theorem 3.3. *Let L_N be a linear operator. The approximation (3.18) is of linear stability, if and only if $R = \infty$, and $c_0(v, N)$ and $c_1(v, N)$ are independent of v and N.*

Proof. If (3.18) is of linear stability, then (3.19) holds for any $N \geq N_0$ and any $v, w \in B_{1,N}$. Thus $R = \infty, c_1(v, R) \equiv c_0$ and $c_0(v, R)$ is arbitrary. Conversely, assume that $R = \infty, c_0(v, R) \equiv \tilde{c}_0$ and $c_1(v, R) \equiv \tilde{c}_1$. For any $v^{(q)} \in B_{1,N}$, let

$$w^{(q)} = \frac{\tilde{c}_0 v^{(q)}}{\|L_N v^{(1)} - L_N v^{(2)}\|_{B_{2,N}}}, \quad q = 1, 2.$$

Then

$$\|L_N w^{(1)} - L_N w^{(2)}\|_{B_{2,N}} = \tilde{c}_0$$

and so by (3.22),

$$\|w^{(1)} - w^{(2)}\|_{B_{1,N}} \leq \tilde{c}_1\|L_N w^{(1)} - L_N w^{(2)}\|_{B_{2,N}}.$$

By the linearity of L_N, we reach (3.19) with $c_0 = \tilde{c}_1$. ∎

Another sufficient and necessary condition of the linear stability for linear problem (3.18) is that $R = \infty$, and $\bar{c}_0(v, N)$ and $\bar{c}_1(v, N)$ are independent of v and N.

We now give a sufficient condition ensuring the generalized uniform weak stability. Let L'_N denote the Frèchet derivative of L_N.

Theorem 3.4. *Assume that*

(i) L_N *is defined and continuous in* $S_{1,N}(v,R)$;

(ii) for all $w \in S_{1,N}(v,R)$ *such that* $L_N w \in S_{2,N}\left(L_N v, c_0^*(v,N)\right)$, *the inverse* $\left(L_N'(w)\right)^{-1}$ *exists and*

$$\| \left(L_N'(w)\right)^{-1} \| \le c_1^*(v,N). \tag{3.25}$$

Then (3.18) is of generalized uniform weak stability for v *in* $S_{1,N}(v,R^*), 0 < R^* \le R$.

Proof. Let

$$r(v,N) = \min\left(c_0^*(v,N), \frac{R}{c_1^*(v,N)}\right).$$

By the Implicit Function Theorem (see Schwartz (1969)), for any $w \in S_{1,N}(v,R)$ such that $L_N w \in S_{2,N}(L_N v, r(v,N))$, the inverse L_N^{-1} exists in an open neighborhood of $L_N w$. Moreover,

$$\left(L_N^{-1}\right)'(L_N w) = \left(L_N'(w)\right)^{-1}.$$

We have from (3.25) that

$$\| \left(L_N^{-1}\right)'(L_N w) \| \le c_1^*(v,N). \tag{3.26}$$

Now let $L_N w^{(q)} \in S_{2,N}\left(L_N v, r(v,N)\right)$, and

$$g(\lambda) = L_N w^{(2)} + \lambda\left(L_N w^{(1)} - L_N w^{(2)}\right), \quad 0 \le \lambda \le 1.$$

Because of the convexity of the ball $S_{2,N}\left(L_N v, r(v,N)\right)$, there exists the inverse $\left(L_N^{-1}\right)'(g(\lambda))$ satisfying (3.26). Therefore it follows from the Mean Value Theorem that

$$\begin{aligned} \|w^{(1)} - w^{(2)}\|_{B_{1,N}} &= \|L_N^{-1}\left(L_N w^{(1)}\right) - L_N^{-1}\left(L_N w^{(2)}\right)\|_{B_{1,N}} \\ &\le c_1^*(v,N)\|L_N w^{(1)} - L_N w^{(2)}\|_{B_{2,N}}. \end{aligned}$$

Thus the conclusion follows. ∎

We next turn to the existence of solutions of (3.18).

Lemma 3.1. *Suppose that*

(i) L_N is defined and continuous in $S_{1,N}(w,R)$;

(ii) for all $v^{(q)} \in S_{1,N}(w,R)$ such that $L_N v^{(q)} \in S_{2,N}(L_N w, r)$, we have

$$\|v^{(1)} - v^{(2)}\|_{B_{1,N}} \leq \sigma \|L_N v^{(1)} - L_N v^{(2)}\|_{B_{2,N}}, \quad \sigma > 0.$$

Let $r_0 = \min\left(r, \dfrac{R}{\sigma}\right)$. Then for any $g \in S_{2,N}(L_N w, r_0)$, the problem

$$L_N v = g$$

possesses a unique solution in the ball $S_{1,N}(w,R)$.

Proof. Since condition (ii) implies the uniqueness of L_N^{-1} in $S_{2,N}(L_N w, r_0) \cap L_N(S_{1,N}(w,R))$, it suffices to prove that $S_{2,N}(L_N w, r_0) \subset L_N(S_{1,N}(w,R))$. Now assume that $g' \in S_{2,N}(L_N w, r_0)$, but $g' \bar{\in} L_N(S_{1,N}(w,R))$. Let

$$g(\lambda) = (1-\lambda)L_N w + \lambda g', \quad \lambda \geq 0 \tag{3.27}$$

and

$$\bar{\lambda} = \begin{cases} \sup\{\lambda' \geq 0 \mid g(\lambda) \in L_N(S_{1,N}(w,R))\}, \\ 0, \quad \text{if the above set is empty.} \end{cases} \tag{3.28}$$

If $\bar{\lambda} > 1$, then $g(1) = g' \in L_N(S_{1,N}(w,R))$ and the conclusion follows. If $\bar{\lambda} = 0$, then $\bar{g} = g(\bar{\lambda}) = L_N w \in L_N(S_{1,N}(w,R))$ and the inverse $\bar{v} = L_N^{-1}\bar{g} = w$. If $0 < \bar{\lambda} \leq 1$, then for any $\varepsilon < \bar{\lambda}$,

$$g\left(\bar{\lambda} - \varepsilon\right) \in S_{2,N}(L_N w, r_0) \cap L_N(S_{1,N}(w,R))$$

and so the element $L_N^{-1}\left(g\left(\bar{\lambda}-\varepsilon\right)\right)$ exists. By condition (ii), there exists the limit $\bar{v}$ such that

$$\bar{v} = \lim_{\varepsilon \to 0} L_N^{-1}\left(g\left(\bar{\lambda}-\varepsilon\right)\right).$$

Moreover, we have from condition (ii) and (3.27) that

$$\begin{aligned} \|L_N^{-1}\left(g\left(\bar{\lambda}-\varepsilon\right)\right) - w\|_{B_{1,N}} &\leq \sigma\|g\left(\bar{\lambda}-\varepsilon\right) - L_N w\|_{B_{2,N}} = \sigma\left(\bar{\lambda}-\varepsilon\right)\|g' - L_N w\|_{B_{2,N}} \\ &< r_0\sigma\left(\bar{\lambda}-\varepsilon\right) \leq (1-\varepsilon)R \end{aligned}$$

and thus $\bar{v} \in S_{1,N}(w,R)$. By the continuity of L_N in the ball $S_{1,N}(w,R)$, we have $L_N(\bar{v}) = \bar{g}$.

Now we choose a closed ball $\bar{S}_{1,N}(\bar{v},\bar{\varepsilon}) \subset S_{1,N}(w,R)$ such that $L_N\left(\bar{S}_{1,N}(\bar{v},\bar{\varepsilon})\right) \subset S_{2,N}(L_N w, r)$. Then L_N is a one-to-one mapping between $\bar{S}_{1,N}(\bar{v},\bar{\varepsilon})$ and $L_N\left(\bar{S}_{1,N}(\bar{v},\bar{\varepsilon})\right)$. Since $\operatorname{Dim}(B_{1,N}) = \operatorname{Dim}(B_{2,N})$, $L_N\left(\bar{S}_{1,N}(\bar{v},\bar{\varepsilon})\right)$ contains an open neighborhood of $\bar{g}$. This is a contradiction to the definition of $\bar{\lambda}$ by (3.28). The proof is complete. ■

Theorem 3.5. *Let U be a solution of (3.1). Assume that*

(i) L_N is defined and continuous in $S_{1,N}(p_{1,N}U, R)$;

(ii) (3.18) is of generalized uniform weak stability for $p_{1,N}U$ in $S_{1,N}(p_{1,N}U, R)$;

(iii) $\|R_N(U)\|_{2,N} < r(N)$ and

$$r(N) = \min\left(\bar{c}_0(p_{1,N}U, N), \frac{R}{\bar{c}_1(p_{1,N}U, N)}\right).$$

Then (3.18) possesses a unique solution u_N in the ball $S_{1,N}(p_{1,N}U, R)$.

Proof. Put $w = p_{1,N}U$ and $r_0 = r(N)$ in Lemma 3.1. Then L_N^{-1} exists in the ball $S_{2,N}(L_N p_{1,N}U, r(N))$. On the other hand, condition (iii) implies

$$\|f_N - L_N p_{1,N}U\|_{2,N} < r(N).$$

Thus $f_N \in S_{2,N}(L_N p_{1,N}U, r(N))$ and so the result follows. ■

We can evaluate the solution of (3.18) by the Newton procedure

$$u_N^{(m+1)} = u_N^{(m)} - \left(L'_N\left(u_N^{(m)}\right)\right)^{-1}\left(L_N u_N^{(m)} - f_N\right), \quad m \geq 0, \tag{3.29}$$

or the simplified Newton procedure

$$u_N^{(m+1)} = u_N^{(m)} - \left(L'_N\left(u_N^{(0)}\right)\right)^{-1}\left(L_N u_N^{(m)} - f_N\right), \quad m \geq 0. \tag{3.30}$$

It can be verified that both (3.29) and (3.30) are convergent with the same limit as the solution of (3.18), if the following conditions are fulfilled:

(i) L'_N exists in $S_{1,N}\left(u_N^{(0)}, \rho\right)$, and for all $v^{(q)} \in S_{1,N}\left(u_N^{(0)}, \rho\right)$ and certain $d_N > 0$,

$$\|L'_N\left(v^{(1)}\right) - L'_N\left(v^{(2)}\right)\| \leq d_N \|v^{(1)} - v^{(2)}\|_{B_{1,N}};$$

(ii) $g_N = \left(L'_N\left(u_N^{(0)}\right)\right)^{-1}$ exists, and

$$\|g_N\| \leq a_N, \quad \|g_N\left(L_N\left(u_N^{(0)}\right) - f_N\right)\| \leq b_N;$$

(iii) $\eta_N = a_N b_N d_N < \dfrac{1}{2}$ and $\rho \geq \dfrac{1-\sqrt{1-2\eta_N}}{a_N d_N}$.

We now explain the relation between the stability defined in the above and the stability in actual computation. For instance, let $\tilde{f}_N$ be the error of f_N which induces the error of u_N, denoted by $\tilde{u}_N$. Then we have

$$L_N\left(u_N + \tilde{u}_N\right) = f_N + \tilde{f}_N.$$

Therefore (3.21) and (3.22) mean that for any $N \geq N_0$ and the disturbed solution $u_N + \tilde{u}_N$ in the ball $S_{1,N}\left(u_N, N\right)$,

$$\|\tilde{u}_N\|_{B_{1,N}} \leq c_1\left(u_N, N\right)\|\tilde{f}_N\|_{B_{2,N}},$$

provided that

$$\|\tilde{f}_N\|_{2,N} \leq c_0\left(u_N, N\right).$$

If L_N^{-1} exists, then (3.18) has a unique solution and $R = \infty$. If, in addition, $c_0\left(u_N, N\right)$ is arbitrary and $c_1\left(u_N, u\right) = c_1(u_N)$, then we get the best stability for usual nonlinear problems. It means that there is no restriction on the data error. In this case, this stability is nearly the same as the linear stability. But they are still different, since $c_1(u_N)$ depends on u_N. If $c_0(u_N, N) = c_0(u_N)$ and $c_1(u_N, N) = c_1(u_N)$, then the error of solution is bounded by the data error provided that the data error does not exceed a critical value. Such stability was also investigated by Stetter (1966). Further, if $c_0(u_N, N) = c_0(N)N^{-s'}$ and $c_1(u_N, N) = c_1(N)$, then it turns out to be the generalized stability of Guo (1982). It is also called g-stability. The infimum of such values s' is called the index of g-stability. On the other hand, if $c_0(u_N, N)$ is arbitrary and $c_1(u_N, N) = cN^q$, then it is a generalization of the weak stability of Strang (1965). Finally if $R < \infty, c_1(u_N, N) = c_1(u_N)$ and $c_0(u_N, N)$ is arbitrary, then it is equivalent to the local stability by Keller (1975). It means that in the ball $S_{1,N}(u_N, R)$, the approximation (3.18) is locally stable.

Under some conditions, the above stability implies the convergence.

Theorem 3.6. *Let U and $u_N \in S_{1,N}(p_{1,N}U, R)$ be the solutions of (3.1) and (3.18) respectively. Suppose that*

(i) (3.18) is of generalized weak stability for $p_{1,N}U$ in $S_{1,N}(p_{1,N}U, R)$;

(ii) $\|R_N(U)\|_{B_{2,N}} \le c_0(p_{1,N}U, N)$.

Then

$$\|u_N - p_{1,N}U\|_{B_{1,N}} \le c_1(p_{1,N}U, N)\|R_N(U)\|_{B_{2,N}}.$$

Proof. Let $\tilde{u}_N = u_N - p_{1,N}U$. Then

$$L_N\left(p_{1,N}U\right) = f_N - R_N(U),$$
$$L_N\left(p_{1,N}U + \tilde{u}_N\right) = f_N.$$

Hence

$$\|\tilde{u}_N\|_{1,N} \le c_1\left(p_{1,N}U, N\right)\|R_N(U)\|_{2,N}.$$

■

We know from Theorem 3.6 that (3.18) is convergent, if $c_1(p_{1,N}U, N)\|R_N(U)\|_{2,N} \to 0$ as $N \to \infty$. In particular, if (3.18) is of the generalized stability with the index s and $\|R_N(U)\|_{B_{2,N}} \le c_2(U)N^{-q}, q > \max(s, 0)$, then

$$\|u_N - p_{1,N}U\|_{B_{1,N}} \le c_2(U)N^{-q}.$$

3.3. Initial-Value Problems

We focus on initial-value problems in this section. Let B_1 and B_2 be two Banach spaces with the norms $\|\cdot\|_{B_1}$ and $\|\cdot\|_{B_2}, B_1 \subseteq B_2$. Q and L are two differential operators with respect to the spatial variables, mapping $v \in B_1$ into B_2. For simplicity, assume that Q is a linear invertible operator. Denote by B_3 the third Banach space with the norm $\|\cdot\|_{B_3}$, whose element is a mapping from the interval $(0, T]$ to $B_2, T > 0$. Let $f(t)$ be a given element in B_3 and U_0 be a given element in B_1. U_0 describes the initial state. Consider the operator equation

$$\begin{cases} \partial_t(QU(t)) = LU(t) + f(t), & 0 < t \le T, \\ U(0) = U_0. \end{cases} \tag{3.31}$$

We define a genuine solution of (3.31) as a one-parameter family $U(t)$ such that $U(t)$ is in the domain of Q and L for each $t \in [0,T]$, and

$$\left\| \frac{1}{s}\left(QU(t+s) - QU(t)\right) - LU(t) - f(t) \right\|_{B_2} \longrightarrow 0, \quad \text{as } s \to 0.$$

Let D_1 and D_3 be such subsets of B_1 and B_3 that for any $U_0 \in D_1$ and $f \in D_3$, (3.31) has a unique solution. Suppose that D_1 and D_3 are dense in B_1 and B_3 respectively. The corresponding subset of solutions is denoted by D_2.

Now let N be any positive integer and $B_{q,N}$ be finite dimensional Banach spaces with the norms $\|\cdot\|_{B_{q,N}}$, approximating the spaces $B_q, q = 1,2,3$. In addition, $\mathrm{Dim}(B_{1,N}) = \mathrm{Dim}(B_{2,N})$. Denote by $p_{q,N}$ the continuous operators from B_q to $B_{q,N}$, satisfying the following conditions:

(i) for all $v \in B_1, p_{1,N}v = p_{2,N}v$;

(ii) for all $v \in B_q, \|p_{q,N}v\| \to \|v\|_q$ as $N \to \infty, q = 1,2,3$.

Next, let $\tau = \tau(N) > 0$ be the mesh spacing of the variable t, and $D_\tau v$ be the forward difference quotient with respect to t, defined in Chapter 1.

We now approximate $QU(t), LU(t), f(t)$ and U_0 by $Q_N u_N(t), L_N(u_N(t), u_N(t+\tau)), f_N(t)$ and $u_{N,0}$ respectively, and then consider the approximate problem

$$\begin{cases} D_\tau\left(Q_N u_N(t)\right) = L_N\left(u_N(t), u_N(t+\tau)\right) + f_N(t), \\ u_N(0) = u_{N,0}. \end{cases} \tag{3.32}$$

Suppose that for each $t, u_N(t)$ is determined uniquely by (3.32). The simplest case is that Q_N^{-1} exists and $L_N\left(u_N(t), u_N(t+\tau)\right)$ is independent of $u_N(t+\tau)$. The approximation error for $t > 0$ is defined as

$$\begin{aligned} R_N(v(t)) &= -p_{2,N}\partial_t Qv(t) + D_\tau Q_N p_{1,N}v(t) + p_{2,N}Lv(t) - L_N\left(p_{1,N}v(t), p_{1,N}v(t+\tau)\right) \\ &\quad + p_{2,N}f(t) - f_N(t). \end{aligned}$$

If $U(t)$ is the solution of (3.31), then $R_N(U(t))$ is reduced to

$$R_N(U(t)) = D_\tau Q_N p_{1,N}U(t) - L_N\left(p_{1,N}U(t), p_{1,N}U(t+\tau)\right) - f_N(t).$$

The approximation (3.32) is said to be consistent with (3.31), if for any $U_0 \in D_1, U \in D_2, f \in D_3$ and $0 \le t \le T$,

$$\|p_{1,N}U_0 - u_{N,0}\|_{B_{1,N}} \longrightarrow 0, \quad \|R_N(U(t))\|_{B_{2,N}} \longrightarrow 0, \quad \text{as } N \to \infty.$$

If for all $U(t)$, the solutions of (3.31) and $u_N(t)$, the corresponding solutions of (3.32),

$$\|p_{1,N}U(t) - u_N(t)\|_{B_{1,N}} \longrightarrow 0, \quad 0 \le t \le T, \quad \text{as } N \to \infty,$$

then we say that the approximation (3.32) is convergent.

We now turn to the stability. We first consider the linear stability. It means that there is a positive constant $c_0(T)$ such that for all $N \ge N_0$ and $0 \le t \le T$,

$$\|u_N^{(1)}(t) - u_N^{(2)}(t)\|_{B_{1,N}} \le c_0(T)\left(\|u_{N,0}^{(1)} - u_{N,0}^{(2)}\|_{B_{1,N}} + \|f_N^{(1)} - f_N^{(2)}\|_{B_{3,N}}\right) \tag{3.33}$$

where $u_N^{(q)}$ are the solutions of (3.32), corresponding to $u_{N,0}^{(q)}$ and $f_N^{(q)}, q = 1, 2$.

Following the same line as in the proof of Theorem 3.1, we can prove the following Lax Theorem (see Richtmyer and Morton (1967)).

Theorem 3.7. *If (3.31) is a well-posed linear problem and (3.32) is consistent with (3.31), then the linear stability is equivalent to the convergence. Moreover for all* $0 \le t \le T$,

$$\|u_N(t) - p_{1,N}U(t)\|_{B_{1,N}} \le c_0(T)\left(\|u_{N,0} - p_{1,N}U_0\|_{B_{1,N}} + \|R_N(U)\|_{B_{3,N}}\right).$$

We now consider nonlinear problems, e.g., see Guo (1994). If there are positive constant $c_0(u_N^{(1)}, N, T)$ and non-negative constant $c_1(u_N^{(1)}, N, T)$ such that for all $N \ge N_0$, the inequality

$$\|u_{N,0}^{(1)} - u_{N,0}^{(2)}\|_{B_{1,N}} + \|f_N^{(1)} - f_N^{(2)}\|_{B_{3,N}} \le c_0\left(u_N^{(1)}, N, T\right) \tag{3.34}$$

implies

$$\begin{aligned} &\|u_N^{(1)}(t) - u_N^{(2)}(t)\|_{B_{1,N}} \\ &\quad \le c_1\left(u_N^{(1)}, N, T\right)\left(\|u_{N,0}^{(1)} - u_{N,0}^{(2)}\|_{B_{1,N}} + \|f_N^{(1)} - f_N^{(2)}\|_{B_{3,N}}\right), \quad 0 \le t \le T, \end{aligned} \tag{3.35}$$

$u_N^{(q)}(t)$ being the solutions of (3.32) related to $u_{N,0}^{(q)}$ and $f^{(q)}, q = 1, 2$, then we say that the approximation (3.32) is of generalized weak stability at $u_N^{(1)}$.

There is a close relationship between the generalized weak stability and the convergence, which extends the result of Lax Theorem.

Theorem 3.8. *Let $U(t)$ and $u_N(t)$ be the solutions of (3.31) and (3.32) respectively. If (3.32) is of generalized weak stability at $p_{1,N}U(t)$ and*

$$\|p_{1,N}U_0 - u_{N,0}\|_{B_{1,N}} + \|R_N(U)\|_{B_{3,N}} \le c_0\left(p_{1,N}U, N, T\right),$$

then for all $0 \le t \le T$,

$$\|p_{1,N}U(t) - u_N(t)\|_{B_{1,N}} \le c_1\left(p_{1,N}U, N, T\right)\left(\|p_{1,N}U_0 - u_{N,0}\|_{B_{1,N}} + \|R_N(U)\|_{B_{3,N}}\right).$$

Proof. We have

$$\begin{cases} D_\tau Q_N p_{1,N}U(t) = L_N\left(p_{1,N}U(t), p_{1,N}U(t+\tau)\right) + f_N(t) + R_N(U(t)), & 0 < t \le T, \\ p_{1,N}U(0) = p_{1,N}U_0. \end{cases} \tag{3.36}$$

By comparing (3.36) with (3.32), the desired result comes immediately from (3.35). ∎

The simplest case of generalized weak stability is also called g-stability, see Griffiths (1982). In this case,

$$c_0\left(u_N, N, T\right) = c_0\left(u_N, T\right) N^{-s'}, \quad c_1\left(u_N, N, T\right) = c_1\left(u_N, T\right).$$

The infimum of such values s' is also called the index of g-stability, denoted by s. Clearly, if $u_{N,0} = p_{1,N}U_0$, $\|p_{1,N}U_0 - U_0\|_{B_{1,N}} = \mathcal{O}\left(N^{-q}\right)$, $\|R_N(U)\|_{B_{3,N}} = \mathcal{O}\left(N^{-q}\right)$, $q > 0$ and $s < q$, then (3.32) is convergent with the accuracy $\mathcal{O}\left(N^{-q}\right)$.

In derivation of generalized stability, we often meet some nonlinear discrete inequalities. Let $R_\tau(T)$ be as in (1.21).

Lemma 3.2. *Let $E(t)$ be a non-negative function and $\rho(t)$ be a non-decreasing function on $R_\tau(T)$. The constants $d > 0, \lambda > 0$ and $c \ge 0$. Assume that*

(i) if $E(s) \leq d$ *for all* $s \in \bar{R}_\tau(t-\tau)$*, then*

$$F_E(t-\tau) = F_E(E(0), E(\tau), \cdots, E(t-\tau)) \leq cE(t-\tau),$$
$$G_E(t) = G_E(E(0), E(\tau), \cdots, E(t)) \geq \lambda E(t);$$

(ii) for certain $t_1 \in R_\tau(T)$ *and all* $t \in R_\tau(t_1)$*,*

$$G_E(t) \leq \rho(t_1) + \tau \sum_{s \in \bar{R}_\tau(t-\tau)} F_E(s);$$

(iii) $\lambda E(0) \leq \rho(t_1) \leq \lambda d e^{-\frac{ct_1}{\lambda}}$.

Then for all $t \in \bar{R}_\tau(t_1)$*,*

$$E(t) \leq \frac{\rho(t_1)}{\lambda} e^{\frac{ct}{\lambda}}.$$

Proof. Let $\psi(t)$ be such a non-negative function that $\psi(0) = \dfrac{\rho(t_1)}{\lambda}$ and for all $t \leq t_1$,

$$\lambda \psi(t) = \rho(t_1) + c\tau \sum_{s \in \bar{R}_\tau(t-\tau)} \psi(s).$$

Then

$$\psi(t) \leq \frac{\rho(t_1)}{\lambda} e^{\frac{ct}{\lambda}} \leq d.$$

Clearly $E(0) \leq \psi(0)$. Assume that $E(t) \leq \psi(t)$ for $t \in \bar{R}_\tau(t_1 - \tau)$. Then for any $t \leq t_1$,

$$\begin{aligned} \lambda E(t) &\leq \rho(t_1) + c\tau \sum_{s \in \bar{R}_\tau(t-\tau)} E(s) \\ &\leq \rho(t_1) + c\tau \sum_{s \in \bar{R}_\tau(t-\tau)} \psi(s) = \lambda \psi(t) \end{aligned}$$

which implies the conclusion. ∎

In many cases, $G_E(t) = E(t)$ and

$$F_E(t) = c_0 \left(1 + \sum_{l=1}^{M} N^{a_l} E^{b_l}(t)\right) E(t) + H(E(0), \cdots, E(t)), \quad b_l \geq 0.$$

In addition, if for $s \leq t$, $E(s) \leq cN^d$ and $c > 0$, then

$$H(E(0), \cdots, E(t)) \leq 0.$$

Therefore, if

$$\rho(t_1)e^{c_0 t_1(M+1)} \leq \min\left(\min_{1\leq l\leq M} N^{-\frac{a_l}{b_l}}, cN^d\right),$$

then for all $t \in \bar{R}_\tau(t_1)$,

$$E(t) \leq \rho(t_1)e^{c_0 t(M+1)}. \tag{3.37}$$

In particular, if $F_E(t) = c_0 E(t)$ and $H(E(0), \cdots, E(t)) \leq 0$, then (3.37) holds for all $\rho(t)$ and $t \in \bar{R}_\tau(T)$.

A more general result is given by Theorem 4.17 of Guo (1988a). In that case, if $E(s) \leq d$ for all $s \in \bar{R}_\tau(t-\tau)$, then

$$G_E(t) \geq \lambda_0 E(t) - \lambda_1 E^2(t), \quad \lambda_0 > 0,\ \lambda_1 \geq 0.$$

It is more suitable for numerical analysis of implicit spectral schemes for evolutionary problems of nonlinear partial differential equations.

CHAPTER 4

SPECTRAL METHODS AND PSEUDOSPECTRAL METHODS

In this chapter, we apply various orthogonal approximations in Sobolev spaces to numerical solutions of partial differential equations. Particular attention is paid to nonlinear problems arising in fluid dynamics, quantum mechanics and some other topics. We use Fourier spectral methods and Fourier pseudospectral methods for periodic problems, Legendre spectral methods, Legendre pseudospectral methods, Chebyshev spectral methods and Chebyshev pseudospectral methods for non-periodic problems. Some filterings are adopted in pseudospectral methods for the improvement of stability. Several spectral penalty methods are introduced for initial-boundary value problems with inhomogeneous data on the boundary. We also describe vanishing viscosity methods for simulating shock waves. Finally, we discuss the spectral approximations to isolated solutions of nonlinear problems.

4.1. Fourier Spectral Methods and Fourier Pseudospectral Methods

The aim of this section is to take Benjamin-Bona-Mahony equation (BBM equation) as an example to show how to construct Fourier spectral schemes and Fourier pseudospectral schemes. We shall also analyze the errors strictly and present some numerical results providing the evidences of theoretical analysis.

BBM equation describes the movement of regularized long waves. It is of the form

$$\begin{cases} \partial_t U(x,t) + d\partial_x U(x,t) + U(x,t)\partial_x U(x,t) & \\ \quad -\delta\partial_t\partial_x^2 U(x,t) = f(x,t), & x \in \mathbb{R}^1,\ 0 < t \le T, \\ U(x,0) = U_0(x), & x \in \mathbb{R}^1, \end{cases} \tag{4.1}$$

where U_0 and f are given functions, and $d \ge 0, \delta > 0$. Moreover U_0 and f possess the period 2π for the variable x, and so does the solution U. Let $\Lambda = (0, 2\pi)$ and denote the space $C(0,T;H_p^r(\Lambda))$ by $C(0,T;H_p^r)$ for simplicity. Also $\|\cdot\|_r = \|\cdot\|_{H^r(\Lambda)}$ etc.. It can be verified that if $U_0 \in H_p^1(\Lambda)$ and $f \in L^2(0,T;L_p^2)$, then (4.1) has a unique solution $U \in L^\infty(0,T;H_p^1)$.

It is commonly admitted that a reasonable numerical algorithm should preserve the features of the genuine solution of the original problem as much as possible. In fact, the solution of (4.1) possesses some conservations, such as

$$\int_\Lambda U(x,t)\,dx = \int_\Lambda U_0(x,t)\,dx + \int_0^t \int_\Lambda f(x,s)\,dxds, \tag{4.2}$$

$$\|U(t)\|^2 + \delta|U(t)|_1^2 = \|U_0\|^2 + \delta|U_0|_1^2 + 2\int_0^t (f(s),U(s))\,ds. \tag{4.3}$$

We are going to construct the Fourier spectral scheme. Let N be any positive integer and V_N be the subspace of all real-valued functions of $\tilde{V}_N$ by (2.17). Denote by P_N the L^2-orthogonal projection from $L_p^2(\Lambda)$ onto V_N. For convenience of analysis, let

$$J(v(x),w(x)) = \frac{1}{3}w(x)\partial_x v(x) + \frac{1}{3}\partial_x\left(w(x)v(x)\right). \tag{4.4}$$

If $v, w, z \in H_p^1(\Lambda)$, then

$$\begin{aligned} (\partial_x(wv), z) + (w\partial_x z, v) &= 0, \\ (w\partial_x v, z) + (\partial_x(wz), v) &= 0. \end{aligned}$$

Hence

$$(J(v,w),z) + (J(z,w),v) = 0. \tag{4.5}$$

In particular,

$$(J(v,w),v) = 0. \tag{4.6}$$

On the other hand,

$$\begin{aligned}(J(v,w),w) &= \frac{1}{3}\left(w^2,\partial_x v\right)-\frac{1}{3}\left(w\partial_x w,v\right),\\(J(v,w),w) &= \frac{2}{3}\left(w^2,\partial_x v\right)+\frac{1}{3}\left(w\partial_x w,v\right)\end{aligned}$$

from which

$$(J(v,w),w)=\frac{1}{2}\left(\partial_x v,w^2\right). \tag{4.7}$$

In addition, for $\phi,\psi\in V_N$,

$$(P_N J(\phi,w),\psi)+(P_N J(\psi,w),\phi)=0, \tag{4.8}$$

$$(P_N J(\phi,w),\phi)=0, \tag{4.9}$$

$$(P_N J(\phi,\psi),\psi)=\frac{1}{2}\left(\partial_x\phi,\psi^2\right). \tag{4.10}$$

For discretization in time t, let τ be the mesh size and use the notations in Chapter 1, such as $R_\tau(T),\bar{R}_\tau(T)$ and $D_\tau v(x,t)$, etc..

Let u_N be the approximation to U, and σ_l be parameters, $0\le\sigma_l\le 1, l=1,2,3$. A Fourier spectral scheme for (4.1) is to find $u_N(x,t)\in V_N$ for all $t\in\bar{R}_\tau(T)$ such that

$$\begin{cases} D_\tau u_N(x,t)+\sigma_1 d\partial_x u_N(x,t+\tau)+(1-\sigma_1)d\partial_x u_N(x,t) \\ \quad +\sigma_2 P_N J\left(u_N(x,t+\tau),u_N(x,t)\right)+(1-\sigma_2)P_N J\left(u_N(x,t),u_N(x,t)\right) \\ \quad -\delta D_\tau\partial_x^2 u_N(x,t)=F_N(x,t), \qquad x\in\mathbb{R}^1,\ t\in R_\tau(T), \\ u_N(x,0)=u_{N,0}(x), \qquad x\in\mathbb{R}^1 \end{cases} \tag{4.11}$$

where $u_{N,0}(x)=P_N U_0(x)$ and

$$F_N(x,t)=\sigma_3 P_N f(x,t+\tau)+(1-\sigma_3)P_N f(x,t).$$

The first formula of (4.11) can be rewritten as

$$\begin{aligned}&D_\tau u_N(x,t)+d\partial_x u_N(x,t)+\sigma_1 d\tau D_\tau\partial_x u_N(x,t)\\&\quad +P_N J(u_N(x,t)+\sigma_2\tau D_\tau u_N(x,t),u_N(x,t))-\delta D_\tau\partial_x^2 u_N(x,t)=F_N(x,t).\end{aligned} \tag{4.12}$$

If $\sigma_1=\sigma_2=\sigma_3=0$, then (4.12) is the explicit spectral scheme in Guo and Manoranjan (1985). If $\sigma_1\neq 0$ and $\sigma_2=0$, it becomes an implicit scheme. But by the orthogonality of Fourier system, we can still evaluate the coefficients of $u_N(x,t)$ explicitly at each time $t \in R_\tau(T)$. This is the advantage of using spectral methods. In (4.11), the nonlinear term is approximated by partially implicit approach and so we only have to solve a linear system for the Fourier coefficients step by step, even if $\sigma_1\sigma_2 \neq 0$. Indeed (4.11) reads

$$\begin{aligned} &u_N(x,t) + \sigma_1 d\tau \partial_x u_N(x,t) \\ &\quad + \sigma_2 \tau P_N J\left(u_N(x,t), u_N(x,t-\tau)\right) - \delta \partial_x^2 u_N(x,t) = G_N(x,t-\tau) \end{aligned} \tag{4.13}$$

where the term $G_N(x,t-\tau)$ depends only on $u_N(x,t-\tau)$ and $\tau F_N(x,t-\tau)$. We take the inner product of both sides of the above equation with $u_N(t)$, i.e., multiplying $u_N(x,t)$ and integrating the resulting equation over Λ. Then we obtain from (4.9) that

$$\|u_N(t)\|^2 + \delta |u_N(t)|_1^2 = \left(G_N(t-\tau), u_N(t)\right).$$

If $G_N(x,t-\tau) \equiv 0$ for $x \in \Lambda$, then $u_N(x,t) \equiv 0$ for $x \in \Lambda$. Since (4.13) is a linear system, the above fact ensures the existence and uniqueness of $u_N(x,t)$ for all $t \in \bar{R}_\tau(T)$.

We check the conservations of the solution of (4.11). We first take $\sigma_2 = 0$, integrate (4.12) over Λ and sum the result for $t \in \bar{R}_\tau(T)$. Then it follows that

$$\int_\Lambda u_N(x,t)\,dx = \int_\Lambda u_{N,0}(x,t)\,dx + \tau \sum_{s\in \bar{R}_\tau(t-\tau)} \int_\Lambda F_N(x,s)\,dx.$$

Next, we let $\sigma_1 = \sigma_2 = \frac{1}{2}$, take the inner product of (4.11) with $u_N(t+\tau) + u_N(t)$ and sum the result for $t \in \bar{R}_\tau(T)$. We obtain from (4.9) that

$$\|u_N(t)\|^2 + \delta |u_N(t)|_1^2 = \|u_{N,0}\|^2 + \delta |u_{N,0}|_1^2 + \tau \sum_{s\in \bar{R}_\tau(t-\tau)} \left(F_N(s), u_N(s+\tau) + u_N(s)\right).$$

Evidently the above two equations are reasonable analogies of the conservations (4.2) and (4.3).

We now turn to analyze the errors. Suppose that $u_{N,0}$ and F_N have the errors $\tilde{u}_{N,0}$ and $\tilde{F}_N$ respectively. Then we get a disturbed solution corresponding to $u_{N,0} + \tilde{u}_{N,0}$ and $F_N + \tilde{F}_N$, denoted by $u_N + \tilde{u}_N$. For simplicity, we denote the errors $\tilde{u}_N, \tilde{u}_{N,0}$ and

$\tilde{F}_N$ by $\tilde{u}, \tilde{u}_0$ and $\tilde{F}$. They satisfy the error equation as follows

$$\begin{aligned}
&D_\tau \tilde{u}(x,t) + d\partial_x \tilde{u}(x,t) + \sigma_1 d\tau D_\tau \partial_x \tilde{u}(x,t) + P_N J\left(u_N(x,t)\right. \\
&\quad \left.+\sigma_2 \tau D_\tau u_N(x,t), \tilde{u}(x,t)\right) + P_N J\left(\tilde{u}(x,t) + \sigma_2 \tau D_\tau \tilde{u}(x,t), u_N(x,t) + \tilde{u}(x,t)\right) \\
&\quad -\delta D_\tau \partial_x^2 \tilde{u}(x,t) = \tilde{F}(x,t), \quad x \in \mathbb{R}^1,\ t \in \bar{R}_\tau(T). \qquad (4.14)
\end{aligned}$$

We take the inner product of (4.14) with $2\tilde{u}(x,t)$ and then obtain from (1.22) and (4.9) that

$$D_\tau \|\tilde{u}(t)\|^2 + \delta D_\tau |\tilde{u}(t)|_1^2 - \tau \|D_\tau \tilde{u}(t)\|^2 - \delta\tau |D_\tau \tilde{u}(t)|_1^2 + \sum_{j=1}^{3} F_j(t) = 2\left(\tilde{F}(t), \tilde{u}(t)\right) \quad (4.15)$$

where

$$\begin{aligned}
F_1(t) &= 2\sigma_1 d\tau \left(D_\tau \partial_x \tilde{u}(t), \tilde{u}(t)\right), \\
F_2(t) &= 2\left(J\left(u_N(t) + \sigma_2 \tau D_\tau u_N(t), \tilde{u}(t)\right), \tilde{u}(t)\right), \\
F_3(t) &= 2\sigma_2 \tau \left(J\left(D_\tau \tilde{u}(t), u_N(t) + \tilde{u}(t)\right), \tilde{u}(t)\right).
\end{aligned}$$

Furthermore, let ξ be an undetermined positive constant and take the inner product of (4.14) with $\xi\tau D_\tau \tilde{u}(x,t)$. We have from (4.9) that

$$\xi\tau \|D_\tau \tilde{u}(t)\|^2 + \xi\delta\tau |D_\tau \tilde{u}(t)|_1^2 + \sum_{j=4}^{6} F_j(t) = \xi\tau \left(\tilde{F}(t), D_\tau \tilde{u}(t)\right) \quad (4.16)$$

where

$$\begin{aligned}
F_4(t) &= d\xi\tau \left(\partial_x \tilde{u}(t), D_\tau \tilde{u}(t)\right), \\
F_5(t) &= \xi\tau \left(J\left(u_N(t) + \sigma_2 \tau D_\tau u_N(t), \tilde{u}(t)\right), D_\tau \tilde{u}(t)\right), \\
F_6(t) &= \xi\tau \left(J\left(\tilde{u}(t), u_N(t) + \tilde{u}(t)\right), D_\tau \tilde{u}(t)\right).
\end{aligned}$$

Let ε be a positive constant. Putting (4.15) and (4.16) together, we obtain that

$$\begin{aligned}
&D_\tau \|\tilde{u}(t)\|^2 + \delta D_\tau |\tilde{u}(t)|_1^2 + \tau(\xi - 1 - \varepsilon)\left(\|D_\tau \tilde{u}(t)\|^2 + \delta |D_\tau \tilde{u}(t)|_1^2\right) \\
&\quad + \textstyle\sum_{j=1}^{6} F_j(t) \le \|\tilde{u}(t)\|^2 + \left(1 + \frac{\tau\xi^2}{4\varepsilon}\right) \|\tilde{F}(t)\|^2.
\end{aligned} \quad (4.17)$$

We begin to estimate $|F_j(t)|$. It is easy to see that

$$\begin{aligned}
|F_1(t) + F_4(t)| &= |d\tau(\xi - 2\sigma_1)\left(\partial_x \tilde{u}(t), D_\tau \tilde{u}(t)\right)| \\
&\le \varepsilon\tau \|D_\tau \tilde{u}(t)\|^2 + \frac{\tau d^2(\xi - 2\sigma_1)^2}{4\varepsilon} |\tilde{u}(t)|_1^2. \qquad (4.18)
\end{aligned}$$

Let $r > \frac{3}{2}$. By (4.10) and imbedding theory,

$$|F_2(t)| \le c\|\partial_x u_N(t)\|_{C(0,T;L^\infty)}\|\tilde{u}(t)\|^2 \le c\|u_N\|_{C(0,T;H^r)}\|\tilde{u}(t)\|^2. \tag{4.19}$$

Similarly

$$|F_5(t)| \le \varepsilon\tau\|D_\tau\tilde{u}(t)\|^2 + \frac{c\tau\xi^2}{\varepsilon}\|u_N\|_{C(0,T;H^r)}\|\tilde{u}(t)\|_1^2. \tag{4.20}$$

Further, by Theorem 2.1,

$$\|\tilde{u}(t)\partial_x\tilde{u}(t)\| \le \|\tilde{u}(t)\|_{L^4}^2\|\partial_x\tilde{u}(t)\|_{L^4}^2 \le cN\|\tilde{u}(t)\|^2|\tilde{u}(t)|_1^2$$

and so by (4.8),

$$\begin{aligned} & |F_3(t) + F_6(t)| \\ = \ & |\tau(\xi - 2\sigma_2)\left(J\left(\tilde{u}(t), u_N(t) + \tilde{u}(t)\right), D_\tau\tilde{u}(t)\right)| \\ \le \ & \varepsilon\tau\|D_\tau\tilde{u}(t)\|^2 + \frac{c\tau(\xi - 2\sigma_2)^2}{\varepsilon}\left(\|u_N\|_{C(0,T;H^r)}\|\tilde{u}(t)\|_1^2 + N\|\tilde{u}(t)\|^2|\tilde{u}|_1^2\right). \end{aligned} \tag{4.21}$$

By substituting (4.18)–(4.21) into (4.17), we find that

$$D_\tau\|\tilde{u}(t)\|^2 + \delta D_\tau|\tilde{u}(t)|_1^2 + \tau(\xi - 1 - 4\varepsilon)\left(\|D_\tau\tilde{u}(t)\|^2 + \delta|D_\tau\tilde{u}(t)|_1^2\right) \le \tilde{R}(t)$$

where

$$\begin{aligned} \tilde{R}(t) \ = \ & c\left(1 + \left(1 + \frac{\tau}{\varepsilon}\left(\xi^2 + (\xi - 2\sigma_2)^2\right)\right)\|u_N\|_{C(0,T;H^r)}\right)\|\tilde{u}(t)\|^2 \\ & + \frac{c\tau}{\varepsilon}\left((\xi - 2\sigma_1)^2 + \left(\xi^2 + (\xi - 2\sigma_2)^2\right)\|u_N\|_{C(0,T;H^r)}\right)|\tilde{u}(t)|_1^2 \\ & + \frac{c\tau N}{\varepsilon}(\xi - 2\sigma_2)^2\|\tilde{u}(t)\|^2|\tilde{u}(t)|_1^2 + \left(1 + \frac{\tau\xi^2}{4\varepsilon}\right)\|\tilde{F}(t)\|^2. \end{aligned}$$

Now let $\xi > 1, 0 < q_0 < \xi - 1$ and $\varepsilon = \frac{1}{4}(\xi - 1 - q_0)$. Then we deduce that

$$D_\tau\|\tilde{u}(t)\|^2 + \delta D_\tau|\tilde{u}(t)|_1^2 + q_0\tau\left(\|D_\tau\tilde{u}(t)\|^2 + \delta|D_\tau\tilde{u}(t)|_1^2\right) \le \tilde{R}(t). \tag{4.22}$$

For describing the average error of numerical solution, we define

$$E(v,t) = \|v(t)\|^2 + \delta|v(t)|_1^2 + q_0\tau^2 \sum_{s\in\bar{R}(t-\tau)}\left(\|D_\tau v(s)\|^2 + \delta|D_\tau v(s)|_1^2\right).$$

Also for the average error of data, set

$$\rho(v,w,t)=\|v\|^2+\delta|v|_1^2+\tau\sum_{s\in\bar{R}(t-\tau)}\|w(s)\|^2.$$

By summing (4.22) for $t\in\bar{R}_\tau(T)$, we obtain

$$E(\tilde{u},t)\le c\rho(\tilde{u}_0,\tilde{F},t)+c\tau\sum_{s\in\bar{R}_\tau(t-\tau)}[(1+\|u_N\|_{C(0,T;H^r)})E(\tilde{u},s)+\tau N(\xi-2\sigma_2)^2E^2(\tilde{u},s)].$$

In particular, if $\sigma_2>\frac{1}{2}$, then we can take $\xi=2\sigma_2$ and so

$$E(\tilde{u},t)\le c\rho(\tilde{u}_0,\tilde{F},t)+c\tau\sum_{s\in\bar{R}_\tau(t-\tau)}\left(1+\|u_N\|_{C(0,T;H^r)}\right)E(\tilde{u},s).$$

Finally we use Lemma 3.2 with $E(t)=E(\tilde{u},t)$ and $\rho(t)=\rho(\tilde{u}_0,\tilde{F},t)$, and obtain the following result.

Theorem 4.1. *There exist positive constants b_1 and b_2 depending only on d,δ and $\|u_N\|_{C(0,T;H^r)}, r>\frac{3}{2}$, such that if for certain $t_1\in R_\tau(T), \rho(\tilde{u}_0,\tilde{F},t_1)\le\frac{b_1}{\tau N}$, then for all $t\in\bar{R}_\tau(t_1)$,*

$$E(\tilde{u},t)\le c\rho(\tilde{u}_0,\tilde{F},t)e^{b_2t}.$$

If in addition $\sigma_2>\frac{1}{2}$, then the above estimate holds for all $t\in\bar{R}_\tau(T)$ and any $\rho(\tilde{u}_0,\tilde{F},t)$.

Theorem 4.1 shows that scheme (4.11) is of the generalized stability with the index $s=1-q$, provided that $\tau=\mathcal{O}(N^{-q})$. If $q=1$, then $s=0$. In this case, the computation is stable when the average error of data does not exceed a critical value cb_1. If $\sigma_2>\frac{1}{2}$, then there is no restriction on $\rho(\tilde{u}_0,\tilde{F},t)$. It means that scheme (4.11) possesses the index of generalized stability $s=-\infty$. We know from the above analysis that the suitable choice of parameter in the approximation of nonlinear term can improve the stability essentially.

We next deal with the convergence of (4.11). Let $U_N=P_NU$ and $\sigma_3=0$ for simplicity. By (4.1),

$$\begin{cases} D_\tau U_N(x,t) + d\partial_x U_N(x,t) + \sigma_1 d\tau D_\tau \partial_x U_N(x,t) \\ \quad + P_N J\left(U_N(x,t) + \sigma_2 \tau D_\tau U_N(x,t), U_N(x,t)\right) - \delta D_\tau \partial_x^2 U_N(x,t) \\ \quad = F_N(x,t) + \sum_{j=1}^{5} G_j(x,t), \qquad x \in \mathbb{R}^1,\ t \in \bar{R}_\tau(T), \\ U_N(x,0) = P_N U_0(x), \quad x \in \mathbb{R}^1, \end{cases}$$

where

$$\begin{aligned} G_1(x,t) &= D_\tau U_N(x,t) - \partial_t U_N(x,t), \\ G_2(x,t) &= \sigma_1 d\tau D_\tau \partial_x U_N(x,t), \\ G_3(x,t) &= P_N J\left(U_N(x,t), U_N(x,t)\right) - P_N J\left(U(x,t), U(x,t)\right), \\ G_4(x,t) &= \sigma_2 \tau P_N J\left(D_\tau U_N(x,t), U_N(x,t)\right), \\ G_5(x,t) &= \delta \partial_t \partial_x^2 U_N(x,t) - \delta D_\tau \partial_x^2 U_N(x,t). \end{aligned}$$

Further, let $\tilde{U}_N(x,t) = u_N(x,t) - U_N(x,t)$, also denoted by $\tilde{U}(x,t)$ for simplicity. By (4.11), it follows that

$$\begin{aligned} & D_\tau \tilde{U}(x,t) + d\partial_x \tilde{U}(x,t) + \sigma_1 d\tau D_\tau \partial_x \tilde{U}(x,t) + P_N J\left(U_N(x,t) + \sigma_2 \tau D_\tau U_N(x,t), \tilde{U}(x,t)\right) \\ & \quad + P_N J\left(\tilde{U}(x,t) + \sigma_2 \tau D_\tau \tilde{U}(x,t), U_N(x,t) + \tilde{U}(x,t)\right) \\ & \quad - \delta D_\tau \partial_x^2 \tilde{U}(x,t) = -\sum_{j=1}^{5} G_j(x,t). \end{aligned} \tag{4.23}$$

In addition, $\tilde{U}(x,0) = 0$. By comparing this error equation to (4.14), we find that it suffices to estimate the terms $G_j(x,t), 1 \le j \le 5$. Since

$$\partial_t v(x,t) - D_\tau v(x,t) = -\frac{1}{\tau} \int_t^{t+\tau} (t+\tau-s) \partial_s^2 v(x,s)\, ds, \tag{4.24}$$

the property of Bochner integral leads to

$$\tau \sum_{s \in \bar{R}(t-\tau)} \|G_1(s)\|^2 \le c\tau^2 \|U\|^2_{H^2(0,t;L^2)}.$$

Clearly,

$$\tau \sum_{s \in \bar{R}(t-\tau)} \|G_2(s)\|^2 \le c\tau^2 \|U\|^2_{H^1(0,t;H^1)}.$$

Moreover for any $v \in H_p^r(\Lambda)$ with $r > \frac{3}{2}$, we deduce from Theorem 2.3 and imbedding theory,

$$\begin{aligned}\|v\partial_x v - P_N v \partial_x P_N v\| &\leq \|v\partial_x v - v\partial_x P_N v\| + \|v\partial_x P_N v - P_N v\partial_x P_N v\| \\ &\leq \|v\|_{L^\infty}|v - P_N v|_1 + \|\partial_x P_N v\|_{L^\infty}\|v - P_N v\| \\ &\leq cN^{1-r}\|v\|_r^2.\end{aligned}$$

Thus

$$\tau \sum_{s\in\bar{R}_\tau(t-\tau)} \|G_3(s)\|^2 \leq cN^{2-2r}\|U\|^4_{C(0,t;H^r)}.$$

It is not difficult to verify that

$$\begin{aligned}\tau \sum_{s\in\bar{R}_\tau(t-\tau)} \|G_4(s)\|^2 &\leq c\tau^2 \left(\|U_N\|^2_{C(0,t;L^\infty)}\|U_N\|^2_{H^1(0,t;H^1)}\right. \\ &\qquad \left.+\|\partial_x U_N\|^2_{C(0,t;L^\infty)}\|U_N\|^2_{H^1(0,t;L^2)}\right) \\ &\leq c\tau^2\|U\|^2_{C(0,t;H^r)}\|U\|^2_{H^1(0,t;H^1)}.\end{aligned}$$

We have

$$\begin{aligned}\left|\tau \sum_{s\in\bar{R}_\tau(t-\tau)} \left(G_5(s), \tilde{U}(s)\right)\right| &\leq \delta\tau \sum_{s\in\bar{R}_\tau(t-\tau)} \left|\left(D_\tau\partial_x U_N(s) - \partial_t\partial_x U_N(s), \partial_x\tilde{U}(s)\right)\right| \\ &\leq \delta\tau \sum_{s\in\bar{R}_\tau(t-\tau)} \left|\tilde{U}(s)\right|_1^2 + c\delta\tau^2\|U\|^2_{H^2(0,t;H^1)}.\end{aligned}$$

It is noted that $\tau^2 + N^{2-2r} = \mathcal{O}\left(\frac{1}{\tau N}\right)$ for $\tau = \mathcal{O}\left(N^{-\frac{1}{3}}\right)$ and $r > \frac{3}{2}$. So by an argument as in the proof of Theorem 4.1, we obtain the following result.

Theorem 4.2. *Let $\sigma_3 = 0$ in (4.11). If $U \in C(0,T;H_p^r) \cap H^2(0,T;H^1), r > \frac{3}{2}$ and $\tau \leq b_3 N^{-\frac{1}{3}}$, then for all $t \in \bar{R}_\tau(T)$,*

$$E\left(U_N - u_N, t\right) \leq b_4 e^{b_5 t}\left(\tau^2 + N^{2-2r}\right)$$

where b_3, b_4 and b_5 are positive constants depending on $d, \delta, \|U\|_{C(0,T;H_p^r)}$ and $\|U\|_{H^2(0,T;H^1)}$. If $\sigma_2 > \frac{1}{2}$, then the above estimate is valid for any $\tau > 0$.

If we take $\sigma_1 = \sigma_3 = \frac{1}{2}$ and approximate the nonlinear term by $\frac{1}{4}P_N J(U_N(x,t+\tau) + U_N(x,t), U_N(x,t+\tau) + U_N(x,t))$, then $E(U_N - u_N, t) = \mathcal{O}\left(\tau^4 + N^{2-2r}\right)$.

We now present some numerical results. For describing the errors, let Λ_N be as in (2.19), and

$$\tilde{E}(v,t) = \max_{x\in\Lambda_N} \frac{|U(x,t)-v(x,t)|}{|U(x,t)|}$$

where U is the solution of (4.1), and v is the corresponding numerical solution. For comparison, we also use a finite difference scheme to solve (4.1). Let $h = \dfrac{2\pi}{2N+1}$ and

$$\begin{aligned}
v_x(x,t) &= \frac{1}{h}\left(v(x+h,t) - v(x,t)\right), \\
v_{\bar{x}}(x,t) &= \frac{1}{h}\left(v(x,t) - v(x-h,t)\right), \\
v_{\hat{x}}(x,t) &= \frac{1}{2}v_x(x,t) + \frac{1}{2}v_{\bar{x}}(x,t).
\end{aligned}$$

Define the finite difference operators as

$$\begin{aligned}
M_h v(x,t) &= \frac{1}{6}v(x-h,t) + \frac{2}{3}v(x,t) + \frac{1}{6}v(x+h,t), \\
J_h(v(x,t),v(x,t)) &= \frac{1}{3}v(x,t)v_{\hat{x}}(x,t) + \frac{1}{3}\left(v^2(x,t)\right)_{\hat{x}}, \\
\Delta_h v(x,t) &= v_{x\bar{x}}(x,t).
\end{aligned}$$

A finite difference scheme for (4.1) is

$$\begin{cases}
D_\tau M_h u_h(x,t) + d u_{h,\hat{x}}(x,t) + J_h\left(u_h(x,t),u_h(x,t)\right) & \\
\quad -\,\delta D_\tau \Delta_h u_h(x,t) = f(x,t), & x\in\Lambda_N,\ t\in\bar{R}_\tau(T), \\
u_h(x,t) = u_h(x+2\pi,t), & x\in\Lambda_N,\ t\in\bar{R}_\tau(T), \\
u_{h,0}(x) = U_0(x), & x\in\Lambda_N.
\end{cases} \tag{4.25}$$

We take the test function

$$U(x,t) = A\exp(B\sin x + \omega t). \tag{4.26}$$

The calculations are carried out with $d=1, \delta=0.001, A=B=0.1, \omega=0.1, N=5$ and $\tau=0.01$. Table 4.1 shows the errors $\tilde{E}(u_N,t)$ for scheme (4.11) with $\sigma_1=\sigma_2=\sigma_3=0$, and the errors $\tilde{E}(u_h,t)$ for scheme (4.25). We find that (4.11) gives quite accurate numerical results even for small N and large τ. It is also shown that (4.11) provides better numerical results than (4.25).

Table 4.1. The errors $\tilde{E}(u_N,t)$ and $\tilde{E}(u_h,t)$.

t	$\tilde{E}(u_N,t)$	$\tilde{E}(u_h,t)$
0.5	$6.719E{-}3$	$8.931E{-}2$
1.0	$8.832E{-}3$	$1.150E{-}2$
1.5	$6.675E{-}3$	$7.574E{-}2$
2.0	$1.027E{-}2$	$7.791E{-}2$

As indicated before, it is not easy to deal with nonlinear terms in spectral methods usually. So we prefer Fourier pseudospectral methods in actual computations. Let $x^{(j)}, \Lambda_N$ and I_N be the Fourier interpolation points, the discrete interval and the Fourier interpolant in (2.19) and (2.20). The discrete inner product $(\cdot,\cdot)_N$ and the discrete norm $\|\cdot\|_N$ are given by (2.22). Clearly

$$I_N\phi = \phi, \qquad \forall\, \phi \in V_N, \tag{4.27}$$

$$(v,w)_N = (I_Nv, I_Nw)_N = (I_Nv, I_Nw), \quad \forall\, v,w \in C_p\left(\bar{\Lambda}\right). \tag{4.28}$$

We can approximate (4.1) directly and derive a scheme similar to (4.11), in which the projection P_N is replaced by I_N. But such a scheme cannot provide good numerical results. This is caused by two things. Firstly, the conservations (4.2) and (4.3) are simulated in spectral scheme (4.11) automatically. However, it is not so in pseudospectral methods, since interpolation is used. In order to preserve the global properties of genuine solution of (4.1), we have to deal with the nonlinear term carefully. For this purpose, we define the operator

$$J_N(v,w)(x) = \frac{1}{3}I_N(w\partial_x v)(x) + \frac{1}{3}\partial_x I_N(wv)(x).$$

Let $\phi, \psi \in V_N$. By (4.27) and (4.28),

$$(J_N(\phi,w),\psi) = \frac{1}{3}(w\partial_x\phi,\psi)_N - \frac{1}{3}(w\phi,\partial_x\psi)_N,$$
$$(J_N(\psi,w),\phi) = \frac{1}{3}(w\partial_x\psi,\phi)_N - \frac{1}{3}(w\psi,\partial_x\phi)_N.$$

Consequently,

$$(J_N(\phi,w),\psi) + (J_N(\psi,w),\phi) = 0, \tag{4.29}$$

$$(J_N(\phi,w),\phi) = 0. \tag{4.30}$$

Therefore $J_N(\phi, w)$ preserves the property of $J(\phi, w)$. This technique is called skew symmetric decomposition of nonlinear convective term. We shall see below that it enables the numerical solution to keep the conservations. On the other hand, the nonlinear computation may destroy the stability. Mainly, this trouble is due to the higher frequency terms. To remedy this, we can use the filtering $R_N(\alpha, \beta)$ in (2.107), denoted by R_N for simplicity. The use of filtering for nonlinear terms brings some difficulties in error analysis. To simplify the statements, we introduce the circle convolution as

$$\phi * \psi = \sum_{|l| \le N} \sum_{|p| \le N} \hat{\phi}_p \hat{\psi}_{l-p} e^{ilx}, \quad \forall\, \phi, \psi \in \tilde{V}_N. \tag{4.31}$$

It is easy to see that $\phi * \psi = \psi * \phi$.

Lemma 4.1. *For any $\phi, \psi \in \tilde{V}_N$ and $\chi \in V_N$,*

$$I_N(\phi\psi) = \phi * \psi,$$
$$(\phi * \chi, \psi) = (\phi, \chi * \psi).$$

Proof. For the first conclusion, it is sufficient to prove that

$$\phi * \psi \left(x^{(j)}\right) = \phi\left(x^{(j)}\right) \psi\left(x^{(j)}\right), \quad 0 \le j \le 2N.$$

In fact,

$$\begin{aligned} \phi * \psi\left(x^{(j)}\right) &= \sum_{|p| \le N} \hat{\phi}_p e^{ipx^{(j)}} \sum_{|l| \le N} \hat{\psi}_{l-p} e^{i(l-p)x^{(j)}} \\ &= \sum_{|p| \le N} \hat{\phi}_l e^{ipx^{(j)}} \psi\left(x^{(j)}\right) = \phi\left(x^{(j)}\right) \psi\left(x^{(j)}\right), \end{aligned}$$

and so the desired result follows. Next, by (4.28) and the first conclusion, we deduce that

$$(\phi * \chi, \psi) = (I_N(\phi\chi), I_N\psi) = (\phi\chi, \psi)_N = (\phi, \chi\psi)_N = (\phi, \chi * \psi).$$

∎

Lemma 4.2. *For any $\phi, \psi \in \tilde{V}_N, w \in H_p^r(\Lambda)$ and $r > \frac{3}{2}$,*

$$|(\partial_x \phi * R_N \psi, w) + (\phi * R_N \partial_x \psi, w)| \le c\alpha\beta \|w\|_r \|\phi\| \|\psi\|.$$

Proof. Let $\hat{w}_l$ be the Fourier coefficients of w. Define

$$\gamma_{l,p} = \begin{cases} l-p+2N+1, & \text{if } l-p<-N, \\ l-p, & \text{if } |l-p|\le N, \\ l-p-2N-1, & \text{if } l-p>N. \end{cases} \tag{4.32}$$

Clearly $\gamma_{-l,-p} = -\gamma_{l,p}$. We have

$$(\partial_x\phi * R_N\psi, w) = (\partial_x\phi * R_N\psi, P_Nw) = 2\pi i \sum_{|l|\le N} \bar{\hat{w}}_l \sum_{|p|\le N} \left(1-\left|\frac{p}{N}\right|^{\alpha}\right)^{\beta} \gamma_{l,p}\hat{\phi}_{l-p}\hat{\psi}_p,$$

$$(\phi * R_N\partial_x\psi, w) = (\phi * R_N\partial_x\psi, P_Nw) = 2\pi i \sum_{|l|\le N} \bar{\hat{w}}_l \sum_{|p|\le N} \left(1-\left|\frac{p}{N}\right|^{\alpha}\right)^{\beta} p\hat{\phi}_{l-p}\hat{\psi}_p.$$

Putting these two formulas together, we get that

$$(\partial_x\phi * R_N\psi, w) + (\phi * R_N\psi, w) = 2\pi i \sum_{|l|\le N} (1+|l|)\bar{\hat{w}}_l \sum_{|p|\le N} \lambda_{l,p}\hat{\phi}_{l-p}\hat{\psi}_p$$

where

$$\lambda_{l,p} = \left(1-\left|\frac{p}{N}\right|^{\alpha}\right)^{\beta}(p+\gamma_{l,p})\,(1+|l|)^{-1}.$$

Let

$$\theta_{l,p} = \left(1-\left|\frac{p}{N}\right|\right)(p+\gamma_{l,p})\,(1+|l|)^{-1}.$$

Set $f(x) = (1-x^{\alpha})^{\beta} - \alpha\beta(1-x)$. Clearly $f(1)=0$ and for $0\le x\le 1$,

$$\partial_x f(x) = \alpha\beta\left(1 - x^{\alpha-1}(1-x^{\alpha})^{\beta-1}\right) \ge 0.$$

Therefore for $0\le x\le 1$,

$$(1-x^{\alpha})^{\beta} \le \alpha\beta(1-x),$$

and so for $|l|, |p| \le N$,

$$|\lambda_{l,p}| \le \alpha\beta|\theta_{l,p}|.$$

We now estimate $|\theta_{l,p}|$. Since $\theta_{-l,-p} = -\theta_{l,p}$, it suffices to consider the case $0\le p\le N$ only. If $|l-p|\le N$, then

$$|\theta_{l,p}| = \frac{|l|\left(1-\frac{p}{N}\right)}{1+|l|} \le 1.$$

If $l-p<-N$ and so $0\le N-p<-l=|l|$, then

$$|\theta_{l,p}| = \frac{(N-p)(2N+1+l)}{N(1+|l|)} \le \frac{|l|(2N+1-|l|)}{N(1+|l|)} \le 2.$$

If $l-p>N$ and so $p\le 0$. It is contrary to $p\ge 0$. Therefore $|\lambda_{l,p}|\le 2\alpha\beta$ for all $|l|,|p|\le N$. Furthermore let ε be a sufficiently small positive constant. We deduce that

$$
\begin{aligned}
&|(\partial_x\phi * R_N\psi, w)+(\phi * R_N\partial_x\psi, w)| \\
&\quad\le c\alpha\beta\sum_{|l|\le N}(1+|l|)|\hat{w}_l|\sum_{|p|\le N}\left|\hat{\phi}_{l-p}\right|\left|\hat{\psi}_p\right| \\
&\quad\le c\alpha\beta\sum_{|l|\le N}(1+|l|)|\hat{w}_l|\left(\sum_{|p|\le N}\left|\hat{\phi}_{l-p}\right|^2\right)^{\frac{1}{2}}\left(\sum_{|p|\le N}\left|\hat{\psi}_p\right|^2\right)^{\frac{1}{2}} \\
&\quad\le c\alpha\beta\|\phi\|\|\psi\|\left(\sum_{|l|\le N}(1+|l|)^{-1-2\varepsilon}\right)^{\frac{1}{2}}\left(\sum_{|l|\le N}(1+|l|)^{3+2\varepsilon}|w_l|^2\right)^{\frac{1}{2}} \\
&\quad\le c\alpha\beta\|w\|_r\|\phi\|\|\psi\|.
\end{aligned}
$$

■

Lemma 4.3. *For any $\phi\in\tilde{V}_N, w\in H_p^r(\Lambda)$ and $r>\frac{3}{2}$,*

$$|(\partial_x\phi * R_N\phi, w)-(\phi * R_N\partial_x\phi, w)|\le c\alpha\beta\|w\|_r\|\phi\|^2.$$

Proof. We have

$$
\begin{aligned}
(\partial_x\phi * R_N\phi, w)=(\partial_x\phi * R_N\phi, P_Nw)&=2\pi i\sum_{|l|\le N}\bar{\hat{w}}_l\sum_{|p|\le N}p\left(1-\left|\frac{\gamma_{l,p}}{N}\right|^\alpha\right)^\beta\hat{\phi}_p\hat{\phi}_{l-p}, \\
(\phi * R_N\partial_x\phi, w)=(\phi * R_N\partial_x\phi, P_Nw)&=2\pi i\sum_{|l|\le N}\bar{\hat{w}}_l\sum_{|p|\le N}p\left(1-\left|\frac{p}{N}\right|^\alpha\right)^\beta\hat{\phi}_p\hat{\phi}_{l-p}.
\end{aligned}
$$

Thus

$$(\partial_x\phi * R_N\phi, w)-(\phi * R_N\partial_x\phi, w)=2\pi i\sum_{|l|\le N}(1+|l|)\bar{\hat{w}}_l\sum_{|p|\le N}\lambda_{l,p}\hat{\phi}_p\hat{\phi}_{l-p}$$

where

$$\lambda_{l,p}=p\left(\left(1-\left|\frac{\gamma_{l,p}}{N}\right|^\alpha\right)^\beta-\left(1-\left|\frac{p}{N}\right|^\alpha\right)^\beta\right)(1+|l|)^{-1}.$$

Let

$$\theta_{l,p}=p\left(\left|\frac{p}{N}\right|-\left|\frac{\gamma_{l,p}}{N}\right|\right)(1+|l|)^{-1}.$$

Then for $|l|, |p| \le N$,

$$|\lambda_{l,p}| \le \alpha\beta |\theta_{l,p}|.$$

We now estimate $|\theta_{l,p}|$. It suffices to consider the case $0 \le p \le N$. If $0 \le p \le l$, then

$$|\theta_{l,p}| \le \left| \left| \frac{p}{N} \right| - \left| \frac{\gamma_{l,p}}{N} \right| \right| \le 1.$$

If $l \le p \le N + l$ and so $0 < p - l \le N$, then

$$|\theta_{l,p}| \le \frac{p|l|}{N(1+|l|)} \le 1.$$

If $p > N + l$ and so $l - p < -N, 0 \le N - p < -l = |l|$, then

$$\begin{aligned} |\theta_{l,p}| &\le \frac{|p - (2N + 1 + l - p)|}{1 + |l|} = \frac{|2(N-p) + 1 - |l||}{1 + |l|} \\ &\le \max\left(\frac{2(N-p)+1}{1+|l|}, \frac{|l|}{1+|l|} \right) \le 2. \end{aligned}$$

Hence $|\lambda_{l,p}| \le \alpha\beta|\theta_{l,p}| \le 2\alpha\beta$. The rest of the proof is clear. ∎

The combination of the above two lemmas implies the following result.

Lemma 4.4. *For any $\phi \in \tilde{V}_N, w \in H_p^r(\Lambda)$ and $r > \frac{3}{2}$,*

$$|(\partial_x \phi * R_N \phi, w)| \le c\alpha\beta \|w\|_r \|\phi\|^2.$$

Lemma 4.1 and Lemma 4.2– Lemma 4.3 with $\beta = 1$ were established in Ma and Guo (1986).

A Fourier pseudospectral scheme for (4.1) is to find $u_N(x,t) \in V_N$ for all $t \in \bar{R}_\tau(T)$ such that

$$\begin{cases} D_\tau u_N(x,t) + dR_N \partial_x u_N(x,t) + \sigma_1 d\tau R_N D_\tau \partial_x u_N(x,t) + R_N J_N \left(R_N u_N(x,t) \right. \\ \quad \left. + \sigma_2 \tau R_N D_\tau u_N(x,t), u_N(x,t) \right) - \delta D_\tau \partial_x^2 u_N(x,t) = F_N(x,t), \quad x \in \mathbb{R}^1, t \in \bar{R}_\tau(T), \\ u_N(x,0) = u_{N,0}(x), \qquad x \in \mathbb{R}^1, \end{cases} \tag{4.33}$$

where $u_{N,0}(x,0) = I_N U_0(x)$ and

$$F_N(x,t) = R_N I_N \left(f(x,t) + \sigma_3 \tau D_\tau f(x,t) \right).$$

Clearly for any $\phi \in V_N$,

$$(R_N \partial_x \phi, \phi) = \left(\partial_x R_N^{\frac{1}{2}} \phi, R_N^{\frac{1}{2}} \phi\right) = 0. \tag{4.34}$$

By (4.29), for any $\phi, \psi \in V_N$,

$$(R_N J_N (R_N \phi, w), \psi) + (R_N J_N (R_N \psi, w), \phi) = 0. \tag{4.35}$$

In particular,

$$(R_N J_N (R_N \phi, w), \phi) = 0. \tag{4.36}$$

If we take $\sigma_2 = 0$ in (4.33), then

$$\int_\Lambda u_N(x,t)\,dx = \int_\Lambda u_{N,0}(x)\,dx + \tau \sum_{s \in \bar{R}_\tau(t-\tau)} \int_\Lambda F_N(x,s)\,dx.$$

If $\sigma_1 = \sigma_2 = \frac{1}{2}$, then it follows from (4.34)–(4.36) that

$$\|u_N(t)\|^2 + \delta |u_N(t)|_1^2 = \|u_{N,0}\|^2 + \delta |u_{N,0}|_1^2 + \tau \sum_{s \in \bar{R}_\tau(t-\tau)} (F_N(s), u_N(s+\tau) + u_N(s)).$$

The above two equalities are reasonable analogies of (4.2) and (4.3).

We next analyze the errors. Let $\tilde{u}_0$ and $\tilde{F}$ be the errors of $u_{N,0}$ and F_N respectively, which induce the error of u_N, denoted by $\tilde{u}$. Then we have the following error equation

$$\begin{aligned} D_\tau \tilde{u}(x,t) &+ d R_N \partial_x \tilde{u}(x,t) + \sigma_1 d\tau R_N D_\tau \partial_x \tilde{u}(x,t) + R_N J_N (R_N u_N(x,t) \\ &+ \sigma_2 \tau R_N D_\tau u_N(x,t), \tilde{u}(x,t)) + R_N J_N (R_N \tilde{u}(x,t) + \sigma_2 \tau R_N D_\tau \tilde{u}(x,t), u_N(x,t) \\ &+ \tilde{u}(x,t)) - \delta D_\tau \partial_x^2 \tilde{u}(x,t) = \tilde{F}(x,t), \qquad x \in \mathbb{R}^1,\ t \in \bar{R}_\tau(T). \end{aligned}$$

By an argument as in the derivation of (4.17), we obtain from (4.34)–(4.36) that for $\xi > 0$,

$$\begin{aligned} D_\tau \|\tilde{u}(t)\|^2 &+ \delta D_\tau |\tilde{u}(t)|_1^2 + \tau(\xi - 1 - \varepsilon)\left(\|D_\tau \tilde{u}(t)\|^2 + \delta |D_\tau \tilde{u}(t)|_1^2\right) \\ &+ \sum_{j=1}^{6} F_j(t) \le \|\tilde{u}(t)\|^2 + \left(1 + \frac{\tau \xi^2}{4\varepsilon}\right) \|\tilde{F}(t)\|^2 \end{aligned} \tag{4.37}$$

where

$$F_1(t) = 2\sigma_1 d\tau \left(R_N D_\tau \partial_x \tilde{u}(t), \tilde{u}(t)\right),$$

$$
\begin{aligned}
F_2(t) &= 2\left(R_N J_N\left(R_N u_N(t)+\sigma_2\tau R_N D_\tau u_N(t), \tilde{u}(t)\right), \tilde{u}(t)\right),\\
F_3(t) &= 2\sigma_2\tau\left(R_N J_N\left(R_N D_\tau \tilde{u}(t), u_N(t)+\tilde{u}(t)\right), \tilde{u}(t)\right),\\
F_4(t) &= d\xi\tau\left(R_N \partial_x \tilde{u}(t), D_\tau \tilde{u}(t)\right),\\
F_5(t) &= \xi\tau\left(R_N J_N\left(R_N u_N(t)+\sigma_2\tau R_N D_\tau u_N(t), \tilde{u}(t)\right), D_\tau \tilde{u}(t)\right),\\
F_6(t) &= \xi\tau\left(R_N J_N\left(R_N \tilde{u}(t), u_N(t)+\tilde{u}(t)\right), D_\tau \tilde{u}(t)\right).
\end{aligned}
$$

By using Lemma 4.1, Lemma 4.4 and (4.34)–(4.36), we can establish the estimates for $F_j(t)$. They are exactly the same as (4.18)–(4.21). Therefore an argument as in the latter part of the proof of Theorem 4.1, yields the following result.

Theorem 4.3. *There exist positive constants b_6 and b_7 depending only on d, δ, α, β and $\|u_N\|_{C(0,T;H^r)}, r > \frac{3}{2}$, such that if for certain $t_1 \in R_\tau(T), \rho(\tilde{u}_0, \tilde{F}, t_1) \le \frac{b_6}{\tau N}$, then for all $t \in \bar{R}(t_1)$,*

$$
E\left(\tilde{u}, t\right) \le c\rho\left(\tilde{u}_0, \tilde{F}, t\right) e^{b_7 t}.
$$

If in addition $\sigma_2 > \dfrac{1}{2}$, then the above estimate holds for all $t \in \bar{R}_\tau(T)$ and any $\rho\left(\tilde{u}_0, \tilde{f}, t\right)$.

We now turn to the convergence of (4.33). Let $U_N = P_N U, \tilde{U} = u_N - U_N$ and $\sigma_3 = 0$ in (4.33). We obtain the error equation as follows

$$
\begin{aligned}
D_\tau \tilde{U}(x,t) &+ dR_N\partial_x \tilde{U}(x,t) + \sigma_1 d\tau R_N D_\tau \partial_x \tilde{U}(x,t) + R_N J_N\left(R_N U_N(x,t)\right.\\
&\left.+\sigma_2\tau R_N D_\tau U_N(x,t), \tilde{U}(x,t)\right) + R_N J_N\left(R_N \tilde{U}(x,t) + \sigma_2\tau R_N D_\tau \tilde{U}(x,t), U_N(x,t)\right.\\
&\left.+\tilde{U}(x,t)\right) - \delta D_\tau \partial_x^2 \tilde{U}(x,t) = -\sum_{j=1}^{6} G_j(x,t), \qquad (4.38)
\end{aligned}
$$

where $G_1(t)$ and $G_5(t)$ are the same as in (4.23), and

$$
\begin{aligned}
G_2(t) &= \sigma_1 d\tau R_N D_\tau \partial_x U_N(x,t),\\
G_3 &= R_N J_N\left(R_N U_N(x,t), U_N(x,t)\right) - P_N J\left(U(x,t), U(x,t)\right),\\
G_4(t) &= \sigma_2\tau R_N J_N\left(R_N D_\tau U_N(x,t), U_N(x,t)\right),\\
G_6 &= P_N f(x,t) - R_N I_N f(x,t).
\end{aligned}
$$

By Theorem 2.4, Theorem 2.26, Lemma 4.4 and imbedding theory, we obtain that for $\alpha \geq r > \frac{3}{2}$,

$$\tau \sum_{s\in \bar{R}_\tau(t-\tau)} \|G_2(s)\|^2 \leq c\tau^2 \|U\|^2_{H^1(0,t;H^1)},$$
$$\tau \sum_{s\in \bar{R}_\tau(t-\tau)} \|G_3(s)\|^2 \leq cN^{2-2r} \|U\|^4_{C(0,T;H^r)},$$
$$\tau \sum_{s\in \bar{R}_\tau(t-\tau)} \|G_4(s)\|^2 \leq c\tau^2 \|U\|^2_{C(0,T;H^r)} \|U\|^2_{H^1(0,t;H^1)},$$
$$\tau \sum_{s\in \bar{R}_\tau(t-\tau)} \|G_6(s)\|^2 \leq cN^{2-2r} \|f\|^2_{C(0,t;H^{r-1})}.$$

In addition, $\tilde{U}(x,0) = I_N U_0(x) - P_N U_0(x)$ and so

$$\|\tilde{U}(0)\|_1^2 \leq cN^{2-2r}\|U_0\|_r^2.$$

Finally by an argument as in the proof of Theorem 4.2, we obtain the following result.

Theorem 4.4. *Let $\sigma_3 = 0$ in (4.33). If $U \in C(0,T;H_p^r) \cap H^2(0,T;H^1), U_0 \in H_p^r(\Lambda), f \in C(0,T;H_p^{r-1}), \alpha \geq r > \frac{3}{2}$ and $\tau \leq b_8 N^{-\frac{1}{3}}$, then for all $t \in \bar{R}_\tau(T)$,*

$$E\left(U_N - u_N, t\right) \leq b_9 e^{b_{10} t}\left(\tau^2 + N^{2-2r}\right)$$

where $b_6 - b_8$ are positive constants depending on $d, \delta, \alpha, \beta, \|U\|_{C(0,T;H_p^r)}, \|U\|_{H^2(0,T;H^1)}, \|U_0\|_r$ and $\|f\|_{C(0,T;H^{r-1})}$. If $\sigma_2 > \frac{1}{2}$, then the conclusion is valid for any $\tau > 0$.

Scheme (4.33) with $\sigma_1 = \sigma_2 = \sigma_3 = 0$ was investigated in Guo and Cao (1988). We have seen in the derivation of (4.37) and (4.38) that since the nonlinear term $U\partial_x U$ is approximated by a skew symmetric operator $R_N J_N\left(R_N u_N, u_N\right)$, the effects of the leading nonlinear error terms in both (4.37) and (4.38) are cancelled, i.e.,

$$\left(R_N J_N\left(R_N \tilde{u}(t), \tilde{u}(t)\right), \tilde{u}(t)\right) = 0,$$
$$\left(R_N J_N\left(R_N \tilde{U}(t), \tilde{U}(t)\right), \tilde{U}(t)\right) = 0.$$

Otherwise, we shall require that $\rho(\tilde{u}_0, \tilde{f}_0, t_1) \leq b_6/N$ in Theorem 4.3. It indicates that the suitable approximation to the nonlinear term, not only enables the scheme to keep the conservations, but also strengthens the stability and raises the accuracy.

It also shows that there is a close relationship between the conservations and the stability. This idea was first proposed for finite difference methods by Guo (1965).

Finally, we present some numerical results. The meaning of $\tilde{E}(v,t)$ is the same as before. We take the test function (4.26) with $A = B$ and $\omega = 0.5$. In calculations, $\sigma_1 = \sigma_2 = \sigma_3 = 0, \beta = 1, N = 8$ and $\tau = 0.01$. We check the effect of the filtering $R_N(\alpha, 1)$, and find the following phenomena.

(i) If the vibrations of genuine solution of (4.1) is small (e.g., $A = 0.1$) and $d = 0$, then the effect of $R_N(\alpha, 1)$ is not clear, see Table 4.2.

(ii) If δ is large (e.g., $\delta = 1$), then the filtering is not important even though the vibration of genuine solution is large, see Table 4.3.

(iii) If the vibration of genuine solution is large and δ is small (e.g., $A = 1, \delta = 10^{-4}, 10^{-6}$), then the effect of filtering is clear, see Tables 4.4 and 4.5.

(iv) The smaller the parameter δ, the more important the linear term $d\partial_x U$. In this case, the filtering plays an important role. Tables 4.2 and 4.5 tell us that its effect is not clear for $d = 0, A = 0.1$, but very clear for $d = 2, A = 0.1$.

(v) The value of α in the filtering must be chosen suitably. For every large α, the filtering is weak. But for too small α, the accuracy is lowered. So far, there is no criterion for the best choice of α. Numerical experiments show that a better choice is $3 \le \alpha \le 10$. But the best ones depend on N. The best one in Table 4.5 is $\alpha = 3$.

Table 4.2. The errors $\tilde{E}(u_N, 5.0), d = 0, A = 0.1$.

	$\alpha = 5$	$\alpha = 10$	$\alpha = 50$
$\delta = 1.0$	$2.36857E{-3}$	$2.36493E{-3}$	$2.36577E{-3}$
$\delta = 10^{-2}$	$2.29817E{-3}$	$2.29622E{-3}$	$2.29651E{-3}$
$\delta = 10^{-4}$	$2.28399E{-3}$	$2.28281E{-3}$	$2.28183E{-3}$
$\delta = 10^{-6}$	$2.28242E{-3}$	$2.28203E{-3}$	$2.28193E{-3}$

Table 4.3. The errors $\tilde{E}(u_N, 5.0), \delta = 1, A = 1.0$.

	$\alpha = 5$	$\alpha = 10$	$\alpha = 50$
$d = 0$	$2.31011E{-3}$	$2.15183E{-3}$	$2.16537E{-3}$
$d = 2$	$1.44118E{-3}$	$1.18726E{-3}$	$1.15333E{-3}$

Table 4.4. The errors $\tilde{E}(u_N, 1.0), d = 0, A = 1.0$.

0	$\alpha = 5$	$\alpha = 10$	$\alpha = 50$
$\delta = 10^{-4}$	$3.29499E{-3}$	$3.55039E{-3}$	$2.35286E{-1}$
$\delta = 10^{-6}$	$2.83234E{-2}$	$1.24055E{-2}$	$1.70894E{-1}$

Table 4.5. The errors $\tilde{E}(u_N, 2.0), d = 2, A = 0.1$.

	$\alpha = 3$	$\alpha = 5$	$\alpha = 10$	$\alpha = 50$
$\delta = 10^{-4}$	$1.55333E{-3}$	$1.63698E{-3}$	$2.43057E{-2}$	$7.02116E{-1}$
$\delta = 10^{-6}$	$1.56715E{-3}$	$1.86465E{-2}$	$1.11043E{-2}$	> 10

Scheme (4.11) and scheme (4.33) are of first order in time. In order to match the spectral accuracy in the space, we should use more accurate temporal discretizations as discussed in Gazdag (1976).

The Fourier spectral methods and the Fourier pseudospectral methods have been widely used for numerical solutions of various partial differential equations. We refer to some early work. For instance, Shamel and Elsässer (1976), Fornberg and Whitham (1978), and Abe and Inou (1980) used them for Korteweg-de Vries equation (KdV equation), while Guo (1981a) proposed the spectral schemes for KdV-Burgers equation with numerical analysis. Pasciak (1982) considered Fourier spectral scheme for generalized BBM equation. Kriess and Oliger (1979) applied them to hyperbolic equations with filterings. Cornille (1982) used Fourier approximation in numerical study of shock waves. Canosa and Gazdag (1977) used it for KdV-Burgers equation and found shock-like waves.

4.2. Legendre Spectral Methods and Legendre Pseudospectral Methods

For non-periodic problems, it is natural to use Legendre spectral methods or Legendre pseudospectral methods. Because of the appearance of the Fast Legendre Transformation, more attention has been paid to these two methods recently. We shall take the nonlinear Klein-Gordan equation (NLKG equation) as an example to describe these two methods and set up the strict error estimations. The numerical results will show their advantages.

NLKG equation plays an important role in quantum mechanics. Let $\Lambda = (-1,1)$. The initial-boundary value problem of NLVG equation is as follows

$$\begin{cases} \partial_t^2 U(x,t) + U(x,t) + U^3(x,t) - \partial_x^2 U(x,t) = f(x,t), & x \in \Lambda,\ 0 < t \leq T, \\ U(-1,t) = U(1,t) = 0, & 0 < t \leq T, \\ \partial_t U(x,0) = U_1(x), & x \in \bar{\Lambda}, \\ U(x,0) = U_0(x), & x \in \bar{\Lambda}. \end{cases} \tag{4.39}$$

It is shown in Lions (1969) that if $U_0 \in H_0^1(\Lambda) \cap L^4(\Lambda), U_1 \in L^2(\Lambda)$ and $f \in L^2(0,T;L^2)$, then (4.39) has a unique solution $U \in L^\infty(0,T;H_0^1 \cap L^4) \cap W^{1,\infty}(0,T;L^2)$. Let

$$E^*(v,t) = \|\partial_t v(t)\|^2 + \|v(t)\|_1^2 + \frac{1}{2}\|v(t)\|_{L^4}^4.$$

The solution U satisfies the conservation

$$E^*(U,t) = E^*(U,0) + 2\int_0^t (f(s), \partial_s U(s))\, ds. \tag{4.40}$$

For Legendre spectral approximation, let N be any positive integer and $\mathbb{P}_N, \mathbb{P}_N^0$ be as in Section 2.3 and (2.56). Denote by P_N the L^2-orthogonal projection from $L^2(\Lambda)$ onto $\mathbb{P}_N$. The H_0^1-orthogonal projection $P_N^{1,0}$ is given by (2.57). Also let τ be the mesh size in t and use the notations $R_\tau(T), \bar{R}_\tau(T), D_\tau v(x,t), \bar{D}_\tau v(x,t)$ and $\hat{D}_{\tau\tau} v(x,t)$ in Chapter 1.

We try to construct a scheme preserving the conservation (4.40). To do this, define

$$G(v(x,t)) = \frac{1}{4}\sum_{j=0}^{3} v^j(x,t+\tau)v^{3-j}(x,t-\tau). \tag{4.41}$$

Clearly

$$\hat{D}_\tau v(x,t)G(v(x,t)) = \frac{1}{8\tau}\left(|v(x,t+\tau)|^4 - |v(x,t-\tau)|^4\right)$$

and so

$$\left(G(v(t)), \hat{D}_\tau v(t)\right) = \frac{1}{4}\hat{D}_\tau \|v(t)\|_{L^4}^4. \tag{4.42}$$

Let u_N be the approximation to U. We approximate the nonlinear term U^3 by $P_N G(u_N)$ instead of $P_N u_N^3$. A Legendre spectral scheme for (4.39) is to find $u_N(x,t) \in \mathbb{P}_N^0$ for all $t \in \bar{R}_\tau(T)$ such that

$$\begin{cases} \left(\hat{D}_{\tau\tau} u_N(t) + \hat{u}_N(t) + G(u_N(t)), \phi\right) \\ \quad + (\partial_x \hat{u}_N(t), \partial_x \phi) = \left(\hat{f}(t), \phi\right), \quad \forall\, \phi \in \mathbb{P}_N^0,\ t \in R_\tau(T), \\ D_\tau u_N(0) = u_{N,1}, \\ u_N(0) = u_{N,0} \end{cases} \tag{4.43}$$

where $u_{N,0} = P_N U_0$ and

$$u_{N,1} = P_N U_1 + \frac{1}{2}\tau P_N\left(\partial_x^2 U_0 - U_0 - U_0^3 + f(0)\right).$$

The first formula of (4.43) stands for

$$\hat{D}_{\tau\tau} u_N(x,t) + \hat{u}_N(t) + P_N G(u_N(x,t)) - P_N \partial_x^2 \hat{u}_N(x,t) = P_N f(x,t).$$

We check the conservation. Putting $v = 2\hat{D}_\tau u_N(t)$ in (4.43), we obtain from (1.23), (1.24) and (4.42) that

$$D_\tau \left\|\bar{D}_\tau u_N(t)\right\|^2 + \hat{D}_\tau \|u_N(t)\|_1^2 + \frac{1}{2}\hat{D}_\tau \|u_N(t)\|_{L^4}^4 = 2\left(\hat{f}(t), \hat{D}_\tau u_N(t)\right). \tag{4.44}$$

Let

$$E_\tau^*(v,t) = \left\|\bar{D}_\tau v(t)\right\|^2 + \frac{1}{2}\|v(t)\|_1^2 + \frac{1}{2}\|v(t-\tau)\|_1^2 + \frac{1}{4}\|v(t)\|_{L^4}^4 + \frac{1}{4}\|v(t-\tau)\|_{L^4}^4$$

and

$$F^*(v,t) = \tau \sum_{s \in R_\tau(t-\tau)} \left(\hat{f}(s), \hat{D}_\tau u_N(s)\right).$$

The summation of (4.44) for $t \in \bar{R}_\tau(T)$ yields

$$E_\tau^*(u_N, t) = E_\tau^*(u_N, \tau) + 2F^*(u_N, t). \tag{4.45}$$

Clearly it is a reasonable analogy of (4.40).

We next derive a priori estimation for $u_N(x,t)$. Assume that $q = \tau N^2 < \infty$. By (4.45), it suffices to bound the initial energy $E^*_\tau(u_N,\tau)$ and estimate the summation $|F^*(u_N,t)|$. From (4.43) and the proof of Theorem 2.8,

$$\|u_N(\tau)\|^2 \leq 2\|u_{N,0}\|^2 + 2\tau^2\|u_{N,1}\|^2,$$
$$|u_N(\tau)|_1^2 \leq 2\|u_{N,0}\|_1^2 + 2a_N^2\tau^2 N^4\|u_{N,1}\|^2 \leq 2\|u_{N,0}\|_1^2 + \frac{9}{2}q^2\|u_{N,1}\|^2.$$

Furthermore

$$\|u_N(\tau)\|_{L^4}^4 \leq 8\left(\|u_{N,0}\|_{L^4}^4 + \tau^4\|u_{N,1}\|_{L^4}^4\right).$$

By Theorem 2.7,

$$\tau^4\|u_{N,1}\|_{L^4}^4 \leq c\tau^4 N^2\|u_{N,1}\|^4 = cq\tau^3\|u_{N,1}\|^4.$$

It is easy to see that

$$|F^*(u_N,t)| \leq \tau\|\bar{D}_\tau u_N(t)\|^2 + 2\tau \sum_{s\in R_\tau(t-\tau)}\left(\|\bar{D}_\tau u_N(s)\|^2 + \|\hat{f}(s)\|^2\right).$$

Hence (4.45) implies that

$$\begin{aligned}(1-\tau)\|\bar{D}_\tau u_N(t)\|^2 + \frac{1}{2}\|u_N(t)\|_1^2 + \frac{1}{2}\|u_N(t-\tau)\|_1^2 + \frac{1}{4}\|u_N(t)\|_{L^4}^4 \\ +\frac{1}{4}\|u_N(t-\tau)\|_{L^4}^4 \leq \rho_0(u_{N,0},u_{N,1},f,t) + 2\tau\sum_{s\in R_\tau(t-\tau)}\|\bar{D}_\tau u_N(s)\|^2 \end{aligned} \quad (4.46)$$

where

$$\rho_0(u_{N,0},u_{N,1},f,t) = c\left(\|u_{N,0}\|_1^2 + \|u_{N,0}\|_{L^4}^4 + \|u_{N,1}\|^2 + \tau^3\|u_{N,1}\|^4\right) + c\tau\sum_{s\in\bar{R}_\tau(t)}\|f(s)\|^2.$$

Let

$$E(v,t) = \|\bar{D}_\tau v(t)\|^2 + \|v(t)\|_1^2 + \|v(t)\|_{L^4}^4.$$

By applying Lemma 3.2 to (4.46), we know that $E(u_N,t) \leq \rho_0(u_{N,0},u_{N,1},f,t)$.

(4.43) is an implicit scheme. Let

$$a(v,w) = (\partial_x v,\partial_x w) + \left(1+\frac{2}{\tau^2}\right)(v,w),$$
$$b(v,w) = \frac{1}{2}\left(v^3,w\right),$$
$$d(v,g,w) = \frac{1}{2}\left(gv^2,w\right) + \frac{1}{2}\left(g^2v,w\right).$$

Then at each time $t \in R_\tau(T)$, we have to solve a nonlinear equation

$$a\left(u_N(t),\phi\right)+b\left(u_N(t),\phi\right)+d\left(u_N(t),u_N(t-2\tau),\phi\right)=\left(F(t),\phi\right), \quad \forall\, \phi \in \mathbb{P}_N^0 \quad (4.47)$$

where $F(t)$ depends only on $u_N(t-2\tau)$ and $\hat{f}(t-\tau)$. We now investigate the existence and uniqueness of solution of (4.47).

Lemma 4.5. *Let $\mathcal{H}$ be an m-dimensional Hilbert space with the inner product $(\cdot,\cdot)_{\mathcal{H}}$ and the norm $\|\cdot\|_{\mathcal{H}}$. A is a continuous operator from $\mathcal{H}$ into $\mathcal{H}$, and $(A\xi,\xi)_{\mathcal{H}} \geq 0$ for all ξ such that $\|\xi\|_{\mathcal{H}} = r > 0$. Then there exists $\xi_0 \in \mathcal{H}$ such that $\|\xi_0\|_{\mathcal{H}} \leq r$ and $A\xi_0 = 0$.*

Proof. Let $S(0,r) = \{\xi \mid \|\xi\|_{\mathcal{H}} < r\}$ and

$$A^*\xi = \frac{-rA\xi}{\|A\xi\|_{\mathcal{H}}}, \qquad \xi \in \bar{S}(0,r).$$

If $A\xi \neq 0$ in the closed ball $\bar{S}(0,r)$, then A^* is a continuous operator and $\|A^*\xi\|_{\mathcal{H}} = r$. By Brouwer Theorem, there exists a fixed point for A^*, say ξ_0. So we have

$$\frac{-rA\xi_0}{\|A\xi_0\|_{\mathcal{H}}} = \xi_0$$

from which

$$(A\xi_0,\xi_0)_{\mathcal{H}} = -r\|A\xi_0\|_{\mathcal{H}} < 0.$$

It is contrary to the assumption and so the proof is complete. ∎

We now use Lemma 4.5 for (4.47). Let $\{\phi_l\}$ be the normalized orthogonal system in $\mathbb{P}_N^0$. The space $\mathcal{H}$ consists of all vectors $\xi = \{\xi_0,\ldots,\xi_{N-2}\}$ such that

$$u_N(x,t) = \sum_{l=0}^{N-2} \xi_l \phi_l(x).$$

Let $A\xi = \{\eta_l\}$ and

$$\eta_l = a\left(u_N(t),\phi_l\right)+b\left(u_N(t),\phi_l\right)+d\left(u_N(t),u_N(t-2\tau),\phi_l\right)-\left(F(t),\phi_l\right).$$

It is easy to see that A is a continuous operator. Moreover,

$$\begin{aligned}(A\xi,\xi)_{\mathcal{H}} = {} & \|u_N(t)\|_1^2+\frac{2}{\tau^2}\|u_N(t)\|^2+\frac{1}{2}\|u_N(t)\|_{L^4}^4 \\ & +d\left(u_N(t),u_N(t-2\tau),u_N(t)\right)-\left(F(t),u_N(t)\right).\end{aligned}$$

According to the a priori estimate, it follows from Theorem 2.7 that

$$\begin{aligned}
&\left|\left(u_N^3(t), u_N(t-\tau)\right)\right| \le \|u_N(t)\|_{L^4}^3 \|u_N(t-2\tau)\|_{L^4} \le \tilde{c}\|u_N(t)\|_{L^4} \le \tilde{c}N^{\frac{1}{2}}\|u_N(t)\|, \\
&\left|\left(u_N^2(t), u_N^2(t-2\tau)\right)\right| \le \|u_N(t)\|_{L^4}^2 \|u_N(t-2\tau)\|_{L^4}^2 \le \tilde{c}N\|u_N(t)\|
\end{aligned}$$

where $\tilde{c}$ is a positive constant depending on the a priori estimate. Clearly,

$$|(F(t), u_N(t))| \le \|F(t)\|\|u_N(t)\|.$$

Since $\tau = qN^{-2}$, we know that for suitably large N and $\|u_N(t)\| = r$, we have $(A\xi, \xi)_{\mathcal{H}} > 0$. This implies the existence of solution of (4.47). If (4.47) has two solutions $u_N^{(1)}(t)$ and $u_N^{(2)}(t)$, then let $\tilde{u}_N(t) = u_N^{(1)}(t) - u_N^{(2)}(t)$. By (4.47),

$$\begin{aligned}
&\|\tilde{u}_N(t)\|_1^2 + \tfrac{2}{\tau^2}\|\tilde{u}_N(t)\|^2 + b\left(u_N^{(1)}(t), \tilde{u}_N(t)\right) - b\left(u_N^{(2)}(t), \tilde{u}_N(t)\right) \\
&\quad + \tilde{d}\left(u_N^{(1)}(t), u_N^{(2)}(t), u_N(t-\tau)\right) = 0
\end{aligned} \tag{4.48}$$

where

$$\begin{aligned}
&\tilde{d}\left(u_N^{(1)}(t), u_N^{(2)}(t), u_N(t-2\tau)\right) \\
&\quad = d\left(u_N^{(1)}(t), u_N(t-2\tau), \tilde{u}_N(t)\right) - d\left(u_N^{(2)}(t), u_N(t-2\tau), \tilde{u}_N(t)\right).
\end{aligned}$$

Clearly

$$b\left(u_N^{(1)}(t), \tilde{u}_N(t)\right) - b\left(u_N^{(2)}(t), \tilde{u}_N(t)\right) \ge 0.$$

Moreover, it can be verified that

$$\left|\tilde{d}\left(u_N^{(1)}(t), u_N^{(2)}(t), u_N(t-2\tau)\right)\right| \le \tilde{c}N\left\|\tilde{u}_N(t)\right\|^2.$$

Finally, (4.48) leads to $\|\tilde{u}_N(t)\| = 0$ for suitably large N. Thus the solution of (4.47) is unique.

We can use the Newton method or the simplified Newton method in Section 3.2 to resolve (4.47).

We now discuss the choice of basis. As we know, Legendre polynomials are orthogonal in $L^2(\Lambda)$. But it is not so for the basis of $\mathbb{P}_N^0$. We could take the basis of $\mathbb{P}_N^0$ as the set of $\phi_l(x), 2 \le l \le N$ where

$$\phi_l(x) = \begin{cases} L_l(x) - L_0(x), & \text{if } l \text{ is even}, \\ L_l(x) - L_1(x), & \text{if } l \text{ is odd}. \end{cases}$$

Unfortunately this basis leads to a linear system with a full matrix whose elements are $M_{l,m} = (\phi_l, \phi_m)$. Moreover, the number condition of the matrix M^* with the elements $M^*_{l,m} = (\partial_x \phi_l, \partial_x \phi_m)$ is very bad. Shen (1994) proposed another basis, i.e.,

$$\psi_l(x) = d_l \left(L_l(x) - L_{l+2}(x) \right), \quad d_l = \frac{1}{\sqrt{4l+6}}, \quad 0 \leq l \leq N-2.$$

It can be checked that

$$(\psi_l, \psi_m) = (\psi_m, \psi_l) = \begin{cases} 2d_l^2 \left(\dfrac{1}{2l+1} + \dfrac{1}{2l+5} \right), & l = m, \\ -\dfrac{2d_l d_{l-2}}{2l+1}, & l = m+2, \\ 0, & \text{otherwise.} \end{cases}$$

In particular,

$$(\partial_x \psi_l, \partial_x \psi_m) = \delta_{l,m}.$$

Therefore this basis leads to a triangular matrix and an identity matrix. It saves a lot of work in computation and improve the condition number of the related matrix, which in turn strengthens the stability of algorithms.

For error estimations, we need some preparations. For simplicity, let $v(t) = v(x,t)$ in the derivations of the following lemmas.

Lemma 4.6. *We have*

$$\|\hat{v}(t) - v(t)\| \leq c\tau^{\frac{3}{2}} \|v\|_{H^2(t-\tau,t+\tau;L^2)},$$
$$\|\hat{D}_{\tau\tau} v(t) - \partial_t^2 v(t)\| \leq c\tau^{\frac{3}{2}} \|v\|_{H^4(t-\tau,t+\tau;L^2)},$$
$$\|D_\tau v(t) - \partial_t v(t) - \frac{\tau}{2} \partial_t^2 v(t)\| \leq c\tau^{\frac{3}{2}} \|v\|_{H^3(t-\tau,t+\tau;L^2)},$$

provided that the related norms exist.

Proof. The results come from some expressions like (4.24) and the property of Bochner integration. ∎

Lemma 4.7. *If* $v \in C(0,T;C) \cap H^1(0,T;L^2)$, *then*

$$\left\| G(v(t)) - \frac{1}{2} v^3(t+\tau) - \frac{1}{2} v^3(t-\tau) \right\| \leq c\tau^{\frac{3}{2}} \|v\|^2_{C(0,T;C)} \|v\|_{H^1(t-\tau,t+\tau;L^2)}.$$

Proof. We have

$$
\begin{aligned}
&G(v(t)) - \frac{1}{2}v^3(t+\tau) - \frac{1}{2}v^3(t-\tau) \\
&\quad = \frac{1}{4}v^2(t+\tau)(v(t-\tau) - v(t+\tau)) + \frac{1}{4}v^2(t-\tau)(v(t+\tau) - v(t-\tau))
\end{aligned}
$$

from which the conclusion comes. ∎

Lemma 4.8. *If $v \in C(0,T;H_0^1)$, then*

$$
\left\|G\left(P_N^{1,0}v(t)\right) - G(v(t))\right\| \le c\left\|P_N^{1,0}v - v\right\|_{C(t-\tau,t+\tau;L^2)} \|v\|^2_{C(t-\tau,t+\tau;H^1)}.
$$

If, in addition, $v \in C(0,T;H_0^1 \cap H^r)$ and $r \ge 1$, then

$$
\left\|G\left(P_N^{1,0}v(t)\right) - G(v(t))\right\| \le cN^{-r}\|v\|^3_{C(t-\tau,t+\tau;H^r)}.
$$

Proof. We have

$$
\begin{aligned}
&4G\left(P_N^{1,0}v(t)\right) - 4G(v(t)) \\
&= \left(\left(P_N^{1,0}(t+\tau)\right)^3 - v^3(t+\tau)\right) + \left(\left(P_N^{1,0}v(t+\tau)\right)^2 P_N^{1,0}v(t-\tau) - v(t+\tau)^2 v(t-\tau)\right) \\
&+ \left(P_N^{1,0}(t+\tau)\left(P_N^{1,0}(t-\tau)\right)^2 - v(t+\tau)v^2(t-\tau)\right) + \left(\left(P_N^{1,0}(t-\tau)\right)^3 - v^3(t-\tau)\right).
\end{aligned}
$$

By Theorem 2.11, $\left\|P_N^{1,0}v\right\|_1 \le \|v\|_1$. Hence by imbedding theory,

$$
\begin{aligned}
&\left\|\left(P_N^{1,0}(t+\tau)\right)^3 - v^3(t+\tau)\right\| \\
&\quad \le c\left\|P_N^{1,0}(t+\tau) - v(t+\tau)\right\|\left(\|v(t+\tau)\|^2_{L^\infty} + \left\|P_N^{1,0}v(t+\tau)\right\|^2_{L^\infty}\right). \\
&\quad \le c\left\|P_N^{1,0}v(t+\tau) - v(t+\tau)\right\| \|v(t+\tau)\|^2_{H^1}.
\end{aligned}
$$

If $v \in C(0,T;H_0^1 \cap H^r)$ with $r \ge 1$, then using Theorem 2.11 again, yields

$$
\begin{aligned}
\left\|\left(P_N^{1,0}(t+\tau)\right)^3 - v^3(t+\tau)\right\| &\le c\left\|P_N^{1,0}(t+\tau) - v(t+\tau)\right\| \|v(t+\tau)\|^2_{H^1} \\
&\le cN^{-r}\|v(t+\tau)\|^3_{H^r}.
\end{aligned}
$$

We can estimate other terms similarly and complete the proof. ∎

Lemma 4.9. *If* $v, w \in C(0,T;H^1)$, *then*

$$G(v(x,t)+w(x,t)) = G(v(x,t)) + G(w(x,t)) + R(x,t),$$
$$\|R(t)\|^2 \le d(v)\left(\|w(t+\tau)\|_1^2 + \|w(t-\tau)\|_1^2 + \|w(t+\tau)\|_{L^4}^4 + \|w(t-\tau)\|_{L^4}^4\right),$$

$d(v)$ being certain positive constant depending only on $\|v\|_{C(t-\tau,t+\tau;H^1)}$.

Proof. We have

$$4G(v(x,t)+w(x,t)) = 4G(v(x,t)) + 4G(w(x,t)) + \sum_{j=1}^{4} R_j(x,t)$$

where

$$\begin{aligned}
R_1(x,t) &= 3v^2(x,t+\tau)w(x,t+\tau) + 3v(x,t+\tau)w^2(x,t+\tau),\\
R_2(x,t) &= v(x,t-\tau)w(x,t+\tau)(2v(x,t+\tau)+w(x,t+\tau))\\
&\quad + v(x,t+\tau)w(t-\tau)(v(x,t+\tau)+2w(x,t+\tau)),\\
R_3(x,t) &= v(x,t+\tau)w(x,t-\tau)(2v(x,t-\tau)+w(x,t-\tau))\\
&\quad + v(x,t-\tau)w(x,t+\tau)(v(x,t-\tau)+2w(x,t-\tau)),\\
R_4(x,t) &= 3v^2(x,t-\tau)w(x,t-\tau) + 3v(x,t-\tau)w^2(x,t-\tau).
\end{aligned}$$

Since

$$\begin{aligned}
&\left\|v(t+\tau)w^2(t+\tau)\right\|^2\\
&\le \|v\|^2_{C(t-\tau,t+\tau;L^\infty)}\|w(t+\tau)\|^4_{L^4} \le c\|v\|^2_{C(t-\tau,t+\tau;H^1)}\|w(t+\tau)\|^4_{L^4}, \text{ etc..}
\end{aligned}$$

We derive the conclusion. ∎

It is noted that for any $v \in H_0^1(\Lambda)$, the Poincaré inequality is of the form

$$\|v\|^2 \le \frac{4}{\pi^2}|v|_1^2. \tag{4.49}$$

We now analyze the generalized stability of (4.43). Suppose that $u_{N,0}, u_{N,1}$ and $P_N f$ have the errors $\tilde{u}_0, \tilde{u}_1$ and $\tilde{f}$ respectively, which induce the error of u_N, denoted by $\tilde{u}_N$ or $\tilde{u}$ for simplicity. Then

$$\begin{cases}
\left(\hat{D}_{\tau\tau}\tilde{u}(t) + \hat{\tilde{u}}(t) + \tilde{G}(t), \phi\right) + \left(\partial_x\hat{\tilde{u}}(t), \partial_x\phi\right) = \left(\hat{\tilde{f}}(t), \phi\right), \quad \forall\, \phi \in \mathbb{P}_N^0,\ t \in R_\tau(T),\\
D_\tau\tilde{u}(0) = \tilde{u}_1,\\
\tilde{u}(0) = \tilde{u}_0
\end{cases} \tag{4.50}$$

where

$$\tilde{G}(x,t) = G\left(u_N(x,t) + \tilde{u}(x,t)\right) - G\left(u_N(x,t)\right) = G(\tilde{u}(x,t)) + \tilde{R}(x,t),$$

and by Lemma 4.9,

$$\left\|\tilde{R}(t)\right\|^2 \leq d\left(u_N\right)\left(\|\tilde{u}(t+\tau)\|_1^2 + \|\tilde{u}(t-\tau)\|_1^2 + \|\tilde{u}(t+\tau)\|_{L^4}^4 + \|\tilde{u}(t-\tau)\|_{L^4}^4\right). \tag{4.51}$$

According to the a priori estimate, $d(u_N)$ is bounded. Now, we take $\phi = 2\hat{D}_\tau\tilde{u}$ in (4.50) and sum up the resulting equality for $t \in R_\tau(T)$. We obtain that

$$\begin{aligned}&\left\|\bar{D}_\tau\tilde{u}(t)\right\|^2 + \frac{1}{2}\|\tilde{u}(t)\|_1^2 + \frac{1}{2}\|\tilde{u}(t-\tau)\|_1^2 + \frac{1}{4}\|\tilde{u}(t)\|_{L^4}^4 + \frac{1}{4}\|\tilde{u}(t-\tau)\|_{L^4}^4 \\ &+2\tau\sum_{s\in R_\tau(t-\tau)}\left(\tilde{R}(s), \hat{D}_\tau\tilde{u}(s)\right) = \|\tilde{u}_1\|^2 + \frac{1}{2}\|\tilde{u}(\tau)\|_1^2 + \frac{1}{2}\|\tilde{u}_0\|_1^2 + \frac{1}{4}\|\tilde{u}(\tau)\|_{L^4}^4 \\ &+\frac{1}{4}\|\tilde{u}_0\|_{L^4}^4 + 2\tau\sum_{s\in R_\tau(t-\tau)}\left(\hat{\tilde{f}}(s), \hat{D}_\tau\tilde{u}(s)\right).\end{aligned}$$

By (4.51), the above equality leads to

$$\begin{aligned}&(1-\tau)\left\|\bar{D}_\tau\tilde{u}(t)\right\|^2 + \left(\tfrac{1}{2} - \tau d\left(u_N\right)\right)|\tilde{u}(t)|_1^2 + \tfrac{1}{2}\|\tilde{u}(t)\|^2 + \left(\tfrac{1}{4} - \tau d\left(u_N\right)\right)\|\tilde{u}(t)\|_{L^4}^4 \\ &\leq \rho_1(t) + \tau\left(c + d\left(u_N\right)\right)\sum_{s\in R_\tau(t-\tau)}\left(\|D_\tau\tilde{u}(t)\|^2 + \|\tilde{u}(s)\|_1^2 + \|\tilde{u}(s)\|_{L^4}^4\right)\end{aligned} \tag{4.52}$$

where

$$\rho_1(t) = \left(1 + \tau d\left(u_{N,0}\right)\right)\left(\|\tilde{u}_0\|_1^2 + \|\tilde{u}_1\|^2 + \|\tilde{u}_0\|_{L^4}^4 + \tau^4\|\tilde{u}_1\|_{L^4}^4\right) + 2\tau\sum_{s\in\bar{R}_\tau(t)}\left\|\tilde{f}(s)\right\|^2.$$

Finally we apply Lemma 3.2 to (4.52), and obtain the following result.

Theorem 4.5. *Let $q = \tau N^2 < \infty$ and N be suitably large. Then for all $t \in \bar{R}_\tau(T)$,*

$$E\left(\tilde{u}, t\right) \leq c\rho_1(t)e^{b_1 t}$$

where b_1 is a positive constant depending only on $\|u_N\|_{C(0,T;H^1)}$.

Scheme (4.43) possesses the index of generalized stability $s = -\infty$.

We next discuss the convergence of (4.43). For the Fourier approximation, the L^2-orthogonal projection is also the best approximation to the norm $\|\cdot\|_1$. But it is not true for Legendre approximation. If we compare the numerical solution u_N to $P_N U$, then the leading term in the error equation is $(\partial_x(U - P_N U), \partial_x \phi)$. It does not vanish, since $\partial_x P_N \neq P_N \partial_x$. This term lowers the convergence rate essentially. Thus, in order to derive the optimal error estimate, we should compare u_N to $P_N^{1,0}U$, the H_0^1-orthogonal projection of U upon $\mathbb{P}_N^0$. To do this, let $U_N = P_N^{1,0}U$ and then from (4.39),

$$\begin{cases} \left(\hat{D}_{\tau\tau}U_N(t) + \hat{U}_N(t) + G\left(U_N(t)\right), \phi\right) + \left(\partial_x \hat{U}_N(t), \partial_x \phi\right) \\ \quad = \left(\hat{f}(t) + \sum_{j=1}^{4} G_j(t), \phi\right), \quad \forall\, \phi \in \mathbb{P}_N^0,\ t \in R_\tau(T), \\ D_\tau U_N(0) = P_N^{1,0}U_1 + \frac{\tau}{2} P_N^{1,0}\left(\partial_x^2 U_0 - U_0 - U_0^3 + f(0)\right) + G_5, \end{cases}$$

where

$$\begin{aligned} G_1(x,t) &= \hat{D}_{\tau\tau}U_N(x,t) - \partial_t^2 \hat{U}(x,t), \\ G_2(x,t) &= \hat{U}_N(x,t) - \hat{U}(x,t), \\ G_3(x,t) &= G\left(U_N(x,t)\right) - G(U(x,t)), \\ G_4(x,t) &= G(U(x,t)) - \frac{1}{2}U^3(x,t+\tau) - \frac{1}{2}U^3(x,t-\tau), \\ G_5 &= D_\tau U_N(x,0) - \partial_t U_N(x,0) - \frac{\tau}{2}\partial_t^2 U_N(x,0). \end{aligned}$$

Setting $\tilde{U} = u_N - U_N$, we have from (4.43) that

$$\begin{cases} \left(\hat{D}_{\tau\tau}\tilde{U}(t) + \hat{\tilde{U}}(t) + G\left(U_N(t) + \tilde{U}(t)\right) - G\left(U_N(t)\right), \phi\right) \\ \quad + \left(\partial_x \hat{\tilde{U}}(t), \partial_x \phi\right) = -\sum_{j=1}^{4} (G_j(t), \phi), \quad \forall\, \phi \in \mathbb{P}_N^0,\ t \in R_\tau(T), \\ D_\tau \tilde{U}(0) = P_N\left(U_1 + \frac{\tau}{2}\partial_t^2 U(0)\right) - P_N^{1,0}\left(U_1 + \frac{\tau}{2}\partial_t^2 U(0)\right) - G_5, \\ \tilde{U}(0) = P_N U_0 - P_N^{1,0}U_0. \end{cases}$$

By comparing this error equation to (4.50), we deduce a result similar to Theorem 4.5. But the terms $\|\bar{D}_\tau \tilde{u}(t)\|, \|\tilde{u}(t)\|_1$ and $\|\tilde{u}(t)\|_{L^4}$ in that theorem are replaced by $\|\bar{D}_\tau \tilde{U}(t)\|, \|\tilde{U}(t)\|_1$ and $\|\tilde{U}(t)\|_{L^4}$ respectively, while $\rho_1(t)$ is replaced by

$$\begin{aligned} \rho_2(t) &= c\left(\tau + d\left(U_N(0)\right)\right)\left(\left\|\tilde{U}(0)\right\|_1^2 + \left\|D_\tau \tilde{U}(0)\right\|^2 + \|\tilde{U}(0)\|_{L^4}^4 + \tau^4 \|D_\tau \tilde{U}(0)\|_{L^4}^4\right) \\ &\quad + c\tau \sum_{j=1}^{5} \sum_{s \in \bar{R}_\tau(t-\tau)} \|G_j(s)\|^2. \end{aligned}$$

For the convergence rate, we only need to bound $\rho_2(t)$. Let $r \geq 1$. We obtain from Theorem 2.11, Lemmas 4.6 –4.8 and imbedding theory that

$$\begin{aligned}
\|G_1(t)\|^2 &\leq cN^{-2r}\|U\|^2_{C^2(0,T;H^r)} + c\tau^3\|U(t)\|^2_{H^4(t-\tau,t+\tau;L^2)},\\
\|G_2(t)\|^2 &\leq cN^{-2r}\|U\|^2_{C(0,T;H^r)},\\
\|G_3(t)\|^2 &\leq cN^{-2r}\|U\|^6_{C(0,T;H^r)},\\
\|G_4(t)\|^2 &\leq c\tau^4\|U\|^4_{C(0,T;H^r)}\|U\|^2_{C^1(0,T;L^2)}.
\end{aligned}$$

Also by Theorems 2.7, 2.9 and 2.11, and Lemma 4.6, we find that

$$\begin{aligned}
\tau^4\|D_\tau\tilde{U}(0)\|^4_{L^4} &\leq c\tau^4N^2\|D_\tau\tilde{U}(0)\|^4 \leq c\tau^4N^{2-4r}\|U_1\|^4_r\\
&\quad + c\tau^8N^{-2}\|\partial_t^2U(0)\|^4_1 + c\tau^{12}N^2\|U\|^4_{C^3(0,T;L^2)},\\
\|\tilde{U}(0)\|^2_1 &\leq cN^{-2r}\|U_0\|^2_{r+\frac{3}{2}}, \text{ etc..}
\end{aligned}$$

Thus

$$\rho_2(t) = \mathcal{O}\left(\tau^4 + \tau^{12}N^2 + \tau^8N^{-2} + \tau^4N^{2-4r} + N^{-2r}\right).$$

Finally we get the following conclusion.

Theorem 4.6. *Let $\tau N^2 < \infty$ and $r \geq 1$. If $U \in C^2(0,T;H_0^1\cap H^r)\cap H^4(0,T;L^2)$, $U_0 \in H^{r+\frac{3}{2}}(\Lambda)$ and $U_1 \in H^r(\Lambda)$, then for all $t \in \bar{R}_\tau(T)$,*

$$E\left(U_N - u_N, t\right) \leq b_2e^{b_3t}\left(\tau^4 + \tau^{12}N^2 + \tau^8N^{-2} + \tau^4N^{2-4r} + N^{-2r}\right)$$

where b_2 and b_3 are positive constants depending on the norms of U and f in the spaces mentioned above.

If we take $u_{N,0} = P_N^{1,0}U_0$, and

$$u_{N,1} = P_N^{1,0}U_1 + \frac{\tau}{2}P_N^{1,0}\left(\partial_x^2U_0 - U_0 - U_0^3 + f(0)\right),$$

then we can remove the restrictions on U_0 and U_1 in Theorem 4.6, and $E\left(U_N - u_N, t\right) = \mathcal{O}\left(\tau^4 + N^{-2r}\right)$.

We now present some numerical results. For describing the numerical errors of scheme (4.43), let

$$\tilde{E}(v,t) = \frac{\|U(t) - v(t)\|}{\|U(t)\|} \tag{4.53}$$

where U is the solution of (4.39), and v is the corresponding numerical solution. For comparison, we also use a finite difference scheme and a finite element scheme to solve (4.39). Let $h = \frac{1}{N}$ and

$$\Lambda_h = \{x = jh,\ -N+1 \le j \le N-1\}, \quad \bar{\Lambda}_h = \Lambda_h \cup \{-1,1\}. \tag{4.54}$$

The finite difference scheme is

$$\begin{cases} \hat{D}_{\tau\tau}u_h(x,t) - \hat{u}_h(x,t) + G(u_h(x,t)) - \Delta_h \hat{u}_h(x,t) = \hat{f}(x,t), & x \in \Lambda_h,\ t \in R_\tau(T), \\ u_h(-1,t) = u_h(1,t) = 0, & t \in R_\tau(T), \\ D_\tau u_h(x,0) = u_{h,1}(x), & x \in \bar{\Lambda}_h, \\ u_h(x,0) = U_0(x), & x \in \bar{\Lambda}_h \end{cases} \tag{4.55}$$

where

$$u_{h,1}(x) = U_1(x) + \frac{\tau}{2}\left(\partial_x^2 U_0(x) - U_0(x) - U_0^3(x) + f(0)\right).$$

Next, let $\Lambda_{h,j} = \{x \mid jh \le x \le jh + h, -N \le j \le N-1\}$,

$$\tilde{S}_h = \{v \mid v \text{ is a linear function in } \Lambda_{h,j}, -N \le j \le N-1\}.$$

Set $S_h = \tilde{S}_h \cap H_0^1(\Lambda)$. The finite element scheme is

$$\left(\hat{D}_{\tau\tau}u^h(t) + \hat{u}^h(t) + G\left(u^h(t)\right), v\right) + \left(\partial_x \hat{u}^h(t), \partial_x v\right) = \left(\hat{f}(t), v\right), \quad \forall\, v \in S_h. \tag{4.56}$$

The values $D_\tau u^h(x,0)$ and $u^h(x,0)$ are evaluated as in (4.55). For describing the errors of (4.55) and (4.56), define

$$\tilde{E}_h(v,t) = \frac{\left(\sum_{x\in\Lambda_h} |U(x,t) - v(x,t)|^2\right)^{\frac{1}{2}}}{\left(\sum_{x\in\Lambda_h} |U(x,t)|^2\right)^{\frac{1}{2}}}.$$

The meanings of U and v are the same as in (4.53).

We take the test function

$$U(x,t) = A\left(x^2 - 1\right)\cos(Bx + Bt)e^{\lambda t}. \tag{4.57}$$

Table 4.6 shows the errors of the scheme (4.43) with $A = 0.5$ and $B = \lambda = 1.0$. Obviously, this approach serves very accurate numerical results even for small N. It can be seen that if N increases and τ decreases proportionally, then the errors decay quickly. This agrees with the theoretical analysis. Table 4.7 lists the errors $\tilde{E}\,(u_N, t)$ for scheme (4.43), $\tilde{E}(u_h, t)$ for scheme (4.55) and $\tilde{E}(u^h, t)$ for scheme (4.56). It is indicated that scheme (4.43) gives much better numerical results than schemes (4.55) and (4.56).

Table 4.6. The errors $\tilde{E}(u_N, t)$.

	$t = 0.2$	$t = 0.6$	$t = 1.0$
$N = 8, \tau = 0.005$	$9.50890E{-7}$	$1.39561E{-6}$	$6.31079E{-6}$
$N = 16, \tau = 0.005$	$9.41326E{-7}$	$1.31907E{-6}$	$6.22905E{-6}$
$N = 8, \tau = 0.001$	$1.64741E{-7}$	$4.94624E{-7}$	$1.03551E{-7}$
$N = 16, \tau = 0.001$	$9.08240E{-8}$	$1.08537E{-7}$	$8.43561E{-8}$

Table 4.7. The errors of schemes (4.43), (4.55) and (4.56).

		$A = B = 1, \lambda = 1$	$A = B = 1, \lambda = 2$	$A = B = 2, \lambda = 2$
$\tilde{E}(u_N, 1.0)$	$N = 8, \tau = 0.005$	$2.53742E{-6}$	$2.97078E{-5}$	$2.17966E{-4}$
	$N = 16, \tau = 0.001$	$3.39940E{-7}$	$1.33918E{-6}$	$9.98687E{-6}$
$\tilde{E}(u_h, 1.0)$	$N = 8, \tau = 0.005$	$1.34663E{-2}$	$2.65475E{-3}$	$3.87233E{-3}$
	$N = 16, \tau = 0.001$	$3.38277E{-3}$	$6.58751E{-4}$	$1.10229E{-3}$
$\tilde{E}(u^h, 1.0)$	$N = 8, \tau = 0.005$	$1.32584E{-3}$	$1.80903E{-2}$	$3.02362E{-2}$
	$N = 16, \tau = 0.001$	$3.28798E{-3}$	$4.42965E{-3}$	$7.71182E{-3}$

The Legendre spectral methods for the generalized NLKG equation were considered in Guo, Li and Vazquez (1996).

Legendre pseudospectral methods are much easier to be implemented. Let $x^{(j)}$ and $\omega^{(j)}$ be the Legendre–Gauss–Lobatto interpolation points and weights in Section

2.4. Denote by $\bar{\Lambda}_N$ the set of $x^{(j)}, 0 \leq j \leq N$. We introduce the discrete L^p-norm associated with the interpolation points as

$$\|v\|_{L^p,N} = \begin{cases} \left(\sum_{j=0}^{N} \left| v^{(x_j)} \right|^p \omega^{(j)} \right)^{\frac{1}{p}}, & \text{for } 1 \leq p < \infty, \\ \max_{x \in \bar{\Lambda}_N} |v(x)|, & \text{for } p = \infty. \end{cases}$$

Also define the discrete L^2-inner product by

$$(v, w)_N = \sum_{j=0}^{N} v\left(x^{(j)}\right) w\left(x^{(j)}\right) \omega^{(j)}.$$

There is a close relationship between the discrete norms and the usual norms. The first relation between the norms $\|\cdot\|$ and $\|\cdot\|_N$ is (2.60). Some other results are given in Li and Guo (1996).

Lemma 4.10. *For any $\phi \in \mathbb{P}_N$ and $p \geq 2$,*

$$\|\phi\|_{L^p,N} \leq \sqrt{3} N^{1-\frac{2}{p}} \|\phi\|.$$

Proof. By virtue of Theorem 2.7,

$$\|\phi\|_{L^\infty} \leq \sqrt{3} N \|\phi\|.$$

By (2.60),

$$\|\phi\|^p_{L^p,N} \leq \|\phi\|^{p-2}_{L^\infty,N} \|\phi\|^2_N \leq \left(2 + \frac{1}{N}\right) \|\phi\|^{p-2}_{L^\infty} \|\phi\|^2 \leq 3^{\frac{p}{2}} N^{p-2} \|\phi\|^p.$$

∎

Lemma 4.11. *For any $\phi \in P^0_N$ and $1 \leq p < \infty$,*

$$\|\phi\|_{L^p,N} \leq c(p) \|\phi\|_{L^p},$$

$c(p)$ is a positive constant dependent of p.

Proof. Let M be any positive integer and $\mathbb{P}_M$ be the set of all polynomials on Λ of degree at most M. Let $\mu(x)$ be a Jacobi weight. $\xi^{(j)}$ and $\rho^{(j)}$ are the nodes and the weights of Gauss quadrature associated with the Jacobi weight $\mu(x)$. Let m be a positive integer. Nevai (1979) (also see Szabados and Vèrtesi (1992)) proved that for any $\psi \in \mathbb{P}_{mM}$, any Jacobi weight $f(x)$ and $0 < p < \infty$,

$$\sum_{j=1}^{M} \left|\psi\left(\xi^{(j)}\right)\right|^p f\left(\xi^{(j)}\right) \rho^{(j)} \leq c(p) \int_\Lambda |\psi(x)|^p f(x)\mu(x)\,dx. \tag{4.58}$$

On the other hand, (2.44) implies that

$$\int_\Lambda \partial_x L_l(x) \partial_x L_n(x) \left(1 - x^2\right) dx = l(l+1) \int_\Lambda L_l(x) L_n(x)\,dx = l(l+1)\delta_{l,n}.$$

So $\{\partial_x L_l(x)\}$ is an orthogonal system with respect to the weight $\mu(x) = 1 - x^2$. As a result, we have that the interior nodes $x^{(j)}$ of the Legendre–Gauss–Lobatto quadrature with $N+1$ nodes coincide with the nodes $\xi^{(j)}, 1 \leq j \leq N-1$ of the Gauss quadrature with $N-1$ nodes, related to the weight $\mu(x)$, i.e., $x^{(j)} = \xi^{(j)}, \quad 1 \leq j \leq N-1$. Furthermore, for any $\phi \in \mathbb{P}_{2N-3}$, we have $\phi(x)(1-x^2) \in \mathbb{P}_{2N-1}$. By (2.37),

$$\sum_{j=1}^{N-1} \phi(\xi^{(j)})\rho^{(j)} = \int_\Lambda \phi(x)(1-x^2)dx = \sum_{j=0}^{N} \phi(x^{(j)})(1-(x^{(j)})^2)\omega^{(j)}$$

whence

$$\omega^{(j)} = \left(1 - \left(\xi^{(j)}\right)^2\right)^{-1} \rho^{(j)}, \quad 1 \leq j \leq N-1.$$

Let $f(x) = (1-x^2)^{-1}$ and $\mu(x) = 1 - x^2$ in (4.58). Then we find that

$$\begin{aligned}
\|\phi\|^p_{L^p,N} &= \sum_{j=0}^{N} \left|\phi\left(x^{(j)}\right)\right|^p \omega^{(j)} = \sum_{j=1}^{N-1} \left|\phi\left(x^{(j)}\right)\right|^p \omega^{(j)} \\
&= \sum_{j=1}^{N} \left|\phi\left(x^{(j)}\right)\right|^p \left(1 - \left(\xi^{(j)}\right)^2\right)^{-1} \rho^{(j)} \\
&\leq c(p) \int_\Lambda |\phi(x)|^p \left(1-x^2\right)^{-1} \left(1-x^2\right) dx = c(p)\|\phi\|^p_{L^p}.
\end{aligned}$$

■

It is noted that by (2.60) and (4.49),

$$\|\phi\|^2_N \leq \frac{4}{\pi^2}\left(2 + \frac{1}{N}\right) \|\partial_x \phi\|^2_N, \quad \forall\, \phi \in \mathbb{P}^0_N. \tag{4.59}$$

Let $I_N v$ be the Legendre interpolant of $v \in C\left(\bar{\Lambda}\right)$, related to the Legendre–Gauss–Lobatto interpolation points. Clearly for all $\phi \in \mathbb{P}_N, I_N\phi = \phi$, and for all $v, w \in C\left(\bar{\Lambda}\right), (v,w)_N = (I_N v, I_N w)_N$. Let $G(v)$ be the same as before. It is easy to see that

$$\left(G(v(t)), \hat{D}_\tau v(t)\right)_N = \frac{1}{4}\hat{D}_\tau \|v(t)\|^4_{L^4,N}. \tag{4.60}$$

Now, we construct the Legendre pseudospectral scheme for (4.39). Let u_N approximate U, and $I_N G(u_N)$ approximate U^3 instead of $I_N u_N^3$. The scheme is to find $u_N(x,t) \in \mathbb{P}_N^0$ for all $t \in \bar{R}_\tau(T)$ such that

$$\begin{cases} \left(\hat{D}_{\tau\tau}u_N(t) + \hat{u}_N(t) + G\left(u_N(t)\right), \phi\right)_N \\ \qquad + \left(\partial_x u_N(t), \partial_x\phi\right) = \left(\hat{f}(t), \phi\right)_N, \quad \forall\, \phi \in \mathbb{P}_N^0,\ t \in R_\tau(T), \\ D_\tau u_N(0) = u_{N,1}, \\ u_N(0) = u_{N,0} \end{cases} \tag{4.61}$$

where $u_{N,0} = I_N U_0$ and

$$u_{N,1} = I_N U_1 + \frac{\tau}{2} I_N\left(\partial_x^2 U_0 - U_0 - U_0^3 + f(0)\right).$$

By (2.37), $(\partial_x u_N(t), \partial_x\phi) = (\partial_x u_N(t), \partial_x\phi)_N$ in (4.61). Thus this scheme stands for

$$\hat{D}_{\tau\tau}u_N(t) + \hat{u}_N(t) + I_N G\left(u_N(t)\right) - I_N\partial_x^2 u_N(t) = I_N f(t).$$

We next check the conservation. Let

$$\begin{aligned} E^*_{\tau,N}(v,t) &= \|\bar{D}_\tau v(t)\|_N^2 + \frac{1}{2}|v(t)|_1^2 + \frac{1}{2}|v(t-\tau)|_1^2 \\ &\quad + \frac{1}{2}\|v(t)\|_N^2 + \frac{1}{2}\|v(t-\tau)\|_N^2 + \frac{1}{4}\|v(t)\|^4_{L^4,N} + \frac{1}{4}\|v(t-\tau)\|^4_{L^4,N}. \end{aligned}$$

We derive from (4.60) and (4.61) that

$$E^*_{\tau,N}\left(u_N,t\right) = E^*_{\tau,N}\left(u_N,\tau\right) + 2\tau \sum_{s\in R_\tau(t-\tau)} \left(\hat{f}(s), \hat{D}_\tau u_N(s)\right)_N$$

which is also a reasonable analogy of (4.40).

For error analysis, we need some preparations. It is not difficult to prove the following lemmas.

Lemma 4.12. *We have*

$$\|\hat{v}(t) - v(t)\|_N \le c\tau^{\frac{3}{2}} \|v\|_{H^2(t-\tau,t+\tau;C)},$$
$$\|\hat{D}_{\tau\tau} v(t) - \partial_t^2 v(t)\|_N \le c\tau^{\frac{3}{2}} \|v\|_{H^4(t-\tau,t+\tau;C)},$$
$$\|D_\tau v(t) - \partial_t v(t) - \frac{\tau}{2}\partial_t^2 v(t)\|_N \le c\tau^{\frac{3}{2}} \|v\|_{H^3(t-\tau,t+\tau;C)},$$

provided that the related norms exist.

Lemma 4.13. *For any $v \in C(0,T;C) \cap H^1(0,T;C)$,*

$$\|G(v(t)) - \frac{1}{2}v^3(t+\tau) - \frac{1}{2}v^3(t-\tau)\|_N \le c\tau^2 \|v\|^2_{C(t-\tau,t+\tau;C)} \|v\|_{H^1(t-\tau,t+\tau;C)}.$$

Lemma 4.14. *For any $v, w \in C(0,T;\mathbb{P}_N)$,*

$$G(v(x,t) + w(x,t)) = G(v(x,t)) + G(w(x,t)) + R(x,t)$$

with

$$\|R(t)\|_N^2 \le d(v)\left(\|\partial_x w(t+\tau)\|_N^2 + \|\partial_x w(t-\tau)\|_N^2 + \|w(t+\tau)\|^4_{L^4,N} + \|w(t-\tau)\|^4_{L^4,N}\right),$$

$d(v)$ being the same as in Lemma 4.9.

We now turn to the generalized stability of (4.61). Suppose that $u_{N,0}, u_{N,1}$ and $I_N f$ have the errors $\tilde{u}_0, \tilde{u}_1$ and $\tilde{f}$ respectively, which cause the error of u_N, denoted by $\tilde{u}$. Then

$$\begin{cases} \left(\hat{D}_{\tau\tau}\tilde{u}(t) + \hat{\tilde{u}}(t) + \tilde{G}(t), \phi\right)_N + \left(\partial_x \hat{\tilde{u}}(t), \partial_x \phi\right) = \left(\hat{\tilde{f}}(t), \phi\right)_N, & \forall \phi \in \mathbb{P}_N^0,\ t \in R_\tau(T), \\ D_\tau \tilde{u}(0) = \tilde{u}_1, \\ \tilde{u}(0) = \tilde{u}_0 \end{cases} \tag{4.62}$$

where

$$\tilde{G}(x,t) = G(\tilde{u}(x,t)) + \tilde{R}(x,t)$$

and

$$\left\|\tilde{R}(t)\right\|_N^2 \le d(u_N)\left(\|\partial_x \tilde{u}(t+\tau)\|_N^2 + \|\partial_x \tilde{u}(t-\tau)\|_N^2 + \|\tilde{u}(t+\tau)\|^4_{L^4,N} + \|\tilde{u}(t-\tau)\|^4_{L^4,N}\right).$$

By an argument as in the derivation of a priori estimate for the solution of (4.43), we know that $d(u_N)$ is bounded. By putting $\phi = \hat{D}_\tau \tilde{u}(t)$ in (4.62) and summing the resulting equality for $t \in R_\tau(T)$, we obtain

$$\begin{aligned}&\left\|\bar{D}_\tau \tilde{u}(t)\right\|_N^2 + \left(\frac{1}{4} - \tau d(u_N)\right)\left(\|\tilde{u}(t)\|_N^2 + |\tilde{u}(t)|_1^2 + \|\tilde{u}(t)\|_{L^4,N}^4\right) \\ &\quad \leq \rho_3(t) + \tau\left(c + d(u_N)\right) \sum_{s \in R_\tau(t-\tau)} \left(\|\bar{D}_\tau \tilde{u}(s)\|_N^2 + |\tilde{u}(s)|_1^2 + \|\tilde{u}(s)\|_{L^4,N}^4\right)\end{aligned}$$

with

$$\begin{aligned}\rho_3(t) &= (c + \tau d(u_{N,0}))\left(\|\tilde{u}_0\|_1^2 + \|\tilde{u}_1\|^2\right) + \left(\frac{9}{4} + \tau d\left(u_{N,0}\right)\right)\|\tilde{u}_0\|_{L^4,N}^4 \\ &\quad + 2\tau^4 \|\tilde{u}_1\|_{L^4,N}^4 + 2\tau \sum_{s \in \bar{R}_\tau(t)} \|\tilde{f}(s)\|^2.\end{aligned}$$

Let

$$E_N(v,t) = \|\bar{D}_\tau v(t)\|_N^2 + |v(t)|_1^2 + \|v(t)\|_{L^4,N}^4.$$

Following the same line as in the latter part of the proof of Theorem 4.5, we reach the following result.

Theorem 4.7. *Let $q = \tau N^2 < \infty$ and N be suitably large. Then for all $t \in \bar{R}_\tau(T)$,*

$$E_N\left(\tilde{u}, t\right) \leq c\rho_3(t) e^{b_4 t},$$

b_4 being a positive constant depending only on $\|u_N\|_{C(0,T;H^1)}$.

We now consider the convergence. We could compare u_N with $I_N U$. In this case, the optimal error estimate fails, since $I_N \partial_x^2 \neq \partial_x^2 I_N$. But the derivation of error bound is simpler. Moreover the technique used is available also for more complicated problems, for instance, the nonlinear term U^3 is replaced by $|u|^\alpha u, \alpha$ being any non-negative constant. For this reason, we first let $U_N = I_N U$. By (4.39),

$$\begin{cases}\left(\hat{D}_{\tau\tau} U_N(t) + \hat{U}_N(t) + G\left(U_N(t)\right), \phi\right)_N + \left(\partial_x \hat{U}_N(t), \partial_x \phi\right) \\ \quad = \sum_{j=1}^{3} (G_j(t), \phi)_N + \left(\hat{f}(t), \phi\right)_N, \ \forall\, \phi \in \mathbb{P}_N^0,\ t \in R_\tau(T), \\ D_\tau U_N(0) = I_N U_1 + \frac{\tau}{2} I_N\left(\partial_x^2 U_0 - U_0 - U_0^3 + f(0)\right) + G_4, \\ U_N(0) = I_N U_0\end{cases}$$

where

$$\begin{aligned}
G_1(x,t) &= \hat{D}_{\tau\tau}U_N(x,t) - \partial_t^2\hat{U}_N(x,t),\\
G_2(x,t) &= I_N\left(G\left(U_N(x,t)\right) - \frac{1}{2}U^3(x,t+\tau) - \frac{1}{2}U^3(x,t-\tau)\right),\\
G_3(x,t) &= I_N\partial_x^2\hat{U}(x,t) - \partial_x^2 I_N\hat{U}(x,t),\\
G_4(x,t) &= D_\tau U_N(x,0) - \partial_t U_N(x,0) - \frac{\tau}{2}\partial_t^2 U_N(0).
\end{aligned}$$

Put $\tilde{U} = u_N - U_N$. Then (4.61) leads to

$$\begin{cases}
\left(\hat{D}_{\tau\tau}\tilde{U}(t) + \hat{\tilde{U}}(t) + G\left(U_N(t) + \tilde{U}(t)\right) - G\left(U_N(t)\right), \phi\right)_N \\
\qquad + \left(\partial_x\hat{\tilde{U}}(t), \partial_x\phi\right) = -\sum_{j=1}^{3}\left(G_j(t),\phi\right)_N, \quad \forall\, \phi \in \mathbb{P}_N^0,\ t \in R_\tau(T),\\
D_\tau\tilde{U}(0) = -G_4,\\
\tilde{U}(0) = 0.
\end{cases}$$

Evidently we could derive an estimate for $\tilde{U}(t)$, similar to Theorem 4.7 in which the norms $\|\bar{D}_\tau\tilde{u}(t)\|_N, |\tilde{u}(t)|_1$ and $\|\tilde{u}(t)\|_{L^4,N}$ are replaced by $\|\bar{D}_\tau\tilde{U}(t)\|_N, |\tilde{U}(t)|_1$ and $\|\tilde{U}(t)\|_{L^4,N}$ respectively. Also $\rho_3(t)$ is replaced by

$$\rho_4(t) = c\|G_4\|^2 + 2\tau^4\|G_4\|_{L^4,N}^4 + c\tau\sum_{j=1}^{3}\sum_{s\in R_\tau(t-\tau)}\|G_j(s)\|_N^2.$$

According to Lemmas 4.12 and 4.13,

$$\begin{aligned}
\|G_1(t)\|_N^2 &\le c\tau^3\|U_N\|_{H^4(t-\tau,t+\tau;C)}^2 \le c\tau^3\|U\|_{H^4(t-\tau,t+\tau;C)}^2,\\
\|G_2(t)\|_N^2 &\le c\tau^4\|U\|_{C^1(0,T;C)}^6.
\end{aligned}$$

By virtue of (2.60), (2.63) and Theorem 2.8,

$$\begin{aligned}
\|G_3(t)\|_N \le c\|G_3(t)\| &\le c\|I_N\partial_x^2\hat{U}(t) - \partial_x^2\hat{U}(t)\|^2 + c\|\partial_x^2\hat{U}(t) - \partial_x^2 I_N\hat{U}(t)\|^2\\
&\le cN^{-2r}\|\hat{U}(t)\|_{r+2}^2 + cN^4\|\hat{U}(t) - I_N\hat{U}(t)\|_1^2\\
&\le cN^{-2r}\|U\|_{C(0,T;H^{r+3})}^2.
\end{aligned}$$

By Lemma 4.12,

$$\|G_4\|_N^2 = \left\|D_\tau U(x,0) - \partial_t U(x,0) - \frac{\tau}{2}\partial_t^2 U(0)\right\|_N^2 \le c\tau^4\|U\|_{C^3(0,T;C)}^2.$$

Thanks to Lemmas 4.11 and 4.12, Theorem 2.7 and (2.60),

$$\|G_4\|^4_{L^4,N} \le c\|G_4\|^4_{L^4} \le cN^2\|G_4\|^4 \le cN^2\|G_4\|^4_N \le c\tau^4N^2\|U\|^4_{C^3(0,T;C)}.$$

Hence $\rho_4(t) = \mathcal{O}\left(\tau^4 + N^{-2r}\right)$ and we have the following result.

Theorem 4.8. *Let $\tau N^2 < \infty$ and $r \ge 1$. If $U \in C(0,T;H^1_0 \cap H^{r+3}) \cap H^4(0,T;C)$, then for all $t \in \bar{R}_\tau(T)$,*

$$E_N\left(U_N - u_{N,t}\right) \le b_5 e^{b_6 t}\left(\tau^4 + N^{-2r}\right),$$

b_5 and b_6 being positive constants depending on the norms of U in the spaces mentioned above.

We next compare u_N with $P^{1,0}_N U$. Put $U_N = P^{1,0}_N U$. In this case, (4.39) yields

$$\begin{cases} \left(\hat{D}_{\tau\tau}U_N(t) + \hat{U}_N + G\left(U_N(t)\right), \phi\right)_N + \left(\partial_x\hat{U}_N(t), \partial_x\phi\right) \\ \quad = \sum\limits_{j=1}^{9} G_j(t) + \left(\hat{f}(t), \phi\right)_N, \quad \forall\, \phi \in \mathbb{P}^0_N,\ t \in R_\tau(T), \\ D_\tau U_N(0) = P^{1,0}_N\left(U_1 + \frac{\tau}{2}\partial^2_t U(0)\right) + G_{10}, \\ U_N(0) = P^{1,0}_N U_0 \end{cases}$$

where

$$\begin{aligned}
G_1(t) &= \left(\hat{D}_{\tau\tau}U_N(t) - \hat{D}_{\tau\tau}U(t), \phi\right)_N, \\
G_2(t) &= \left(\hat{D}_{\tau\tau}U(t), \phi\right)_N - \left(\hat{D}_{\tau\tau}U(t), \phi\right), \\
G_3(t) &= \left(\hat{D}_{\tau\tau}U(t) - \partial^2_t\hat{U}(t), \phi\right), \\
G_4(t) &= \left(\hat{U}_N(t) - \hat{U}(t), \phi\right)_N, \\
G_5(t) &= \left(\hat{U}(t), \phi\right)_N - \left(\hat{U}(t), \phi\right), \\
G_6(t) &= \left(G\left(U_N(t)\right) - \frac{1}{2}U^3_N(t+\tau) - \frac{1}{2}U^3_N(t-\tau), \phi\right)_N, \\
G_7(t) &= \frac{1}{2}\left(U^3_N(t+\tau) + U^3_N(t-\tau) - U^3(t+\tau) - U^3(t-\tau), \phi\right)_N, \\
G_8(t) &= \frac{1}{2}\left(U^3(t+\tau) + U^3(t-\tau), \phi\right)_N - \frac{1}{2}\left(U^3(t+\tau) + U^3(t-\tau), \phi\right), \\
G_9(t) &= \left(\hat{f}(t), \phi\right) - \left(\hat{f}(t), \phi\right)_N, \\
G_{10}(x) &= D_\tau U_N(x,0) - \partial_t U_N(x,0) - \frac{\tau}{2}\partial^2_t U_N(x,0).
\end{aligned}$$

Setting $\tilde{U} = u_N - U_N$, we obtain from (4.61) that

$$\begin{cases} \left(\hat{D}_{\tau\tau}\tilde{U}(t) + \hat{\tilde{U}}(t) + G\left(U_N(t) + \hat{U}(t)\right) - G\left(U_N(t)\right), \phi\right)_N \\ \quad + \left(\partial_x \hat{\tilde{U}}(t), \partial_x \phi\right) + \sum_{j=1}^{9} G_j(t) = 0, \ \forall\, \phi \in \mathbb{P}_N^0, \ t \in R_\tau(T), \\ D_\tau \tilde{U}(0) = I_N\left(U_1 + \frac{\tau}{2}\partial_t^2 U(0)\right) - P_N^{1,0}\left(U_1 + \frac{\tau}{2}\partial_t^2 U(0)\right) - G_{10}, \\ \tilde{U}(0) = I_N U_0 - P_N^{1,0} U_0. \end{cases}$$

Comparing this error equation to (4.62), it suffices to deal with $G_j(t)$. Let $r \geq 1$ and $\varepsilon > 0$. Since $\left(\hat{D}_{\tau\tau}U(t), \phi\right)_N = \left(I_N \hat{D}_{\tau\tau}U(t), \phi\right)_N$, we know from (2.60), (2.63) and Theorem 2.11 that

$$\begin{aligned} |G_1(t)| &\leq \left\|\hat{D}_{\tau\tau}U_N(t) - \hat{D}_{\tau\tau}I_N U(t)\right\|_N \|\phi\|_N \\ &\leq \varepsilon\|\phi\|^2 + \frac{c}{\varepsilon}\left\|\hat{D}_{\tau\tau}\left(P_N^{1,0}U(t) - U(t)\right)\right\|^2 + \frac{c}{3}\left\|\hat{D}_{\tau\tau}\left(I_N U(t) - U(t)\right)\right\|^2 \\ &\leq \varepsilon\|\phi\|^2 + \frac{c}{\varepsilon}N^{-2r}\|U\|_{C^2(t-\tau,t+\tau;H^r)}. \end{aligned}$$

By (2.64),

$$|G_2(t)| \leq \varepsilon\|\phi\|^2 + \frac{c}{\varepsilon}N^{-2r}\|U\|^2_{C(t-\tau,t+\tau;H^r)}.$$

By Lemma 4.6,

$$|G_3(t)| \leq \varepsilon\|\phi\|^2 + \frac{c}{\varepsilon}\tau^3\|U\|^2_{H^4(t-\tau,t+\tau;L^2)}.$$

It is not difficult to verify that

$$\begin{aligned} |G_4(t)| &\leq \left\|\hat{U}_N(t) - I_N\hat{U}(t)\right\|_N \|\phi\|_N \leq \varepsilon\|\phi\|^2 + \frac{c}{\varepsilon}N^{-2r}\|U\|^2_{C(t-\tau,t+\tau;H^r)}, \\ |G_5(t)| &\leq \varepsilon\|\phi\|^2 + \frac{c}{\varepsilon}N^{-2r}\|U\|^2_{C(t-\tau,t+\tau;H^r)}. \end{aligned}$$

Furthermore, Lemma 4.13, Theorem 2.11 and imbedding theory lead to

$$\begin{aligned} |G_6(t)| &\leq \varepsilon\|\phi\|^2 + \frac{c}{\varepsilon}\tau^4\|U_N\|^4_{C(t-\tau,t+\tau;C)}\|U_N\|^2_{H^1(t-\tau,t+\tau;C)} \\ &\leq \varepsilon\|\phi\|^2 + \frac{c}{\varepsilon}\tau^4\|U\|^6_{H^1(t-\tau,t+\tau;H^r)}. \end{aligned}$$

Imbedding theory, Theorem 2.11 and (2.63) yield

$$\begin{aligned} |G_7(t)| &\leq \varepsilon\|\phi\|^2 + \frac{c}{\varepsilon}\left(\|U^3(t+\tau) - U_N^3(t+\tau)\|_N^2 + \|U^3(t-\tau) - U_N^3(t-\tau)\|_N^2\right) \\ &\leq \varepsilon\|\phi\|^2 + \frac{c}{\varepsilon}(\|U\|^4_{C(t-\tau,t+\tau;C)} + \|U_N\|^4_{C(t-\tau,t+\tau;C)})\left(\|I_N U(t+\tau) - U_N(t+\tau)\|_N^2\right. \\ &\quad \left. + \|I_N U(t-\tau) - U_N(t-\tau)\|_N^2\right) \\ &\leq \varepsilon\|\phi\|^2 + \frac{c}{\varepsilon}N^{-2r}\|U\|^6_{C(t-\tau,t+\tau;H^r)}. \end{aligned}$$

By using imbedding theory and (2.64) again,

$$|G_8(t)| \leq \varepsilon\|\phi\|^2 + \frac{c}{\varepsilon}N^{-2r}\|U^3\|^2_{C(t-\tau,t+\tau;H^r)} \leq \varepsilon\|\phi\|^2 + \frac{c}{\varepsilon}N^{-2r}\|U\|^6_{C(t-\tau,t+\tau,H^r)}.$$

Clearly

$$|G_9(t)| \leq \varepsilon\|\phi\|^2 + \frac{c}{\varepsilon}N^{-2r}\left(\|f(t+\tau)\|^2_r + \|f(t-\tau)\|^2_r\right).$$

By Theorems 2.7 and 2.11, Lemmas 4.6, 4.11 and 4.12, (2.60), (2.63) and imbedding theory,

$$\begin{aligned} \left\|D_\tau\tilde{U}(0)\right\|^2 &= c\tau^4\|U\|^2_{C^3(0,T;C)}, \\ \left\|D_\tau\tilde{U}(0)\right\|^4_{L^4,N} &\leq c\|D_\tau\tilde{U}(0)\|^4_{L^4} \leq cN^2\|D_\tau\tilde{U}(0)\|^4 \leq cN^{2-4r}(\|U_1\|^4_r + \|U\|^4_{C^1(0,T;H^r)}) \\ &\quad + c\tau^4N^{-2}\|U\|^4_{C^2(0,T;H^1)} + c\tau^8N^2\|U\|^4_{C^3(0,T;C^3)}. \end{aligned}$$

Besides, by Theorem 2.11 and (2.63),

$$\left\|\tilde{U}(0)\right\|^2_1 \leq cN^{-2r}\|U_0\|^2_{r+1}.$$

The above statements lead to the following convergence result.

Theorem 4.9. *Let $\tau N^2 < \infty$ and $r \geq 1$. If $U \in C^2(0,T;H^r) \cap C^3(0,T;C) \cap H^4(0,T;L^2)$,*
$U_0 \in H^{r+1}(\Lambda), U_1 \in H^r(\Lambda)$ and $f \in L^2(0,T;H^r)$, then for all $t \in \bar{R}_\tau(T)$,

$$E_N(U_N - u_N, t) \leq b_7 e^{b_8 t}\left(\tau^4 + \tau^{12}N^2 + \tau^8N^{-2} + \tau^4N^{2-4r} + N^{-2r}\right),$$

b_7 and b_8 being positive constants depending on the norms of U, U_0 and f in the spaces mentioned above.

Finally, we present some numerical results. For describing the errors of scheme (4.61), let

$$\tilde{E}_N(u_N, t) = \frac{\|U(t) - u_N(t)\|_N}{\|U(t)\|_N}.$$

We also use scheme (4.55) with the relative error $\tilde{E}_h(u_h, t)$ as before. The test function is given by (4.57). In Table 4.8, the calculation is carried out with $A = 0.5, B = \lambda = 1.0, N = 8$ and $\tau = 0.005$. The errors $\tilde{E}_N(u_N, t)$ of scheme (4.61) are compared to the errors $\tilde{E}_h(u_h, t)$ of scheme (4.55). Clearly (4.61) provides better numerical results than scheme (4.55), and has very high accuracy even for small N. Table 4.9 shows the numerical errors with $A = B = 1.0$ and $\lambda = 2.0$. We find that if N increases and τ decreases proportionally, then the errors decay very quickly, and that the solution of (4.61) converges faster than that of (4.55).

Table 4.8. The errors of schemes (4.55) and (4.61).

	$\tilde{E}_N(u_N, t)$	$\tilde{E}_h(u_h, t)$
$t = 0.2$	$9.21281E{-7}$	$1.23154E{-3}$
$t = 0.4$	$5.91656E{-7}$	$4.34661E{-3}$
$t = 0.6$	$1.52947E{-6}$	$8.66063E{-3}$
$t = 0.8$	$3.82931E{-6}$	$1.36452E{-2}$
$t = 1.0$	$6.61214E{-6}$	$1.88353E{-2}$

Table 4.9. The errors of schemes (4.55) and (4.61).

	$\tilde{E}_N(u_N, 1.0)$			$\tilde{E}_h(u_h, 1.0)$		
N	$\tau = 0.005$	$\tau = 0.001$	$\tau = 0.0005$	$\tau = 0.005$	$\tau = 0.001$	$\tau = 0.0005$
4	$1.52081E{-3}$	$1.54167E{-3}$	$1.54159E{-3}$	$8.79735E{-3}$	$8.78729E{-3}$	$8.78698E{-3}$
8	$2.95840E{-5}$	$1.16568E{-6}$	$2.97163E{-7}$	$2.65475E{-3}$	$2.63820E{-3}$	$2.63768E{-3}$
16	$2.94326E{-5}$	$1.1641E{-6}$	$2.93173E{-7}$	$6.76090E{-4}$	$6.58751E{-4}$	$6.58222E{-4}$

The generalized NLKG equation is of the form

$$\partial_t^2 U + U + g(U) - \partial_x^2 U = f$$

where $g(z) = |z|^\alpha \alpha, \alpha \geq 0$. In this case, define

$$G(v(x,t)) = \int_0^1 g(yv(x, t+\tau) + (1-y)v(t-\tau))\, dy,$$

and approximate $g(U)$ by $I_N G(u_N)$, see Li and Guo (1997). The numerical solution u_N is compared to $I_N U$ when α is not an integer.

Legendre spectral methods and Legendre pseudospectral methods have been used for numerical studies in various fields. For instance, Maday and Quarteroni (1981) used it for the steady Burgers' equation. Canuto and Quarteroni (1982b) studied the stability of semi-discrete Legendre spectral and Legendre pseudospectral schemes for linear hyperbolic equations. They also considered Jacobi spectral approximations in that paper.

4.3. Chebyshev Spectral Methods and Chebyshev Pseudospectral Methods

Chebyshev spectral methods and Chebyshev pseudospectral methods are also used widely for non-periodic problems. Since Chebyshev polynomials can be changed into trigonometric polynomials by a transformation of independent variable, the Fast Fourier Transformation can be used in actual computations. But the existence of the weight function brings some difficulties in theoretical analysis. In this section, we take the Burgers' equation as an example to show how to build Chebyshev spectral schemes and Chebyshev pseudospectral scheme, and how to analyze the errors precisely. The filterings are applied to improving the stability. The numerical results presented show the effects of filterings.

Burgers' equation is one of the important models describing the movement of unsteady one-dimensional fluid flow. Let $\Lambda = (-1, 1)$. The initial-boundary value problem of Burgers' equation is the following

$$\begin{cases} \partial_t U(x,t) + U(x,t)\partial_x U(x,t) - \mu\partial_x^2 U(x,t) = f(x,t), & x \in \Lambda,\, 0 < t \leq T, \\ U(-1,t) = U(1,t) = 0, & 0 < t \leq T, \\ U(x,0) = U_0(x), & x \in \bar{\Lambda} \end{cases} \tag{4.63}$$

where $U_0(x), f(x,t)$ and $\mu > 0$ are the initial state, the body force and the kinetic viscosity respectively. It is known that if $U_0 \in L^2(\Lambda)$ and $f \in L^2(0,T;L^2)$, then (4.63) has a unique solution $U \in L^2(0,T;H^1) \cap L^\infty(0,T;L^2)$. The solution of (4.63) possesses the conservation

$$\|U(t)\|^2 + 2\mu \int_0^t |U(s)|_1^2\, ds = \|U_0\|^2 + 2\int_0^t (f(s), U(s))\, ds. \tag{4.64}$$

Let $\omega(x) = (1-x^2)^{-\frac{1}{2}}$. The space $L^2_\omega(\Lambda)$ is defined as in Section 2.5, equipped with the inner product $(\cdot,\cdot)_\omega$ and the norm $\|\cdot\|_\omega$. The spaces $L^p_\omega(\Lambda), H^m_\omega(\Lambda), H^m_{0,\omega}(\Lambda)$ and their norms $\|\cdot\|_{L^p_\omega}, \|\cdot\|_{m,\omega}$ are also the same as before. Let N be any positive integer and P_N be the L^2_ω-orthogonal projection from $L^2_\omega(\Lambda)$ onto $\mathbb{P}^0_N$. The H^1_ω-orthogonal projections $P^1_N, \tilde{P}^{1,0}_N$ and $P^{1,0}_N$ are given in Section 2.5. The mesh size in time t is taken to be τ. The notations $R_\tau(T), \hat{v}(x,t)$ and $\hat{D}_\tau v(x,t)$ are the same as before.

When we approach (4.63) by the Legendre spectral approximations, we can get a scheme whose solution fulfills a conservation simulating (4.64). But this property is destroyed by the weight function in Chebyshev spectral approximations. In Section 4.1, we constructed the schemes with the accuracy of first order in time discretization, unless the nonlinear term is approximated by a fully implicit technique. If we use three-level scheme, then we can obtain the accuracy of second order in temporal discretization. While the unknown solution can still be evaluated explicitly by the variable transformation and the orthogonality of Fourier system. We shall adopt this trick.

Let u_N be the approximation to U. A Chebyshev spectral scheme for (4.63) is to find $u_N(x,t) \in \mathbb{P}_N^0$ for all $t \in \bar{R}_\tau(T)$ such that

$$\begin{cases} \left(\hat{D}_\tau u_N(t), \phi\right)_\omega - \frac{1}{2}\left(u_N^2(t), \partial_x(\phi\omega)\right) \\ \qquad +\mu\left(\partial_x \hat{u}_N(t), \partial_x(\phi\omega)\right) = \left(\hat{f}(t), \phi\right)_\omega, \ \forall\, \phi \in \mathbb{P}_N^0,\ t \in R_\tau(T), \\ u_N(\tau) = u_{N,1}, \\ u_N(0) = u_{N,0} \end{cases} \tag{4.65}$$

where $u_{N,0} = P_N U_0$ and

$$u_{N,1} = P_N\left(U_0 + \tau\partial_t U(0)\right) = P_N U_0 + \tau P_N\left(\mu\partial_x^2 U_0 - \frac{1}{2}\partial_x U_0^2 + f(0)\right).$$

The scheme (4.65) stands for

$$\hat{D}_\tau u_N(x,t) + \frac{1}{2}P_N\partial_x u_N^2(x,t) - \mu P_N \partial_x^2 \hat{u}_N(x,t) = P_N \hat{f}(x,t)$$

which is easy to be implemented in practice. (4.65) is an implicit scheme. At each time $t \in R_\tau(T)$, we need to solve a linear system

$$\left(u_N(t), \phi\right)_\omega + 2\tau\mu\left(\partial_x u_N(t), \partial_x(\phi\omega)\right) = \left(F(t), \phi\right)_\omega$$

where $F(t)$ depends on $u_N(t-\tau), u_N(t-2\tau)$ and $\hat{f}(x,t)$. By Lemma 2.1, it possesses a unique solution. Furthermore, $u_N(t)$ can be calculated explicitly as explained above.

In actual calculations, we can take the basis functions as

$$\phi_l(x) = \begin{cases} T_l(x) - T_0(x), & l \text{ is even}, \\ T_l(x) - T_1(x), & l \text{ is odd}. \end{cases}$$

But these functions lead to a linear system with a full matrix. A better choice (see Shen (1995)) is to take

$$\psi_l(x) = T_l(x) - T_{l+2}(x).$$

Then

$$(\psi_m, \psi_l)_\omega = (\psi_l, \psi_m)_\omega = \begin{cases} \frac{\pi}{2}(c_m + 1), & l = m, \\ -\frac{\pi}{2}, & l = m - 2 \text{ or } m + 2, \\ 0, & \text{otherwise,} \end{cases}$$

and

$$(\partial_x \psi_l, \partial_x(\psi_m \omega)) = \begin{cases} 2\pi(m+1)(m+2), & l = m, \\ 4\pi(m+1), & l = m+2, m+4, \dots, \\ 0, & l < m \text{ or } l + m \text{ odd.} \end{cases}$$

We now consider the generalized stability of (4.65). Let $\tilde{u}_0, \tilde{u}_1$ and $\tilde{f}$ be the errors of $u_{N,0}, u_{N,1}$ and $P_N \hat{f}$, which induce the error of u_N, say $\tilde{u}$. From (4.65),

$$\begin{aligned} &\left(\hat{D}_\tau \tilde{u}(t), \phi\right)_\omega - \left(F\left(u_N(t), \tilde{u}(t)\right), \partial_x(\phi\omega)\right) \\ &+\mu\left(\partial_x \hat{\tilde{u}}(t), \partial_x(\phi\omega)\right) = \left(\tilde{f}(t), \phi\right), \ \forall \, \phi \in \mathbb{P}_N^0, \, t \in R_\tau(T), \end{aligned} \tag{4.66}$$

where

$$F(v(x,t), w(x,t)) = v(x,t)w(x,t) + \frac{1}{2}w^2(x,t).$$

By taking $\phi = \hat{\tilde{u}}$ in (4.66), we get from (1.23) and Lemma 2.8 that

$$\frac{1}{2}\hat{D}_\tau \|\tilde{u}(t)\|_\omega^2 + \frac{\mu}{4}\|\hat{\tilde{u}}(t)\|_{1,\omega}^2 \le \left(\tilde{f}(t), \hat{\tilde{u}}(t)\right)_\omega + \left(F\left(u_N(t), \tilde{u}(t)\right), \partial_x\left(\hat{\tilde{u}}(t)\omega\right)\right). \tag{4.67}$$

We next deal with the right-hand side of (4.67). First of all, we know from Theorem 2.13 and the proof of Lemma 2.8 that

$$\begin{aligned} &\left|4\left(F\left(u_N(t), \tilde{u}(t)\right), \partial_x\left(\hat{\tilde{u}}(t)\omega\right)\right)\right| \\ &\le \frac{\mu}{2}\left\|\hat{\tilde{u}}(t)\right\|_{1,\omega}^2 + \frac{32}{\mu}\left\|F\left(u_N(t), \tilde{u}(t)\right)\right\|_\omega^2 \\ &\le \frac{\mu}{2}\left\|\hat{\tilde{u}}(t)\right\|_{1,\omega}^2 + d_1(u_N, \mu)\left(\|\tilde{u}(t)\|_\omega^2 + N\|\tilde{u}(t)\|_\omega^4\right) \end{aligned} \tag{4.68}$$

where $d_1(v,u)$ is a certain positive constant depending only on $\|v\|_{C(0,T;L^\infty)}$ and μ. Clearly,

$$\left|4\left(\tilde{f}(t), \hat{\tilde{u}}(t)\right)_\omega\right| \le \|\tilde{u}(t+\tau)\|_\omega^2 + \|\tilde{u}(t-\tau)\|_\omega^2 + 2\left\|\hat{\tilde{f}}(t)\right\|_\omega^2. \tag{4.69}$$

Let $\tau < \frac{1}{2}$ and

$$E(v,t) = \|v(t)\|_\omega^2 + \frac{1}{2}\mu\tau \sum_{s \in R_\tau(t-\tau)} \left\|\hat{\tilde{u}}(s)\right\|_{1,\omega}^2,$$

$$\rho_1(t) = 2\|\tilde{u}_0\|_\omega^2 + 2\|\tilde{u}_1\|_\omega^2 + 4\tau \sum_{s \in \bar{R}_\tau(t)} \left\|\tilde{f}(s)\right\|_\omega^2.$$

By substituting (4.68) and (4.69) into (4.67) and summing up the result for $t \in R_\tau(T)$, we obtain

$$E(\tilde{u},t) \le \rho_1(t) + \tau \sum_{s\in R_\tau(t-\tau)} d(u_N,\mu)\left(E(\tilde{u},s) + NE^2(\tilde{u},s)\right).$$

Finally we use Lemma 3.2 to conclude that

Theorem 4.10. *Let τ be suitably small. There exist positive constants b_1 and b_2 depending only on $\|u_N\|_{C(0,T;L^\infty)}$ and μ, such that if $\rho_1(t_1) \le \dfrac{b_1}{N}$ for certain $t_1 \in R_\tau(T)$, then for all $t \in \bar{R}_\tau(t_1)$,*

$$E(\tilde{u},t) \le \rho_1(t)e^{b_2 t}.$$

We now deal with the convergence. We can compare u_N to $P_N U, P_N^1 U$ or $\tilde{P}_N^{1,0} U$. But in these cases, the leading error terms cannot be cancelled. It means that the terms $(\partial_x(U - P_N U), \partial_x(\phi\omega)), (\partial_x(U - P_N^1 U), \partial_x(\phi\omega))$ and $\left(\partial_x\left(U - \tilde{P}_N^{1,0}U\right), \partial_x(\phi\omega)\right)$ do not vanish for $\phi \in \mathbb{P}_N^0$. They shall lower the convergence rate. Therefore we compare u_N to $U_N = P_N^{1,0}U$, where the H_0^1-projection $P_N^{1,0}$ is defined as in (2.82). By (4.63),

$$\begin{aligned}&\left(\hat{D}_\tau U_N(t),\phi\right)_\omega - \frac{1}{2}\left(U_N^2(t),\partial_x(\phi\omega)\right) + \mu\left(\partial_x \hat{U}_N(t),\partial_x(\phi\omega)\right)\\ &\quad = \left(G_1(t) + \hat{f}(t),\phi\right)_\omega + \left(G_2(t)+G_3(t),\partial_x(\phi\omega)\right), \quad \forall\, \phi \in \mathbb{P}_N^0,\ t\in R_\tau(T),\end{aligned}$$

with

$$\begin{aligned}G_1(x,t) &= \hat{D}_\tau U_N(x,t) - \partial_t \hat{U}(x,t),\\ G_2(x,t) &= \frac{1}{2}U^2(x,t) - \frac{1}{2}U_N^2(x,t),\\ G_3(x,t) &= \frac{1}{4}\tau^2 \hat{D}_{\tau\tau}U^2(x,t).\end{aligned}$$

Putting $\tilde{U} = u_N - U_N$, we obtain from (4.65) that

$$\begin{aligned}&\left(\hat{D}_\tau \tilde{U}(t),\phi\right)_\omega - \left(F\left(U_N(t),\tilde{U}(t)\right),\partial_x(\phi\omega)\right) + \mu\left(\partial_x\tilde{U}(t),\partial_x(\phi\omega)\right).\\ &\quad + (G_1(t),\phi)_\omega + (G_2(t)+G_3(t),\partial_x(\phi\omega)) = 0, \quad \forall\,\phi\in \mathbb{P}_N^0,\ t\in R_\tau(T). \quad (4.70)\end{aligned}$$

Let $r \ge 1$. By Theorem 2.18,

$$\begin{aligned}\tau \sum_{s\in R_\tau(t-\tau)} \|G_1(s)\|_\omega^2 &\le c\tau^4 \|U\|^2_{H^3(0,T;L^2_\omega)} + cN^{-2r}\|U\|^2_{H^1(0,T;H^r_\omega)},\\ \|G_2(t)\|^2_\omega &\le d_2(U)N^{-2r}\|U(t)\|^2_{r,\omega},\\ \|G_3(t)\|^2_\omega &\le \tau^4 d_3(U)\left(\|U\|^2_{C^2(0,T;L^2_\omega)} + 1\right)\end{aligned}$$

where $d_2(v)$ is a certain positive constant depending only on $\|v\|_{C(0,T;H^1_\omega)}$, and $d_3(v)$ is a certain positive constant depending only on $\|v\|_{C^1(0,T;L^\infty)}$. On the other hand,

$$\left\|\tilde{U}(0)\right\|_\omega \leq 2\|P_N U_0 - U_0\|_\omega + 2\|P_N^{1,0} U_0 - U_0\|_\omega \leq cN^{-r}\|U_0\|_{r,\omega}.$$

Let I be the identity operator. Since

$$\tilde{U}(\tau) = (P_N - I)(U_0 + \tau\partial_t U(0)) + (I - P_N^{1,0})U(\tau) + U_0 + \tau\partial_t U(0) - U(\tau),$$

we assert that

$$\|\tilde{U}(\tau)\|_\omega \leq CN^{-r}(\|U_0\|_{r,\omega} + \|U(\tau)\|_{r,\omega} + \tau\|\partial_t U(0)\|_{r,\omega}) + c\tau^2\|P_N^{1,0}\partial_t^2 U\|_{C(0,T;L^2_\omega)}.$$

Finally we apply Theorem 4.10 to (4.70).

Theorem 4.11. *Let $\tau \leq b_3 N^{-\frac{1}{4}}$. If $U \in C^1(0,T;H^1_{0,\omega} \cap H^r_\omega) \cap H^3(0,T;L^2_\omega)$ with $r \geq 1$, then*

$$E\left(U_N - u_N, t\right) \leq b_4 e^{b_5 t}\left(\tau^4 + N^{-2r}\right),$$

b_3, b_4 and b_5 being positive constants depending on μ and the norms of U in the spaces mentioned above.

Chebyshev pseudospectral methods are more preferable in practical cases. We shall use this approach for (4.63). Let $x^{(j)}$ and $\omega^{(j)}$ be the Chebyshev–Gauss–Lobatto interpolation points and weights as in Section 2.5. Set $\bar{\Lambda}_N = \left\{x^{(j)} \mid 0 \leq j \leq N\right\}$. Denote by $I_N v$ the Chebyshev interpolant of $v \in C(\bar{\Lambda})$, associated with $\bar{\Lambda}_N$. The discrete inner product $(\cdot,\cdot)_{N,\omega}$ and the discrete norm $\|\cdot\|_{N,\omega}$ are defined by (2.36), related to the Chebyshev–Gauss–Lobatto points and weights. Clearly, $I_N\phi = \phi$ for any $\phi \in \mathbb{P}_N$, and $(v,w)_{N,\omega} = (I_N v, I_N w)_{N,\omega}$ for all $v, w \in C(\bar{\Lambda})$.

Let u_N be the approximation to U. In order to strengthen the stability, we use the filtering $R_N = R_N(\alpha,\beta), \alpha,\beta \geq 1$, defined by (2.136). The Chebyshev pseudospectral scheme for (4.63) is to find $u_N(x,t) \in \mathbb{P}_N^0$ for all $t \in \bar{R}_\tau(T)$ such that

$$\begin{cases} \left(\hat{D}_\tau u_N(t), \phi\right)_{N,\omega} - \frac{1}{2}\left(R_N I_N u_N^2(t), \partial_x(\phi\omega)\right) \\ \quad +\mu\left(\partial_x u_N(t), \partial_x(\phi\omega)\right) = \left(R_N I_N \hat{f}(t), \phi\right)_{N,\omega}, \quad \forall\, \phi \in \mathbb{P}_N^0,\ t \in R_\tau(T), \\ u_N(\tau) = u_{N,1}, \\ u_N(0) = u_{N,0} \end{cases} \tag{4.71}$$

where $u_{N,0} = I_N U_0$ and

$$u_{N,1} = I_N (U_0 + \tau \partial_t U(0)) = I_N U_0 + \tau I_N \left(\mu \partial_x^2 U_0 - \frac{1}{2} \partial_x U_0^2 + f(0) \right).$$

Scheme (4.71) stands for

$$\hat{D}_\tau u_N(x,t) + \frac{1}{2} I_N \partial_x R_N I_N u_N^2(x,t) - \mu I_N \partial_x^2 u_N(x,t) = I_N R_N I_N \hat{f}(x,t).$$

This expression is easier to be performed in actual computations.

Now suppose that $u_{N,0}, u_{N,1}$ and $R_N I_N \hat{f}$ have the errors $\tilde{u}_0, \tilde{u}_1$ and $\tilde{f}$ respcetively, which cause $\tilde{u}_N$, the error of u_N. From (4.71),

$$\begin{aligned} &\left(\hat{D}_\tau \tilde{u}(t), \phi \right)_{N,\omega} - \left(R_N I_N F\left(u_N(t), \tilde{u}(t) \right), \partial_x(\phi\omega) \right) \\ &\quad + \mu \left(\partial_x \hat{\tilde{u}}(t), \partial_x(\phi\omega) \right) = \left(\tilde{f}(t), \phi \right)_{N,\omega}, \quad \forall\, \phi \in \mathbb{P}_N^0,\ t \in R_\tau(T), \end{aligned} \tag{4.72}$$

where $F(v,w)$ is the same as in (4.66). We take $\phi = \hat{\tilde{u}}(t)$ in the above equation, and then by (1.23) and Lemma 2.8,

$$2\hat{D}_\tau \|\tilde{u}(t)\|_{N,\omega}^2 + \mu \|\hat{\tilde{u}}(t)\|_{1,\omega}^2 \le 4 \left(\tilde{f}(t), \hat{\tilde{u}}(t) \right)_{N,\omega} + 4 \left(R_N I_N F(t), \partial_x \left(\hat{\tilde{u}}(t)\omega \right) \right). \tag{4.73}$$

We now estimate the right-hand side of (4.73). By Lemma 2.8, Theorems 2.19 and 2.27, and (2.86),

$$\left| 4 \left(R_N I_N F(t), \partial_x \left(\hat{\tilde{u}}(t)\omega \right) \right) \right| \le 8 \left\| I_N F(t) \right\|_\omega \left| \hat{\tilde{u}}(t) \right|_{1,\omega} \le \frac{\mu}{2} \left\| \hat{\tilde{u}}(t) \right\|_{1,\omega}^2 + \frac{32}{\mu} \|F(t)\|_\omega^2.$$

Hence it is bounded by the right-hand side of (4.68). Clearly

$$\left| 4 \left(\tilde{f}(t), \hat{\tilde{u}}(t) \right)_{N,\omega} \right| \le \|\tilde{u}(t+\tau)\|_\omega^2 + \|\tilde{u}(t-\tau)\|_\omega^2 + 2 \left\| \tilde{f}(t) \right\|_\omega^2.$$

Let $\tau < \frac{1}{2}$, and $E(v,t)$ and $d(u_N,\mu)$ have the same meaning as before. Set

$$\rho_2(t) = c \left(\|\tilde{u}_0\|_\omega^2 + \|\tilde{u}_1\|_\omega^2 + \tau \sum_{s \in \bar{R}_\tau(t)} \left\| \tilde{f}(s) \right\|_\omega^2 \right).$$

By putting the above two estimates into (4.73) and summing the result for $t \in R_\tau(T)$, we find that

$$E\left(\tilde{u}, t \right) \le \rho_2(t) + \tau \sum_{s \in R_\tau(t-\tau)} d(u_N, \mu) \left(E(\tilde{u}, s) + N E^2(\tilde{u}, s) \right).$$

Applying Lemma 3.2 to the above inequality, we obtain the generalized stability expressed as follows.

Theorem 4.12. *Let τ be suitably small. There exist positive constants b_6 and b_7 depending only on $\|u_N\|_{C(0,T;L^\infty)}$ and μ, such that if $\rho_2(t_1) \le \frac{b_6}{N}$ for some $t_1 \in R_\tau(T)$, then for all $t \in \bar{R}_\tau(t_1)$,*

$$E(\tilde{u}, t) \le \rho_2(t)e^{b_7 t}.$$

We next state the convergence of (4.71). For the optimal error estimate, we also compare u_N to $U_N = P_N^{1,0}U$. Let $\tilde{U} = u_N - U_N$. Then (4.63) and (4.71) produce that

$$\begin{aligned}&\left(\hat{D}_\tau \tilde{U}(t), \phi\right)_{N,\omega} - \left(R_N I_N F\left(U_N(t), \tilde{U}(t)\right), \partial_x(\phi\omega)\right) + \mu\left(\partial_x \hat{\tilde{U}}(t), \partial_x(\phi\omega)\right)\\ &\quad + G_1(t) + (G_2(t) + G_3(t), \partial_x(\phi\omega)) + G_4(t) = 0, \quad \forall\ \phi \in \mathbb{P}_N^0,\ t \in R_\tau(T), \end{aligned} \tag{4.74}$$

where $F(v,w)$ and $G_3(t)$ are the same as in (4.70), and

$$\begin{aligned} G_1(t) &= \left(\hat{D}_\tau U_N(t), \phi\right)_{N,\omega} - \left(\partial_t \hat{U}(t), \phi\right)_\omega, \\ G_2(x,t) &= \frac{1}{2}U^2(x,t) - \frac{1}{2}R_N I_N U_N^2(x,t), \\ G_4(t) &= \left(\hat{f}(t), \phi\right)_\omega - \left(R_N I_N \hat{f}(t), \phi\right)_{N,\omega}. \end{aligned}$$

Let $r \ge 1$. We have

$$\begin{aligned} |G_1(t)| \;\le\; & \left|\left(\hat{D}_\tau U(t) - \partial_t U(t), \phi\right)_\omega\right| + \left|\left(\hat{D}_\tau U(t), \phi\right)_\omega - \left(\hat{D}_\tau U(t), \phi\right)_{N,\omega}\right| \\ & + \left|\left(\hat{D}_\tau U(t) - \hat{D}_\tau U_N(t), \phi\right)_{N,\omega}\right|. \end{aligned}$$

According to Theorem 2.18 and (2.87),

$$|G_1(t)| \le A(t) + c\|\phi\|_\omega^2$$

with

$$A(t) = c\left\|\hat{D}_\tau U(t) - \partial_t U(t)\right\|_\omega^2 + cN^{-2r}\left\|\hat{D}_\tau U(t)\right\|_{r,\omega}^2.$$

Moreover,

$$\tau \sum_{s \in R_\tau(t-\tau)} A(s) \le c\tau^4 \|U\|^2_{H^3(0,T;L^2_\omega)} + cN^{-2r}\|U\|^2_{C^1(0,T;H^r_\omega)}.$$

We have

$$\|G_2(t)\|_\omega \le \|R_N I_N(U_N^2(t) - U^2(t))\|_\omega + \|R_N(I_N U^2(t) - U^2(t))\|_\omega + \|R_N U^2(t) - U^2(t)\|_\omega\,.$$

By Theorems 2.19 and 2.27,

$$\|G_2(t)\|_\omega \le d_4(U)N^{-r}$$

where $d_4(v)$ is a positive constant depending only on $\|v\|_{C(0,T;H^r_\omega)}$. We also have

$$\|G_4(t)\|_\omega \le cN^{-2r}\|f\|^2_{C(0,T;H^r_\omega)} + c\|\phi\|^2_\omega.$$

Besides,

$$\begin{aligned}
\left\|\tilde{U}(0)\right\|_\omega &\le cN^{-r}\|U_0\|_{r,\omega},\\
\left\|\tilde{U}(\tau)\right\|_\omega &\le cN^{-r}(\|U(0)\|_{r,\omega} + \|U(\tau)\|_{r,\omega} + \tau\|\partial_t U(0)\|_{r,\omega}) + c\tau^2\|U\|_{C^2(0,T;L^2_\omega)}.
\end{aligned}$$

Finally, we apply Theorem 4.12 to (4.74).

Theorem 4.13. *Let* $\tau \le b_8 N^{-\frac{1}{4}}$. *If* $U \in C^1(0,T;H^1_{0,\omega}\cap H^r_\omega)\cap H^3(0,T;L^2_\omega), f \in C(0,T;H^r_\omega)$ *and* $1 \le r \le \alpha$, *then*

$$E\left(U_N - u_N, t\right) \le b_9 e^{b_{10}t}\left(\tau^4 + N^{-2r}\right),$$

b_8, b_9 *and* b_{10} *being positive constants depending on* μ *and the norms of* U *and* f *in the spaces mentioned above.*

In order to save work, we sometimes use explicit Chebyshev pseudospectral schemes for (4.63). For instance, let $R^0_N = R^0_N(\alpha,\beta), \alpha,\beta \ge 1$ be the filter defined by (2.137), and consider the following Chebyshev pseudospectral scheme

$$D_\tau u_N(x,t) + \frac{1}{2}I_N R_N \partial_x\left(R_N I_N u^2_N(x,t)\right) - \mu I_N R_N \partial^2_x R^0_N u_N(x,t) = I_N R_N I_N f(x,t). \tag{4.75}$$

If $\tau = \mathcal{O}\left(N^{-4}\right)$, then we can derive the same result of stability and convergence as in Theorems 4.12 and 4.13. The condition on τ is much more severe than the condition for the stability in the corresponding Fourier pseudospectral schemes, in which $\tau = \mathcal{O}\left(N^{-2}\right)$. However, the usefulness of the filters R_N and R^0_N can improve the results in practical computations. We check the effects of the filters in (4.75) by numerical experiments. For description of the accuracy, let

$$\begin{aligned}
\tilde{E}_2(u_N,t) &= \left(\frac{1}{N+1}\sum_{x\in\bar{\Lambda}_N}|U(x,t)-u_N(x,t)|^2\right)^{\frac{1}{2}},\\
\tilde{E}_\infty(u_N,t) &= \max_{x\in\bar{\Lambda}_N}|U(x,t)-u_N(x,t)|.
\end{aligned}$$

We take $f = 0$ and so (4.63) has a solution

$$U(x,t) = \frac{1}{2}\left(1 - \tanh\frac{2x-t}{8\mu}\right).$$

Set $\mu = 0.005, \tau = 0.0005$ and $N = 64$. To test the effects of the filters, various values of α and β are carried out. For $\alpha = \infty$ or $\beta = 0$, (4.75) becomes the usual Chebyshev pseudospectral scheme. In this case, $\tilde{E}_2(u_N, 0.25) = 1.302$ and $\tilde{E}_\infty(u_N, 0.25) = 3.907$. The numerical errors with other choices of α and β are presented in Tables 4.10 and 4.11. Evidently, the filters improve the accuracy. For N=64, a better choice appears to be $\alpha = 6 \sim 8, \beta = 2 \sim 3$.

Table 4.10. The errors $\tilde{E}_2(u_N, 0.25)$.

	$\alpha = 10$	$\alpha = 8$	$\alpha = 6$
$\beta = 1$	0.883	0.787	0.023
$\beta = 2$	0.393	0.016	0.018
$\beta = 3$	0.016	0.018	0.021

Table 4.11. The errors $\tilde{E}_\infty(u_N, 0.25)$.

	$\alpha = 10$	$\alpha = 8$	$\alpha = 6$
$\beta = 1$	3.409	3.283	0.102
$\beta = 2$	2.596	0.065	0.073
$\beta = 3$	0.065	0.071	0.082

The Chebyshev spectral methods and Chebyshev pseudospectral methods for generalized Burgers' equation are given in Ma and Guo (1988). The filtered schemes can be found in Ma and Guo (1994).

The Chebyshev spectral methods and Chebyshev pseudospectral methods have been used successfully for various partial differential equations. For instance, Maday and Quarteroni (1981) used it for steady Burgers' equation. Gottlieb (1981) first applied the properties of Gauss-type quadrature to studying the stability of Chebyshev pseudospectral methods for linear hyperbolic and parabolic equations. Gottlieb, Orszag and Turkel (1981) considered hyperbolic equations with variable coefficients.

4.4. Spectral Penalty Methods

We can deal with inhomogeneous boundary conditions in several ways. One of them is the spectral penalty method in which the boundary conditions become a part of equation, see Funaro and Gottlieb (1988,1991). In the last section, we introduced the Chebyshev pseudospectral method associated with the Chebyshev–Gauss–Lobatto interpolation points. Those points are chosen, because they allow the use of **FFT** in computations. For inhomogeneous boundary conditions, we can use Chebyshev penalty method as described below. Consequently, we adopt the weighted norm in numerical analysis. However, this is not a natural norm and complicates the analysis. A more natural way is to use Legendre–Gauss–Lobatto interpolation and the corresponding penalty method. As those points are not given explicitly, their evaluation for large N is not robust due to roundoff errors. In this section, we focus on a new approach, implementing Legendre method on Chebyshev points, enjoying the advantages of both Legendre and Chebyshev methods. It is named the Chebyshev–Legendre method by Don and Gottlieb (1994).

We begin with a simple problem. Let $\Lambda = (-1,1)$ and consider the hyperbolic equation

$$\begin{cases} \partial_t U(x,t) - \partial_x U(x,t) = f(x,t), & x \in \Lambda,\ t > 0, \\ U(1,t) = g(t), & t > 0, \\ U(x,0) = U_0(x), & x \in \bar{\Lambda}. \end{cases} \tag{4.76}$$

Let $x^{(j)}$ and $\omega^{(j)}$ be the Chebyshev–Gauss–Lobatto interpolation points and weights. $\bar{\Lambda}_N = \left\{x^{(j)} \mid 0 \le j \le N\right\}$. The usual semi-discrete Chebyshev pseudospectral scheme is to find $u_N(x,t) \in \mathbb{P}_N$ for $t \ge 0$ such that

$$\begin{cases} \partial_t u_N(x,t) - \partial_x u_N(x,t) = f(x,t), & x \in \bar{\Lambda}_N/\{1\},\ t > 0, \\ u_N(1,t) = g(t), & t > 0, \\ u_N(x,0) = U_0(x), & x \in \bar{\Lambda}_N. \end{cases} \tag{4.77}$$

Funaro and Gottlieb (1988) proposed a penalty method in which, instead of the boundary condition in (4.77), we require that

$$\partial_t u_N(1,t) - \partial_x u_N(1,t) + \lambda\left(u_N(1,t) - g(t)\right) = f(1,t)$$

where λ is determined from stability considerations. This approach is stable for the initial data as long as $\lambda \ge \frac{1}{2}N^2$. Since $x^{(j)}$ are the zeros of the polynomial

$(1-x^2)\,\partial_x T_N(x)$, this scheme can be written as

$$\begin{cases} \partial_t u_N(x,t) - \partial_x u_N(x,t) + \dfrac{\lambda(1+x)\partial_x T_N(x)}{2\partial_x T_N(1)}(u_N(1,t)-g(t)) = f(x,t), & x \in \bar{\Lambda}_N, t>0, \\ u_N(x,0) = U_0(x), & x \in \bar{\Lambda}_N. \end{cases} \tag{4.78}$$

The main feature of (4.78) is that the numerical solution does not fulfill the boundary condition exactly, but only in the limit as $N \to \infty$. In fact, the boundary condition is a part of the equation. Another penalty method based on Legendre–Gauss–Lobatto interpolation was served by Funaro and Gottlieb (1991). Let $\tilde{x}^{(j)}$ and $\tilde{\omega}^{(j)}$ be the nodes and weights of this interpolation. $\tilde{\Lambda}_N = \left\{ y^{(j)} = \tilde{x}^{(N-j)} \mid 0 \le j \le N \right\}$. Since $y^{(j)}$ are the zeros of the polynomial $(1-x^2)\,\partial_x L_N(x)$, the penalty scheme for (4.76) is to find $u_N(x,t) \in \mathbb{P}_N$ for $t \ge 0$ such that

$$\begin{cases} \partial_t u_N(x,t) - \partial_x u_N(x,t) + \dfrac{\lambda(1+x)\partial_x L_N(x)}{2\partial_x L_N(1)}(u_N(1,t)-g(t)) = f(x,t), & x \in \tilde{\Lambda}_N, t>0, \\ u_N(x,0) = U_0(x), & x \in \tilde{\Lambda}_N. \end{cases} \tag{4.79}$$

This scheme is also stable for the initial data when $\lambda \ge \frac{1}{2}N(N+1)$.

The Chebyshev–Legendre penalty scheme for (4.76) is to find $u_N(x,t) \in \mathbb{P}_N$ for $t \ge 0$ such that

$$\begin{cases} \partial_t u_N(x,t) - \partial_x u_N(x,t) + \dfrac{\lambda(1+x)\partial_x L_N(x)}{2\partial_x L_N(1)}(u_N(1,t)-g(t)) = f(x,t), & x \in \bar{\Lambda}_N, t>0, \\ u_N(x,0) = U_0(x), & x \in \bar{\Lambda}_N. \end{cases} \tag{4.80}$$

We now explore the relation between (4.79) and (4.80). The insight can be gained if one compares the differentiation matrix induced by (4.79) and the differentiation matrix induced by (4.80). As in the proof of Theorem 2.6, put

$$q_C(x) = \left(1-x^2\right)\partial_x T_N(x),$$
$$q_L(x) = \left(1-x^2\right)\partial_x L_N(x).$$

Define the Lagrange polynomials

$$g_l(x) = \frac{q_C(x)}{\left(x-x^{(l)}\right)\partial_x q_C\left(x^{(l)}\right)}, \quad 0 \le l \le N,$$
$$h_l(x) = \frac{q_L(x)}{\left(x-y^{(l)}\right)\partial_x q_L\left(y^{(l)}\right)}, \quad 0 \le l \le N.$$

For $v \in C(\bar{\Lambda})$, the Chebyshev interpolant $I_N v$ and the Legendre interpolant $\tilde{I}_N v$ are as follows

$$I_N v(x) = \sum_{l=0}^{N} v\left(x^{(l)}\right) g_l(x), \tag{4.81}$$

$$\tilde{I}_N v(x) = \sum_{l=0}^{N} v\left(y^{(l)}\right) h_l(x). \tag{4.82}$$

Denote by D_C the differentiation matrix in Chebyshev pseudospectral approximation, and by D_L the differentiation matrix in Legendre pseudospectral approximation. It means that

$$\partial_x I_N v\left(x^{(l)}\right) = \sum_{m=0}^{N} D_{C,l,m} v\left(x^{(m)}\right), \quad 0 \le l \le N, \tag{4.83}$$

$$\partial_x \tilde{I}_N v\left(y^{(l)}\right) = \sum_{m=0}^{N} D_{L,l,m} v\left(y^{(m)}\right), \quad 0 \le l \le N. \tag{4.84}$$

By virtue of (4.81) and (4.82),

$$D_{C,l,m} = \partial_x g_m\left(x^{(l)}\right), \quad D_{L,l,m} = \partial_x h_m\left(y^{(l)}\right), \quad 0 \le l,\, m \le N.$$

Let A and B be the $(N+1) \times (N+1)$ matrices with the elements

$$A_{l,m} = h_m\left(x^{(l)}\right), \quad B_{l,m} = g_m\left(y^{(l)}\right), \quad 0 \le l,\, m \le N.$$

Furthermore, denote by F the matrix induced by the differentiation and the penalty in (4.80), and denote by G the matrix induced by the differentiation and the penalty in (4.79). We have the following results.

Lemma 4.15. *We have $AB = I$ and $D_C = AD_LB$.*

Proof. Since $g_m(x) \in \mathbb{P}_N$, it can be expressed as

$$g_m(x) = \sum_{j=0}^{N} g_m\left(y^{(j)}\right) h_j(x). \tag{4.85}$$

Because $g_m\left(x^{(l)}\right) = \delta_{l,m}$,

$$\delta_{l,m} = \sum_{j=0}^{N} g_m\left(y^{(j)}\right) h_j\left(x^{(l)}\right) = \sum_{j=0}^{N} A_{l,j} B_{j,m}$$

whence $AB = I$. Next, differentiating (4.85) yields

$$\partial_x g_m(x) = \sum_{j=0}^{N} g_m\left(y^{(j)}\right)\partial_x h_j(x).$$

Since $\partial_x h_j(x) \in \mathbb{P}_N$, it can be expressed as

$$\partial_x h_j(x) = \sum_{k=0}^{N} \partial_x h_j\left(y^{(k)}\right) h_k(x).$$

Therefore

$$\begin{aligned} D_{C,l,m} = \partial_x g_m\left(x^{(l)}\right) &= \sum_{j=0}^{N}\sum_{k=0}^{N} g_m\left(y^{(j)}\right)\partial_x h_j\left(y^{(k)}\right) h_k\left(x^{(l)}\right) \\ &= \sum_{j=0}^{N}\sum_{k=0}^{N} A_{l,k} D_{L,k,j} B_{j,m} = (AD_L B)_{l,m}. \end{aligned}$$

■

Theorem 4.14. *We have $F = AGB$.*

Proof. The differentiation matrix G is essentially the matrix D_L introduced in (4.84), modified to take into account the boundary conditions, imposed via penalty in (4.79). Thus

$$G_{l,m} = D_{L,l,m} - \lambda\delta_{l,0}\delta_{m,0}.$$

In the same manner, we obtain from (4.80) that

$$F_{l,m} = D_{C,l,m} - \frac{\lambda\left(1 + x^{(l)}\right)\partial_x L_N\left(x^{(l)}\right)}{2\partial_x L_N(1)}\delta_{m,0}. \tag{4.86}$$

By Lemma 4.15, we deduce that

$$\begin{aligned} (AGB)_{l,m} &= (AD_L B)_{l,m} - \lambda\sum_{j=0}^{N}\sum_{k=0}^{N}\delta_{j,0}\delta_{k,0}h_j\left(x^{(l)}\right) g_m\left(y^{(k)}\right) \\ &= D_{C,l,m} - \lambda h_0\left(x^{(l)}\right) g_m\left(y^{(0)}\right). \end{aligned} \tag{4.87}$$

According to the definition of $h_0(x)$, a computation produces that

$$h_0\left(x^{(l)}\right) = \frac{\left(1 + x^{(l)}\right)\partial_x L_N\left(x^{(l)}\right)}{2\partial_x L_N(1)}.$$

Since $y^{(0)} = x^{(0)} = 1, g_m\left(y^{(0)}\right) = \delta_{m,0}$. Hence (4.87) is equal to the right-hand side of (4.86). Then the desired result follows. ■

We now turn to analyze the stability of scheme (4.80). By the theory in Chapter 3, it suffices to verify that the norm of u_N depends on the norms of data continuously. Let

$$(v,w)_N = \sum_{j=0}^{N} v\left(y^{(j)}\right) w\left(y^{(j)}\right) \tilde{\omega}^{(j)},$$

and $\|v\|_N = (v,v)_N^{\frac{1}{2}}$. Also set

$$\|v\|_N^* = \sum_{l=0}^{N}\sum_{m=0}^{N} H_{l,m} v\left(x^{(l)}\right) w\left(x^{(m)}\right)$$

where

$$H_{l,m} = \sum_{j=0}^{N} g_l\left(y^{(j)}\right) g_m\left(y^{(j)}\right) \tilde{\omega}^{(j)}.$$

For any $\phi \in \mathbb{P}_N$, we have

$$\phi(x) = I_N\phi(x) = \sum_{l=0}^{N} \phi\left(x^{(l)}\right) g_l(x)$$

and so

$$\phi\left(y^{(m)}\right) = \sum_{l=0}^{N} \phi\left(x^{(l)}\right) g_l\left(y^{(m)}\right).$$

Consequently,

$$\|\phi\|_N^* = \|\phi\|_N. \tag{4.88}$$

Theorem 4.15. *If* $\frac{1}{4}(1+\varepsilon)\left(N^2+N\right) \le \lambda \le c_1\left(N^2+N\right)$ *and* $\varepsilon > 0$, *then*

$$\|u_N(t)\|_N^2 \le e^{\varepsilon t}\left(\|u_N(0)\|_N^2 + \frac{1}{\varepsilon}\int_0^t \left(\|I_N f(s)\|_N^2 + 4c_1^2 g^2(s)\right) ds\right).$$

Proof. Since $u_N(x,t) \in \mathbb{P}_N$ for $t \ge 0$, (4.80) implies that for all $x \in \bar{\Lambda}$ and $t > 0$,

$$\partial_t u_N(x,t) - \partial_x u_N(x,t) + \frac{\lambda(1+x)\partial_x L_N(x)}{2\partial_x L_N(1)}\left(u_N(1,t) - g(t)\right) = I_N f(x,t). \tag{4.89}$$

By setting $x = y^{(j)}$ in (4.89), multiplying it by $2u_N\left(y^{(j)},t\right)\tilde{\omega}^{(j)}$ and summing the result for $0 \le j \le N$, we obtain that

$$\begin{aligned}&\partial_t\|u_N(t)\|_N^2 - 2\left(u_N(t), \partial_x u_N(t)\right)_N + 2\lambda\tilde{\omega}^{(0)} u_N^2(1,t)\\ &= 2\lambda\tilde{\omega}^{(0)} u_N(1,t) g(t) + 2\left(I_N f(t), u_N(t)\right)_N.\end{aligned}$$

By (2.37),

$$\begin{aligned} 2\left(u_N(t), \partial_x u_N(t)\right)_N &= 2\int_\Lambda u_N(x,t)\partial_x u_N(x,t)\,dx \\ &= u_N^2(1,t) - u_N^2(-1,t) \end{aligned}$$

and then

$$\begin{aligned} &\partial_t \|u_N(t)\|_N^2 + \left(2\lambda\tilde{\omega}^{(0)} - 1 - \varepsilon\right) u_N^2(1,t) + u_N^2(-1,t) \\ &\quad \le \varepsilon \|u_N(t)\|_N^2 + \frac{1}{\varepsilon}\|I_N f(t)\|_N^2 + \frac{\left(\lambda\tilde{\omega}^{(0)}\right)^2}{\varepsilon} g^2(t). \end{aligned}$$

Since $\tilde{\omega}^{(0)} = \dfrac{2}{N(N+1)}$, we integrate the above inequality to obtain

$$\|u_N(t)\|_N^2 \le \|u_N(0)\|_N^2 + \int_0^t \left(\varepsilon\|u_N(s)\|_N^2 + \frac{1}{\varepsilon}\|I_N f(s)\|^2 + \frac{4c_1^2}{\varepsilon} g^2(s)\right) ds$$

from which the conclusion follows. ∎

Theorem 4.15 shows the linear stability of scheme (4.80). In particular, if $f(x,t) = g(t) = 0$, then

$$\|u_N(t)\|_N^2 + \int_0^t \left(\left(\frac{4\lambda}{N(N+1)} - 1\right) u_N^2(1,s) + u_N^2(-1,s)\right) ds = \|u_N(0)\|_N^2.$$

Next, we consider the convergence of (4.80). Let $U_N = \tilde{I}_N U$. We get from (4.76) that

$$\begin{cases} \partial_t U_N(x,t) - \partial_x U_N(x,t) + \dfrac{\lambda(1+x)\partial_x L_N(x)}{2\partial_x L_N(1)}\left(U_N(1,t) - g(t)\right) \\ \qquad = G_1(x,t) + G_2(x,t) + I_N f(x,t), \quad x \in \bar{\Lambda},\ t > 0, \\ U_N(x,0) = I_N U_0(x) + G_3(x), \quad x \in \bar{\Lambda} \end{cases} \tag{4.90}$$

where

$$\begin{aligned} G_1(x,t) &= \tilde{I}_N \partial_x U(x,t) - \partial_x \tilde{I}_N U(x,t), \\ G_2(x,t) &= \tilde{I}_N f(x,t) - I_N f(x,t), \\ G_3(x,t) &= \tilde{I}_N U_0(x) - I_N U_0(x). \end{aligned}$$

Put $\tilde{U} = u_N - U_N$. Then we have from (4.80) and (4.90) that for all $x \in \bar{\Lambda}$ and $t > 0$,

$$\partial_t \tilde{U}(x,t) - \partial_x \tilde{U}(x,t) + \frac{\lambda(1+x)\partial_x L_N(x)}{2\partial_x L_N(1)}\tilde{U}(1,t) = -G_1(x,t) - G_2(x,t). \tag{4.91}$$

Since $\tilde{I}_N\partial_x U \in \mathbb{P}_N$ and $\partial_x\tilde{I}_N U \in \mathbb{P}_{N-1}$, we have from (2.60) that

$$\|G_1(t)\|_N^2 \le c\|G_1(t)\|^2 \le c\|\tilde{I}_N\partial_x U(t) - \partial_x U(t)\|^2 + c\|\partial_x\tilde{I}_N U(t) - \partial_x U(t)\|^2.$$

Furthermore, by using (2.63),

$$\|G_1(t)\|_N^2 \le cN^{-2r}\|U\|_{r+1}^2, \quad r \ge 0.$$

Similarly,

$$\|G_2(t)\|_N^2 \le c\|G_2(t)\|^2 \le c\|\tilde{I}_N f(t) - f(t)\|^2 + c\|I_N f(t) - f(t)\|_\omega^2.$$

By Theorem 2.19 and (2.63),

$$\|G_2(t)\|_N^2 \le cN^{-2r}\|f\|_{r,\omega}^2, \quad r > \frac{1}{2}.$$

Also, $\tilde{U}(x,0) = -G_3(x)$, and

$$\|G_3\|_N^2 \le cN^{-2r}\|U_0\|_{r,\omega}^2.$$

Finally we get the following result by comparing (4.91) with (4.89).

Theorem 4.16. *Let $\lambda \ge \frac{1}{4}N(N+1)$ and $r > \frac{1}{2}$. If $U \in L^2(0,T;H^{r+1}), U_0 \in H_\omega^r(\Lambda)$ and $f \in L^2(0,T;H_\omega^r)$, then for all $t \le T$,*

$$\|U_N - u_N\| \le b_1 N^{-r}$$

where b_1 is a positive constant depending only on the norms of U, U_0 and f in the mentioned spaces.

Now we apply the Chebyshev–Legendre penalty method to parabolic equations. Consider the problem

$$\begin{cases} \partial_t U(x,t) - \partial_x^2 U(x,t) = f(x,t), & x \in \Lambda,\ t > 0, \\ \alpha\partial_x U(1,t) + \beta U(1,t) = g^+(t), & t > 0, \\ -\gamma\partial_x U(-1,t) + \delta U(-1,t) = g^-(t), & t > 0, \\ U(x,0) = U_0(x), & x \in \bar{\Lambda} \end{cases} \tag{4.92}$$

where $\alpha, \beta, \delta, \gamma \ge 0, \alpha + \beta > 0$ and $\gamma + \delta > 0$. For the Chebyshev pseudospectral scheme, Gottlieb (1981) proved the stability with $\alpha = \gamma = 0$, and Gottlieb, Hussiani and Orszag (1984) proved the stability with $\beta = \delta = 0$. In order to describe the Chebyshev–Legendre penalty method, we introduce the operator A as

$$Av(x,t) = -\partial_x^2 v(x,t) + R(v,x,t)$$

where

$$\begin{aligned} R(v,x,t) &= \lambda_0 Q_0(x)\left(\alpha\partial_x v(1,t) + \beta v(1,t) - g^+(t)\right) \\ &\quad + \lambda_N Q_N(x)\left(-\gamma\partial_x v(-1,t) + \delta v(-1,t) - g^-(t)\right) \end{aligned}$$

with

$$Q_0(x) = \frac{(1+x)\partial_x L_N(x)}{2\partial_x L_N(1)}, \qquad Q_N(x) = \frac{(1-x)\partial_x L_N(x)}{2\partial_x L_N(-1)}.$$

The Chebyshev-Legendre penalty scheme for (4.92) is to seek $u_N(x,t) \in \mathbb{P}_N$ for $t \geq 0$ such that

$$\begin{cases} \partial_t u_N(x,t) - \partial_x^2 u_N(x,t) + R\left(u_N,x,t\right) = f(x,t), & x \in \bar{\Lambda}_N,\ t > 0, \\ u_N(x,0) = U_0(x), & x \in \bar{\Lambda}_N. \end{cases} \tag{4.93}$$

We first consider the stability. For simplifying the statements, let $\kappa = \dfrac{b\tilde{\omega}^{(0)}}{a}$ and

$$\begin{aligned} c(a,b) &= \frac{1}{a\tilde{\omega}^{(0)}}\left(1 + 2\kappa + 2\sqrt{\kappa^2+\kappa}\right), \\ d(a,b) &= \frac{1}{a\tilde{\omega}^{(0)}}\left(1 + 2\kappa - 2\sqrt{\kappa^2+\kappa}\right). \end{aligned}$$

Lemma 4.16. *Let $d(\alpha,\beta) \leq \lambda_0 \leq c(\alpha,\beta)$ and $d(\gamma,\delta) \leq \lambda_N \leq c(\gamma,\delta)$. Then for all $\phi \in \mathbb{P}_N$,*

$$(A\phi,\phi)_N \geq \sum_{j=1}^{N-1}\left(\partial_x\phi\left(y^{(j)}\right)\right)^2\tilde{\omega}^{(j)} + \lambda_0\tilde{\omega}^{(0)}\phi(1)g^+(t) + \lambda_N\tilde{\omega}^{(N)}\phi(-1)g^-(t).$$

Proof. Using (2.37), we derive that

$$\begin{aligned} -\sum_{j=0}^{N}\phi\left(y^{(j)}\right)\partial_x^2\phi\left(y^{(j)}\right)\tilde{\omega}^{(j)} &= -\int_\Lambda \phi(x)\partial_x^2\phi(x)\,dx \\ &= \|\partial_x\phi\|^2 - \phi(1)\partial_x\phi(1) + \phi(-1)\partial_x\phi(-1) \\ &= \sum_{j=1}^{N-1}\left(\partial_x\phi\left(y^{(j)}\right)\right)^2\tilde{\omega}^{(j)} + \left(\partial_x\phi(1)\right)^2\tilde{\omega}^{(0)} \\ &\quad + \left(\partial_x\phi(-1)\right)^2\tilde{\omega}^{(N)} - \phi(1)\partial_x\phi(1) + \phi(-1)\partial_x\phi(-1). \end{aligned}$$

Therefore

$$
\begin{aligned}
(A\phi,\phi)_N &= \sum_{j=1}^{N-1}\left(\partial_x\phi\left(y^{(j)}\right)\right)^2\tilde{\omega}^{(j)} + F(1,\alpha,\beta,0) + F(-1,\gamma,\delta,N)\\
&\quad - \lambda_0\tilde{\omega}^{(0)}\phi(1)g^+(t) - \lambda_N\tilde{\omega}^{(N)}\phi(-1)g^-(t)
\end{aligned}
$$

where

$$F(x,a,b,j) = x\left(a\lambda_j\tilde{\omega}^{(j)} - 1\right)\phi(x)\partial_x\phi(x) + b\lambda_j\tilde{\omega}^{(j)}\phi^2(x) + \tilde{\omega}^{(j)}\left(\partial_x\phi(x)\right)^2.$$

For $F(1,\alpha,\beta,0)$ to be non-negative, we need

$$\left(\alpha\lambda_0\tilde{\omega}^{(0)} - 1\right)^2 \le 4\beta\lambda_0\left(\tilde{\omega}^{(0)}\right)^2$$

or

$$\alpha^2\left(\tilde{\omega}^{(0)}\right)^2\lambda_0^2 - 2\left(\alpha + 2\beta\tilde{\omega}^{(0)}\right)\tilde{\omega}^{(0)}\lambda_0 + 1 \le 0,$$

i.e., $d(\alpha,\beta) \le \lambda_0 \le c(\alpha,\beta)$. Similarly $F(-1,\gamma,\delta,N) \ge 0$, if $d(\gamma,\delta) \le \lambda_N \le c(\gamma,\delta)$. The proof is complete. ∎

If $\alpha = 0, \beta = 1$, then the condition in Lemma 4.16 is reduced to that $\lambda_0 \ge \frac{1}{16}N^2(N+1)^2$. While for $\alpha = 1, \beta = 0$, we require that $\lambda_0 = \frac{1}{2}N(N+1)$. We now turn to the stability. For simplicity, we only consider the case $g^+(t) = g^-(t) = 0$.

Theorem 4.17. *Let λ_0 and λ_N satisfy the conditions in Lemma 4.16. If $g^+(t) = g^-(t) = 0$, then for any $\varepsilon > 0$,*

$$\|u_N(t)\|_N^2 + 2\int_0^t\sum_{j=1}^{N-1}\left(\partial_x u_N\left(y^{(j)},s\right)\right)^2\tilde{\omega}^{(j)}\,ds \le e^{\varepsilon t}\left(\|u_N(0)\|_N^2 + \frac{1}{\varepsilon}\int_0^t\|I_Nf(s)\|_N^2\right)ds.$$

Proof. Since $u_N(x,t) \in \mathbb{P}_N$ for $t \ge 0$, (4.93) implies that for all $x \in \Lambda$ and $t > 0$,

$$\partial_t u_N(x,t) - \partial_x^2 u_N(x,t) + R\left(u_N,x,t\right) = I_Nf(x,t), \tag{4.94}$$

i.e.,

$$\partial_t u_N(x,t) + Au_N(x,t) = I_Nf(x,t).$$

By Lemma 4.16,

$$\partial_t\|u_N(t)\|_N^2 + 2\sum_{j=1}^{N-1}\left(\partial_x u_N\left(y^{(j)},t\right)\right)^2\tilde{\omega}^{(j)} \le \varepsilon\|u_N(t)\|_N^2 + \frac{1}{\varepsilon}\|I_Nf(t)\|_N^2$$

from which the conclusion comes. ■

We next consider the convergence of scheme (4.93). For simplicity let $\alpha = \gamma = 0, \beta = \delta = 1$. Let $\mathbb{P}_N^{1,0}$ be the H_0^1-orthogonal projection upon $\mathbb{P}_N^0$ defined by (2.57), and $U_N = P_N^{1,0}U$. Then by (4.92) and (2.37),

$$\begin{cases} (\partial_t U_N(t), \phi)_N + (\partial_x U_N(t), \partial_x \phi)_N + (R(U_N, t), \phi)_N \\ \qquad = G_1(t) + G_2(t) + (I_N f(t), \phi)_N, & \forall\, \phi \in \mathbb{P}_N^0,\ t > 0, \\ U_N(0) = P_N^{1,0} U_0, & x \in \bar{\Lambda} \end{cases}$$

where

$$\begin{aligned} G_1(t) &= (\partial_t U_N(t), \phi)_N - (\partial_t U(t), \phi), \\ G_2(t) &= (f(t), \phi) - (f(t), \phi)_N + (f(t) - I_N f(t), \phi)_N. \end{aligned}$$

On the other hand, (4.94) and (2.37) imply that

$$(\partial_t u_N(t), \phi)_N + (\partial_x u_N(t), \partial_x \phi)_N + (R(u_N, t), \phi)_N = (I_N f(t), \phi)_N.$$

Set $\tilde{U} = u_N - U_N$. We obtain that

$$\begin{cases} \left(\partial_t \tilde{U}(t), \phi\right)_N + \left(\partial_x \tilde{U}(t), \partial_x \phi\right)_N + \left(R\left(\tilde{U}, t\right), \phi\right)_N = -G_1(t) - G_2(t), \\ \tilde{U}(0) = I_N U_0 - P_N^{1,0} U_0. \end{cases} \tag{4.95}$$

By Theorem 2.11, Theorem 2.19, (2.60) and (2.64), we know that for $r \geq 1$,

$$\begin{aligned} |G_1(t)| &\leq |(\partial_t U_N(t), \phi)_N - (\partial_t U_N(t), \phi)| + |(\partial_t U_N(t) - \partial_t U(t), \phi)| \\ &\leq cN^{-r} \|\phi\| \|\partial_t U(t)\|_r \leq \varepsilon \|\phi\|_N^2 + \frac{c}{\varepsilon} N^{-2r} \|\partial_t U(t)\|_r^2, \\ |G_2(t)| &\leq cN^{-r} \|\phi\| \|f(t)\|_{r,\omega} \leq \varepsilon \|\phi\|_N^2 + \frac{c}{\varepsilon} N^{-2r} \|f(t)\|_{r,\omega}^2. \end{aligned}$$

Also by Theorem 2.11, Theorem 2.19 and (2.60),

$$\begin{aligned} \left\|\tilde{U}(0)\right\|_N \leq c\left\|\tilde{U}(0)\right\| &\leq c\left(\|U_0 - I_N U_0\| + \|U_0 - P_N^{1,0} U_0\|\right) \\ &\leq cN^{-r} \|U_0\|_r. \end{aligned}$$

By taking $\phi = \tilde{U}$ in (4.95) and an argument as in the proof of Theorem 4.17, we reach the following aim.

Theorem 4.18. *Let $\alpha = \gamma = 0, \beta = \delta = 1, r \geq 1$ and $\lambda_0, \lambda_N \geq \frac{1}{16} N^2 (N+1)^2$. If $U \in H^1(0, t; H^r), U_0 \in H^r(\Lambda)$ and $f \in L^2(0, T; H_\omega^r)$, then for all $t \leq T$,*

$$\|U_N(t) - u_N(t)\| \leq b_2 N^{-r}$$

where b_2 is a positive constant depending only on the norms of U, U_0 and f in the mentioned spaces.

The Chebyshev–Legendre penalty method can be used for nonlinear problems. Let $f(z)$ be a continuous convex-function and $f'(z) = \partial_z f(z)$. Consider the problem

$$\partial_t U(x,t) + \partial_x f(U(x,t)) - \mu\partial_x^2 U(x,t) = f(x,t), \quad x \in \Lambda,\, 0 < t \le T$$

with $\mu \ge 0$ and the boundary conditions

$$\begin{array}{ll} U(1,t) = g^+(t), & \text{if } \mu > 0 \text{ or } \mu = 0, f'\left(g^+(t)\right) \le 0, \\ U(-1,t) = g^-(t), & \text{if } \mu > 0 \text{ or } \mu = 0, f'\left(g^-(t)\right) \ge 0. \end{array}$$

The corresponding Chebyshev–Legendre penalty method is as follows

$$\partial_t u_N(x,t) + \partial_x I_N f\left(u_N(x,t)\right) - \mu\partial_x^2 u_N(x,t) + R\left(u_N, x, t\right) = f(x,t), \quad x \in \bar{\Lambda}_N,\, t > 0$$

with

$$R(v,x,t) = b_0(t)\lambda_0 Q_0(x)\left(v(1,t) - g^+(t)\right) + b_N(t)\lambda_N Q_N(x)\left(v(-1,t) - g^-(t)\right)$$

where $b_0(t) = 1$ for $\mu > 0$ or $\mu = 0, f'\left(g^+(t)\right) \le 0$, and $b_0(t) = 0$ otherwise; and $b_N(t) = 1$ for $\mu > 0$ or $\mu = 0, f'\left(g^-(t)\right) \ge 0$, and $b_N(t) = 0$ otherwise.

Finally we list some numerical results. Let τ be the mesh size in t. We discretize (4.78)–(4.80) by the third order Heun–Runge–Kutta method in time. Then for $x \in \Lambda_N^*$ and $t \in \bar{R}_\tau(T)$,

$$\begin{aligned} u_N^*(x,t) &= u_N(x,t) + \frac{\tau}{3}\left(Du_N(x,t) - \lambda Q(x)\left(u_N(x,t) - g(t)\right)\right), \\ u_N^{**}(x,t) &= u_N(x,t) + \frac{2\tau}{3}\left(Du_N^*(x,t) - \lambda Q(x)\left(u_N^*(x,t) - g(t) - \frac{\tau}{3}\partial_t g(t)\right)\right), \\ u_N(x,t+\tau) &= \frac{1}{4}u_N(x,t) + \frac{3}{4}u^*(x,t) + \frac{3\tau}{4}\left(Du_N^{**}(x,t) - \lambda Q(x)\left(u_N^{**}(x,t) \right.\right. \\ &\qquad \left.\left. - g(t) - \frac{\tau}{3}\partial_t g(t) - \frac{2}{9}\tau^2\partial_t^2 g(t)\right)\right) \end{aligned}$$

where $\Lambda_N^* = \Lambda_N, D = D_C$ for (4.78) and (4.80), and $\Lambda_N^* = \tilde{\Lambda}_N, D = D_L$ for (7.79), and

$$Q(x) = \begin{cases} \dfrac{(1+x)\partial_x T_N(x)}{2\partial_x T_N(1)}, & \text{for (7.78)}, \\ \dfrac{(1+x)\partial_x L_N(x)}{2\partial_x L_N(1)}, & \text{for (7.79) and (7.80)}. \end{cases}$$

For scheme (4.77), $\lambda = 0$ and $u_N(1,t) = g(t)$. In all calculations, we take $q = \tau N^2, \lambda = 2\chi\tilde{\omega}^{(0)}$. The test function is

$$U(x,t) = \sin(2\pi(x+t)).$$

Table 4.12 shows the L^2-errors between $U(x,t)$ and the numerical solution of (4.77). Clearly large time step τ is allowed. Table 4.13 shows the L^2-errors between $U(x,t)$ and the solution of (4.80) with $\chi = 2$. It indicates the advantage of scheme (4.80). Table 4.14 shows the L^2-errors between $U(x,t)$ and the solution of (4.80) with different values of χ. It is found that the accuracy with $\chi \geq 1$ is very good. While the computation is unstable for $\chi < 1$. It confirms the theoretical prediction.

Table 4.12. The L^2-errors of scheme (4.77).

	$q=8$	$q=4$	$q=1$
$N=16$	$8.2E-4$	$1.0E-4$	$2.9E-6$
$N=32$	$1.5E-5$	$1.8E-6$	$2.8E-8$
$N=64$	$4.2E-7$	$4.9E-8$	$7.2E-10$
$N=128$	$1.7E-8$	$1.9E-9$	$2.8E-11$

Table 4.13. The L^2-errors of scheme (4.80) with $\chi = 2$.

	$q=8$	$q=4$	$q=1$
$N=16$	$4.7E-4$	$6.0E-5$	$2.8E-6$
$N=32$	$7.4E-6$	$9.3E-7$	$1.5E-8$
$N=64$	$1.2E-7$	$1.5E-8$	$2.3E-10$
$N=128$	$1.8E-9$	$2.3E-10$	$3.6E-12$

Table 4.14. The L^2-errors of scheme (4.80) with $q = 1$.

	$\chi=8$	$\chi=2$	$\chi=1$	$\chi=0.9$
$N=16$	$3.1E-6$	$2.8E-6$	$6.5E-6$	$8.3E-6$
$N=32$	$1.5E-8$	$1.5E-8$	$1.5E-8$	$1.5E-8$
$N=64$	$2.3E-10$	$2.3E-10$	$2.3E-10$	$2.3E-7$
$N=128$	$3.6E-12$	$3.6E-12$	$3.6E-12$	unstable

4.5. Spectral Vanishing Viscosity Methods

In this section, we study the spectral methods for nonlinear conservation laws whose solutions may develop spontaneous jump discontinuities, i.e., shock waves. Hence weak solutions must be admitted. There are many weak solutions for the same problem usually. So we have to single out the unique "physically relevant" solution satisfying the entropy condition. We shall describe the spectral vanishing viscosity methods for nonlinear conservation laws such that the numerical solutions converge to weak solutions fulfilling the entropy condition. Such solutions are called the entropy solutions.

Let $\mathbb{R}_x = \{x \mid -\infty < x < \infty\}, \mathbb{R}_t^+ = \{t \mid t > 0\}$ and $Q = \mathbb{R}_x \times \mathbb{R}_t^+$. Let D be a set in $\mathbb{R}^2$. If $\bar{D}_1$ is a compact set in D, then we say that $D_1 \subset\subset D$. Moreover, if $v(x,t)$ is defined on D almost everywhere, then we say that $v(x,t)$ is defined on $D, a.e.$. If a function $v(x,t)$ is defined on $D, a.e.$, and for any measurable subset $D_1 \subset\subset D, v \in L^q(D_1), 1 \le q \le \infty$, then it is denoted by $v \in L^q_{\rm loc}(D)$. We can define the spaces $H^r_{\rm loc}(D)$ and $W^{r,q}_{\rm loc}(D)$ similarly. Furthermore, if v_l is a sequence such that

$$\lim_{l\to\infty} \iint_D v_l(x,t)w(x,t)\,dxdt = \iint_D v(x,t)w(x,t)\,dxdt, \quad \forall\, w \in L^1(D),$$

then we say that v_l converges to v in $L^\infty(D)$ weak-star. Besides, let $\mathcal{D}(Q)$ be the space involving all infinitely differentiable functions vanishing for sufficiently large $|x| + t$.

We now discuss the nonlinear conservation laws. Let $f(z)$ be a convex function in $C^1(\mathbb{R})$, and $f'(z) = \partial_z f(z)$. Consider the problem

$$\begin{cases} \partial_t U(x,t) + \partial_x f(U(x,t)) = 0, & (x,t) \in Q, \\ U(x,0) = U_0(x), & x \in \mathbb{R}_x. \end{cases} \tag{4.96}$$

A weak solution of (4.96) is a bounded measurable function $U(x,t)$ such that

$$\iint_{\bar{Q}} \left(U(x,t)\partial_t w(x,t) + f(U(x,t))\partial_x w(x,t)\right) dxdt + \int_{\mathbb{R}_x} U_0(x)w(x,0)\,dx = 0, \quad \forall\, w \in \mathcal{D}(Q). \tag{4.97}$$

Next let $E(z)$ be certain smooth function in $\mathbb{R}^1$, and

$$F(z) = \int E'(z)f'(z)\,dz.$$

If for any smooth solution of (4.96), there holds

$$\partial_t E(U(x,t)) + \partial_x F(U(x,t)) = 0, \quad (x,t) \in Q,$$

then we say that E and F are a pair of entropy and entropy flux for (4.96). If for any strictly convex entropy $E(z)$, a weak solution $U(x,t)$ satisfies

$$\partial_t E(U(x,t)) + \partial_x F(U(x,t)) \leq 0, \quad (x,t) \in Q, \tag{4.98}$$

then we say that $U(x,t)$ is an admissible solution. It can be verified that if both $U(x,t)$ and $V(x,t)$ are admissible solutions of (4.96), and

$$\lim_{t\to 0^+} U(x,t) = \lim_{t\to 0^+} V(x,t), \quad \text{in } L^1_{\text{loc}}(Q),$$

then $U(x,t) = V(x,t)$ in $\bar{Q}$, *a.e.*.

There are fruitful results for the existence of weak solutions, see Lax (1972). Tartar (1975) used compensated compactness to study (4.96). The key point is to explore the conditions with which a sequence $\{v_l\}$ tends to a weak solution as it tends to v in $L^\infty_{\text{loc}}(Q)$ weak-star, or in other words, $f(v_l)$ tends to $f(v)$. For this purpose, the following lemma plays an important role, see Tartar (1975). Let $M(D)$ be a space of Borel measures.

Lemma 4.17. *Let D be a bounded open set in $\mathbb{R}^2$ and $f(z) \in C^1(\mathbb{R}^1)$. If $v_l \to v$ in $L^\infty(D)$ weak-star, and for all convex function $E(z), \partial_t E(v_l) + \partial_x F(v_l)$ are in a compact set of $H^{-1}(D)$ or a bounded set of $M(D)$, then*

$$\begin{aligned} f(v_l) &\longrightarrow f(v), \quad \text{in } L^\infty(D) \text{ weak-star}, \\ f'(v_l) &\longrightarrow f'(v), \quad \text{in } L^q(D) \text{ strongly for all } 1 \leq q < \infty. \end{aligned}$$

Moreover, if there is no interval on which f is affine, then

$$v_l \longrightarrow v, \quad \text{in } L^q(D) \text{ strongly for all } 1 \leq q < \infty.$$

We can take $L^1(D)$ as a space of measures. Other two commonly used results from compensated compactness are due to Murat and Tartar, see Tartar (1975).

Lemma 4.18. *Let D be a bounded open set in $\mathbb{R}^2$, and the sequence $\{v_l\}$ be in a bounded set of $W^{-1,\infty}(D)$. If $\{v_l\}$ is also in a union of a compact set of $H^{-1}(D)$ and a bounded set of $M(D)$, then it belongs to a compact set of $H^{-1}_{\text{loc}}(D)$.*

Let $\boldsymbol{v}$ be a function vector and $\boldsymbol{v} = \left(v^{(1)}, v^{(2)}\right)$. Set $\operatorname{div} \boldsymbol{v} = \partial_1 v^{(1)} + \partial_2 v^{(2)}$ and $\operatorname{curl} \boldsymbol{v} = \partial_1 v^{(2)} - \partial_2 v^{(1)}$.

Lemma 4.19. *Let D be a bounded open set in $\mathbb{R}^2$. If $\boldsymbol{v}_l \to \boldsymbol{v}, \boldsymbol{w}_l \to \boldsymbol{w}$ in $L^2(D) \times L^2(D)$ weakly, and $\operatorname{div} \boldsymbol{v}_l \to \operatorname{div} \boldsymbol{v}, \operatorname{curl} \boldsymbol{w}_l \to \operatorname{curl} \boldsymbol{w}$ in $H^{-1}(D) \times H^{-1}(D)$ strongly, then $\boldsymbol{v}_l \cdot \boldsymbol{w}_l \to \boldsymbol{v} \cdot \boldsymbol{w}$ in $\mathcal{D}'(D)$.*

We first consider periodic problems. Let $\Lambda = (0, 2\pi)$ and $Q = \Lambda \times \mathbb{R}_t^+$. Denote by $\mathcal{D}_p(Q)$ the space involving all infinitely differentiable functions with the period 2π for the variable x, vanishing for sufficiently large t. Let $D = \Lambda \times \{t \mid 0 < t_1 < t < t_2\}$. Define the spaces $L_p^\infty(Q), L_{p,\mathrm{loc}}^\infty(Q)$ and $H_{p,\mathrm{loc}}^r(Q)$ similarly. The periodic nonlinear conservation laws are of the form

$$\begin{cases} \partial_t U_(x,t) + \partial_x f(U(x,t)) = 0, & (x,t) \in Q, \\ U(x,t) = U(x + 2\pi, t), & (x,t) \in Q, \\ U(x,0) = U_0(x), & x \in \bar{\Lambda}. \end{cases} \tag{4.99}$$

The weak solutions of (4.99) are bounded measurable functions such that

$$\iint_{\bar{Q}} (U(x,t)\partial_t w(x,t) + f(U(x,t))\partial_x w(x,t))dxdt + \int_\Lambda U_0(x) w(x,0) = 0, \quad \forall\, w \in \mathcal{D}_p(Q). \tag{4.100}$$

The admissible solutions are also identified by (4.98). In this case, Lemmas 4.17 –4.19 are also valid with slight modifications. For instance, $L^p(D)$ and $W^{r,q}(D)$ are replaced by $L_p^q(D)$ and $W_p^{r,q}(D)$, etc..

We are going to construct the Fourier spectral scheme. Let P_N be the L^2-orthogonal projection upon V_N as defined in Section 2.2. The standard Fourier spectral scheme is to find $u_N(x,t) \in V_N$ for $t \geq 0$ such that

$$\begin{cases} \partial_t u_N(x,t) + \partial_x P_N f(u_N(x,t)) = 0, & (x,t) \in Q, \\ u_N(x,0) = P_N U_0(x), & x \in \bar{\Lambda}. \end{cases}$$

Let I be the identity operator, then this scheme stands for

$$\partial_t u_N(x,t) + \partial_x f(u_N(x,t)) = (I - P_N)\partial_x f(u_N(x,t)).$$

Let us multiply the above equation by $u_N(x,t)$. We find after integration that

$$\|u_N(t)\|^2 = \|u_N(0)\|^2 \leq \|U_0\|^2. \tag{4.101}$$

This yields the existence of a weak solution U, which is the weak limit of u_N in $L^2(0,T;L^2)$ for any $t>0$. Unfortunately it does not solve our problem. In fact, if U is a weak solution, then $P_N f(u_N)$, and hence $f(u_N)$ should tend to $f(U)$ weakly. If in addition $f(z)=\frac{1}{2}z^2$, then u_N^2 tends to U^2 weakly and so u_N converges to U in $L^2(0,T;L^2)$ strongly. Therefore by (4.101), $\|U(t)\|^2$ is also conserved in time. However the inequality (4.98) is valid strictly at the locations of shock waves. It leads to a contradiction. Hence the standard Fourier spectral methods do not work for the problems with shock waves. This fact motivates us to modify the above scheme. Tadmor (1989) proposed the following Fourier spectral vanishing viscosity scheme

$$\partial_t u_N(x,t)+\partial_x P_N f\left(u_N(x,t)\right)=\varepsilon\partial_x\left((I-P_M)\partial_x u_N(x,t)\right) \tag{4.102}$$

where $\varepsilon=\varepsilon(N)\to 0$ as $N\to\infty$, and $M=M(N)<N, M(N)\to\infty$ as $N\to\infty$.

We shall establish some a priori estimates for u_N. Let

$$\varepsilon(N)\sim cN^{-\alpha},\quad M(N)\sim cN^{\beta},\quad 0<2\beta<\alpha\le 1. \tag{4.103}$$

We multiply (4.102) by u_N and integrate over Λ. Since

$$\begin{aligned}\left(\partial_x P_N f\left(u_N(x,t)\right),u_N(t)\right) &= -\left(P_N f\left(u_N(x,t)\right),\partial_x u_N(t)\right)\\ &= -\left(f\left(u_N(x,t)\right),\partial_x u_N(t)\right)=0,\end{aligned}$$

we obtain

$$\partial_t\|u_N(t)\|^2+2\varepsilon\|(I-P_M)\partial_x u_N(t)\|^2=0$$

and so

$$\|u_N(t)\|^2+2\varepsilon\int_0^t\|(I-P_M)\partial_x u_N(s)\|^2\,ds=\|u_N(0)\|^2\le\|U(0)\|^2.$$

Let c_0 be a positive constant depending only on $\|U(0)\|^2$. Then for any fixed $T>0$,

$$\|u_N\|_{L^\infty(0,T;L^2)}\le c_0,\quad \varepsilon\|(I-P_M)\partial_x u_N\|_{L^2(0,T;L^2)}\le c_0. \tag{4.104}$$

Furthermore, $P_M\partial_x u_N=\partial_x P_M u_N$ and $P_M u_N\in\mathbb{P}_M$. So by Theorem 2.2,

$$\begin{aligned}\|\partial_x u_N(t)\|^2 &= \|(I-P_M)\partial_x u_N(t)\|^2+\|P_M\partial_x u_N(t)\|^2\\ &\le \|(I-P_M)\partial_x u_N(t)\|^2+M^2\|u_N(t)\|^2\\ &\le \|(I-P_M)\partial_x u_N(t)\|^2+cN^{2\beta}\|u_N(t)\|^2.\end{aligned}$$

By using (4.103) and (4.104), we assert that

$$\|\partial_x u_N\|^2_{L^2(0,T;L^2)} \le \frac{cc_0}{\varepsilon}. \tag{4.105}$$

Next, multiplying (4.102) by $\partial_t u_N$ and integrating the result over Λ, we get that

$$\begin{aligned}\|\partial_t u_N(t)\|^2 + \frac{\varepsilon}{2}\partial_t\|(I-P_M)\partial_x u_N(t)\|^2 &\le \frac{1}{2}\|\partial_t u_N(t)\|^2 + \frac{1}{2}\|\partial_x P_N f(u_N(t))\|^2\\ &\le \frac{1}{2}\|\partial_t u_N(t)\|^2 + \frac{1}{2}\|f'(u_N(t))\,\partial_x u_N(t)\|^2.\end{aligned}$$

Then the integration of the above formula from 0 to t yields

$$\begin{aligned}&\|\partial_t u_N\|^2_{L^2(0,t;L^2)} + \varepsilon\|(I-P_M)\partial_x u_N(t)\|^2\\ &\le \varepsilon\|(I-P_M)\partial_x u_N(0)\|^2 + \|f'(u_N)\|^2_{L^\infty(0,t;\Lambda)}\int_0^t \|\partial_x u_N(s)\|^2\,ds.\end{aligned} \tag{4.106}$$

Now assume that there exists a positive constant such that

$$\varepsilon\|\partial_x u_N(0)\|^2 \le A, \quad \|u_N\|_{L^\infty(0,T;L^\infty)} \le A, \tag{4.107}$$

and denote by c_A a positive constant depending on A. Then (4.105)–(4.107) lead to

$$\|\partial_t u_N\|^2_{L^2(0,T;L^2)} \le c_A\left(\frac{c_0}{\varepsilon}+1\right). \tag{4.108}$$

We are going to derive the convergence of scheme (4.102) by a compensated compactness argument. Let $D = \Lambda\times(0,T)$ and

$$(v,w)_D = \int_0^T\int_\Lambda v(x,t)w(x,t)\,dxdt, \quad \|v\|_D = (v,v)_D^{\frac{1}{2}}.$$

Then by (4.102), for any $w\in H_0^1(0,T;H_p^1)$,

$$(\partial_t u_N + \partial_x f(u_N), w)_D = G_1(w) + G_2(w)$$

where

$$\begin{aligned}G_1(w) &= (\partial_x(I-P_N)f(u_N), w)_D\,,\\ G_2(w) &= \varepsilon\,(\partial_x(I-P_M)\partial_x u_N, w)_D\,.\end{aligned}$$

By Theorem 2.3, (4.103) and (4.105),

$$\begin{aligned}|G_1(w)| &= |((I-P_N)f(u_N),\partial_x w)_D| \le \frac{c}{N}\|f'(u_N)\partial_x u_N\|_D\|\partial_x w\|_D\\ &\le \frac{c_0 c_A}{\sqrt{\varepsilon}N}\|\partial_x w\|_D \le c_0 c_A N^{\frac{\alpha}{2}-1}\|\partial_x w\|_D.\end{aligned}$$

On the other hand, (4.104) leads to

$$\begin{aligned} |G_2(w)| &\leq \varepsilon\,|((I-P_M)\partial_x u_N, \partial_x w)_D| \leq \varepsilon\|(I-P_M)\partial_x u_N\|_D\|\partial_x w\|_D \\ &\leq c_0\sqrt{\varepsilon}\|\partial_x w\|_D \leq c_0 N^{-\frac{\alpha}{2}}\|\partial_x w\|_D. \end{aligned}$$

Therefore

$$|(\partial_t u_N + \partial_x f(u_N), w)_D| \leq c_0\left(c_A N^{\frac{\alpha}{2}-1} + N^{-\frac{\alpha}{2}}\right)\|\partial_x w\|_D. \qquad (4.109)$$

Since $\alpha \leq 1$, the right-hand side of the above inequality tends to zero as $N \to \infty$. It implies that $\partial_t u_N + \partial_x f(u_N)$ belongs to a compact subset of $H^{-1}_{\text{loc}}(Q)$.

Next, let $E(z)$ be any strictly convex entropy and $F(z)$ be the corresponding entropy flux. We repeat the above procedure with $E'(u_N)w$. Then we deduce from (4.109) that for any $w \in H^1_0(0,T;H^1_p)$,

$$\begin{aligned} & |(\partial_t E(u_N) + \partial_x F(u_N), w)_D| \\ = & |(\partial_t u_N + \partial_x f(u_N), E'(u_N)w)_D| \\ \leq & c_0\left(c_A N^{\frac{\alpha}{2}-1} + N^{-\frac{\alpha}{2}}\right)\|\partial_x(E'(u_N)w)\|_D \\ \leq & c_0\left(c_A N^{\frac{\alpha}{2}-1} + N^{-\frac{\alpha}{2}}\right)\left(\|E''(u_N)\partial_x u_N\|_D\|w\|_{L^\infty(D)} + \|E'(u_N)\partial_x w\|_D\right). \end{aligned}$$

By using (4.103) and (4.105) again, we find that

$$|(\partial_t E(u_N) + \partial_x F(u_N), w)_D| \leq B_1 + B_2$$

where

$$\begin{aligned} B_1 &= c_0 c_A\left(c_A N^{\alpha-1} + c_0\right)\|w\|_{L^\infty(D)}, \\ B_2 &= c_0 c_A\left(c_A N^{\frac{\alpha}{2}-1} + N^{-\frac{\alpha}{2}}\right)\|\partial_x w\|_D. \end{aligned}$$

Since $\alpha \leq 1, B_1$ is bounded. Clearly $B_2 \to 0$ as $N \to \infty$. Thus $\partial_t E(u_N) + \partial_x F(u_N)$ is composed of two terms, each of them is in a compact subset of $H^{-1}_{\text{loc}}(Q)$ or a bounded subset of $L^1_{\text{loc}}(Q)$. By Lemma 4.18, it is in a compact subset of $H^{-1}_{\text{loc}}(Q)$. Furthermore Lemma 4.17 ensures that u_N converges to U^*, a weak solution of (4.99) in $L^q_{\text{loc}}(Q)$ strongly for all $1 \leq q < \infty$.

Now, we check the entropy condition. Let $\alpha < 1$. We first have that

$$(\partial_t E(u_N) + \partial_x F(u_N), w)_D = G_3(w) + G_4(w)$$

where

$$G_3(w) = -\left((I - P_N)f(u_N), \partial_x\left(E'(u_N)w\right)\right)_D,$$
$$G_4(w) = -\varepsilon\left((I - P_M)\partial_x u_N, \partial_x\left(E'(u_N)w\right)\right)_D.$$

By Theorem 2.3, (4.103) and (4.105), we get that

$$\begin{aligned}|G_3(w)| &\le \frac{c_A}{N}\|\partial_x u_N\|_D\|\partial_x(E'(u_N)w)\|_D \le \frac{c_0 c_A}{\sqrt{\varepsilon}N}\|\partial_x(E'(u_N)w)\|_D \\ &\le c_0 c_A N^{\frac{\alpha}{2}-1}\left(\|\partial_x u_N\|_D\|w\|_{L^\infty(D)} + \|E'(u_N)\|_{L^\infty(D)}\|\partial_x w\|_D\right).\end{aligned}$$

Thus $G_3 \to 0$, as $N \to \infty$. Furthermore, we have

$$G_4(w) = G_4^{(1)}(w) + G_4^{(2)} + G_4^{(3)}(w)$$

with

$$G_4^{(1)}(w) = -\varepsilon\left(\partial_x u_N, E''(u_N)w\partial_x u_N\right)_D,$$
$$G_4^{(2)}(w) = -\varepsilon\left(\partial_x u_N, E'(u_N)\partial_x w\right)_D,$$
$$G_4^{(3)}(w) = \varepsilon\left(P_M\partial_x u_N, \partial_x\left(E'(u_N)w\right)\right)_D.$$

By taking $w \ge 0$ and using the convexity of E, we know that $G_4^{(1)}(w) \le 0$. By (4.105),

$$\left|G_4^{(2)}(w)\right| \le c_A\sqrt{\varepsilon}\|\partial_x w\|_D \longrightarrow 0, \quad \text{as } N \longrightarrow \infty.$$

Because $P_M u_N \in \mathbb{P}_M$, we have from Theorem 2.2 and (4.105) that

$$\begin{aligned}\left|G_4^{(3)}(w)\right| &\le \varepsilon\|\partial_x P_M u_N\|_D\left(\|E''(u_N)w\partial_x u_N\|_D + \|E'(u_N)\partial_x w\|_D\right) \\ &\le c_A\varepsilon M\left(\frac{1}{\sqrt{\varepsilon}}\|w\|_{L^\infty(D)} + \|\partial_x w\|_D\right) \\ &\le c_A N^{\beta-\frac{\alpha}{2}}\left(\|w\|_{L^\infty(D)} + \|\partial_x w\|_D\right).\end{aligned}$$

Since $2\beta < \alpha$, $\left|G_4^{(3)}(w)\right| \to 0$ as $N \to \infty$. Therefore U^* fulfills (4.98) in the sense of distribution. The above statements lead to the following conclusion.

Theorem 4.19. *Let (4.103) and (4.107) hold. Then for $1 \le q < \infty$, the solution of (4.102) tends to a weak solution of (4.99) in $L^q_{\mathrm{loc}}(\bar{Q})$ strongly. If in addition $\alpha < 1$, then it is the unique entropy solution.*

For the existence of weak solution of (4.99) with $f(z) = \frac{1}{2}z^2$, we only need to check that both $\partial_t u_N + \frac{1}{2}\partial_x u_N^2$ and $\frac{1}{2}\partial_t u_N^2 + \frac{1}{3}\partial_x u_N^3$ belong to a union of a compact subset of $H_{\mathrm{loc}}^{-1}(Q)$ and a bounded subset of $L_{\mathrm{loc}}^1(Q)$. In this case, let U_j^* be the weak limits of u_N^j. Then by Lemma 4.19,

$$U_4^* = 4U_1^*U_3^* - 3\left(U_2^*\right)^2. \tag{4.110}$$

Moreover

$$\left(u_N - U_1^*\right)^4 \longrightarrow U_4^* - 4U_3^*U_1^* + 6U_2^*\left(U_1^*\right)^2 - 4\left(U_1^*\right)^4 + \left(U_1^*\right)^4, \text{ weakly,}$$

from which and (4.110),

$$\left(u_N - U_1^*\right)^4 \longrightarrow -3\left(U_2^* - \left(U_1^*\right)^2\right)^2 \leq 0, \text{ weakly.}$$

So $U_2^* = \left(U_1^*\right)^2$. Consequently, u_N converges to U_1^* in $L_{\mathrm{loc}}^q(\bar{Q})$ strongly and so it is a weak solution of (4.99) with $f(z) = \frac{1}{2}z^2$, see Tadmor (1989).

In actual computations, we can take the vanishing viscosity term as $\varepsilon\partial_x\left(q_N\partial_x u_N\right)$ where

$$q_N(x,t) = \sum_{0\leq|l|\leq N} \hat{q}_l(t)e^{ilx}$$

and $\varepsilon = \varepsilon(N) \to 0, M = M(N) \to \infty$ as $N \to \infty$. For instance, $M \sim 2N^{\frac{1}{2}}$ and

$$\hat{q}_l = \begin{cases} 0, & |l| \leq M, \\ 1, & |l| > M. \end{cases} \tag{4.111}$$

Tadmor (1990) solved (4.99) with $f(z) = \frac{1}{2}z^2$ and $U_0(x) = \sin x$ by the Fourier spectral method and the Fourier vanishing viscosity method with $M = 8, N = 32, \varepsilon = 0.31$ and (4.111). They are integrated in time up to $t = 1.5$, using the fourth order Runge–Kutta procedure. In the figures, the real lines present the same exact solution of (4.99), while the dotted curves present the numerical ones. Figure 4.1 shows the results of Fourier spectral method, and Figure 4.2 shows the results of Fourier vanishing viscosity method. Obviously the latter provides much better numerical approximation to the shock waves. In Figure 4.3, the improved results are obtained by using C^∞-viscosity coefficients $\hat{q}_l$ connecting the wave numbers in the inviscid region $l < M(N)$ and the highest wave numbers $|l| \sim N$. This technique prevents the propagation of the Gibbs phenomenon into the whole computational domain. Figure 4.4 shows the numerical solutions of Fourier vanishing viscosity scheme after it is post-processed by the spectrally accurate smoothing procedure discussed in Gottlieb and Tadmor (1985).

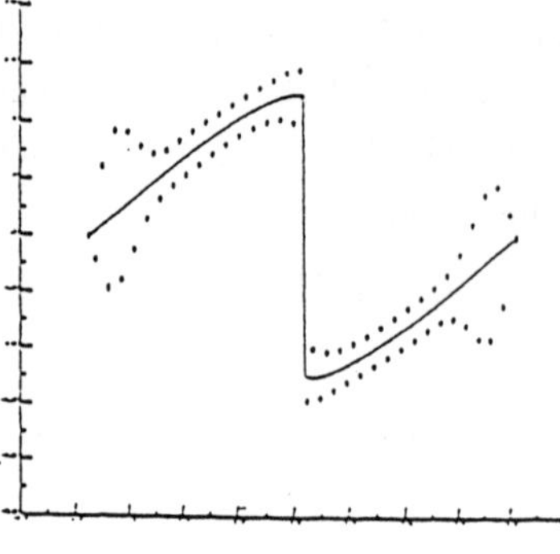

Figure 4.1.

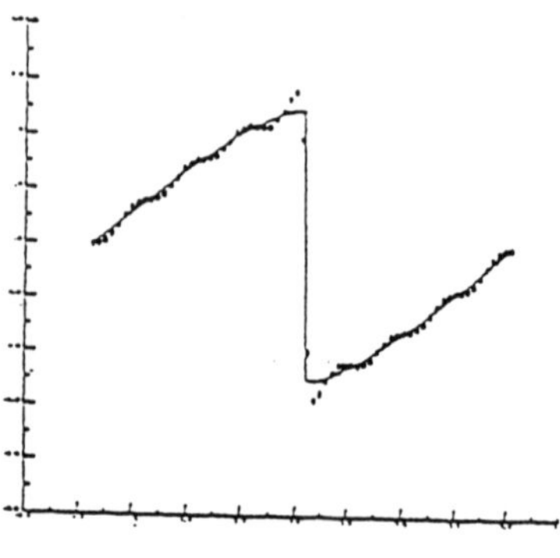

Figure 4.2.

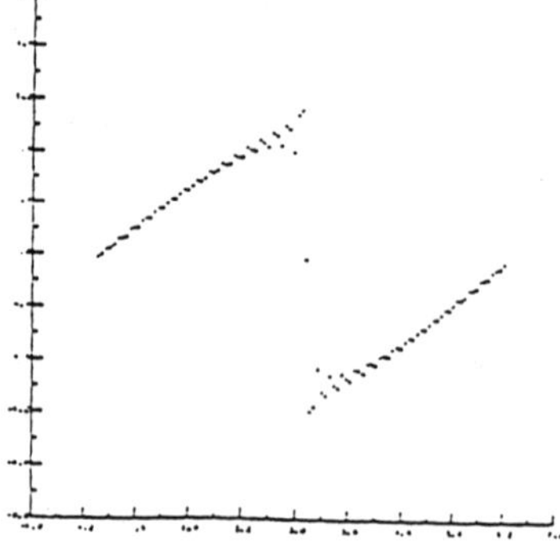

Figure 4.3.

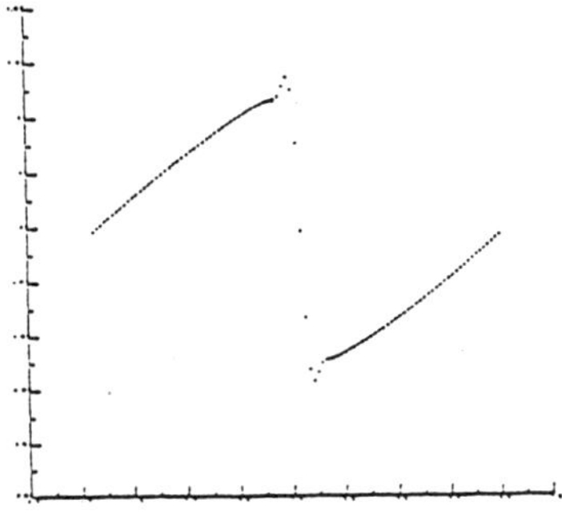

Figure 4.4.

The total variations of Fourier viscosity schemes and their stabilities and convergences are discussed in Maday and Tadmor (1989), and Tadmor (1991,1993a). In particular, Chen, Du and Tadmor (1993) considered multi-dimensional scalar conservation laws, and proved that for certain Fourier vanishing viscosity schemes, the temporal L^∞-norms of the solutions depend on the L^∞-norms of the initial data continuously. Thus, the assumption $\|u_N\|_{L^\infty(0,T;L^\infty)} \leq A$ is reasonable. The first assumption in (4.107) can also be fulfilled once the initial data satisfies certain Lipschitz condition. Chen, Du and Tadmor (1993) also showed that for fixed $t > 0$, the norm $\|U(t) - u_N(t)\|_{L^1(\Lambda)}$ is bounded by $\sqrt{\varepsilon(N)}$.

Now, let $\Lambda = (-1,1), Q = \Lambda \times \mathbb{R}_t^+$ and consider the initial-boundary value problem of (4.96) with appropriate data $g(t) \in H^1_{\text{loc}}(\mathbb{R}_t^+)$, prescribed at $x = \pm 1$. Let P_N and I_N be the L^2-orthogonal projection upon $\mathbb{P}_N$ and the interpolation associated with Legendre–Gauss–Lobatto interpolation points $x^{(j)}$ and weights $\omega^{(j)}$. Let $v \in L^2(\Lambda)$ and $\hat{v}_l$ be the coefficients of its Legendre expansion. A viscosity operator q is defined as

$$qv(x) = \sum_{l=0}^{N} \hat{q}_l \hat{v}_l L_l(x)$$

with $M = M(N)$ and

$$\begin{cases} \hat{q}_l = 0, & \text{for } l \leq M, \\ \hat{q}_l \geq 1 - \dfrac{M^4}{l^4}, & \text{for } M < l \leq N. \end{cases}$$

Let $\varepsilon = \varepsilon(N)$ be the small parameter. The free pair (ε, M) will be chosen later, such that $\varepsilon \to 0, M \to \infty$ as $N \to \infty$. On the other hand, let

$$B\left(u_N(t)\right) = \left(\lambda(t)(1-x) + \mu(t)(1+x)\right) \partial_x L_N(x).$$

The pair (λ,μ) is chosen to match the inflow boundary data prescribed at $x=1$ whenever $f'(u_N(1,t))<0$, and at $x=-1$ whenever $f'(u_N(-1,t))>0$. A Legendre vanishing viscosity penalty scheme for the initial-boundary value problem of (4.96) was given by Maday, Kaber and Tadmor (1993). It is to find $u_N(x,t)\in \mathbb{P}_N$ for $t\geq 0$ such that

$$\begin{cases} (\partial_t u_N(t)+\partial_x I_N f(u_N(t)),\phi)_N+\varepsilon\,(q\partial_x u_N(t),\partial_x\phi)_N \\ \qquad =(B\,(u_N(t))\,,\phi)_N\,, \quad \forall\,\phi\in\mathbb{P}_N,\, t>0, \\ u_N(x,t)=g(t), \quad \text{at the inflow boundary points, } t>0, \\ u_N(0,t)=I_N U_0(x), \qquad x\in\bar{\Lambda}. \end{cases} \tag{4.112}$$

We shall establish some a priori estimates for u_N. First of all, let σ be a filter such that for any $\phi\in\mathbb{P}_N$,

$$\sigma\phi(x)=\sum_{l=0}^{N}\hat{\sigma}_l\hat{\phi}_l L_l(x).$$

Also let $\|v\|_\sigma^2=(\sigma v,v)$.

Lemma 4.20. *For any $\phi\in\mathbb{P}_N$,*

$$\|\partial_x\phi\|_\sigma^2\leq cN^2\sum_{l=1}^{N}l\hat{\sigma}_l\|\phi\|^2.$$

Proof. Let

$$J_{l,N}=\{j\mid l+1\leq j\leq N,\ l+j \text{ odd}\}.$$

Then by (2.51),

$$\partial_x\phi(x)=\sum_{l=0}^{N-1}\hat{\phi}_l^{(1)}L_l(x),\quad \hat{\phi}_l^{(1)}=(2l+1)\sum_{j\in J_{l,N}}\hat{\phi}_j.$$

By (2.48),

$$\begin{aligned}(\sigma\partial_x\phi,\partial_x\phi) &= \sum_{l=0}^{N-1}\hat{\sigma}_l\left|\hat{\phi}_l^{(1)}\right|^2\|L_l\|^2=\sum_{l=0}^{N-1}\hat{\sigma}_l(2l+1)^2\left|\sum_{j\in J_{l,N}}\hat{\phi}_j\right|^2\|L_l\|^2 \\ &\leq 2\sum_{l=0}^{N-1}\hat{\sigma}_l(2l+1)\left(\sum_{j\in J_{l,N}}\left|\hat{\phi}_j\right|^2\|L_j\|^2\cdot\sum_{j\in J_{l,N}}\frac{1}{\|L_j\|^2}\right) \\ &\leq c\sum_{j=0}^{N}\left|\hat{\phi}_j\right|^2\|L_j\|^2\sum_{l=0}^{N}l\hat{\sigma}_l\left(N^2-l^2\right)\leq cN^2\sum_{l=0}^{N}l\hat{\sigma}_l\|\phi\|^2.\end{aligned}$$

∎

It is noted that when $\hat{\sigma}_l\equiv 1$, Lemma 4.20 is reduced to Theorem 2.8.

Lemma 4.21. *For any $\phi \in \mathbb{P}_N$,*

$$\|\partial_x \phi\|^2 \le \|\partial_x \phi\|_q^2 + cM^4 \ln N \|\phi\|^2.$$

Proof. Let $\hat{\sigma}_l = 1 - \hat{q}_l$. Then $\hat{\sigma}_l = 1$ for $l \le M$ and $\hat{\sigma}_l \le \frac{M^4}{l^4}$ for $M < l \le N$. Since $\|\partial_x \phi\|^2 = \|\partial_x \phi\|_q^2 + \|\partial_x \phi\|_\sigma^2$, it remains to bound $\|\partial_x \phi\|_\sigma^2$. To this end, let $J = \log_2 \frac{N}{M}$ and decompose the function ϕ as

$$\phi(x) = \sum_{l=0}^{M} \hat{\phi}_l L_l(x) + \sum_{j=1}^{J} \phi_j(x), \quad \phi_j(x) = \sum_{l > 2^{j-1}M}^{2^j M} \hat{\phi}_l L_l(x).$$

By Lemma 4.20,

$$\|\partial_x \phi_j\|_\sigma^2 \le c\left(2^j M\right)^2 \sum_{l>2^{j-1}M}^{2^j M} \frac{M^4}{l^3} \|\phi_j\|^2 \le 2cM^4 \|\phi_j\|^2.$$

Therefore

$$\begin{aligned} \|\partial_x \phi\|_\sigma^2 &\le 2\left\|\partial_x \sum_{l=0}^{M} \hat{\phi}_l L_l\right\|_\sigma^2 + 2J \sum_{j=1}^{J} \|\partial_x \phi_j\|_\sigma^2 \\ &\le 2cM^4 \left\|\sum_{l=0}^{M} \hat{\phi}_l L_l\right\|^2 + 4cJM^4 \sum_{j=1}^{J} \|\phi_j\|^2 \le cM^4 \ln N \|\phi\|^2. \end{aligned}$$

∎

To simplify the presentation, we shall deal with the prototype case where one boundary, say $x = -1$, is an inflow boundary, while $x = 1$ is an outflow one. Then

$$B\left(u_N(t)\right) = \lambda(t)(1-x)\partial_x L_N(x).$$

As we know, $x^{(j)}$ are the zeros of $\partial_x L_N(x), 1 \le j \le N-1$ and $\omega^{(0)} = \dfrac{2}{N(N+1)}$, $\partial_x L_N(-1) = \dfrac{1}{2}(-1)^{N+1} N(N+1)$. Thus

$$\left(B\left(u_N(t)\right), v\right)_N = 2(-1)^{N+1} \lambda(t) v(-1,t). \tag{4.113}$$

Let $\phi \equiv 1$ in (4.112). Since $I_N f\left(u_N(x,t)\right) \in \mathbb{P}_N$, we deduce from (2.37) that

$$\partial_t \left(u_N(t), 1\right) + f\left(u_N(1,t)\right) - f\left(u_N(-1,t)\right) = 2(-1)^{N+1} \lambda(t).$$

Consequently,

$$|\lambda(t)| \le \frac{1}{\sqrt{2}} \|\partial_t u_N(t)\| + \max\left(|f(u_N(1,t))|, |f(u_N(-1,t))|\right). \tag{4.114}$$

Further, set

$$\psi(t) = \int_0^t \lambda(s)\, ds,$$

and assume that for all N,

$$\varepsilon\|\partial_x u_N(0)\|^2 \le A, \quad \max_{0\le t\le T} \|u_N(t)\|_{L^\infty(\Lambda)} \le A. \tag{4.115}$$

Then for $t \le T$,

$$|\psi(t)| \le \frac{1}{\sqrt{2}}\|u_N(t)\| + \frac{1}{\sqrt{2}}\|u_N(0)\| + t \max_{|z|\le A} |f(z)|. \tag{4.116}$$

We now derive a priori estimates. Let c_0 be a positive constant depending on $\|U_0\|$, c_A be a positive constant depending on A, and

$$F^*(z) = \int_0^z \xi f'(\xi)\, d\xi.$$

Assume that

$$\varepsilon \sim cN^{-\alpha}, \quad M \sim cN^{\beta}, \quad 0 < 4\beta < \alpha \le 1. \tag{4.117}$$

Set $\phi = u_N$ in (4.112). In view of (2.37), (2.60), (2.63) and Lemma 4.21, we find that

$$\begin{aligned}
&\frac{1}{2}\partial_t\|u_N(t)\|_N^2 + F^*(u_N(1,t)) - F^*(u_N(-1,t)) \\
&\quad + \varepsilon(q\partial_x u_N(t), \partial_x u_N(t)) + 2(-1)^N \lambda(t) u_N(-1,t) \\
&\quad = (\partial_x f(u_N(t)), u_N(t)) - (\partial_x I_N f(u_N(t)), u_N(t))_N \\
&\quad = -((I - I_N) f(u_N(t)), \partial_x u_N(t)) \\
&\quad \le \frac{c}{N}\|\partial_x f(u_N(t))\|\|\partial_x u_N(t)\| \le \frac{c_A}{N}\left(\|\partial_x u_N(t)\|_q^2 + M^4 \ln N \|u_N(t)\|_N^2\right).
\end{aligned}$$

Thus for any $t \le T$,

$$\begin{aligned}
&\|u_N(t)\|_N^2 + 2\left(\varepsilon - \frac{c_A}{N}\right)\int_0^t \|\partial_x u_N(s)\|_q^2\, ds \\
&\quad \le c_0 + 2\int_0^t \left(\frac{c_A M^4 \ln N}{N}\|u_N(s)\|^2 + 2\max_{|z|\le A}|F^*(z)| - 2(-1)^N \lambda(s) g(s)\right) ds.
\end{aligned}$$

By (4.116),

$$\begin{aligned}
\left|\int_0^t \lambda(s) g(s)\, ds\right| &= \left|g(t)\psi(t) - g(0)\psi(0) - \int_0^t \partial_s g(s)\psi(s)\, ds\right| \\
&\le \|g(t)\|\left(\|u_N(t)\| + c_0 + c_A\right) + \int_0^t |\partial_s g(s)|\left(\|u_N(s)\| + c_0 + c_A\right) ds.
\end{aligned}$$

Putting the above two estimates together, we have from (4.117) that

$$\|u_N(t)\|^2 + \varepsilon \int_0^t \|\partial_x u_N(s)\|_q^2 \, ds \le c_A \left(c + \|g\|^2_{H^1_{\rm loc}(0,t)} \right) + c_0.$$

By using Lemma 4.21 and (4.117) again,

$$\varepsilon \|\partial_x u_N\|^2_{L^2_{\rm loc}(Q)} \le c_A \left(1 + \|g\|^2_{H^1_{\rm loc}(\mathbb{R}_t^+)} \right) + c_0. \tag{4.118}$$

Next, we set $\phi = \partial_t u_N$ in (4.112). By Maday (1991), $|I_N v|_1 \le c|v|_1$ for any $v \in H^1(\Lambda)$. So by (2.37), (2.60) and (4.114),

$$\|\partial_t u_N(t)\|^2 + \frac{\varepsilon}{2} \partial_t \|\partial_x u_N(t)\|_q^2 \le c_A \|\partial_x u_N(t)\|^2 + \frac{1}{2}\|\partial_t u_N(t)\|^2 + c_A \left|\partial_t g(t)\right|^2 + c_A.$$

Temporal integration of the above inequality followed by (4.118), implies

$$\|\partial_t u_N\|^2_{L^2_{\rm loc}(Q)} \le c_A \left(c_0 + \|\partial_x u_N\|^2_{L^2_{\rm loc}(Q)} + \|g\|^2_{H^1_{\rm loc}(\mathbb{R}_t^+)} + 1 \right) \le \frac{c_A}{\varepsilon} \left(c_0 + \|g\|^2_{H^1_{\rm loc}(\mathbb{R}_t^+)} + 1 \right). \tag{4.119}$$

We now prove the existence of weak solutions. Let $D = \Lambda \times (0,T)$ and define $(\cdot,\cdot)_D$ and $\|\cdot\|_D$ as before. Let

$$(v,w)_{D,N} = \int_0^T (v(t), w(t))_N \, dt.$$

For any $w \in H_0^1(D)$, define

$$w_N(x,t) = \int_{-1}^x P_{N-1} \partial_y w(y,t) \, dy.$$

Clearly $w_N \in \mathbb{P}_N$ and $w_N(-1,t) = 0$. Let $\sigma = 1 - q$ and $\phi = w_N$ in (4.112). We have from (2.37) and (4.113) that

$$(\partial_t u_N, \partial_x f(u_N), w)_D = \sum_{j=1}^5 G_j(w), \quad \forall\, w \in H_0^1(D),$$

where

$$\begin{aligned}
G_1(w) &= (\partial_t u_N + \partial_x f(u_N), w - w_N)_D, \\
G_2(w) &= (\partial_x f(u_N) - \partial_x I_N f(u_N), w_N)_D, \\
G_3(w) &= (\partial_t u_N + \partial_x I_N f(u_N), w_N)_D - (\partial_t u_N + \partial_x I_N f(u_N), w_N)_{D,N}, \\
G_4(w) &= \varepsilon(\sigma \partial_x u_N, \partial_x w_N)_D, \\
G_5(w) &= -\varepsilon(\partial_x u_N, \partial_x w_N)_D.
\end{aligned}$$

By (4.118) and (4.119),

$$|G_1(w)| \leq \frac{c_A}{\sqrt{\varepsilon}}\|w - w_N\|_D \leq \frac{c_A}{\sqrt{\varepsilon}N}\|\partial_x w\|_D.$$

According to (2.63) and (4.118),

$$|G_2(w)| = |(f(u_N) - I_N f(u_N), \partial_x w_N)_D| \leq \frac{c}{N}\|\partial_x f(u_N)\|_D\|\partial_x w_N\|_D \leq \frac{c_A}{\sqrt{\varepsilon}N}\|\partial_x w_N\|_D.$$

By virtue of (2.64) and (4.119),

$$|G_3(w)| = |(\partial_t u_N, w_N)_D - (\partial_t u_N, w_N)_{D,N}| \leq \frac{c_A}{N}\|\partial_t u_N\|_D\|\partial_x w_N\|_D \leq \frac{c_A}{\sqrt{\varepsilon}N}\|\partial_x w_N\|_D.$$

We see from the proof of Lemma 4.21 that

$$|G_4(w)| \leq \varepsilon c_A M^2\sqrt{\ln N}\|u_N\|_D\|\partial_x w_N\|_D \leq \varepsilon c_A M^2\sqrt{\ln N}\|\partial_x w_N\|_D. \qquad (4.120)$$

Clearly,

$$|G_5(w)| \leq \sqrt{\varepsilon}c_A\|\partial_x w_N\|_D.$$

The above statements with (4.117) imply that

$$\|\partial_t u_N + \partial_x f(u_N)\|_{H^{-1}_{\mathrm{loc}}(Q)} \longrightarrow 0, \quad \text{as } N \longrightarrow \infty.$$

It means that $\partial_t u_N + \partial_x f(u_N)$ belongs to a compact subset of $H^{-1}_{\mathrm{loc}}(Q)$. Furthermore let E be any strictly convex entropy and F be the corresponding entropy flux. We have

$$(\partial_t E(u_N) + \partial_x F(u_N), w)_D = \sum_{j=1}^{5} G_j\left(E'(u_N)w\right). \qquad (4.121)$$

The previous statements tell us that

$$\begin{aligned}
&\left|\sum_{j=1}^{5} G_j\left(E'(u_N)w\right)_D\right| \\
\leq\ & c_A\left(\frac{1}{N\sqrt{\varepsilon}} + \varepsilon M^2\sqrt{\ln N} + \sqrt{\varepsilon}\right)\left(\|\partial_x u_N\|_D\|w\|_{L^\infty(D)} + c_A\|\partial_x w\|_D\right) \\
\leq\ & c_A\left(\|w\|_{L^\infty(D)} + \left(\frac{1}{N\sqrt{\varepsilon}} + \varepsilon M^2\sqrt{\ln N} + \sqrt{\varepsilon}\right)\|\partial_x w\|_D\right).
\end{aligned}$$

Thus the term $\partial_t E(u_N) + \partial_x F(u_N)$ can be written as a sum of two terms which belong to a compact subset of $H^{-1}_{\mathrm{loc}}(Q)$ and a bounded set of $L^1_{\mathrm{loc}}(Q)$. Then by Lemma 4.18, it belongs to a compact set of $H^{-1}_{\mathrm{loc}}(Q)$. Further, Lemma 4.17 ensures that for $1 \le q < \infty, u_N$ converges to U^*, a weak solution, in $L^q_{\rm loc}(Q)$ strongly. Finally, we check the entropy condition. Let $\alpha < 1$. It is easy to see that

$$\begin{aligned}
\sum_{j=1}^{3} |G_j(E'(u_N)w)| &\le \frac{c_A}{\sqrt{\varepsilon}N} \|\partial_x (E'(u_N)w)_N\|_D \\
&\le \frac{c_A}{\sqrt{\varepsilon}N} \left(\|\partial_x u_N\|_D \|w\|_{L^\infty(D)} + \|u_N\|_{L^\infty(D)} \|\partial_x w\|_D\right) \\
&\le c_A \left(N^{\alpha-1}\|w\|_{L^\infty(D)} + N^{\frac{\alpha}{2}-1}\|\partial_x w\|_D\right) \longrightarrow 0.
\end{aligned}$$

By (4.117), (4.118) and (4.120),

$$\begin{aligned}
|G_4(E'(u_N)w)| &\le c_A \varepsilon M^2 \sqrt{\ln N}\, \|\partial_x (E'(u_N)w)\|_D \\
&\le c_A \sqrt{\varepsilon} M^2 \sqrt{\ln N}\, (\|w\|_D + \|\partial_x w\|_D) \longrightarrow 0.
\end{aligned}$$

Moreover, for $w \ge 0$,

$$\begin{aligned}
G_5(E'(u_N)w) &= -\varepsilon \left(\partial_x u_N, P_{N-1}\partial_x (E'(u_N)w)\right)_D \\
&= -\varepsilon \left(\partial_x u_N, E''(u_N) w \partial_x u_N\right)_D - \varepsilon \left(\partial_x u_N, E'(u_N)\partial_x w\right)_D \\
&\le -\varepsilon \left(\partial_x u_N, E'(u_N)\partial_x w\right)_D \le c_A \sqrt{\varepsilon} \|\partial_x w\|_D \longrightarrow 0.
\end{aligned}$$

Thus by (4.121), U^* satisfies (4.98) in the sense of distribution, and so it is the unique entropy solution.

Theorem 4.20. *Let (4.115) and (4.117) hold. Then the solution of (4.112) tends to a weak solution of the initial-boundary value problem of (4.96) in $L^q_{\rm loc}(Q)$ strongly, $1 \le q < \infty$. If in addition $\alpha < 1$, then it is the unique entropy solution.*

We now present some numerical results. We consider the Hopf equation with $U_0(x) = 1 + \frac{1}{2}\sin \pi x$ and $U(-1,t) = 1$. For time discretization, we use the second-order Adams-Bashforth method with the step size $\tau = 10^{-5}$. In all calculations, $N = 128$. The standard Legendre pseudospectral scheme corresponds to $\varepsilon = 0$, while the Legendre vanishing viscosity scheme is implemented with $\varepsilon \sim N^{-1}, M \sim 5\sqrt{N}$ and

$$\hat{q}_l = \exp\left(-\frac{(l-N)^2}{(l-M)^2}\right), \quad M < l \le N.$$

Figure 4.5 shows the numerical results of Legendre pseudospectral scheme before post-processing, and Figure 4.6 shows the corresponding ones after post-processing. Figure 4.7 presents the results of Legendre pseudospectral vanishing viscosity scheme before post-processing. Figure 4.8 is for the corresponding results after post-processing. In all figures, the real lines present the same exact solution of Hopf equation, and the dotted curves correspond to the numerical solutions. Obviously the Legendre pseudospectral vanishing viscosity method provides much better results.

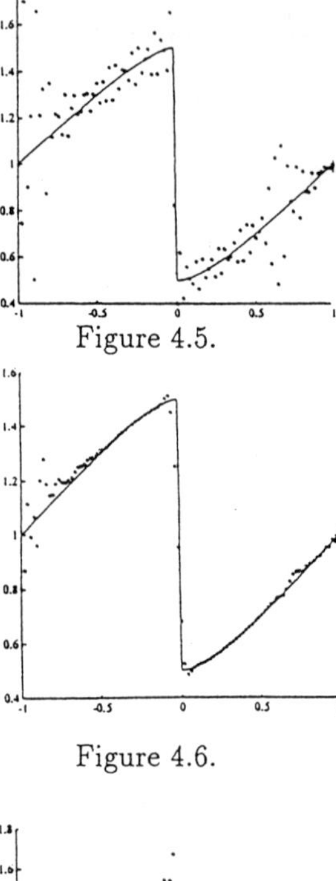

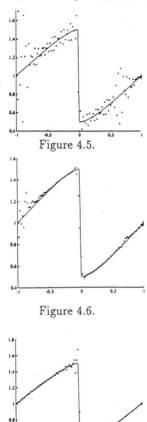

Figure 4.5.

Figure 4.6.

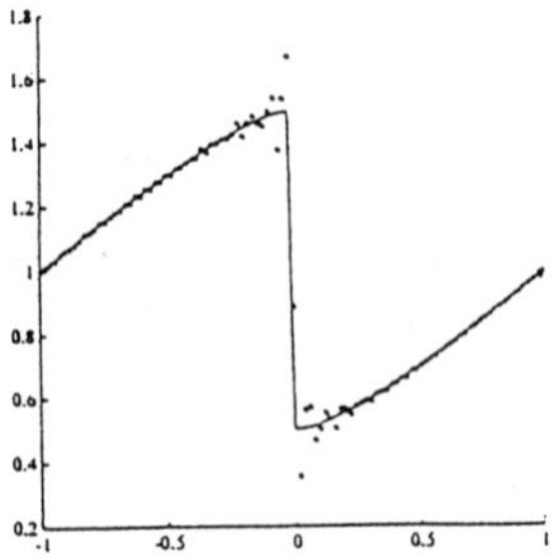

Figure 4.7.

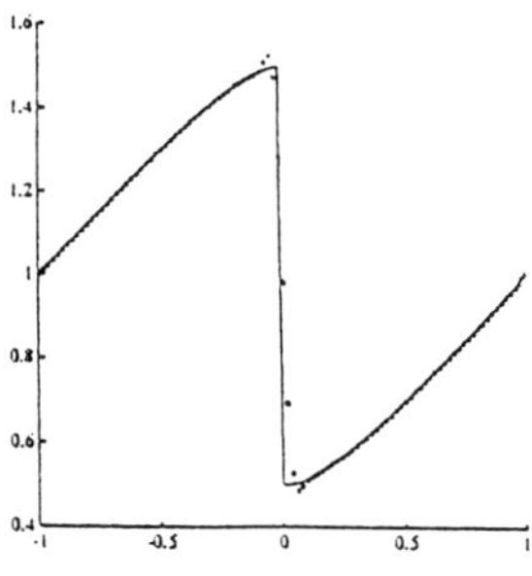

Figure 4.8.

Recently, Ma (1996a) developed a Chebyshev–Legendre pseudospectral vanishing viscosity method for nonlinear conservation laws which is a combination of Chebyshev–Legendre penalty method in Section 4.4 and vanishing viscosity method in this section. He also developed a new method in which the viscosity term is taken to be $\varepsilon P_N((1-x^2)^{-\frac{1}{2}}D^2(qu_N)), D=\sqrt{1-x^2}\partial_x$. It was proved that the solution of both schemes converge to the unique entropy solution provided that certain reasonable conditions are fulfilled. Ma (1996b) also served a Chebyshev–Legendre super viscosity method in which the viscosity term is of the form $(-1)^{s+1}\varepsilon P_N\left((1-x^2)^{-\frac{1}{2}}D^{2s}u_N\right), s\geq 1$. Its solution also converges to the unique entropy solution. The super viscosity technique for the Fourier spectral methods can be found in Tadmor (1993b).

4.6. Spectral Approximations of Isolated Solutions

In the previous sections, we studied various spectral methods for nonlinear revolutionary problems. Generally speaking, all of those methods are applicable to the corresponding steady problems. But for nonlinear elliptic equations, the situations are more complicated. The main feature is that they may possess several solutions. In this section, we take the steady Burgers' equation as an example to describe the spectral approximations of isolated solutions.

Let $\Lambda=(-1,1)$ and $\mu>0$. $f(x)$ is a given function. Consider the problem

$$\begin{cases} U\partial_x U-\mu\partial_x^2 U=f, & x\in\Lambda,\\ U=0, & x=-1,1. \end{cases} \tag{4.122}$$

Set

$$a(v,w) = \int_\Lambda \partial_x v(x)\partial_x w(x)\,dx, \qquad \forall\, v,w \in H^1(\Lambda),$$
$$b(v,w,z) = \tfrac{1}{2}\int_\Lambda v(x)w(x)\partial_x z\,dx, \quad \forall\, v,w \in L^4(\Lambda),\ z \in H^1(\Lambda).$$

The solution of (4.122) is defined as a function $U \in H_0^1(\Lambda)$, satisfying

$$\mu a(U,v) - b(U,U,v) = (f,v), \quad \forall\, v \in H_0^1(\Lambda). \tag{4.123}$$

If (f,v) is a linear continuous functional in $H_0^1(\Lambda)$, then by Fixed Point Theorem, (4.123) has at least one solution. Letting $v = U$ in (4.123) and

$$|||f||| = \sup_{\substack{v\in H_0^1(\Lambda)\\ v\neq 0}} \frac{|(f,v)|}{|v|_1},$$

we get that $|U|_1 \le \mu^{-1}|||f|||$. Assume that (4.123) has the solutions U and U^*, and $\tilde{U} = U^* - U$. Then

$$\mu a\left(\tilde{U},v\right) - b\left(\tilde{U},U+\tilde{U},v\right) - b\left(U,\tilde{U},v\right) = 0, \quad \forall\, v \in H_0^1(\Lambda).$$

Putting $v = \tilde{U}$, we deduce that

$$\mu\left|\tilde{U}\right|_1^2 \le \frac{1}{2}\|U\|_{L^4}\left\|\tilde{U}\right\|_{L^4}\left|\tilde{U}\right|_1.$$

We know that $\|v\|_{L^4} \le \sqrt{2}|v|_1$ for any $v \in H_0^1(\Lambda)$, and thus

$$\mu\left|\tilde{U}\right|_1^2 \le \mu^{-1}|||f|||\left|\tilde{U}\right|_1^2.$$

Hence (4.123) has only one solution as long as $|||f||| < \mu^2$.

If (4.123) has a unique solution, then it is easy to solve it by various spectral approximations as presented in Sections 2.4 and 2.5. Here we are more interested in some problems with multi-solutions. In this case, it is better to consider (4.123) in an abstract framework. To do this, we need some preparations. For convenience, we shall adopt a unified notation as in Section 2.3, i.e., $\omega(x) \equiv 1$ for the Legendre approximation or $\omega(x) = (1-x^2)^{-\frac{1}{2}}$ for the Chebyshev approximation.

Lemma 4.22. *The imbedding $H_{0,\omega}^1(\Lambda) \subset L_\omega^2(\Lambda)$ is compact.*

Proof. For $\omega(x) \equiv 1$, the conclusion comes from the well-known imbedding theory. Now, let $\omega(x) = (1-x^2)^{-\frac{1}{2}}$. It suffices to verify that $v \in L_\omega^2(\Lambda) \to v\omega^{\frac{1}{2}} \in L^2(\Lambda)$ is an isomorphism, and that $v \in H^1_{0,\omega}(\Lambda) \to v\omega^{\frac{1}{2}} \in H^1_0(\Lambda)$ is a continuous mapping. The first fact is trivial. We now prove the second one. Let $v \in H^1_{0,\omega}(\Lambda)$. We have

$$\partial_x \left(v\omega^{\frac{1}{2}}\right) = \partial_x v\omega^{\frac{1}{2}} + \frac{1}{2}xv\omega^{\frac{5}{2}}.$$

Clearly $\partial_x v\omega^{\frac{1}{2}} \in L^2(\Lambda)$. Moreover by Lemma 2.7, $v\omega^{\frac{5}{2}} \in L^2(\Lambda)$. Hence $\partial_x \left(v\omega^{\frac{1}{2}}\right) \in H^1(\Lambda)$.

Next, the set $\mathcal{D}(\Lambda)$ is dense in $H^1_\omega(\Lambda)$ and so for any $v \in H^1_{0,\omega}(\Lambda)$, there exists a sequence $z_l \in \mathcal{D}(\Lambda), l = 1, 2, \ldots,$ such that $z_l \to v$ in $H^1_\omega(\Lambda)$. Then $v\omega^{\frac{1}{2}}$ is the limit in $H^1(\Lambda)$ of $z_l\omega^{\frac{1}{2}}$ which vanishes at $x = -1, 1$. Therefore $v\omega^{\frac{1}{2}} \in H^1_0(\Lambda)$. It implies the continuity of the mapping $v \in H^1_{0,\omega}(\Lambda) \to v\omega^{\frac{1}{2}} \in H^1_0(\Lambda)$. ∎

Lemma 4.23. *For any $0 \le \mu < r$, the imbedding $H^r_\omega(\Lambda) \subset H^\mu_\omega(\Lambda)$ is compact.*

Proof. It suffices to consider the case with $\omega = (1 - x^2)^{-\frac{1}{2}}$. Firstly, let $r = 1, \mu = 0$, and v_l be a bounded sequence of $H^1_\omega(\Lambda)$. For each v_l, there is $v_l^0 \in H^1_{0,\omega}(\Lambda)$ and $w_l \in \mathbb{P}_1$ such that $v_l = v_l^0 + w_l$ and

$$\left\|v_l^0\right\|_{1,\omega} + \|w_l\|_{L^\infty(\Lambda)} \le c\|v_l\|_{1,\omega}.$$

By Lemma 4.22 and the equivalence of norms on $\mathbb{P}_1$, we can extract subsequences $v^0_{l,m}$ and $w_{l,m}$ such that $v_{l,m} \to v$ in $L^2_\omega(\Lambda)$ and $w_{l,m} \to w$ in $L^\infty(\Lambda)$. Thus $v^0_{l,m} + w_{l,m} \to v^0 + w$ in $L^2_\omega(\Lambda)$, and this completes the proof for $r = 1$ and $\mu = 0$. The general result follows from an inductive procedure and the space interpolation. ∎

Lemma 4.24. *For any $r > \frac{1}{2}, H^r_\omega(\Lambda) \subset L^\infty(\Lambda)$, and for any $r \ge 1, H^r_\omega(\Lambda)$ is an algebra.*

Proof. We only have to verify the conclusions with $\omega(x) = (1 - x^2)^{-\frac{1}{2}}$. Since $\omega(x) \ge 1, H^r_\omega(\Lambda) \subset H^r(\Lambda) \subset L^\infty(\Lambda)$ for $r > \frac{1}{2}$. We next prove the second result. It suffices to verify that for $r \ge 1$,

$$\|vw\|_{r,\omega} \le c\|v\|_{r,\omega}\|w\|_{r,\omega}, \quad \forall\, v, w \in H^r_\omega(\Lambda).$$

Indeed, for any integer $r \ge 1$,

$$\begin{aligned} \|\partial_x^r(vw)\|_\omega &= \left(\int_\Lambda \left(\partial_x^r(vw)\right)^2 \omega\, dx\right)^{\frac{1}{2}} \\ &\le \sum_{\mu=0}^{r-1} C_r^\mu \|v\|_{\mu+1,\omega} \left\|\partial_x^{r-\mu} w\right\|_\omega + \|\partial_x^r v\|_\omega \|w\|_{1,\omega} \le c\|v\|_{r,\omega}\|w\|_{r,\omega}. \end{aligned}$$

Then the space interpolation extends this result to any real numbers $r \geq 1$. ■

Now, let

$$a_\omega(v, w) = \int_\Lambda \partial_x v(x) \partial_x (w(x)\omega(x))\, dx$$

and define the linear mapping $A : \left(H^1_{0,\omega}(\Lambda)\right)' \to H^1_{0,\omega}(\Lambda)$ by

$$a_\omega(Av, w) = \langle v, w \rangle, \quad \forall\, w \in H^1_{0,\omega}(\Lambda), \tag{4.124}$$

where $\langle \cdot, \cdot \rangle$ denotes the duality pairing between $\left(H^1_{0,\omega}(\Lambda)\right)'$ and $H^1_{0,\omega}(\Lambda)$.

Lemma 4.25. *For any $r \geq 0, A$ is a linear, bounded and continuous mapping from $H^r_\omega(\Lambda)$ into $H^{r+2}_\omega(\Lambda) \cap H^1_{0,\omega}(\Lambda)$. For any $-1 \leq r < 0, A$ is continuous from $\left(H^{-r}_{0,\omega}(\Lambda)\right)'$ into $H^{r+2}_\omega(\Lambda) \cap H^1_{0,\omega}(\Lambda)$.*

Proof. If r is a non-negative integer, then the integration by parts in (4.124) yields

$$-\partial_x^2(Av) = v$$

and so $\|Av\|_{r+2,\omega} \leq c\|v\|_{r,\omega}$. For real number $r > 0$, the same result follows from the space interpolation. Finally for $-1 \leq r < 0$, the result comes as in Corollary 4.5.2 of Bergh and Löfström (1976). ■

Now by Lemma 2.8, A is a bounded mapping on $H^1_{0,\omega}(\Lambda)$. Moreover, Lemmas 4.22 and 4.25 imply that A is a compact mapping from $L^2_\omega(\Lambda)$ into $H^1_{0,\omega}(\Lambda)$. Let $\lambda = \frac{1}{\mu}$ and define the mapping $G : \mathbb{R}^1 \times H^1_{0,\omega}(\Lambda) \to L^2_\omega(\Lambda)$ by

$$G(\lambda, v) = \lambda\left(v \partial_x v - f\right). \tag{4.125}$$

Clearly G is infinitely differentiable. Let DG be the Frèchet derivative of G. Then D^2G is bounded on any bounded subset of $\mathbb{R}^1 \times H^1_{0,\omega}(\Lambda)$. Moreover, since $H^1_{0,\omega}(\Lambda) \subset L^\infty(\Lambda)$, we get

$$\|G(\lambda, v)\|_\omega \leq c|\lambda| \left(\|v\|_{L^\infty} \|v\|_{1,\omega} + \|f\|_\omega\right) \leq c|\lambda| \left(\|v\|^2_{1,\omega} + \|f\|_\omega\right). \tag{4.126}$$

Finally, define the mapping $F : \mathbb{R}^1 \times H^1_{0,\omega}(\Lambda) \to H^1_{0,\omega}(\Lambda)$, by

$$F(\lambda, v) = v + AG(\lambda, v).$$

Then problem (4.123) can be written equivalently as follows: find $v \in H^1_{0,\omega}(\Lambda)$ such that

$$F(\lambda, U) = 0. \tag{4.127}$$

Let Λ^* be any compact interval in $\mathbb{R}^+_\lambda = \{\lambda \mid \lambda > 0\}$. Using the techniques in Lions (1969), we can prove that there exists a branch of isolated solutions of (4.123), denoted by $(\lambda, U(\lambda))$. According to the description in Section 3.2, it means that there exists a positive constant δ, such that

$$\|(I + ADG(\lambda, U(\lambda)))v\|_{1,\omega} \geq \delta\|v\|_{1,\omega}, \quad \forall\, \lambda \in \Lambda^*,\, v \in H^1_{0,\omega}(\Lambda). \tag{4.128}$$

We begin to approximate the isolated solutions of (4.127). Let P_N be the L^2_ω-orthogonal projection and $P^{1,0}_N$ be the $H^1_{0,\omega}$-orthogonal projection, i.e.,

$$a_\omega\left(v - P^{1,0}_N v, \phi\right) = 0, \quad \forall\, \phi \in H^1_{0,\omega}(\Lambda). \tag{4.129}$$

Set $A_N = P^{1,0}_N A$. Thanks to (4.124) and (4.129), for any $v \in \left(H^1_{0,\omega}(\Lambda)\right)'$,

$$a_\omega(A_N v, \phi) = a_\omega(Av, \phi) = \langle v, \phi\rangle, \quad \forall\, \phi \in \mathbb{P}^0_N.$$

Define

$$F_N(\lambda, v) = v + A_N G(\lambda, v).$$

Maday and Quarteroni (1981) proposed a spectral approximation to (4.127). It is to find $u_N \in \mathbb{P}^0_N$ such that

$$F_N(\lambda, u_N) = 0. \tag{4.130}$$

We can use the Newton procedure or the modified Newton procedure to solve (4.130). For instance, let $u_N^{(m)}$ be the m-th iteration of u_N, and

$$DF_N\left(\lambda, u_N^{(m)}\right) u_N^{(m+1)} = DF_N\left(\lambda, u_N^{(m)}\right) u_N^{(m)} - F_N\left(\lambda, u_N^{(m)}\right), \quad m \geq 0.$$

We are going to analyze the convergence. To do this, we introduce the result of Brezzi, Rappaz and Raviart (1980). Let B and W be two Banach spaces. Λ^* is a compact interval in $\mathbb{R}^1$. $\mathcal{G}$ is a C^1 mapping from $\Lambda^* \times W$ into B, and $\mathcal{A}$ is a linear continuous mapping from B into W. Suppose that for all $\lambda \in \Lambda^*$ and $v \in W$, the operator $\mathcal{A}D\mathcal{G}(\lambda, v) \in \mathcal{L}(W, W)$ is compact. Let $\mathbb{F}(\lambda, v) = v + \mathcal{A}\mathcal{G}(\lambda, v)$ and consider the problem

$$\mathbb{F}(\lambda, U) = 0. \tag{4.131}$$

Next let W_N be the finite dimensional subspaces of $W, N = 0, 1, \ldots$. Let $\mathcal{A}_N \in \mathcal{L}(B, W_N)$ and $\mathbb{F}_N(\lambda, v) = v + \mathcal{A}_N\mathcal{G}(\lambda, v)$ for $\lambda \in \Lambda^*, v \in W_N$.

Lemma 4.26. *Assume that*

(i) $\mathcal{G}$ *is a* C^{m+1} *mapping from* $\Lambda^* \times W$ *into* B, *and* $D^{m+1}\mathcal{G}$ *is bounded over any bounded subset of* $\Lambda^* \times W, m \geq 1$;

(ii) $\sqcap_N \in \mathcal{L}(W, W_N)$ *and*

$$\lim_{N\to\infty} \|\sqcap_N v - v\|_W = 0, \quad \forall\, v \in W; \tag{4.132}$$

(iii) $\mathcal{A}_N \in \mathcal{L}(W', W_N)$ *and*

$$\lim_{N\to\infty} \|\mathcal{A} - \mathcal{A}_N\|_{\mathcal{L}(B,W)} = 0; \tag{4.133}$$

(iv) $\{(\lambda, U(\lambda)), \lambda \in \Lambda^*\}$ *is a branch of isolated solutions of (4.131).*

Then there exists a neighborhood Ⓗ of the origin in W, and for $N \geq N_0$ large enough, a unique C^{m+1} mapping $\lambda \in \Lambda^* \to u_N(\lambda) \in W_N$ such that

$$\mathbb{F}_N(\lambda, u_N(\lambda)) = 0, \quad U(\lambda) - u_N(\lambda) \in Ⓗ, \quad \forall\, \lambda \in \Lambda^*.$$

Moreover there exists a positive constant c_1 independent of λ and N, such that for any $\lambda \in \Lambda^*$,

$$\|U(\lambda) - u_N(\lambda)\|_W \leq c_1 \left(\|U(\lambda) - \sqcap_N U(\lambda)\|_W + \|(\mathcal{A} - \mathcal{A}_N)\mathcal{G}(\lambda, U(\lambda))\|_W\right). \tag{4.134}$$

Following the line as in Corollary 1.1 of Maday and Quarteroni (1982a), we have the following lemma.

Lemma 4.27. *Let* H *be a Banach space and* $K \subseteq W \subseteq H \subseteq B$. *Assume that*

(i) the conditions of Lemma 4.26 hold;

(ii) the mapping $v \in K \to D\mathcal{G}(\lambda, v) \in \mathcal{L}(H, W')$ *is continuous and*

$$\lim_{N\to\infty} \|\mathcal{A} - \mathcal{A}_N\|_{\mathcal{L}(W',H)} = 0; \tag{4.135}$$

(iii) if $v \in H$ *satisfies* $v + \mathcal{A}D\mathcal{G}(\lambda, U(\lambda))v = 0$, *then* $v \in W$.

(iv) $U(\lambda), u_N(\lambda) \in K$ *and* $\|U(\lambda) - u_N(\lambda)\|_K$ *tends to zero as* N *goes to infinity.*

Then for $N \geq N_0$ *large enough,*

$$\|U(\lambda) - u_N(\lambda)\|_H \leq c\|\mathbb{F}_N(\lambda, U(\lambda))\|_H. \tag{4.136}$$

Using the above two lemmas, we can state the following result.

Theorem 4.21. *Let* $\{(\lambda, U(\lambda)), \lambda \in \Lambda^*\}$ *be a branch of isolated solutions of (4.127). There exists a neighborhood* Ⓗ *of the origin in* $H^1_{0,\omega}(\Lambda)$ *and, for* $N \geq N_0$, *a unique* C^∞ *mapping:* $\lambda \in \Lambda^* \to u_N(\lambda) \in \mathbb{P}^0_N$ *such that for any* $\lambda \in \Lambda^*, u_N(\lambda)$ *solves (4.130). Furthermore, if the mapping* $\lambda \in \Lambda^* \to U(\lambda) \in H^r_\omega(\Lambda) \cap H^1_{0,\omega}(\Lambda)$ *is continuous for some* $r > 1$, *then for any* $\lambda \in \Lambda^*$, *there exists a constant* b, *depending on* $\|U(\lambda)\|_{r,\omega}$ *such that*

$$\|U(\lambda) - u_N(\lambda)\|_{1,\omega} + N\|U(\lambda) - u_N(\lambda)\|_\omega \leq bN^{1-r}. \tag{4.137}$$

Proof. Take $B = L^2_\omega(\Lambda)$, $W = H^1_{0,\omega}(\Lambda), \sqcap_N = P^{1,0}_N, \mathcal{A} = A$ and $\mathcal{G}(\lambda, v) = G(\lambda, v)$ in Lemma 4.26. By Theorems 2.11 and 2.18 and a density argument, (4.132) is valid. Since $(A - A_N)v = (I - P^{1,0}_N)Av$, we have from Theorems 2.11 and 2.18, and Lemma 4.25 that

$$\|(A - A_N)v\|_{1,\omega} \leq cN^{-1}\|Av\|_{2,\omega} \leq cN^{-1}\|v\|_\omega$$

and so (4.133) holds. On use of Lemma 4.26, the first part of this theorem comes with

$$\|U(\lambda) - u_N(\lambda)\|_{1,\omega} \leq c\left(\|U(\lambda) - P^{1,0}_N U(\lambda)\|_{1,\omega} + \|(A - A_N)G(\lambda, U(\lambda))\|_{1,\omega}\right). \tag{4.138}$$

Since $U(\lambda)$ is a solution of (4.127), $AG(\lambda, U(\lambda)) = -U(\lambda)$. Thus Theorems 2.11 and 2.18, and (4.138) lead to

$$\|U(\lambda) - u_N(\lambda)\|_{1,\omega} \leq cN^{1-r}\|U(\lambda)\|_{r,\omega}.$$

To obtain the L^2_ω-estimate, we check the hypotheses of Lemma 4.27. By (4.125), $DG(\lambda, v)w = \lambda\partial_x(vw)$. Let $H = B$ and $K = W$. Then by Lemma 4.24, the first part of condition (ii) in Lemma 4.27 is fulfilled. Using Theorems 2.11 and 2.18, and Lemma 4.25 again, we deduce (4.135). Furthermore, condition (iii) of Lemma 4.27 is satisfied by Lemma 4.25 and condition (ii) of Lemma 4.27. Clearly condition (iv) is valid. Therefore (4.136) leads to

$$\|U(\lambda) - u_N(\lambda)\|_\omega \leq c\|F_N(\lambda, U(\lambda))\|_\omega.$$

Since

$$F_N(\lambda, U(\lambda)) = (A_N - A)G(\lambda, U(\lambda)) = U(\lambda) - P^{1,0}_N U(\lambda),$$

we have

$$\|U(\lambda) - u_N(\lambda)\|_\omega \leq cN^{-r}\|U(\lambda)\|_{r,\omega}$$

which completes the proof. ■

CHAPTER 5

SPECTRAL METHODS FOR MULTI-DIMENSIONAL AND HIGH ORDER PROBLEMS

In this chapter, we study the spectral methods in several spatial dimensions. We first list some results of orthogonal approximations in several dimensions, which are the generalizations of the results in Chapter 2. Next, we take the periodic problem of three-dimensional steady incompressible fluid flow and the stream function form of two-dimensional unsteady flow as two examples to describe the spectral methods for multi-dimensional problems and high order problems. The final part of this chapter is devoted to the spectral domain decomposition methods and the spectral multigrid methods. These techniques make the spectral methods more effective. In this chapter, the Fourier approximations, the Legendre approximations and the Chebyshev approximations in several dimensions are used for elliptic, parabolic and hyperbolic equations respectively. These examples will help readers to handle the related techniques.

5.1. Orthogonal Approximations in Several Dimensions

This section is for orthogonal approximations in several dimensions. In principle, all results in Chapter 2 could be generalized to those in several dimensions.

We first consider the Fourier approximations. Let $x = (x_1, \dots, x_n)$ and $\Omega = (0, 2\pi)^n$. Let l_p be integers and $l = (l_1, l_2, \dots, l_n)$. Set $|l| = \max\limits_{1 \le q \le n} |l_q|$ and $l \cdot x = l_1 x_1 + \cdots + l_n x_l$. The set of functions $e^{il \cdot x}$ is an orthogonal system in $L^2(\Omega)$. The

Fourier transformation of a function $v \in L^2(\Omega)$ is as

$$Sv(x) = \sum_{|l|=0}^{\infty} \hat{v}_l e^{il\cdot x}$$

with the Fourier coefficients

$$\hat{v}_l = \left(\frac{1}{2\pi}\right)^n \int_\Omega v(x) e^{-il\cdot x}\, dx, \quad |l| = 0, 1, \ldots.$$

Let N be any positive integer, and

$$\tilde{V}_N = \text{span}\left\{e^{il\cdot x} \mid |l| \le N\right\}.$$

V_N denotes the subspace of all real-valued functions in $\tilde{V}_N$. In this subspace, some inverse inequalities are valid. Firstly, for any $\phi \in \tilde{V}_N$ and $1 \le p \le q \le \infty$,

$$\|\phi\|_{L^q} \le cN^{\frac{n}{p}-\frac{n}{q}} \|\phi\|_{L^p}. \tag{5.1}$$

Next, let m be a non-negative integer. For any $\phi \in \tilde{V}_N$ and $|k| = m$,

$$\left\|\partial_x^k \phi\right\|_{L^p} \le (2N)^m \|\phi\|_{L^p}, \tag{5.2}$$

and for any $r \ge 0$,

$$\|\phi\|_r \le cN^r \|\phi\|. \tag{5.3}$$

The L^2-orthogonal projection of a function $v \in L^2(\Omega)$ is

$$P_N v(x) = \sum_{|l| \le N} \hat{v}_l e^{il\cdot x}.$$

Let $H_p^r(\Omega)$ be the subspace of $H^r(\Omega)$ containing all functions with period 2π for all variables. If $0 \le \mu \le r$, then for any $v \in H_p^r(\Omega)$,

$$\|v - P_N v\|_\mu \le cN^{r-\mu} \|v\|_r. \tag{5.4}$$

$P_N v$ is also the H^r-orthogonal projection of $v \in H_p^r(\Omega)$. Now let $j = (j_1, j_2, \ldots, j_n)$, $x^{(j)} = (x_1^{(j_1)}, \ldots, x_n^{(j_n)})$ and

$$\Omega_N = \left\{x^{(j)} \,\middle|\, x_q^{(j_q)} = \frac{2\pi j_q}{2N+1}, 0 \le j_q \le 2N, 1 \le q \le n\right\}.$$

The Fourier interpolant $I_N v(x)$ of a function $v \in C(\bar{\Omega})$ is such a trigonometric function in $\tilde{V}_N$ that

$$I_N v(x) = v(x), \quad x \in \Omega_N.$$

If $v \in H_p^r(\Omega), 0 \le \mu \le r$ and $r > \frac{n}{2}$, then

$$\|v - I_N v\|_\mu \le cN^{r-\mu}\|v\|_r. \tag{5.5}$$

We next consider the Legendre approximations. Let $\Omega = (-1,1)^n$. The Legendre polynomial of degree l is

$$L_l(x) = \prod_{q=1}^{n} L_{l_q}(x_q).$$

The Legendre transformation of a function $v \in L^2(\Omega)$ is as

$$Sv(x) = \sum_{|l|=0}^{\infty} \hat{v}_l L_l(x)$$

with the Legendre coefficients

$$\hat{v}_l = \prod_{q=1}^{n}\left(l_q + \frac{1}{2}\right)\int_\Omega v(x)L_l(x)\,dx, \quad |l| = 0,1,\ldots.$$

Let $\mathbb{P}_N$ be the set of all polynomials of degree at most N in all variables. Several inverse inequalities exist in $\mathbb{P}_N$. For any $\phi \in \mathbb{P}_N$ and $1 \le p \le q \le \infty$, we have

$$\|\phi\|_{L^q} \le cN^{2n\left(\frac{1}{p}-\frac{1}{q}\right)}\|\phi\|_{L^p}. \tag{5.6}$$

On the other hand, for any $\phi \in \mathbb{P}_N, 2 \le p \le \infty$ and $|k| = m$,

$$\left\|\partial_x^k \phi\right\|_{L^p} \le cN^{2m}\|\phi\|_{L^p}, \tag{5.7}$$

and for any $r \ge 0$,

$$\|\phi\|_r \le cN^{2r}\|\phi\|. \tag{5.8}$$

The L^2-orthogonal projection of a function $v \in L^2(\Omega)$ is

$$P_N v(x) = \sum_{|l|=0}^{N} \hat{v}_l L_l(x).$$

If $r \ge 0$ and $\mu \le r$, then for any $v \in H^r(\Omega)$,

$$\|v - P_N v\|_\mu \le cN^{\sigma(\mu,r)}\|v\|_r \tag{5.9}$$

where $\sigma(\mu, r)$ is given in (2.53). We introduce the inner product of $H^m(\Omega)$, i.e.,

$$(v,w)_m = \sum_{0\le|k|\le m}\left(\partial_x^k v, \partial_x^k w\right).$$

The H^m-orthogonal projection $P_N^m : H^m(\Omega) \to \mathbb{P}_N$ is such a mapping that for any $v \in H^m(\Omega)$,

$$\left(v - P_N^m v, \phi\right)_m = 0, \quad \forall \phi \in \mathbb{P}_N. \tag{5.10}$$

If $v \in H^r(\Omega)$ and $0 \le \mu \le m \le r$, then

$$\|v - P_N^m v\|_\mu \le cN^{r-\mu}\|v\|_r. \tag{5.11}$$

We also need another kind of orthogonal projection in the numerical analysis of spectral methods. Let

$$\begin{aligned} \mathbb{P}_N^{m,0} &= \left\{\phi \in \mathbb{P}_N \,\middle|\, \partial_x^k \phi(x) = 0 \text{ on } \partial\Omega, |k| \le m-1\right\}, \\ a_m(v,w) &= \sum_{|k|=m} \left(\partial_x^k v, \partial_x^k w\right). \end{aligned}$$

The H_0^m-orthogonal projection $P_N^{m,0} : H_0^m(\Omega) \to \mathbb{P}_N^{m,0}$ is such a mapping that for any $v \in H_0^m(\Omega)$,

$$a_m\left(v - P_N^{m,0} v, \phi\right) = 0, \quad \forall\, \phi \in \mathbb{P}_N^{m,0}.$$

If $v \in H^r(\Omega) \cap H_0^m(\Omega)$ and $0 \le \mu \le m \le r$, then

$$\left\|v - P_N^{m,0} v\right\|_\mu \le cN^{\mu-r}\|v\|_r. \tag{5.12}$$

Let $x^{(j)} = \left(x_1^{(j_1)}, \ldots, x_n^{(j_n)}\right), 0 \le j_q \le N, x_q^{(j_q)}$ being the Legendre-Gauss type interpolation points. Ω_N is the set of all $x^{(j)}$. The Legendre interpolant $I_N v(x) \in \mathbb{P}_N$ of a function $v \in C(\bar{\Omega})$ is defined by

$$I_N v\left(x^{(j)}\right) = v\left(x^{(j)}\right), \quad x \in \Omega_N.$$

If $v \in H^r(\Omega), r > \dfrac{n}{2}$ and $0 \le \mu \le r$, then

$$\|v - I_N v\|_\mu \le cN^{2\mu + \frac{n}{2} - r}\|v\|_r. \tag{5.13}$$

In some special cases, we can improve the above result. For Legendre-Gauss interpolation and $r > \dfrac{n}{2}$,

$$\|v - I_N v\| \le cN^{-r}\|v\|_r.$$

For Legendre-Gauss-Lobatto interpolation and $0 \le \mu \le \min(1, 2r - n)$,

$$\|v - I_N v\|_\mu \le cN^{\mu-r}\|v\|_r.$$

For the Chebyshev approximations in several dimensions, let $\Omega = (-1,1)^n$ and $\omega(x) = \prod_{q=1}^{n} \left(1 - x_q^2\right)^{-\frac{1}{2}}$. The Chebyshev polynomial of degree l is

$$T_l(x) = \prod_{q=1}^{n} T_{l_q}(x_q).$$

The Chebyshev transformation of a function $v \in L^2_\omega(\Omega)$ is as

$$Sv(x) = \sum_{|l|=0}^{\infty} \hat{v}_l T_l(x)$$

with the Chebyshev coefficients

$$\hat{v}_l = \left(\frac{2}{\pi}\right)^n \prod_{q=1}^{n} \frac{1}{c_{l_q}} \int_\Omega v(x) T_l(x) \omega(x)\, dx, \quad |l| = 0, 1, \ldots,$$

where for $1 \le q \le n, c_{l_q} = 2$ for $l_q = 0$ and $c_{l_q} = 1$ for $l_q \ge 1$. For any $\phi \in \mathbb{P}_N$ and $1 \le p \le q \le \infty$.

$$\|\phi\|_{L^q_\omega} \le cN^{\frac{n}{p} - \frac{n}{q}} \|\phi\|_{L^p_\omega}. \tag{5.14}$$

Also for any $\phi \in \mathbb{P}_N, 2 \le p \le \infty$ and $|k| = m$,

$$\left\|\partial_x^k \phi\right\|_{L^p_\omega} \le cN^{2m} \|\phi\|_{L^p_\omega} \tag{5.15}$$

and for any $r \ge 0$,

$$\|\phi\|_{r,\omega} \le cN^{2r} \|\phi\|_\omega. \tag{5.16}$$

The L^2_ω-orthogonal projection of a function $v \in L^2_\omega(\Omega)$ is

$$P_N v(x) = \sum_{|l|=0}^{N} \hat{v}_l T_l(x).$$

If $r \ge 0$ and $\mu \le r$, then for any $v \in H^r_\omega(\Omega)$,

$$\|v - P_N v\|_{\mu,\omega} \le cN^{\sigma(\mu,r)} \|v\|_r, \tag{5.17}$$

$\sigma(\mu, r)$ is given in (2.53). Furthermore, we define the inner product of $H^m_\omega(\Omega)$ by

$$(v, w)_{m,\omega} = \sum_{0 \le |k| \le m} \left(\partial_x^k v, \partial_x^k w\right)_\omega.$$

The H^m_ω-orthogonal projection $P^m_N : H^m_\omega(\Omega) \to \mathbb{P}_N$ is such a mapping that for any $v \in H^m_\omega(\Omega)$,

$$(v - P^m_N v, \phi)_{m,\omega} = 0, \quad \forall\, \phi \in \mathbb{P}_N.$$

If $v \in H^r_\omega(\Omega)$ and $0 \le \mu \le 1 \le r$, then

$$\left\|v - P^1_N v\right\|_{\mu,\omega} \le cN^{\mu-r}\|v\|_{r,\omega}. \tag{5.18}$$

Moreover, let

$$\tilde{a}_{m,\omega}(v,w) = \sum_{|k|=m} \left(\partial^k_x v, \partial^k_x w\right)_\omega \tag{5.19}$$

and

$$a_{m,\omega}(v,w) = \sum_{|k|=m} \left(\partial^k_x v, \partial^k_x (w\omega)\right). \tag{5.20}$$

In particular, $\tilde{a}_\omega(v,w) = \tilde{a}_{1,\omega}(v,\omega)$ and $a_\omega(v,w) = a_{1,\omega}(v,w)$. The $H^1_{0,\omega}$-orthogonal projection $\tilde{P}^{1,0}_N : H^1_{0,\omega}(\Omega) \to \mathbb{P}^0_N$ is such a mapping that for any $v \in H^1_{0,\omega}(\Omega)$,

$$\tilde{a}_\omega\left(v - \tilde{P}^{1,0}_N v, \phi\right) = 0, \quad \forall\, \phi \in \mathbb{P}^0_N.$$

If $v \in H^r_\omega(\Omega) \cap H^1_{0,\omega}(\Omega)$ and $r \ge 1$, then

$$\left\|v - \tilde{P}^1_N v\right\|_{1,\omega} \le cN^{1-r}\|v\|_{r,\omega}.$$

The other $H^1_{0,\omega}$-orthogonal projection $P^{1,0}_N : H^1_{0,\omega}(\Omega) \to \mathbb{P}^0_N$ is such a mapping that for any $v \in H^1_{0,\omega}(\Omega)$,

$$a_\omega\left(v - P^{1,0}_N v, \phi\right) = 0, \quad \forall\, \phi \in \mathbb{P}^0_N. \tag{5.21}$$

If $v \in H^r_\omega(\Omega) \cap H^1_{0,\omega}(\Omega)$ and $0 \le \mu \le 1 \le r$, then

$$\left\|v - P^{1,0}_N v\right\|_{\mu,\omega} \le cN^{\mu-r}\|v\|_{r,\omega}. \tag{5.22}$$

Let $x^{(j)} = \left(x^{(j_1)}_1, \ldots, x^{(j_n)}_n\right), 0 \le j_q \le N, x^{(j_q)}_q$ be the Chebyshev-Gauss type interpolation points. Ω_N is the set of all $x^{(j)}$. The Chebyshev interpolant $I_N v(x) \in \mathbb{P}_N$ of a function $v \in C(\bar{\Omega})$ is defined by

$$I_N v\left(x^{(j)}\right) = v\left(x^{(j)}\right), \quad x^{(j)} \in \Omega_N.$$

If $v \in H^r_\omega(\Omega), r > \dfrac{n}{2}$ and $0 \le \mu \le r$, then

$$\|v - I_N v\|_{\mu,\omega} \le cN^{2\mu-r}\|v\|_{r,\omega}. \tag{5.23}$$

We now turn to the orthogonal approximations on unbounded domains. First of all, let $\Omega = (0,\infty)^n$ and $\omega(x) = \exp\left(-\sum_{q=1}^{n} x_q\right)$. The Laguerre polynomial of degree l is

$$\mathcal{L}_l(x) = \prod_{q=1}^{n} \mathcal{L}_{l_q}(x_q).$$

The Laguerre transformation of a function $v \in L^2_\omega(\Omega)$ is

$$Sv(x) = \sum_{|l|=0}^{\infty} \hat{v}_l \mathcal{L}_l(x)$$

with the Laguerre coefficients

$$\hat{v}_l = \int_\Omega v(x)\mathcal{L}_l(x)\omega(x)\,dx, \quad |l| = 0,1,\ldots.$$

For any $\phi \in \mathbb{P}_N$ and $1 \le p \le q \le \infty$,

$$\|\phi\|_{L^q_\omega} \le cN^{\frac{n}{p}-\frac{n}{q}}\|\phi\|_{L^p_\omega}. \tag{5.24}$$

Also for $1 \le p \le n$,

$$\|\partial_p\phi\|_\omega \le cN\|\phi\|_\omega. \tag{5.25}$$

The L^2_ω-orthogonal projection of a function $v \in L^2_\omega(\Omega)$ is

$$P_N v(x) = \sum_{|l|=0}^{N} \hat{v}_l \mathcal{L}_l(x).$$

Let $\alpha > 0$ and

$$H^r_\omega(\Omega,\alpha) = \left\{ v \in H^r_\omega(\Omega) \,\middle|\, v\prod_{q=1}^{n} x_q^{\frac{\alpha}{2}} \in H^r_\omega(\Omega) \right\}$$

with the norm

$$\|v\|_{r,\omega,\alpha} = \left\| v\prod_{q=1}^{n}(1+x_q)^{\frac{\alpha}{2}} \right\|_{r,\omega}.$$

For any $v \in H^r_\omega(\Omega,\alpha)$ and $0 \le \mu \le r$,

$$\|v - P_N v\|_{\mu,\omega} \le cN^{\mu-\frac{r}{2}}\|v\|_{r,\omega,\alpha} \tag{5.26}$$

where α is the largest integer such that $\alpha < r+1$.

Finally let $\Omega = (-\infty,\infty)^n$ and $\omega(x) = \exp\left(-\sum_{q=1}^{n} x_q^2\right)$. The Hermite polynomial of degree l is

$$H_l(x) = \prod_{q=1}^{n} H_{l_q}(x_q).$$

The Hermite transformation of a function $v \in L^2_\omega(\Omega)$ is as

$$v(x) = \sum_{|l|=0}^{\infty} \hat{v}_l H_l(x)$$

with the Hermite coefficients

$$\hat{v}_l = (\pi)^{-\frac{n}{2}} \left(\prod_{q=1}^{n} 2^{l_q} (l_q)! \right)^{-1} \int_\Omega v(x) H_l(x) \omega(x)\, dx, \quad |l| = 0, 1, \ldots.$$

For any $\phi \in \mathbb{P}_N$ and $1 \le p \le q \le \infty$,

$$\|\phi\|_{L^q_\omega} \le cN^{\frac{5n}{6}\left(\frac{1}{p}-\frac{1}{q}\right)} \|\phi\|_{L^p_\omega}. \tag{5.27}$$

Also for $1 \le p \le n$,

$$\|\partial_p \phi\|_\omega \le c\sqrt{N} \|\phi\|_\omega. \tag{5.28}$$

The L^2_ω-orthogonal projection of $v \in L^2_\omega(\Omega)$ is

$$P_N v(x) = \sum_{|l|=0}^{N} \hat{v}_l H_l(x).$$

For any $v \in H^r_\omega(\Omega)$ and $0 \le \mu \le r$, we have

$$\|v - P_N v\|_{\mu,\omega} \le cN^{\frac{\mu}{2}-\frac{r}{2}} \|v\|_{r,\omega}. \tag{5.29}$$

The above results can be proven along the same lines as in Sections 2.2– 2.6. Some of them can be found in Canuto and Quarteroni (1982a), and Bernardi and Maday (1992a).

The filterings and the techniques for recovering the spectral accuracy in Section 2.7 can be generalized to the orthogonal approximations in several dimensions. For example, let $\Omega = (0, 2\pi)^n$ and $|l|_2 = \left(\sum_{q=1}^{n} l_q^2 \right)^{\frac{1}{2}}$. The filtering $R_N(\alpha, \beta)$ can be defined as

$$R_N(\alpha, \beta)\phi(x) = \sum_{|l| \le N} \prod_{q=1}^{n} \left(1 - \frac{|l|_2^\alpha}{N^\alpha} \right)^\beta \hat{\phi}_l e^{il\cdot x}, \quad \alpha, \beta \ge 1, \phi \in V_N.$$

For any $\phi \in \tilde{V}_N$ and $0 \le r - \mu \le \alpha$,

$$\|R_N(\alpha, \beta)\phi - \phi\|_\mu \le c\beta N^{\mu - r} \|\phi\|_r.$$

If in addition $\mu \geq 0$, then

$$\|R_N(\alpha,\beta)\phi - \phi\|_\mu \leq c\beta N^{\mu-r}|\phi|_r. \tag{5.30}$$

The L^∞-error estimates between $R_N(\alpha,\beta)P_N v(x)$ and $v(x)$ also exist. For instance, for any $\phi \in V_N$, we define the filtering $R_N(\alpha,\beta)$ by

$$R_N(\alpha,\beta)\phi(x) = \sum_{|l|_2\leq N}\left(1-\frac{|l|_2^\alpha}{N^\alpha}\right)^\beta \hat{\phi}_l e^{il\cdot x}, \quad \beta = \left[\frac{n-1}{2}\right]+1.$$

Set

$$\omega(v;\rho) = \max_{|x-x'|\leq\rho}|v(x)-v(x')|, \quad \omega_m(v;\rho) = \max_{|k|=m}\omega\left(\partial_x^k v;\rho\right).$$

Cheng and Chen (1956) proved that if $\alpha > m+1$ or $\alpha = m+1, m$ being even, then for any $v \in C_p^m(\bar{\Omega})$,

$$R_N(\alpha,\beta)P_N v(x) - v(x) = \mathcal{O}\left(\frac{1}{N^m}\omega_m\left(v;\frac{1}{N}\right)\right),$$

and if $\alpha = m+1, m$ being odd, then for any $v \in C_p^m(\bar{\Omega})$,

$$R_N(\alpha,\beta)P_N v(x) - v(x) = \mathcal{O}\left(\frac{\ln N}{N^m}\omega_m\left(v;\frac{1}{N}\right)\right).$$

For the sake of simplicity, we still use the notations $L^\infty(\Omega), L_\omega^q(\Omega), W_\omega^{q,r}(\Omega)$ for the spaces of vector functions in the following discussions.

5.2. Spectral Methods for Multi-Dimensional Nonlinear Systems

The aim of this section is to present the spectral methods for multi-dimensional nonlinear systems. In particular, we are interested in spectral approximations of isolated solutions. For simplicity, we focus on the three-dimensional periodic problem of steady Navier-Stokes equation. As in Section 4.6, we shall rewrite it as an abstract operator equation in certain space. Then we use the Fourier spectral methods and Fourier pseudospectral method to solve it numerically. Finally the convergences of both schemes are given.

Navier-Stokes equation is the fundamental equation describing incompressible fluid flows. Let $\Omega = (0,2\pi)^3$ and $L_0^2(\Omega)$ be the subspace of $L^2(\Omega)$, containing all functions with the zero average over Ω. Denote by $U(x), P(x)$ and $\mu > 0$ the speed vector, the pressure and the kinetic viscosity, respectively. The components of $U(x)$ are $U^{(j)}(x), 1 \leq j \leq 3$. Moreover $f(x)$ stands for the body force, and $f(x) = \left(f^{(1)}(x), f^{(2)}(x), f^{(3)}(x)\right)$. The steady Navier-Stokes equation is of the form

$$\begin{cases} (U \cdot \nabla)U - \mu \Delta U + \nabla p = f, & x \in \Omega, \\ \nabla \cdot U = 0, & x \in \Omega. \end{cases} \tag{5.31}$$

In this section, we assume that all functions have period 2π for all variables. For fixing the values of P, we require that $P \in L_0^2(\Omega)$. Let $\mathcal{V} = \{v \mid v \in \mathcal{D}_p(\Omega), \nabla \cdot v = 0\}$ and V be the closure of $\mathcal{V}$ in $H^1(\Omega)$. Set

$$a(v, w) = \int_\Omega \nabla v(x) \cdot \nabla w(x)\, dx,$$

$$\tilde{b}(v, w, z) = \int_\Omega v(x) \cdot (w(x) \cdot \nabla) z(x)\, dx.$$

A weak solution of (5.31) is defined as a pair $(U, P) \in V \times L_0^2(\Omega)$, satisfying

$$\mu a(U, v) - \tilde{b}(U, U, v) = (f, v), \quad \forall\, v \in H_p^1(\Omega). \tag{5.32}$$

If (f, v) is a linear continuous functional in V, then by the Fixed Point Theorem, (5.32) has at least one weak solution.

If (5.32) has a unique solution, then it is not difficult to approximate it directly by the Fourier spectral method or the Fourier pseudospectral method. But in those cases, the trial functions ϕ_l have to satisfy the incompressibility condition, i.e., $\nabla \cdot \phi_l = 0$. It is not convenient in actual computations. We can also approximate (5.31) with some trial functions whose divergences do not vanish. However, we need to evaluate U coupled with P. In order to remove this problem, we try to eliminate the pressure P in the first formula of (5.31) by the technique of Orszag (1971b). To do this, let

$$\phi_l(x) = \left(\frac{1}{\sqrt{2\pi}}\right)^3 e^{il \cdot x}$$

and for any $v \in L^2(\Omega)$,

$$\hat{v}_l = (v, \phi_l), \quad |l| = 0, 1, \ldots.$$

For $l \neq 0$, we multiply (5.31) by $\overline{\nabla \phi_l}$ and integrate the result over Ω. Then it follows that

$$\int_\Omega (-\mu \Delta U(x) + (U(x) \cdot \nabla) U(x) - f(x)) \cdot \overline{\nabla \phi_l}(x)\, dx = -\int_\Omega \nabla P(x) \cdot \overline{\nabla \phi_l}(x)\, dx.$$

Integrating by parts and using $\nabla \cdot U = 0$, we get

$$-\int_\Omega \Delta U(x) \cdot \overline{\nabla \phi_l}(x)\,dx = \int_\Omega (\nabla \cdot U(x))\overline{\Delta \phi_l}(x)\,dx = 0.$$

Therefore

$$\int_\Omega ((U(x)\cdot\nabla)U(x) - f(x)) \cdot \overline{\nabla\phi_l}(x)\,dx = \int_\Omega P(x)\overline{\Delta\phi_l}(x)\,dx = -|l|_2^2 \int_\Omega P(x)\bar{\phi}_l(x)\,dx \tag{5.33}$$

where $|l|_2^2 = \sum_{q=1}^{n} l_q^2$. Since $P \in L_0^2(\Omega)$, $\hat{P}_0 = 0$. Thus for any $v \in H_p^1(\Omega)$,

$$\begin{aligned}
\int_\Omega \nabla P(x)\cdot\bar{v}(x)\,dx &= -\int_\Omega P(x)\overline{\nabla\cdot v}(x)\,dx = i\sum_l \left(l\cdot\bar{\hat{v}}_l\right)(P,\phi_l) \\
&= -i\sum_{l\neq 0}\frac{1}{|l|_2^2}\left(l\cdot\bar{\hat{v}}_l\right)\int_\Omega ((U(x)\cdot\nabla)U(x) - f(x))\cdot\overline{\nabla\phi_l}(x)\,dx \\
&= -\sum_{l\neq 0}\frac{1}{|l|_2^2}\left(l\cdot\bar{\hat{v}}_l\right)\int_\Omega [l\cdot((U(x)\cdot\nabla)U(x) - f(x))]\,\bar{\phi}_l(x)\,dx
\end{aligned}$$

from which and (5.32), we obtain that

$$\begin{aligned}
&-\mu\int_\Omega \Delta U(x)\cdot\bar{v}(x)\,dx + \int_\Omega (U(x)\cdot\nabla)U(x)\cdot\bar{v}(x)\,dx \\
&\quad -\sum_{l\neq 0}\frac{1}{|l|_2^2}\left(l\cdot\bar{\hat{v}}_l\right)\int_\Omega [l\cdot((U(x)\cdot\nabla)U(x) - f(x))]\,\bar{\phi}_l(x)\,dx = \int_\Omega f(x)\cdot\bar{v}(x)\,dx.
\end{aligned}$$

By (5.33) and $P \in L_0^2(\Omega)$,

$$\begin{cases} (P,\phi_0) = 0, \\ (P,\phi_l) = -\dfrac{1}{|l|_2^2}\displaystyle\int_\Omega ((U(x)\cdot\nabla)U(x) - f(x))\cdot\overline{\nabla\phi_l}(x)\,dx, \quad \text{for } l\neq 0. \end{cases} \tag{5.34}$$

Set

$$\begin{aligned}
b(v,w,z) = &\int_\Omega [(w(x)\cdot\nabla)v(x)]\cdot\bar{z}(x)\,dx \\
&-\sum_{l\neq 0}\frac{l\cdot\bar{\hat{z}}_l}{|l|_2^2}\int_\Omega (l\cdot(w(x)\cdot\nabla)v(x))\bar{\phi}_l(x)\,dx,
\end{aligned} \tag{5.35}$$

$$\mathcal{F}(z) = \int_\Omega f(x)\cdot\bar{z}(x)\,dx - \sum_{l\neq 0}\frac{l\cdot\bar{\hat{z}}_l}{|l|_2^2}(l\cdot f,\phi_l). \tag{5.36}$$

Then we can derive a new representation of (5.32). It is to find $U \in H_p^1(\Omega)$ such that

$$\mu a(U,v) + b(U,U,v) = \mathcal{F}(v), \quad \forall\, v \in H_p^1(\Omega). \tag{5.37}$$

While P is determined by (5.34).

We can rewrite (5.32) in another way. To do this, let

$$\begin{aligned} b^0(v,w,z) &= i\sum_{l\neq 0}\int_\Omega (w(x)\cdot l)\left(v(x)\cdot\bar{\hat{z}}_l\right)\bar{\phi}_l(x)\,dx \\ &\quad - i\sum_{l\neq 0}\frac{l\cdot\bar{\hat{z}}_l}{|l|_2^2}\int_\Omega (w(x)\cdot l)(v(x)\cdot l)\bar{\phi}_l(x)\,dx, \qquad (5.38)\\ b^1(v,w,z) &= -\int_\Omega (\nabla\cdot w(x))(v(x)\cdot\bar{z}(x))dx + \sum_{l\neq 0}\frac{l\cdot\bar{\hat{z}}_l}{|l|_2^2}(\nabla\cdot w,(\bar{v}(x)\cdot l)\phi_l). \qquad (5.39)\end{aligned}$$

By integrating by parts in (5.35), we obtain that

$$b(v,w,z) = b^0(v,w,z) + b^1(v,w,z), \quad \forall\, v,w\in H_p^1(\Omega). \tag{5.40}$$

Moreover, if in addition $w\in V$, then $b^1(v,w,z)=0$. Thus (5.32) can be presented as follows. It is to find $U\in H_p^1(\Omega)$, such that

$$\mu a(U,v) + b^0(U,U,v) = \mathcal{F}(v), \quad \forall\, v\in H_p^1(\Omega). \tag{5.41}$$

We now verify the equivalence of the representations (5.31), (5.37) with (5.34) and (5.41) with (5.34).

Lemma 5.1. *For any* $v,w\in H_p^1(\Omega)$,

$$|a(v,w)| \le |v|_1|w|_1,$$

and for any $w\in L^2(\Omega)$,

$$|\mathcal{F}(w)| \le c\|f\|\|w\|.$$

Proof. The first result is clear. Next, we have

$$|\mathcal{F}(w)| \le \|f\|\|w\| + \sum_{l\neq 0}|\hat{w}_l|\,|(f,\phi_l)| \le 2\|f\|\|w\|.$$

∎

Lemma 5.2. *For any* $v,w,z\in H_p^1(\Omega)$,

$$|b(v,w,z)| \le c|v|_1\|w\|_1\|z\|_{3/4},$$
$$|b(v,w,z)| \le c\|v\|_1\|w\|_1\|z\|_1.$$

Proof. We have from (5.35) that

$$b(v,w,z)=\int_\Omega((w(x)\cdot\nabla)v(x))\,\bar{z}(x)\,dx-\sum_{j,m=1}^{3}\int_\Omega w^{(j)}(x)\partial_j v^{(m)}(x)\bar{\sigma}^{(m)}(x)\,dx \quad (5.42)$$

where

$$\sigma^{(m)}(x)=\sum_{l\neq 0}\frac{l_m}{|l|_2^2}\,(l\cdot\hat{z}_l)\,\phi_l(x),\quad 1\le m\le 3.$$

Obviously $|\hat{\sigma}_l|\le|\hat{z}_l|$ which implies that for any $r\ge 0$,

$$\|\sigma\|_r\le c\|z\|_r.$$

Hence by imbedding theory, (5.42) reads

$$\begin{aligned}|b(v,w,z)| &\le c|v|_1\|w\|_{L^4}\,(\|z\|_{L^4}+\|\sigma\|_{L^4})\\ &\le c|v|_1\|w\|_{\frac{3}{4}}\|z\|_{\frac{3}{4}}\le c|v|_1\|w\|_1\|z\|_{\frac{3}{4}}.\end{aligned}$$

The rest of the proof is clear. ∎

Lemma 5.3. *Let $v,w,z\in H_p^1(\Omega)$. We have that for $v\in W^{1,4}(\Omega)$,*

$$|b(v,w,z)|\le c|v|_{W^{1,4}}\|w\|\|z\|_1,$$

and for $v\in H^{\frac{7}{4}}(\Omega)$,

$$|b(v,w,z)|\le c\|v\|_{\frac{7}{4}}\|w\|\|z\|_1.$$

Proof. For any $v\in W^{1,4}(\Omega)$, we deduce from (5.42) that

$$|b(v,w,z)|\le c|\nabla v|_{L^4}\|w\|\,(\|z\|_{L^4}+\|\sigma\|_{L^4})\le c|v|_{W^{1,4}}\|w\|\|z\|_1.$$

The second result follows from the fact that $H^{\frac{7}{4}}(\Omega)\subset W^{1,4}(\Omega)$. ∎

Lemma 5.4. *Let $v,w,z\in H_p^1(\Omega)$. If in addition $v\in W^{1,\frac{12}{5}}(\Omega)$ and $w\in L^{12}(\Omega)$, then*

$$|b(v,w,z)|\le c\|v\|_{W^{1,\frac{12}{5}}}\|w\|_{L^{12}}\|z\|.$$

If $w\in H^{\frac{7}{4}}(\Omega)\cap V$, then

$$|b(v,w,z)|\le c\|v\|\|w\|_{\frac{7}{4}}\|z\|_1.$$

Proof. By (5.42),

$$|b(v,w,z)| \le c\|\nabla v\|_{L^{\frac{12}{5}}}\|w\|_{L^{12}}(\|z\|+\|\sigma\|) \le c\|v\|_{W^{1,\frac{12}{5}}}\|w\|_{L^{12}}\|z\|.$$

If $w \in H^{\frac{7}{4}}(\Omega)\cap V$, then by (5.38),

$$|b(v,w,z)| = \left|b^0(v,w,z)\right| \le \sum_{j,m=1}^{3}\left|\sum_{l\neq 0}|l|_2 A_l^{j,m}\left(\int_\Omega w^{(j)}(x)v^{(m)}(x)\bar{\phi}_l(x)\,dx\right)\right| \tag{5.43}$$

with

$$A_l^{j,m} = \frac{l_j}{|l|_2}\bar{\hat{z}}_l^{(m)} + \frac{l_j l_m}{|l|_2^3}\left(l\cdot\bar{\hat{z}}_l\right).$$

Clearly $\left|A_l^{j,m}\right| \le 2\,|\hat{z}_l|$ and so

$$\sum_{j,m=1}^{3}\sum_{l\neq 0}|l|^{2r}\left|A_l^{j,m}\right|^2 \le c\|z\|_r^2, \quad r\ge 0.$$

Thus (5.43) implies that

$$\left|b^0(v,w,z)\right| \le c\|v\|\|w\|_{L^\infty}\|z\|_1 \le c\|v\|\|w\|_{\frac{7}{4}}\|z\|_1. \tag{5.44}$$

■

Lemma 5.5. *For any $v,w,z\in H_p^1(\Omega)$,*

$$\left|b^0(v,w,z)\right| \le c\|v\|_1\|w\|_1\|z\|_{\frac{3}{4}}.$$

Proof. Let $A_l^{j,m}$ be the same as before, and $v,w,z\in \mathcal{D}(\Omega)$. By (5.43),

$$\begin{aligned}\left|b^0(v,w,z)\right| &\le \sum_{j,m=1}^{3}\left(\sum_{l\neq 0}|l|_2^{2r}\left|A_l^{j,m}\right|^2\right)^{\frac{1}{2}}\left(\sum_{l\neq 0}|l|_2^{2-2r}\left(\int_\Omega w^{(j)}(x)v^{(m)}(x)\bar{\phi}_l(x)\,dx\right)^2\right)^{\frac{1}{2}}\\ &\le c\|z\|_r\sum_{j,m=1}^{3}\left\|w^{(j)}v^{(m)}\right\|_{1-r}.\end{aligned}$$

By putting $r=0,1$, we obtain from imbedding theory that

$$\left|b^0(v,w,z)\right| \le c\|z\|\|v\|_{W^{1,4}}\|w\|_{W^{1,4}} \le c\|z\|\|v\|_{\frac{7}{4}}\|w\|_{\frac{7}{4}},$$
$$\left|b^0(v,w,z)\right| \le c\|z\|_1\|v\|_{L^4}\|w\|_{L^4} \le c\|z\|_1\|v\|_{\frac{3}{4}}\|w\|_{\frac{3}{4}}.$$

By the space interpolation, we assert that

$$\left|b^0(v,w,z)\right| \le c\|z\|_{\frac{3}{4}}\|v\|_1\|w\|_1.$$

■

Lemma 5.6. *Let $v, w, z \in H_p^1(\Omega)$. We have that*

$$\left|b^0(v,w,z)\right| \le c\|v\|_{W^{1,4}}\|w\|\|z\|_1, \quad \text{for } v \in W^{1,4}(\Omega)$$

and

$$\left|b^0(v,w,z)\right| \le c\|v\|_{\frac{7}{4}}\|w\|\|z\|_1, \quad \text{for } v \in H^{\frac{7}{4}}(\Omega).$$

On the other hand, if $w \in H^{\frac{7}{4}}(\Omega)$, then

$$\left|b^0(v,w,z)\right| \le c\|v\|\|w\|_{\frac{7}{4}}\|z\|_1.$$

Proof. It is easy to derive the above estimates. For instance, the last one comes immediately from (5.44). ∎

Theorem 5.1. *The representation (5.31) is equivalent to (5.37) with (5.34), or to (5.41) with (5.34).*

Proof. It is clear that a solution of (5.31) is also a solution of (5.37) with (5.34). Now, suppose that (U, P) is a solution of (5.37) with (5.34). We have from Lemma 5.1, Lemma 5.2 and (5.37) that for any $v \in \mathcal{D}(\Omega)$,

$$|\mu a(U,v)| \le c\left(\|U\|_1|U|_1 + \|f\|\right)\|v\|_{\frac{3}{4}}.$$

Moreover by Green's formula, $\Delta U \in H^{-\frac{3}{4}}(\Omega)$ and so $U \in H_p^1(\Omega) \cap H^{\frac{5}{4}}(\Omega)$. By imbedding theory, $U \in L^{12}(\Omega) \cap W^{1,\frac{12}{5}}(\Omega)$. By virtue of (5.37) and Lemma 5.4,

$$|\mu a(U,v)| \le c\left(\|U\|_{W^{1,\frac{12}{5}}}\|U\|_{L^{12}} + \|f\|\right)\|v\|.$$

By Green's formula again, $\Delta U \in L^2(\Omega)$ and so $U \in H_p^1(\Omega) \cap H^2(\Omega)$. The integration by parts in (5.37) leads to (5.31). A standard argument tells us that the normal derivative of U is also periodic and so $U \in H_p^2(\Omega)$. Furthermore by (5.34),

$$\sum_l |l|_2^2\,|(P,\phi_l)|^2 \le \sum_l \sum_{j=1}^{3}\left|\left((U\cdot\nabla)U^{(j)} - f^{(j)}, \phi_l\right)\right|^2.$$

Since $(U\cdot\nabla)U - f \in L^2(\Omega)$, the above series converges and so $P \in H_p^1(\Omega)$. Using (5.34) again, we have

$$\int_\Omega P(x)\,dx = (\sqrt{2\pi})^3\,(P,\phi_0) = 0$$

and so $P \in H_p^1(\Omega) \cap L_0^2(\Omega)$. Finally it follows from (5.35)–(5.37) that $\mu a(U, \nabla\phi_l) = 0$. On the other hand, by integrating by parts, we obtain that for $l \neq 0$,

$$\begin{aligned} a(U, \nabla\phi_l) &= \sum_{j,m=1}^{3} \int_\Omega \partial_m U^{(j)}(x) \partial_m \partial_j \bar{\phi}_l(x)\, dx = -\sum_{j,m=1}^{3} l_m l_j \int_\Omega \partial_m U^{(j)}(x) \bar{\phi}_l(x)\, dx \\ &= i|l|_2^2 \sum_{j=1}^{3} l_j \hat{U}_l^{(j)} = i|l|_2^2 (\nabla \cdot U, \phi_l). \end{aligned} \tag{5.45}$$

In addition, $(\nabla \cdot U, \phi_0) = 0$ and so $\nabla \cdot U = 0$. The above statements lead to that (U, P) is a solution of (5.31).

We now turn to the equivalence of (5.37) with (5.34) and (5.41) with (5.34). If U is a solution of (5.37), then for any $v \in H_p^1(\Omega)$, $b(U,U,v) = b^0(U,U,v)$ and so it is also a solution of (5.41). Conversely, if U is a solution of (5.41), then by taking $v = \nabla\phi_l, l \neq 0$ in (5.41) and arguing as in the previous paragraph, we can deduce that $\nabla \cdot U = 0$. Hence $U \in V$ and for any $v \in H_p^1(\Omega)$, $b(U,U,v) = b^0(U,U,v)$. This means that U is a solution of (5.37). This completes the proof. ∎

We are going to construct the Fourier-spectral scheme. For this purpose, we put (5.37) in an abstract framework. Let $\lambda = \dfrac{1}{\mu}$ and $E(\Omega) = \left(H_p^{\frac{3}{4}}(\Omega)\right)'$. Since $H_0^{\frac{3}{4}}(\Omega) \subset H_p^{\frac{3}{4}}(\Omega)$, $E(\Omega) \subset H^{-\frac{3}{4}}(\Omega)$ and also $E(\Omega) \subset \left(H_p^1(\Omega)\right)'$. We define an inner product on $H_p^1(\Omega)$ by

$$(v, w)_1 = a(v, w) + (v, w), \quad \forall\, v, w \in H_p^1(\Omega), \tag{5.46}$$

and a linear mapping $A : \left(H_p^1(\Omega)\right)' \to H_p^1(\Omega)$ by

$$(Av, w)_1 = \langle v, w \rangle, \quad \forall\, w \in H_p^1(\Omega). \tag{5.47}$$

Moreover, by the theory of linear elliptic equations, A is a continuous mapping from $H^r(\Omega)$ with $r \geq -1$ into $H^{r+2}(\Omega) \cap H_p^1(\Omega)$. Since $E(\Omega) \subset H^{-\frac{3}{4}}(\Omega)$, A is continuous from $E(\Omega)$ into $H^{\frac{5}{4}}(\Omega) \cap H_p^1(\Omega)$. But the latter space is compactly contained in $H_p^1(\Omega)$, whence A is a compact mapping from $E(\Omega)$ into $H_p^1(\Omega)$. Define $G : \mathbb{R}^1 \times H_p^1(\Omega) \to E(\Omega)$ by

$$\langle G(\lambda, v), w \rangle = \lambda(b(v, v, w) - \mathcal{F}(w)) - (v, w). \tag{5.48}$$

From Lemmas 5.1 and 5.2, we know that for any $\lambda \in \mathbb{R}^1$ and $v \in H_p^1(\Omega)$,

$$\|G(\lambda, v)\|_E \leq c|\lambda| \left(\|v\|_1^2 + \|f\|\right).$$

Moreover G is a C^∞ mapping. Let D be the Fréchet derivative of G. Then for any positive integer $m, D^m G$ is bounded over any bounded subset of $\mathbb{R}^1 \times H_p^1(\Omega)$. According to (5.46) and (5.48), (5.37) is equivalent to

$$(U,v)_1 + \langle G(\lambda,U),v\rangle = 0, \quad \forall\, v \in H_p^1(\Omega).$$

Setting $F : \mathbb{R}^1 \times H_p^1(\Omega) \to H_p^1(\Omega)$,

$$F(\lambda,v) = v + AG(\lambda,v). \tag{5.49}$$

By (5.46)–(5.48), we get an equivalent form of (5.37). It is to find $U \in H_p^1(\Omega)$ such that

$$F(\lambda,U) = 0. \tag{5.50}$$

Let Λ^* be a compact interval in $\mathbb{R}^1$, and assume that $(\lambda, U(\lambda))$ is a branch of isolated solutions of (5.50). So there exists $\delta > 0$ such that for all $\lambda \in \Lambda^*$ and $v \in H_p^1(\Omega)$,

$$\|(I + ADG(\lambda,U(\lambda)))v\|_1 \geq \delta\|v\|_1\,. \tag{5.51}$$

Now let V_N be the same as in Section 5.1, and P_N be the L^2-orthogonal projection upon V_N. It can be checked that for any $v \in H_p^1(\Omega)$,

$$(P_N v - v, \phi)_1 = 0, \quad \forall\, \phi \in V_N. \tag{5.52}$$

(5.4) asserts that P_N is a continuous mapping in $H_p^r(\Omega)$ with $r \geq 0$. Let $A_N = P_N A$. By (5.47) and (5.52), $A_N \in \mathcal{L}\left(\left(H_p^1(\Omega)\right)', V_N\right)$ and

$$(A_N v, \phi)_1 = \langle v, \phi\rangle, \quad \forall\, \phi \in V_N. \tag{5.53}$$

Setting $F_N : \mathbb{R}^1 \times V_N \to V_N$,

$$F_N(\lambda,v) = v + A_N G(\lambda,v).$$

Maday and Quarteroni (1982b) provided a Fourier-spectral approximation to (5.50). It is to find $u_N \in V_N$ such that

$$F_N(\lambda,u_N) = 0. \tag{5.54}$$

By (5.53), (5.54) is equivalent to finding $u_N \in V_N$ such that

$$a(u_N,\phi) + \lambda b(u_N,u_N,\phi) = \lambda\mathcal{F}(\phi), \quad \forall\, \phi \in V_N.$$

By arguing as in the proof of Theorem 5.1, it can be shown that $\nabla \cdot u_N = 0$.

Theorem 5.2. *Let $\{(\lambda, U(\lambda)), \lambda \in \Lambda^*\}$ be a branch of isolated solutions of (5.50). There exists a neighborhood Ⓗ of the origin in $H^1_p(\Omega)$, and for N sufficiently large, a unique C^∞ mapping $\lambda \in \Lambda^* \to u_N(\lambda) \in V_N$ such that for any $\lambda \in \Lambda^*, F_N(\lambda, u_N(\lambda)) = 0$ and $U(\lambda) - u_N(\lambda) \in$ Ⓗ. Moreover, if the mapping $\lambda \in \Lambda^* \to U(\lambda) \in H^r_p(\Omega)$ is continuous for some $r \geq 1$, then for any λ, there is a positive constant b_1 depending on $\|U(\lambda)\|_r$ such that*

$$\|U(\lambda) - u_N(\lambda)\|_1 + N\|U(\lambda) - u_N(\lambda)\| \leq b_1 N^{1-r}.$$

Proof. We first use Lemma 4.26, with $B = E(\Omega), W = H^1_p(\Omega), \sqcap_N = P_N, \mathcal{A} = A$ and $\mathcal{G}(\lambda, v) = G(\lambda, v)$. Clearly condition (i) of Lemma 4.26 is satisfied. Further, by (5.4) and a density argument,

$$\lim_{N\to\infty} \|P_N v - v\|_1 = 0, \quad \forall\, v \in H^1_p(\Omega). \tag{5.55}$$

Since A is a compact mapping from $E(\Omega)$ into $H^1_p(\Omega)$, we derive that

$$\lim_{N\to\infty} \|A_N - A\|_{\mathcal{L}(E,H^1_p)} = \lim_{N\to\infty} \|(P_N - I)A\|_{\mathcal{L}(E,H^1_p)} = 0. \tag{5.56}$$

Therefore both (4.132) and (4.133) are also fulfilled. So there exists a unique branch $\{(\lambda, u_N(\lambda)), \lambda \in \Lambda^*\}$. The combination of (4.134), (5.4) and (5.50) yields that

$$\begin{aligned}
\|U(\lambda) - u_N(\lambda)\|_1 &\leq c\,(\|U(\lambda) - P_N U(\lambda)\|_1 + \|(A - A_N)G(\lambda, U(\lambda))\|_1) \\
&\leq c\,(\|U(\lambda) - P_N U(\lambda)\|_1 + \|(I - P_N)AG(\lambda, U(\lambda))\|_1) \\
&\leq c\|U(\lambda) - P_N U(\lambda)\|_1 \leq cN^{1-r}\|U(\lambda)\|_r.
\end{aligned} \tag{5.57}$$

For the L^2-error estimate, we put

$$H = L^2(\Omega), \quad K = \left\{v \in H^{\frac{7}{4}}(\Omega)\,|\,\nabla \cdot v = 0\right\}.$$

Lemmas 5.3 and 5.4 ensure the first part of condition (ii) in Lemma 4.27. By $A \in \mathcal{L}\left(\left(H^1_p\right)', H^1_p\right)$, condition (iii) of Lemma 4.27 is also satisfied. Since $A - A_N = (I - P_N)A$, (5.4) ensures condition (4.135). Finally by (5.3), (5.4) and (5.57),

$$\begin{aligned}
\|U(\lambda) - u_N(\lambda)\|_K &\leq \|U(\lambda) - P_N U(\lambda)\|_K + \|P_N U(\lambda) - u_N(\lambda)\|_K \\
&\leq \|U(\lambda) - P_N U(\lambda)\|_K \\
&\quad + N^{\frac{3}{4}}\,(\|U(\lambda) - P_N U(\lambda)\|_1 + \|U(\lambda) - u_N(\lambda)\|_1) \\
&\leq cN^{-\frac{1}{4}}\|U(\lambda)\|_2.
\end{aligned}$$

Hence condition (iv) of Lemma 4.27 is also fulfilled. Now, using (4.136) and (5.4), we obtain that

$$\begin{aligned}\|U(\lambda)-u_N(\lambda)\| &\le c\|F_N(\lambda,U(\lambda))\| = \|(A-A_N)G(\lambda,U(\lambda))\| \\ &= \|(P_N-I)AG(\lambda,U(\lambda))\| = \|(P_N-I)U(\lambda)\| \le cN^{-r}\|U(\lambda)\|_r\end{aligned}$$

which completes the proof. ∎

In actual computations, we can use the Newton procedure to evaluate $u_N(\lambda)$. For instance, let $u_N^{(m)}(\lambda)$ be the m-th iterated value of $u_N(\lambda)$. Then the iteration is as follows

$$\begin{aligned}&\left(I+A_N DG\left(\lambda,u_N^{(m)}(\lambda)\right)\right)\left(u_N^{(m+1)}(\lambda)-u_N^{(m)}(\lambda)\right)\\ &\quad=-\left(u_N^{(m)}(\lambda)+A_N G\left(\lambda,u_N^{(m)}(\lambda)\right)\right),\quad m\ge 0.\end{aligned}$$

On the other hand, the approximate pressure $p_N(\lambda)=p_N(x,\lambda)$ could be determined by

$$p_N(\lambda)=\sum_{|l|\le N}\hat{p}_{N,l}(\lambda)\phi_l(\lambda)$$

with

$$\hat{p}_{N,0}(\lambda)=0,\quad \hat{p}_{N,l}(\lambda)=\frac{i}{|l|_2^2}\left(l\cdot\left(\left(u_N(\lambda)\cdot\nabla\right)u_N(\lambda)-f\right),\phi_l\right),\quad l\ne 0. \tag{5.58}$$

Theorem 5.3. *Assume that the hypotheses of Theorem 5.2 hold, $P(\lambda)\in H_p^r(\Omega)\cap L_0^2(\Omega)$, and $r\ge 1$. Then there is a positive constant b_2 depending on $\|U(\lambda)\|_r$ such that*

$$\|P(\lambda)-p_N(\lambda)\|+N^{-1}\|P(\lambda)-p_N(\lambda)\|_1\le b_2N^{-r}.$$

Proof. Let $P_N(\lambda)=P_NP(\lambda)$ and $\tilde{P}(\lambda)=p_N(\lambda)-P_N(\lambda)$. By (5.34) and (5.58), $\left(\tilde{P}(\lambda),\phi_0\right)=0$ and for $l\ne 0$,

$$\begin{aligned}\left(\tilde{P}(\lambda),\phi_l\right) &= \frac{i}{|l|_2^2}\left(l\cdot\left[\left(u_N(\lambda)\cdot\nabla\right)u_N(\lambda)-(U(\lambda)\cdot\nabla)U(\lambda)\right],\phi_l\right)\\ &= \frac{i}{|l|_2^2}\left(l\cdot\left[\left(\left(u_N(\lambda)-U(\lambda)\right)\cdot\nabla\right)u_N(\lambda)+\left(U(\lambda)\cdot\nabla\right)\left(u_N(\lambda)-U(\lambda)\right)\right],\phi_l\right).\end{aligned}$$

By integrating by parts and using $\nabla\cdot U(\lambda)=\nabla\cdot u_N(\lambda)=0$, we obtain that for $l\ne 0$,

$$\begin{aligned}\left|\left(\tilde{P}(\lambda),\phi_l\right)\right| &= \left|\int_\Omega\left[\left(U(\lambda)-u_N(\lambda)\right)\cdot\frac{l}{|l|_2}\right]\left[\left(U(\lambda)+u_N(\lambda)\right)\cdot\frac{l}{|l|_2}\right]\bar{\phi}_l\,dx\right|\\ &\le \sum_{j,m=1}^{3}\left|\left(\left(U^{(j)}(\lambda)-u_N^{(j)}(\lambda)\right)\left(U^{(m)}(\lambda)+u_N^{(m)}(\lambda)\right),\bar{\phi}_l\right)\right|.\end{aligned}$$

Consequently,

$$\begin{aligned}
\left\|\tilde{P}(\lambda)\right\| &\leq \sum_{j,m=1}^{3}\left\|\left(U^{(j)}(\lambda)-u_N^{(j)}(\lambda)\right)\left(U^{(m)}(\lambda)+u_N^{(m)}(\lambda)\right)\right\|^2 \\
&\leq \sum_{j=1}^{3}\left\|U^{(j)}(\lambda)-u_N^{(j)}(\lambda)\right\|^2\sum_{m=1}^{3}\left\|U^{(m)}(\lambda)+u_N^{(m)}(\lambda)\right\|_{L^\infty} \\
&\leq c\left\|U(\lambda)-u_N(\lambda)\right\|^2\left\|U(\lambda)+u_N(\lambda)\right\|_2^2 \\
&\leq c\|U(\lambda)\|_2^2\left\|U(\lambda)-u_N(\lambda)\right\|^2,
\end{aligned}$$

from which together with Theoerem 5.2 and (5.4), we find that

$$\|P(\lambda)-p_N(\lambda)\| \leq \|\tilde{P}(\lambda)\| + \|P(\lambda)-P_N(\lambda)\| \leq b_2N^{-r}.$$

■

We now consider the Fourier-pseudospectral approximation. Define the mapping $G: \mathbb{R}^1 \times H_p^1(\Omega) \to E(\Omega)$ by

$$\langle G(\lambda,v),w\rangle = \lambda\left(b^0(v,v,w)-\mathcal{F}(w)\right)-(v,w). \tag{5.59}$$

Due to Lemmas 5.1 and 5.5, b^0 possesses all the continuity properties of b, except the first estimate in Lemma 5.4, which was used only in order to prove the regularity of $U(\lambda)$ in Theorem 5.1. So the mapping defined by (5.59) has the properties of G defined by (5.48). In particular, (5.41) can be equivalently written in the form (5.50) where G is given by (5.59).

Now, let Ω_N be the set of all Fourier interpolation points as in Section 5.1. For any $v, w \in C(\bar{\Omega})$, the discrete inner product is given by

$$(v,w)_N = \left(\frac{2\pi}{2N+1}\right)^3 \sum_{x\in\Omega_N} v(x)\bar{w}(x).$$

Denote by $I_N v$ and $\tilde{v}_l$ the Fourier interpolant of $v \in C(\bar{\Omega})$ and the Fourier coefficients. Assume that $f \in C(\bar{\Omega})$. Define $b_N^0(v,w,z): V_N \to C(\bar{\Omega})$ and $\mathcal{F}_N(w): V_N \to C(\bar{\Omega})$ by

$$b_N^0(v,w,z) = i\sum_{|l|\leq N, l\neq 0}\left[\left((w\cdot l)\left(v\cdot\bar{\tilde{z}}_l\right),\phi_l\right)_N - \frac{l\cdot\bar{\tilde{z}}_l}{\|l\|_2^2}\left((w\cdot l)(v\cdot l),\phi_l\right)_N\right],$$

$$\mathcal{F}_N(w) = (f,w)_N - \sum_{|l|\leq N, l\neq 0}\frac{1}{\|l\|_2^2}\left(l\cdot\bar{\tilde{w}}_l\right)(l\cdot f,\phi_l)_N.$$

Further, define $\bar{G}_N : \mathbb{R}^1 \times V_N \to V_N'$ as

$$\langle \bar{G}_N(\lambda, v), w\rangle = \lambda \left(b_N^0(v,v,w) - \mathcal{F}(w)\right) - (v,w)_N, \tag{5.60}$$

and let

$$\langle G_N(\lambda, v), w\rangle = \langle \bar{G}_N(\lambda, v), P_N w\rangle, \quad \forall\, \lambda \in \mathbb{R}^1,\, v \in V_N,\, w \in H_p^1(\Omega). \tag{5.61}$$

Let A_N be the same as in (5.53) and set $F_N : \mathbb{R}^1 \times V_N \to V_N$,

$$F_N(\lambda, v) = v + A_N G_N(\lambda, v).$$

A Fourier-pseudospectral approximation of (5.41) is to find $u_N \in V_N$ such that

$$F_N(\lambda, u_N) = 0. \tag{5.62}$$

By (5.53), (5.60) and (5.61), (5.62) is equivalent to finding $u_N \in V_N$ such that

$$a(u_N, \phi) + \lambda b_N^0(u_N, u_N, \phi) = \lambda \mathcal{F}_N(\phi), \quad \forall\, \phi \in V_N.$$

By an argument as in the proof of Theorem 5.1, it can be checked that $\nabla \cdot u_N = 0$. We have the following result.

Theorem 5.4. *Let $\{(\lambda, U(\lambda)), \lambda \in \Lambda^*\}$ be a branch of isolated solutions of (5.41), and $r_1 > \frac{3}{2}, r_2 > 2$. If $f \in H_p^{r_1}(\Omega)$ and the mapping $\lambda \in \Lambda^* \to U(\lambda) \in H_p^{r_2}(\Omega)$ is continuous, then for $N \geq N_0$ large enough, there exists a positive constant b_3 depending on $\|U(\lambda)\|_{r_2}$ and $\|f\|_{r_1}$, and a unique C^∞ mapping $\lambda \in \Lambda^* \to u_N(\lambda) \in V_N$ such that for any $\lambda \in \Lambda^*, F_N(\lambda, u_N(\lambda)) = 0$ and*

$$\|U(\lambda) - u_N(\lambda)\|_1 \leq b_3 \left(N^{1-r_2} + N^{-r_1}\right).$$

In order to obtain an improved L^2-error estimate, let $M > N$ be a suitably chosen integer. The corresponding set of Fourier interpolation points, the discrete inner product and the Fourier interpolation are denoted by $\Omega_M, (\cdot,\cdot)_M$ and I_M. If $M > N^{\frac{r}{r-1}}$, then

$$\|f - I_M f\| \leq c M^{1-r} \|f\|_{r-1} \leq c N^{-r} \|f\|_{r-1}.$$

Define the mapping $G_N^*(\lambda, v)$ as

$$G_N^*(\lambda, v) = G_N(\lambda, v) + \lambda \left(\mathcal{F}^*(v) - \mathcal{F}_N(v)\right)$$

where $\mathcal{F}_N^* : V_N \to C(\bar{\Omega})$ is given by

$$\mathcal{F}_N^*(v) = (f, v)_M - \sum_{|l| \le N, l \ne 0} \frac{1}{|l|_2^2} (l \cdot \bar{\tilde{v}}_l)(l \cdot f, \phi_l)_M \,.$$

Let $F_N^*(\lambda, v) = v + A_N G_N^*(\lambda, v)$. A new Fourier pseudospectral approximation is to find $u_N^* \in V_N$ such that

$$F_N^*(\lambda, u_N^*(\lambda)) = 0. \tag{5.63}$$

We have the following result.

Theorem 5.5. *Let $\{(\lambda, U(\lambda)), \lambda \in \Lambda^*\}$ be a branch of isolated solutions of (5.41) and u_N^* be the solution of (5.63). If $f \in H^{r-1}(\Omega)$ with $r > 2$ and the mapping $\lambda \in \Lambda^* \to U(\lambda) \in H_p^1(\Omega) \cap H_p^r(\Omega)$ is continuous, then there exists a positive constant b_4 depending on $\|U(\lambda)\|_r$ and $\|f\|_{r-1}$ such that for any $\lambda \in \Lambda^*$,*

$$\|U(\lambda) - u_N^*(\lambda)\| \le b_4 N^{-r}.$$

In actual computations, we can use the Newton procedure to solve (5.62) or (5.63). For example,

$$\begin{aligned}&\left(I + A_N D G_N^*\left(\lambda, u_N^{*(m)}(\lambda)\right)\right)\left(u_N^{*(m+1)}(\lambda) - u_N^{*(m)}(\lambda)\right)\\ &\quad = -\left(u_N^{*(m)}(\lambda) + A_N G_N^*\left(\lambda, u_N^{*(m)}(\lambda)\right)\right), \quad m \ge 0.\end{aligned}$$

On the other hand, the approximate pressure $p_N^*(\lambda)$ is given by

$$p_N^*(\lambda) = \sum_{|l| \le N} \hat{p}_{N,l}^*(\lambda)\phi_l(x)$$

with $\hat{p}_{N,0}^*(\lambda) = 0$ and for $l \ne 0$,

$$\hat{p}_{N,l}^*(\lambda) = \frac{-1}{|l|_2^2}\left((u_N^*(\lambda) \cdot l)(u_N^*(\lambda) \cdot l) + if \cdot l, \phi_l\right)_M \,.$$

If $M > N^{\frac{r}{r-1}}$, then

$$\|P(\lambda) - p_N^*(\lambda)\| + N^{-1}\|P(\lambda) - p_N^*(\lambda)\| \le b_5 N^{-r} \tag{5.64}$$

where b_5 is a positive constant depending on $\|U(\lambda)\|_r$ and $\|f\|_{r-1}$.

Theorem 5.4, Theorem 5.5 and (5.64) were proved in Maday and Quarteroni (1982b).

For the Fourier approximations of unsteady Navier-Stokes equation, we refer to the early work in Orszag (1971a, 1971b), Fox and Orszag (1973a, 1973b), Fronberg (1977), and Moin and Kim (1980). Hald (1981), Kuo (1983), Guo (1985), Guo and Ma (1987), Ma and Guo (1987), Cao and Guo (1991), and Rushid, Cao and Guo (1994) provided various schemes with numerical analysis for compuational fluid dynamics. The applications of Fourier approximations to other multi-dimensional equations can be found in numerous articles, such as Wengle and Seifeld (1978), Feit, Flack and Steiger (1982), Guo, Cao and Tahira (1993), and Cao and Guo (1993b).

5.3. Spectral Methods for Nonlinear High Order Equations

Many nonlinear partial differential equations of high order occur in science and engineering, such as the Korteweg-de Vries equation possessing solitons, the Laudau-Ginzburg equation in the dynamics of solid-state phase transitions in shape memory alloys, etc.. In this section, we consider the spectral methods for such kind of equations. We shall take the initial-boundary value problem of two-dimensional Navier-Stokes equation in stream function representation as an example to describe the Legendre spectral method for solving initial-boundary value problems of nonlinear revolutionary equations of fourth order. The corresponding numerical solution converges to the weak solution of original problem. The spectral accuracy is obtained for smooth solution. The prediction-correction method is also used for raising the total accuracy of numerical solution. The numerical experiments coincide with the theoretical analysis.

Let $\Omega = (-1,1)^2$ and $\partial\Omega$ be a non-slip wall. The initial-boundary value problem of two-dimensional Navier-Stokes equation is of the form

$$\begin{cases} \partial_t U + (U\cdot\nabla)U - \mu\Delta U + \nabla P = f, & x\in\Omega,\ 0<t\le T, \\ \nabla\cdot U = 0, & x\in\Omega,\ 0\le t\le T, \\ U = 0, & x\in\partial\Omega,\ 0<t\le T, \\ U(x,0) = U_0(x), & x\in\bar\Omega. \end{cases} \tag{5.65}$$

Let $\mathcal{V} = \{v \mid v\in\mathcal{D}(\Omega),\ \nabla\cdot v = 0\}$. The closures of $\mathcal{V}$ in $L^2(\Omega)$ and $H^1(\Omega)$ are denoted by H and V respectively. Let

$$a(v,w) = \int_\Omega \nabla v(x)\cdot\nabla w(x)\,dx,$$

$$b(v,w,z) = \int_\Omega (w(x)\cdot\nabla)v(x)\cdot z(x)\,dx.$$

The weak form of (5.65) is as follows

$$\begin{cases} (\partial_t U, v) + b(U,U,v) + \mu a(U,v) = (f,v), & \forall\, v \in V,\, 0 < t \le T, \\ U(0) = U_0, & x \in \bar{\Omega}. \end{cases} \tag{5.66}$$

It is shown in Lions (1969) that if $U_0 \in H$ and $f \in L^2(0,T;V')$, then (5.66) has a unique solution in $L^2(0,T;V) \cap L^\infty(0,T;H)$. The solution of (5.66) possesses the conservation

$$\|U(t)\|^2 + 2\mu \int_0^t |U(s)|_1^2\, ds = \|U_0\|^2 + 2\int_0^t (f(s), U(s))\, ds. \tag{5.67}$$

For numerical simulations, we can approximate (5.66) directly. But in this case, we should construct a trial function space whose elements satisfy the incompressibility. It is not convenient in actual computations. To avoid it, several alternative expressions are concerned. Firstly, we can take the divergences of both sides of (5.65), and get

$$\partial_t(\nabla\cdot U) + \nabla\cdot((U\cdot\nabla)U) - \mu\Delta(\nabla\cdot U) + \Delta P = \nabla\cdot f.$$

By the incompressibility, we obtain

$$\begin{cases} \partial_t U + (U\cdot\nabla)U - \mu\Delta U + \nabla P = f, & x \in \Omega,\, 0 < t \le T, \\ \Delta P + \Phi(U) = \nabla\cdot f, & x \in \Omega,\, 0 < t \le T \end{cases} \tag{5.68}$$

where

$$\Phi(U) = 2\left(\partial_2 U^{(1)}\partial_1 U^{(2)} - \partial_1 U^{(1)}\partial_2 U^{(2)}\right).$$

Conversely if U and P satisfy (5.68) and $\nabla\cdot U(0) = 0$, then for all $t \ge 0, \nabla\cdot U = 0$. Thus (5.68) is equivalent to (5.65) provided that U and P are suitably smooth. The second formula of (5.68) is a Poisson equation for P at each time t. If we use it to evaluate the pressure, then we need certain boundary conditions for the pressure. As pointed out in Gresho and Sani (1987), we have that

$$\partial_\nu P(x,t) = \nu\cdot(\mu\Delta U(x,t) + f(x,t)), \quad x \in \partial\Omega,\, 0 < t \le T$$

where $\partial_\nu P$ is the outward normal derivative of P. A simplified condition is

$$\partial_\nu P(x,t) = N(x,t), \quad x \in \partial\Omega,\, 0 < t \le T.$$

But the first condition is too complicated in actual calculations, while the simplified one is not physical.

The second way is to introduce the vorticity $H(x,t)$, the stream function $\Psi(x,t)$ and

$$J(v,w) = \partial_2 w \partial_1 v - \partial_1 w \partial_2 v, \quad \forall\, v,w \in H^1(\Omega).$$

Then (5.65) is equivalent to

$$\begin{cases} \partial_t H + J(H,\Psi) - \mu\Delta H = \nabla \times f, & x \in \Omega,\ 0 < t \le T, \\ -\Delta\Psi = H, & x \in \Omega,\ 0 \le t \le T, \\ H(x,0) = H_0(x), & x \in \bar{\Omega}. \end{cases} \tag{5.69}$$

Since the incompressibility is already included, we do not need to construct the trial function space with free-divergence. Clearly $\Psi(x,t) = 0$ on the boundary. But it is not easy to deal with the boundary values of the vorticity. In some literature, it is assumed approximately that $H(x,t) = 0$ on $\partial\Omega$. But it is not physical.

The third representation of Navier-Stokes equation is the stream function form. Let

$$G(v,w) = \partial_2 v \partial_1(\Delta w) - \partial_1 v \partial_2(\Delta w).$$

Then the stream function form is as follows

$$\begin{cases} \partial_t \Delta\Psi + G(\Psi,\Psi) - \mu\Delta^2\Psi = g, & x \in \Omega,\ 0 < t \le T, \\ \Psi = \dfrac{\partial\Psi}{\partial n} = 0, & x \in \Omega,\ 0 < t \le T, \\ \Psi(x,0) = \Psi_0(x), & x \in \bar{\Omega} \end{cases} \tag{5.70}$$

where $g = -\nabla \times f$. The main merits of this expression are remedying the problems mentioned in the above and keeping the physical boundary values naturally. But the appearance of the terms $G(\Psi,\Psi)$ and $\Delta^2\Psi$ brings some difficulties in theoretical analysis.

We shall derive the weak form of (5.70). For any $v,z \in W^{1,4}(\Omega)$ and $w \in H^2(\Omega)$, let

$$J(v,w,z) = (\Delta w, \partial_2 v \partial_1 z - \partial_1 v \partial_2 z).$$

Obviously

$$J(v,w,z) + J(z,w,v) = 0. \tag{5.71}$$

It can be verified that

$$J(v,w,z) = -(G(v,w),z). \tag{5.72}$$

We can estimate $|J(v,w,z)|$ in several ways.

Lemma 5.7. *For any* $v \in H_0^2(\Omega)$,

$$\|\nabla v\|_{L^4} \leq 2^{\frac{1}{4}} \|\nabla v\|^{\frac{1}{2}} \|\Delta v\|^{\frac{1}{2}}.$$

Proof. By a density argument, it suffices to prove the conclusion for $v \in C_0^\infty(\Omega)$. For such a function, we observe that for $\bar{x} = (\bar{x}_1, \bar{x}_2) \in \Omega$,

$$(\partial_1 v(\bar{x}))^2 \leq 2 \int_{-1}^{\bar{x}_1} |\partial_1 v(x_1, \bar{x}_2)| \left|\partial_1^2 v(x_1, \bar{x}_2)\right| dx_1,$$
$$(\partial_1 v(\bar{x}))^2 \leq 2 \int_{-1}^{\bar{x}_2} |\partial_1 v(\bar{x}_1, x_2)| \, |\partial_1 \partial_2 v(\bar{x}_1, x_2)| \, dx_2, \text{ etc.}$$

Thus

$$\|\partial_1 v\|_{L^4}^4 \leq 4 \|\partial_1 v\|^2 \|\partial_1 \partial_2 v\| \left\|\partial_1^2 v\right\| \leq 2 \|\partial_1 v\|^2 \|\Delta v\|^2.$$

Similarly

$$\|\partial_2 v\|_{L^4}^4 \leq 2 \|\partial_2 v\|^2 \|\Delta v\|^2.$$

This completes the proof. ∎

Lemma 5.8. *For any* $v, w \in H_0^2(\Omega)$,

$$|J(v, v, w)| \leq 2\|\nabla v\| \|\Delta v\| \|\Delta w\|.$$

Proof. By integrating by parts, we obtain that

$$\begin{aligned} J(v,v,w) &= -\int_\Omega \left(\partial_1 v \partial_2 v \partial_1^2 w - (\partial_1 v)^2 \partial_1 \partial_2 w + (\partial_2 v)^2 \partial_1 \partial_2 w - \partial_1 v \partial_2 v \partial_2^2 w\right) dx \\ &\quad + \int_\Omega \left(\partial_1 v \partial_2 w \partial_1^2 v - \partial_1 v \partial_1 w \partial_1 \partial_2 v + \partial_2 v \partial_2 w \partial_1 \partial_2 v - \partial_2 v \partial_1 w \partial_2^2 v\right) dx. \end{aligned}$$

Integrating the last integral by parts again, we find that it vanishes. So by Lemma 5.7,

$$\begin{aligned} |J(v,v,w)| &\leq \|\partial_1 v\|_{L^4} \|\partial_2 v\|_{L^4} \left\|\partial_1^2 w\right\| + \|\partial_1 v\|_{L^4}^2 \|\partial_1 \partial_2 w\| \\ &\quad + \|\partial_2 v\|_{L^4}^2 \|\partial_1 \partial_2 w\| + \|\partial_1 v\|_{L^4} \|\partial_2 v\|_{L^4} \left\|\partial_2^2 w\right\| \\ &\leq \left(2 \|\partial_1 v\|_{L^4}^2 \|\partial_2 v\|_{L^4}^2 + \|\partial_1 v\|_{L^4}^4 + \|\partial_2 v\|_{L^4}^4\right)^{\frac{1}{2}} \\ &\quad \cdot \left(2 \|\partial_1 \partial_2 w\|^2 + \left\|\partial_1^2 w\right\|^2 + \left\|\partial_2^2 w\right\|^2\right)^{\frac{1}{2}} \\ &\leq 2\|\nabla v\| \|\Delta v\| \|\Delta w\|. \end{aligned}$$

∎

Lemma 5.9. *If $v, z \in H_0^2(\Omega)$ and $w \in H^2(\Omega)$, then*

$$|J(v,w,z)| \leq 2\|\Delta w\| \left(\|\nabla v\|\|\Delta v\|\|\nabla z\|\|\Delta z\|\right)^{\frac{1}{2}}.$$

Proof. By Lemma 5.7,

$$\begin{aligned} |J(v,w,z)| &\leq \sqrt{2}\|\Delta w\|\|\nabla v\|_{L^4}\|\nabla z\|_{L^4} \\ &\leq 2\|\Delta w\| \left(\|\nabla v\|\|\Delta v\|\|\nabla z\|\|\Delta z\|\right)^{\frac{1}{2}}. \end{aligned}$$

■

Lemma 5.10. *For any $v, w, z \in H_0^2(\Omega)$,*

$$|J(v,w,z)| \leq 2\|\nabla w\|^{\frac{1}{2}}\|\Delta w\|^{\frac{1}{2}} \left(\|\Delta v\|\|\nabla z\|^{\frac{1}{2}}\|\Delta z\|^{\frac{1}{2}} + \|\Delta z\|\|\nabla v\|^{\frac{1}{2}}\|\Delta v\|^{\frac{1}{2}}\right).$$

Proof. By integrating by parts, we have that

$$\begin{aligned} J(v,w,z) &= \int_\Omega \left(\partial_1 w \partial_2 z \partial_1^2 v - \partial_1 w \partial_1 z \partial_1 \partial_2 v + \partial_2 w \partial_2 z \partial_1 \partial_2 v - \partial_2 w \partial_1 z \partial_2^2 v\right) dx \\ &+ \int_\Omega \left(\partial_1 v \partial_1 w \partial_1 \partial_2 z - \partial_2 v \partial_1 w \partial_1^2 z + \partial_1 v \partial_2 w \partial_2^2 z - \partial_2 v \partial_2 w \partial_1 \partial_2 z\right) dx. \quad (5.73) \end{aligned}$$

Then the desired result follows from an argument as before. ■

Lemma 5.11. *Let $v, w, z \in H_0^2(\Omega)$. If in addition $w \in H^r(\Omega)$ with $r > 2$, then*

$$|J(v,w,z)| \leq c\|\Delta v\|\|w\|_r\|\nabla z\|.$$

Proof. We first let $2 < r < \frac{5}{2}$. Set $r = \mu + 1, \alpha = \frac{2}{2-\mu}$ and $\beta = \frac{2}{\mu - 1}$. Then $H^1(\Omega) \hookrightarrow L^\beta(\Omega)$ and $H^{r-2}(\Omega) \hookrightarrow L^\alpha(\Omega)$. Thus

$$\begin{aligned} |J(v,w,z)| &\leq \|\Delta w\|_{L^\alpha} \left(\|\partial_1 v\|_{L^\beta}\|\partial_2 z\| + \|\partial_2 v\|_{L^\beta}\|\partial_1 z\|\right) \\ &\leq c\|\Delta v\|\|w\|_r\|\nabla z\|. \end{aligned}$$

Next, let $r \geq \frac{5}{2}$. Since $H^{\mu-1}(\Omega) \hookrightarrow L^4(\Omega)$, we see that

$$|J(v,w,z)| \leq \sqrt{2}\|\Delta w\|_{L^4}\|\nabla v\|_{L^4}\|\nabla z\| \leq c\|\Delta v\|\|w\|_r\|\nabla z\|.$$

■

Lemma 5.12. *Let $v, w, z \in H_0^2(\Omega)$. If in addition $v \in H^r(\Omega)$ with $r > 2$, then*

$$|J(v,w,z)| \le c\|v\|_r\|\Delta w\|\|\nabla z\|,$$
$$|J(v,w,z)| \le c\|v\|_r\|\nabla w\|\|\Delta z\|.$$

Proof. Since $H^{r-1}(\Omega) \hookrightarrow L^\infty(\Omega)$, we get that

$$\begin{aligned} |J(v,w,z)| &\le \|\partial_1 v\|_{L^\infty}\|\Delta w\|\,\|\partial_2 z\| + \|\partial_2 v\|_{L^\infty}\|\Delta w\|\,\|\partial_1 z\| \\ &\le c\|v\|_r\|\Delta w\|\|\nabla z\|. \end{aligned}$$

On the other hand, we have that

$$\left|\int_\Omega \partial_2 v \partial_1 w \partial_1^2 z\, dx\right| \le \|\partial_2 v\|_{L^\infty}\|\nabla w\|\|\Delta z\|.$$

Moreover for $2 < r < \frac{5}{2}$,

$$\left|\int_\Omega \partial_1\partial_2 v \partial_1 w \partial_1 z\, dx\right| \le \|\partial_1\partial_2 v\|_{L^\alpha}\|\partial_1 w\|_{L^2}\|\partial_1 z\|_{L^\beta},$$

and for $r \ge \frac{5}{2}$,

$$\left|\int_\Omega \partial_1\partial_2 v \partial_1 w \partial_1 z\, dx\right| \le \|\partial_1\partial_2 v\|_{L^4}\|\partial_1 w\|_{L^2}\|\partial_1 z\|_{L^4}.$$

The remaining terms in (5.73) can be estimated similarly. Finally, the second conclusion follows from imbedding theory as in the proof of the previous lemma. ∎

The weak solution of (5.70) is a function $\Psi \in L^2(0,T;H_0^2) \cap L^\infty(0,T;H^1)$ such that

$$\begin{cases} (\partial_t\nabla\Psi(t),\nabla v) + J(\Psi(t),\Psi(t),v) \\ \quad +\mu(\Delta\Psi(t),\Delta v) + \langle g(t),v\rangle_{\mathcal{L}(H^{-2}(\Omega),H_0^2(\Omega))} = 0, & \forall\, v \in H_0^2(\Omega),\ 0 < t \le T, \\ \Psi(x,0) = \Psi_0(x), & x \in \bar\Omega. \end{cases} \tag{5.74}$$

It is shown in Guo, He and Mao (1997) that if $\Psi_0 \in H_0^1(\Omega)$ and $g \in L^2(0,T;H^{-2})$, then (5.74) has a unique solution. The solution satisfies the conservation

$$\|\nabla\Psi(t)\|^2 + 2\mu\int_0^t \|\Delta\Psi(s)\|^2\, ds + s\int_0^t \langle g(s),\Psi(s)\rangle_{\mathcal{L}(H^{-2}(\Omega),H_0^2(\Omega))}\, ds = \|\nabla\Psi_0\|^2. \tag{5.75}$$

We can adopt Legendre approximations or Chebyshev approximations to solve (5.74) numerically. For the sake of simplicity, we focus on the Legendre spectral method. Let $\mathbb{P}_N^{2,0} = \mathbb{P}_N \cap H_0^2(\Omega)$ and

$$(v,w)_1 = (\nabla v, \nabla w), \qquad \forall\, v,w \in H_0^1(\Omega),$$
$$(v,w)_2 = (\Delta v, \Delta w), \qquad \forall\, v,w \in H_0^2(\Omega).$$

It can be checked that in the space $H_0^2(\Omega)$, the inner product $(v,w)_2$ here is equal to the inner product $a_2(v,w)$ in Section 5.1, and $|v|_2 = \|\Delta v\|$. The orthogonal projections $\tilde{P}_N^{m,0} : H_0^m(\Omega) \to \mathbb{P}_N^{2,0}, m = 1,2,$ are two such mappings that for any $v \in H_0^m(\Omega)$,

$$\left(v - \tilde{P}_N^{m,0} v, \phi\right)_m = 0, \quad \forall\, \phi \in \mathbb{P}_N^{2,0},\ m = 1,2.$$

It is not difficult to show that for any $v \in H_0^1(\Omega)$, $\left\|v - \tilde{P}_N^{1,0} v\right\|_1 \to 0$ as $N \to \infty$. Clearly the projection $\tilde{P}^{2,0}$ is the same as the projection $P_N^{2,0}$ in Section 5.1. Now let ψ_N be the approximation to Ψ. A Legendre spectral scheme for (5.74) is to find $\psi_N(x,t) \in \mathbb{P}_N^{2,0}$ for all $t \in \bar{R}_\tau(T)$ such that

$$\begin{cases} (D_\tau \nabla \psi_N(t), \nabla \phi) + J\left(\psi_N(t) + \delta\tau D_\tau \psi_N(t), \psi_N(t), \phi\right) & \\ \qquad + \mu\left(\Delta\left(\psi_N(t) + \sigma\tau D_\tau \psi_N(t)\right), \Delta\phi\right) + (g(t), \phi) = 0, & \forall\, \phi \in \mathbb{P}_N^{2,0},\ t \in \bar{R}_\tau(T), \\ \psi(0) = \psi_{N,0}(x), & x \in \bar{\Omega} \end{cases} \tag{5.76}$$

where σ, δ are parameters, $0 \le \sigma, \delta \le 1$ and

$$\|\Psi_0 - \psi_{N,0}\|_1 \longrightarrow 0, \quad \text{as } N \to \infty. \tag{5.77}$$

For $\Psi_0 \in H_0^1(\Omega)$, we take $\psi_{N,0} = \tilde{P}_N^{1,0}\Psi_0$. While for $\Psi_0 \in H_0^2(\Omega)$, we take $\psi_{N,0} = \tilde{P}_N^{2,0}\Psi_0$. Both of them satisfy (5.77). Thus the approximations on the initial level are well defined. Now, assume that $\psi_N(t-\tau)$ is known. Let τ be suitably small and

$$A(v,w) = (\nabla v, \nabla w) + \mu\sigma\tau(\Delta v, \Delta w) + \delta\tau J\left(v, \psi_N(t-\tau), w\right), \quad \forall\, v,w \in \mathbb{P}_N^{2,0}.$$

By Lemma 5.9, $A(v,w)$ is a continuous bilinear elliptic form on $\mathbb{P}_N^{2,0} \times \mathbb{P}_N^{2,0}$. Hence by Lemma 2.1, the numerical solution at the time t is determined uniquely. So (5.76) has a unique solution as long as $\Psi_0 \in H_0^1(\Omega)$ and $g \in L^2(0,T;L^2(\Omega))$. Especially, if $\sigma = \delta = \frac{1}{2}$, then we deduce from (5.71) and (5.76) that

$$\begin{aligned} &\|\nabla\psi_N(t)\|^2 + \frac{1}{2}\mu\tau \sum_{s\in\bar{R}_\tau(t-\tau)} \|\Delta\psi_N(s) + \Delta\psi_N(s+\tau)\|^2 \\ &\quad + \tau \sum_{s\in\bar{R}_\tau(t-\tau)} (g(s), \psi_N(s) + \psi_N(s+\tau))\, ds = \|\nabla\psi_{N,0}\|^2 \end{aligned}$$

which is a reasonable simulation of (5.75).

In practical computations, a good choice of basis functions is essentially important. Shen (1994) proposed a nice set of basis functions. Let $L_m(y)$ be the Legendre polynomial of degree m. Set

$$\chi_m(y) = d_m\left(L_m(y) - \frac{2(2m+5)}{2m+7}L_{m+2}(y) + \frac{2m+3}{2m+7}L_{m+4}(y)\right),$$

$$d_m = \frac{1}{\sqrt{2(2m+3)^2(2m+5)}}, \qquad 0 \le m \le N-4.$$

Let $l = (l_1, l_2)$ and

$$\phi_l(x) = \chi_{l_1}(x_1)\chi_{l_2}(x_2).$$

Then

$$\mathbb{P}_N^{2,0} = \{\phi_l(x) \mid 0 \le l_1, l_2 \le N-4\}.$$

By the orthogonality of Legendre polynomials, both the matrix with the elements $(\phi_l, \phi_{l'})$ and the matrix with the elements $(\nabla\phi_l, \nabla\phi_{l'})$ are pentadiagonal. Hence this choice provides an efficient algorithm. In particular,

$$\int_{-1}^{1} \partial_y^2\chi_m(y)\partial_y^2\chi_{m'}(y)\,dy = \delta_{m,m'}$$

which simplifies the calculation and improves the condition number of the related matrix essentially.

The numerical solution $\psi_N(x,t)$ given by (5.76) is a discrete approximation to $\Psi(x,t)$. But we can creat a continuous version by the interpolation in time t. Define two step functions $\Psi_N^{(m)}(x,t) : [0,T] \to \mathbb{P}_N^{2,0}, m = 1,2$, such that for $t \in \bar{R}_\tau(T)$,

$$\Psi_N^{(m)}(x,t) = \psi_N(s + (2-m)\tau), \quad t \in [s, s+\tau).$$

Guo and He (1996) proved the following results.

Theorem 5.6. *Let $\Psi_0 \in H_0^1(\Omega), g \in C(0,T;L^2)$ and $\psi_{N,0} = \tilde{P}_N^{1,0}\Psi_0, \tau \le \frac{C_*}{N^4}, C_*$ being a certain positive constant independent of N and τ. Then $\Psi_N^{(m)}, m = 1,2$, converge to the weak solution Ψ in the following sense:*

(i) $\Psi_N^{(m)} \to \Psi$ in $L^2(0,T;H_0^2(\Omega))$ weakly, as $N \to \infty, \tau \to 0$;

(ii) $\Psi_N^{(m)} \to \Psi$ in $L^\infty(0,T;H_0^1(\Omega))$ weak-star, as $N \to \infty$, $\tau \to 0$;

(iii) $\Psi_N^{(m)} \to \Psi$ *in* $L^2(0,T;H_0^1(\Omega))$ *strongly, as* $N \to \infty$, $\tau \to 0$.

If in addition $\tau = o(N^{-4})$, then

(iv) $\Psi_N^{(m)} \to \Psi$ *in* $L^2(0,T;H_0^2(\Omega))$ *strongly, as* $N \to \infty$, $\tau \to 0$.

It is noted that if $\sigma, \delta > \frac{1}{2}$, then the conclusions (i)–(iii) hold for any $\tau = \mathcal{O}(N^{-4})$. If in addition $\Psi_0 \in H_0^2(\Omega)$ and $\psi_{N,0} = P_N^{2,0}\Psi_0$, then the conclusions (i)–(iii) are valid for any $\tau \to 0$. Furthermore, if $\tau = \mathcal{O}(N^{-4})$ or $\Psi_0 \in H_0^2(\Omega)$, $\psi_{N,0} = P_N^{2,0}\Psi_0$, $\tau \to 0$, then the conclusion (iv) holds for $\sigma = \delta = 1$.

If Ψ is smooth, then (5.76) provides the numerical solution with high accuracy. Let $q_0 \geq 0$ and ξ be a positive constant chosen below. Set

$$E(v,t) = \|\nabla v(t)\|^2 + \mu\tau\left(\sigma + \frac{\xi}{2}\right)\|\Delta v(t)\|^2 + \frac{1}{2}\mu\tau \sum_{s\in\bar{R}_\tau(t-\tau)} \left(\|\Delta v(s)\|^2 + q_0\tau\,\|D_\tau \nabla v(s)\|^2\right). \tag{5.78}$$

Theorem 5.7. *Assume that*

(i) $\Psi \in L^2\left(0,T;H^{r_0}\cap H_0^2\right) \cap H^1\left(0,T;H^{r_1}\right) \cap H^2\left(0,T;H^1\right), \Psi_0 \in H_0^2(\Omega), g \in C(0,T;L^2)$ *and* $\psi_{N,0} = P_N^{2,0}\Psi_0$;

(ii) $r_0 \geq r_1 = 2$;

(iii) *one of the following conditions is fulfilled*

(a) $\sigma > \frac{1}{2}, \tau = o\left(N^{-\lambda}\right), \lambda \geq \max\left(\frac{4}{3}, 7 - 2r_0\right)$,

(b) $\sigma = \frac{1}{2}, \tau = \mathcal{O}\left(N^{-4}\right)$,

(c) $\sigma < \frac{1}{2}, \tau N^4 \leq \dfrac{2}{\mu c_p^2(1-2\sigma)}$, c_p *being a positive constant such that for any* $v \in \mathbb{P}_N^{2,0}$, $\|\Delta v\| \leq c_p N^2 \|\nabla v\|$.

Then there exists a positive constant b_1 *depending only on* μ, Ω *and the norms of* Ψ *and* g *in the mentioned spaces, such that for all* $t \in \bar{R}_\tau(T)$,

$$E\left(\Psi - \psi_N, t\right) \leq b_1\left(\tau^2 + N^{3-2r_0} + N^{-4} + \tau N^{-2}\right).$$

If $\sigma > \frac{1}{2}$ *and* δ *is suitably large, then the above result is vaild for any* $\tau \to 0$.

Proof. Let $\Psi_N = P_N^{2,0}\Psi$. We derive from (5.74) that

$$\begin{cases} (D_\tau \nabla \Psi_N(t), \nabla\phi) + J(\Psi_N(t) + \delta\tau D_\tau \Psi_N(t), \Psi_N(t), \phi) + \mu(\Delta(\Psi_N(t) + \sigma\tau D_\tau \Psi_N(t)), \Delta\phi) \\ \quad = -(g(t), v) + B(\Psi(t), \Psi_N(t), \phi), \quad \forall\, \phi \in \mathbb{P}_N^{2,0}, \quad 0 < t \le T, \\ \Psi_N(0) = P_N^{2,0}\Psi_0, \quad x \in \bar{\Omega} \end{cases} \tag{5.79}$$

where

$$\begin{aligned} B(\Psi, \Psi_N, \phi) \;&=\; (\nabla(D_\tau \Psi_N - \partial_t \Psi), \nabla\phi) + J(\Psi_N + \delta\tau D_\tau \Psi_N, \Psi_N, \phi) \\ &\quad -J(\Psi, \Psi, \phi) + \mu\sigma\tau(D_\tau \Delta\Psi, \Delta\phi). \end{aligned}$$

Let $\tilde{\Psi} = \psi_N - \Psi_N$. We obtain from (5.76) and (5.79) that

$$\begin{aligned} &\left(D_\tau \nabla\tilde{\Psi}(t), \nabla\phi\right) + J\left(\tilde{\Psi}(t) + \delta\tau D_\tau \tilde{\Psi}(t), \Psi_N(t) + \tilde{\Psi}(t), \phi\right) \\ &\quad + J\left(\Psi_N(t) + \delta\tau D_\tau \Psi_N(t), \tilde{\Psi}(t), \phi\right) + \mu\left(\Delta\left(\tilde{\Psi}(t) + \sigma\tau D_\tau \tilde{\Psi}(t)\right), \Delta\phi\right) \\ &\quad = -B(\Psi(t), \Psi_N(t), \phi), \quad \forall\, \phi \in \mathbb{P}_N^{2,0}. \end{aligned}$$

By taking $\phi = 2\tilde{\Psi}(t) + \xi\tau D_\tau \tilde{\Psi}(t)$ and using an argument as before, we obtain from (5.71) that

$$\begin{aligned} &D_\tau \|\nabla\tilde{\Psi}(t)\|^2 + \tau(\xi - 1)\left\|D_\tau \nabla\tilde{\Psi}(t)\right\|^2 + 2\mu\|\Delta\tilde{\Psi}(t)\|^2 \\ &\quad + \mu\tau\left(\sigma + \frac{\xi}{2}\right) D_\tau \|\Delta\tilde{\Psi}(t)\|^2 \\ &\quad + \mu\tau^2\left(\xi\sigma - \sigma - \frac{\xi}{2}\right)\|D_\tau \Delta\tilde{\Psi}(t)\|^2 = \sum_{j=1}^{8} G_j(t) \end{aligned} \tag{5.80}$$

where

$$\begin{aligned} G_1(t) &= -\tau(\xi - 2\delta)J\left(\tilde{\Psi}(t), \tilde{\Psi}(t), D_\tau \tilde{\Psi}(t)\right), \\ G_2(t) &= -\tau(\xi - 2\delta)J\left(\tilde{\Psi}(t), \Psi_N(t), D_\tau \tilde{\Psi}(t)\right), \\ G_3(t) &= -2J\left(\Psi_N(t) + \delta\tau D_\tau \Psi_N(t), \tilde{\Psi}(t), \tilde{\Psi}(t)\right), \\ G_4(t) &= -\xi\tau J\left(\Psi_N(t) + \delta\tau D_\tau \Psi_N(t), \tilde{\Psi}(t), D_\tau \tilde{\Psi}(t)\right), \\ G_5(t) &= 2\left(D_\tau \Psi_N(t) - \partial_t \Psi(t), \Delta\tilde{\Psi}(t)\right) - \xi\tau\left(\nabla(\Psi_N(t) - \partial_t \Psi(t)), D_\tau \nabla\tilde{\Psi}(t)\right), \\ G_6(t) &= -2\mu\sigma\tau\left(D_\tau \Delta\Psi(t), \Delta\tilde{\Psi}(t)\right) - \xi\mu\sigma\tau^2\left(D_\tau \Delta\Psi(t), D_\tau \Delta\tilde{\Psi}(t)\right), \\ G_7(t) &= -2J\left(\Psi_N(t) + \delta\tau D_\tau \Psi_N(t), \Psi_N(t), \tilde{\Psi}(t)\right) + 2J\left(\Psi(t), \Psi(t), \tilde{\Psi}(t)\right), \\ G_8(t) &= -\xi\tau J\left(\Psi_N(t) + \delta\tau D_\tau \Psi_N(t), \Psi_N(t), D_\tau \tilde{\Psi}(t)\right) + \xi\tau J\left(\Psi(t), \Psi(t), D_\tau \tilde{\Psi}(t)\right). \end{aligned}$$

We are going to bound $|G_j(t)|$. Let $\varepsilon > 0$ and D be an undetermined positive constant. By (5.71) and Lemmas 5.8–5.10, we have that

$$\begin{aligned}|G_1(t)| &\le 2\tau|\xi-2\delta|\left\|\nabla\tilde{\Psi}(t)\right\|\left\|\Delta\tilde{\Psi}(t)\right\|\left\|D_\tau\Delta\tilde{\Psi}(t)\right\| \\ &\le \varepsilon_0\tau\left\|D_\tau\nabla\tilde{\Psi}(t)\right\|^2 + c_1\tau N^4\left\|\nabla\tilde{\Psi}(t)\right\|^2\left\|\Delta\tilde{\Psi}(t)\right\|^2,\end{aligned}$$

$$\begin{aligned}|G_2(t)| &\le 2\tau|\xi-2\delta|\,\|\Delta\Psi_N(t)\|\left(\left\|\nabla\tilde{\Psi}(t)\right\|\left\|\Delta\tilde{\Psi}(t)\right\|\left\|D_\tau\nabla\tilde{\Psi}(t)\right\|\left\|D_\tau\Delta\tilde{\Psi}(t)\right\|\right)^{\frac{1}{2}} \\ &\le 2\left(\sqrt{\tau}|\xi-2\delta|\|\Delta\Psi_N(t)\|\left\|\Delta\tilde{\Psi}(t)\right\|^{\frac{1}{2}}\left\|D_\tau\nabla\tilde{\Psi}(t)\right\|^{\frac{1}{2}}\right)\left(\tau\left\|\nabla\tilde{\Psi}(t)\right\|\left\|D_\tau\Delta\tilde{\Psi}(t)\right\|\right)^{\frac{1}{2}} \\ &\le \tau\left\|\nabla\tilde{\Psi}(t)\right\|\left\|D_\tau\Delta\tilde{\Psi}(t)\right\| + \tau(\xi-2\delta)^2\|\Delta\Psi_N(t)\|^2\|\Delta\tilde{\Psi}(t)\|\left\|D_\tau\nabla\tilde{\Psi}(t)\right\| \\ &\le \varepsilon_0\mu\tau^2\left\|D_\tau\Delta\tilde{\Psi}(t)\right\|^2 + \frac{1}{4\varepsilon_0\mu}\left\|\nabla\tilde{\Psi}(t)\right\|^2 + \varepsilon_0\tau\left\|D_\tau\nabla\tilde{\Psi}(t)\right\|^2 + c_2\tau\left\|\Delta\tilde{\Psi}(t)\right\|^2,\end{aligned}$$

$$\begin{aligned}|G_3(t)| &\le 2\left|J\left(\tilde{\Psi}(t),\tilde{\Psi}(t),\Psi_N(t)+\delta\tau D_\tau\Psi_N(t)\right)\right| \\ &\le 2D\left\|\nabla\tilde{\Psi}(t)\right\|\left\|\Delta\tilde{\Psi}(t)\right\|\|\Delta\left(\Psi_N(t)+\delta\tau D_\tau\Psi_N(t)\right)\| \\ &\le D\left\|\Delta\tilde{\Psi}(t)\right\|^2 + c_3\left\|\nabla\tilde{\Psi}(t)\right\|^2,\end{aligned}$$

$$\begin{aligned}|G_4(t)| \le\ & 2\xi\tau\left\|\nabla\tilde{\Psi}(t)\right\|^{\frac{1}{2}}\left\|\Delta\tilde{\Psi}(t)\right\|^{\frac{1}{2}}\left(\left\|D_\tau\nabla\tilde{\Psi}(t)\right\|^{\frac{1}{2}}\left\|D_\tau\Delta\tilde{\Psi}(t)\right\|^{\frac{1}{2}}\right. \\ & \cdot\|\Delta\left(\Psi_N(t)+\delta\tau D_\tau\Psi_N(t)\right)\| + \left\|D_\tau\Delta\tilde{\Psi}(t)\right\|\|\nabla\left(\Psi_N(t)+\delta\tau D_\tau\Psi_N(t)\right)\|^{\frac{1}{2}} \\ & \left.\|\Delta\left(\Psi_N(t)+\delta\tau D_\tau\Psi_N(t)\right)\|^{\frac{1}{2}}\right) \\ \le\ & 2\xi\tau\|\Psi\|_{C(0,T;H^2)}\left(\left\|D_\tau\nabla\tilde{\Psi}(t)\right\|^{\frac{1}{2}}\left\|D_\tau\Delta\tilde{\Psi}(t)\right\|^{\frac{1}{2}}\left\|\nabla\tilde{\Psi}(t)\right\|^{\frac{1}{2}}\left\|\Delta\tilde{\Psi}(t)\right\|^{\frac{1}{2}}\right. \\ & \left.+\left\|D_\tau\Delta\tilde{\Psi}(t)\right\|\left\|\nabla\tilde{\Psi}(t)\right\|^{\frac{1}{2}}\left\|\Delta\tilde{\Psi}(t)\right\|^{\frac{1}{2}}\right) \\ \le\ & 2\xi\tau\|\Psi\|_{C(0,T;H^2)}\left\|D_\tau\nabla\tilde{\Psi}(t)\right\|^{\frac{1}{2}}\left\|D_\tau\Delta\tilde{\Psi}(t)\right\|^{\frac{1}{2}}\left\|\nabla\tilde{\Psi}(t)\right\|^{\frac{1}{2}}\left\|\Delta\tilde{\Psi}(t)\right\|^{\frac{1}{2}} \\ & +\varepsilon_0\mu\tau^2\left\|D_\tau\Delta\tilde{\Psi}(t)\right\|^2 + \frac{\xi^2}{\varepsilon_0\mu}\|\Psi\|^2_{C(0,T;H^2)}\left\|\nabla\tilde{\Psi}(t)\right\|\left\|\Delta\tilde{\Psi}(t)\right\| \\ \le\ & 2\xi\tau\|\Psi\|_{C(0,T;H^2)}\left\|D_\tau\nabla\tilde{\Psi}(t)\right\|^{\frac{1}{2}}\left\|D_\tau\Delta\tilde{\Psi}(t)\right\|^{\frac{1}{2}}\left\|\nabla\tilde{\Psi}(t)\right\|^{\frac{1}{2}}\left\|\Delta\tilde{\Psi}(t)\right\|^{\frac{1}{2}} \\ & +\varepsilon_0\mu\tau^2\left\|D_\tau\Delta\tilde{\Psi}(t)\right\|^2 + D\left\|\Delta\tilde{\Psi}(t)\right\|^2 + c_4\left\|\nabla\tilde{\Psi}(t)\right\|^2 \\ \le\ & 2\varepsilon_0\mu\tau^2\left\|D_\tau\Delta\tilde{\Psi}(t)\right\|^2 + \frac{1}{4\varepsilon_0\mu}\left\|\nabla\tilde{\Psi}(t)\right\|^2 + \varepsilon_0\tau\left\|D_\tau\nabla\tilde{\Psi}(t)\right\|^2 \\ & + D\left\|\Delta\tilde{\Psi}(t)\right\|^2 + c_4\left\|\nabla\tilde{\Psi}(t)\right\|^2 + c_4\tau\left\|\Delta\tilde{\Psi}(t)\right\|^2\end{aligned}$$

where

$$c_1 = \frac{1}{\varepsilon_0}c_p^2(\xi - 2\delta)^2, \qquad c_2 = \frac{1}{4\varepsilon_0}(\xi - 2\delta)^4\|\Psi\|^4_{C(0,T;H^2)},$$
$$c_3 = \frac{4}{D}\|\Psi\|^2_{C(0,T;H^2)}, \qquad c_4 = \frac{1}{4\varepsilon_0}\xi^4\|\Psi\|^4_{C(0,T;H^2)} \max\left(1, \frac{1}{\varepsilon_0\mu^2 D}\right).$$

We next estimate $|F_5(t)|$ and $|F_6(t)|$. It can be verified as before that

$$\begin{aligned}\|D_\tau\Psi_N(t) - \partial_t\Psi(t)\|^2 &\le \|D_\tau\Psi_N(t) - D_\tau\Psi(t)\|^2 + \|D_\tau\Psi(t) - \partial_t\Psi(t)\|^2 \\ &\le \frac{2}{\tau}\int_t^{t+\tau} \|\partial_s(\Psi(s) - \Psi_N(s))\|^2\, ds + \frac{2\tau}{3}\int_t^{t+\tau}\left\|\partial_s^2\Psi(s)\right\|^2 ds. \end{aligned} \qquad (5.81)$$

A similar estimate is valid for $\|\nabla(D_\tau\Psi_N(t) - \partial_t\Psi(t))\|^2$. Thus

$$\begin{aligned}|G_5(t)| \le\ & D\left\|\Delta\tilde{\Psi}(t)\right\|^2 + \varepsilon_0\tau\left\|D_\tau\nabla\tilde{\Psi}(t)\right\|^2 + \frac{c_5}{\tau}\int_t^{t+\tau}\|\partial_s(\Psi(s) - \Psi_N(s))\|^2\, ds \\ &+ c_5\int_t^{t+\tau}\|\partial_s(\Psi(s) - \Psi_N(s))\|_1^2\, ds + c_5\tau\int_t^{t+\tau}\left\|\partial_s^2\Psi(s)\right\|^2 ds \\ &+ c_5\tau^2\int_t^{t+\tau}\left\|\partial_s^2\Psi(s)\right\|_1^2 ds\end{aligned}$$

where $c_5 = \max\left(\frac{1}{2\varepsilon_0}\xi^2, \frac{2}{D}\right)$. Clearly

$$|G_6(t)| \le D\left\|\Delta\tilde{\Psi}(t)\right\|^2 + \varepsilon_0\mu\tau^2\left\|D_\tau\Delta\tilde{\Psi}(t)\right\|^2 + c_6\tau\int_t^{t+\tau}\|\partial_s\Psi(s)\|_2^2\, ds$$

where $c_6 = \mu\sigma^2\left(\frac{\mu}{D} + \frac{1}{4\varepsilon_0}\xi^2\tau^2\right)$. Finally, we estimate $|G_7(t)|$ and $|G_8(t)|$. By virtue of Lemma 5.10,

$$\begin{aligned}&2\left|J\left(\Psi_N(t) + \delta\tau D_\tau\Psi_N(t), \Psi(t) - \Psi_N(t), \tilde{\Psi}(t)\right)\right| \\ &\quad\le 8c\|\Psi\|_{C(0,T;H^2)}\|\Psi(t) - \Psi_N(t)\|_1^{\frac{1}{2}}\|\Psi(t) - \Psi_N(t)\|_2^{\frac{1}{2}}\left\|\Delta\tilde{\Psi}(t)\right\| \\ &\quad\le D\left\|\Delta\tilde{\Psi}(t)\right\|^2 + c_{7,1}\|\Psi(t) - \Psi_N(t)\|_1\|\Psi(t) - \Psi_N(t)\|_2\end{aligned} \qquad (5.82)$$

where $c_{7,1} = \frac{16c}{D}\|\Psi\|^2_{C(0,T;H^2)}$. According to Lemma 5.9,

$$\begin{aligned}&2\left|J\left(\Psi_N(t) + \delta\tau D_\tau\Psi_N(t) - \Psi(t), \Psi(t), \tilde{\Psi}(t)\right)\right| \\ &\quad\le 4c\|\Psi\|_{C(0,T;H^2)}\left\|\Delta\tilde{\Psi}(t)\right\|\left(\|\Psi(t) - \Psi_N(t)\|_1^{\frac{1}{2}}\|\Psi(t) - \Psi_N(t)\|_2^{\frac{1}{2}} + \delta\tau\|D_\tau\Psi_N(t)\|_2\right) \\ &\quad\le D\left\|\Delta\tilde{\Psi}(t)\right\|^2 + c_{7,1}\|\Psi(t) - \Psi_N(t)\|_1\|\Psi(t) - \Psi_N(t)\|_2 \\ &\quad\quad + c_{7,2}\tau\int_t^{t+\tau}\|\partial_s\Psi(s)\|_2^2\, ds\end{aligned} \qquad (5.83)$$

where $c_{7,2} = \dfrac{8}{D}\delta^2 c\|\Psi\|^2_{C(0,T;H^2)}$. Let $c_7 = 2c_{7,1} + c_{7,2}$. The combination of (5.82) and (5.83) leads to

$$|G_7(t)| \le 2D\left\|\Delta\tilde{\Psi}(t)\right\|^2 + c_7\|\Psi(t)-\Psi_N(t)\|_1\|\Psi(t)-\Psi_N(t)\|_2 + c_7\tau\int_t^{t+\tau}\|\partial_s\Psi(s)\|_2^2\,ds.$$

Similarly,

$$\begin{aligned}|G_8(t)| \ \le\ & 2\varepsilon_0\mu\tau^2\left\|D_\tau\Delta\tilde{\Psi}(t)\right\|^2 + c_8\|\Psi(t)-\Psi_N(t)\|_1\|\Psi(t)-\Psi_N(t)\|_2\\ & + c_8\tau\int_t^{t+\tau}\|\partial_s\Psi(s)\|_2^2\,ds\end{aligned}$$

where $c_8 = \dfrac{8}{\varepsilon_0\mu}c\xi^2\left(2+\delta^2\right)\|\Psi\|^2_{C(0,T;H^2)}$. We now take $D = \dfrac{\mu}{12}$ and substitute the previous inequalities into (5.80). Then

$$\begin{aligned}& D_\tau\left\|\nabla\tilde{\Psi}(t)\right\|^2 + \tau(\xi-1-4\varepsilon_0)\left\|D_\tau\nabla\tilde{\Psi}(t)\right\|^2 + \frac{\mu}{2}\left\|\Delta\tilde{\Psi}(t)\right\|^2\\ &\quad + \mu\tau\left(\sigma+\frac{\xi}{2}\right)D_\tau\left\|\Delta\tilde{\Psi}(t)\right\|^2 + \mu\tau^2\left(\xi\sigma-\sigma-\frac{\xi}{2}-6\varepsilon_0\right)\left\|D_\tau\Delta\tilde{\Psi}(t)\right\|^2\\ &\le A_1\left\|\nabla\tilde{\Psi}(t)\right\|^2 + A_2(t)\left\|\Delta\tilde{\Psi}(t)\right\|^2 + A_3(t)\end{aligned} \tag{5.84}$$

where

$$\begin{aligned}A_1 &= c_3+c_4+\frac{1}{2\varepsilon_0\mu},\\ A_2(t) &= -\mu+\tau(c_2+c_4)+c_1\tau N^4\left\|\nabla\tilde{\Psi}(t)\right\|^2,\\ A_3(t) &= \frac{c_5}{\tau}\int_t^{t+\tau}\|\partial_s(\Psi(t)-\Psi_N(t))\|^2\,ds + c_5\int_t^{t+\tau}\|\partial_s(\Psi(t)-\Psi_N(t))\|_1^2\,ds\\ &\quad + \tau(c_6+c_7+c_8)\int_t^{t+\tau}\|\partial_s\Psi(s)\|_2^2\,ds + c_5\tau\int_t^{t+\tau}\left\|\partial_s^2\Psi(s)\right\|^2 ds\\ &\quad + c_5\tau^2\int_t^{t+\tau}\left\|\partial_s^2\Psi(s)\right\|_1^2 ds + (c_7+c_8)\|\Psi(t)-\Psi_N(t)\|_1\|\Psi(t)-\Psi_N(t)\|_2.\end{aligned}$$

We next choose the value of ξ as follows. For $\sigma > \frac{1}{2}$, we take

$$\xi = \xi_1 = \max\left(1+4\varepsilon_0+q_0, \frac{2\sigma+12\varepsilon_0}{2\sigma-1}\right).$$

Then $\xi\sigma-\sigma-\frac{1}{2}\xi-6\varepsilon_0 \ge 0$ and so

$$\begin{aligned}&\tau(\xi-1-4\varepsilon_0)\left\|D_\tau\nabla\tilde{\Psi}(t)\right\|^2\\ &\quad+\mu\tau^2\left(\xi\sigma-\sigma-\frac{1}{2}\xi-6\varepsilon_0\right)\left\|D_\tau\Delta\tilde{\Psi}(t)\right\|^2 \ge q_0\tau\left\|D_\tau\nabla\tilde{\Psi}(t)\right\|^2.\end{aligned} \tag{5.85}$$

If $\sigma = \dfrac{1}{2}$ and $\tau = \mathcal{O}(N^{-4})$, then we take

$$\xi \geq \xi = 1 + 4\varepsilon_0 + q_0 + \mu c_p^2 \tau N^4 \left(\frac{1}{2} + 6\varepsilon_0\right)$$

and so (5.85) also holds. Finally for $\sigma < \dfrac{1}{2}$ and $\tau N^4 < \dfrac{2}{\mu c_p^2 (1 - 2\sigma)}$, we take

$$\xi \geq \xi_3 = \left(1 + 4\varepsilon_0 + q_0 + \mu c_p^2 \tau N^4 (\sigma + 6\varepsilon_0)\right) \left(1 + \mu c_p^2 \tau N^4 \left(\sigma - \frac{1}{2}\right)\right)^{-1}$$

and so (5.85) is still true. Let

$$\rho(t) = \tau \sum_{s \in \bar{R}_\tau(t-\tau)} A_3(s).$$

By summing (5.85) for $t \in \bar{R}_\tau(T)$, we obtain that

$$E\left(\tilde{\Psi}, t\right) \leq \rho(t) + \tau \sum_{s \in \bar{R}_\tau(t-\tau)} \left(A_1 E\left(\tilde{\Psi}, s\right) + A_2(s) \left\|\Delta \tilde{\Psi}(s)\right\|^2\right). \tag{5.86}$$

If τ is suitably small, then

$$A_2(t) \leq -\frac{\mu}{2} + c_1 \tau N^4 \left\|\nabla \tilde{\Psi}(t)\right\|^2.$$

By virtue of Lemma 3.2, if for certain $t_1 \leq T$,

$$\rho(t_1) \leq \frac{\mu}{2 c_1 \tau N^4}, \tag{5.87}$$

then for all $t \leq t_1$,

$$E\left(\tilde{\Psi}, t\right) \leq \rho(t) e^{A_1 t}.$$

In order to get the convergence rate, we only need to estimate $\rho(t)$ and to check (5.87). Clearly by (5.12),

$$\rho(t) \leq b_1 \left(\tau^2 + N^{3-2r_0} + N^{-4} + \tau N^{-2}\right).$$

Furthermore if $\sigma \leq \dfrac{1}{2}$ and $\tau = \mathcal{O}(N^{-4})$, then (5.87) holds for $t_1 = T$ and N large enough. If $\sigma > \dfrac{1}{2}$, then the values of ξ and all c_j do not depend on τN^4. By Lemma 3.2, we only require $A_2(t) \leq 0$, i.e.,

$$\tau \leq b_2, \quad \tau N^4 \left(\tau^2 + N^{3-2r_0} + N^{-4} + \tau N^{-2}\right) \leq b_2, \tag{5.88}$$

b_2 being a positive constant independent of N and τ. Since $r_0 \geq 2$, a sufficient condition for (5.88) is that $\tau = o(N^{-\lambda}), \lambda \geq \max\left(\frac{4}{3}, 7 - 2r_0\right)$. Finally, the above statements together with the triangle inequality lead to the first part of this theorem. If in addition $\sigma > \frac{1}{2}$ and $\delta \geq \frac{\xi_1}{2}$, then we can take $\xi = 2\delta$ and so $c_1 = c_2 = 0$. Therefore the condition (a) can be weakened to any $\tau \to 0$. ■

Theorem 5.7 does not provide an optimal estimate. But if Ψ is a little smoother, then the optimal estimate exists, stated as follows.

Theorem 5.8. *Assume that*

(i) condition (i) of Theorem 5.7 holds;

(ii) $r_0 \geq r_1 > 2$, *or* $r_0 > r_1 = 2$ *and* $\Psi \in L^\infty(0,T;H^r)$ *for some* $r \in (2, r_0)$*;*

(iii) one of the following conditions is fulfilled

(a) $\sigma > \frac{1}{2}, \tau = o\left(N^{-\lambda}\right), \lambda \geq \max\left(\frac{4}{3}, 6 - 2r_0\right)$,

(b) $\sigma = \frac{1}{2}, \tau = \mathcal{O}\left(N^{-4}\right)$,

(c) $\sigma < \frac{1}{2}, \tau N^4 \leq \dfrac{2}{\mu c_p^2(1-2\sigma)}$.

Then for all $t \in \bar{R}_\tau(T)$,

$$E\left(\Psi - \psi_N, t\right) \leq b_3\left(\tau^2 + N^{2-2r_0} + N^{-2r_1} + \tau N^{2-2r_1}\right)$$

where b_3 *is a positive constant depending only on* μ, Ω *and the norms of* Ψ *and* g *in the mentioned spaces. If in addition* $\sigma > \frac{1}{2}$ *and* δ *is suitably large, then the above result holds for any* $\tau \to 0$.

Proof. We rewrite (5.80) as

$$\begin{aligned} D_\tau \left\|\nabla\tilde{\Psi}(t)\right\|^2 + \tau(\xi - 1)\left\|D_\tau\nabla\tilde{\Psi}(t)\right\|^2 + 2\mu\left\|\Delta\tilde{\Psi}(t)\right\|^2 + \mu\tau\left(\sigma + \tfrac{\xi}{2}\right)D_\tau\left\|\tilde{\Psi}(t)\right\|^2 \\ +\mu\tau^2\left(\xi\sigma - \sigma - \tfrac{\xi}{2}\right)\left\|D_\tau\Delta\tilde{\Psi}(t)\right\|^2 = R_1(t) + R_2(t) \end{aligned} \tag{5.89}$$

with

$$R_1(t) = \sum_{j=1}^{6} G_j(t), \quad R_2(t) = G_7(t) + G_8(t).$$

Since $|R_1(t)|$ has been handled in Theorem 5.7, we need only to estimate $|R_2(t)|$ more precisely. If $r_1 > 2$, then $\Psi \in L^\infty(0,T;H^r)$ for $r \in (2, r_1)$. We have

$$G_7(t) = \sum_{j=1}^{4} G_{7,j}(t)$$

where

$$\begin{aligned}
G_{7,1}(t) &= 2J\left(\Psi_N(t) - \Psi(t), \Psi(t) - \Psi_N(t), \tilde{\Psi}(t)\right), \\
G_{7,2}(t) &= 2J\left(\Psi(t) + \delta\tau D_\tau \Psi(t), \Psi(t) - \Psi_N(t), \tilde{\Psi}(t)\right), \\
G_{7,3}(t) &= 2J\left(\Psi(t) - \Psi_N(t) - \delta\tau D_\tau \Psi_N(t), \Psi(t), \tilde{\Psi}(t)\right), \\
G_{7,4}(t) &= 2\delta\tau J\left(D_\tau \Psi_N(t) - D_\tau \Psi(t), \Psi(t) - \Psi_N(t), \tilde{\Psi}(t)\right).
\end{aligned}$$

By Lemmas 5.8 and 5.12, we obtain that

$$\begin{aligned}
|G_{7,1}(t)| &\le 4\left\|\Delta\tilde{\Psi}(t)\right\| \left\|\nabla\left(\Psi(t) - \Psi_N(t)\right)\right\| \left\|\Delta\left(\Psi(t) - \Psi_N(t)\right)\right\| \\
&\le D\left\|\Delta\tilde{\Psi}(t)\right\|^2 + c_{7,3}\left\|\Psi(t) - \Psi_N(t)\right\|_1^2
\end{aligned}$$

and

$$\begin{aligned}
|G_{7,2}(t)| &\le 2c\left\|\Psi(t) + \delta\tau D_\tau\Psi(t)\right\|_r \left\|\Psi(t) - \Psi_N(t)\right\|_1 \left\|\Delta\tilde{\Psi}(t)\right\| \\
&\le D\left\|\Delta\tilde{\Psi}(t)\right\|^2 + c_{7,4}\left\|\Psi(t) - \Psi_N(t)\right\|_1^2
\end{aligned}$$

where $c_{7,3} = \dfrac{16}{D}\|\Psi\|^2_{L^\infty(0,T;H^2)}$ and $c_{7,4} = \dfrac{c}{D}\|\Psi\|^2_{L^\infty(0,T;H^r)}$. Moreover by (5.71) and Lemma 5.11,

$$|G_{7,3}(t)| \le 2c\|\Psi(t)\|_r \left\|\Delta\tilde{\Psi}(t)\right\| \left(\left\|\Psi(t) - \Psi_N(t)\right\|_1 + \delta\tau\left\|D_\tau\Psi_N(t)\right\|_1\right).$$

Since

$$\left\|D_\tau\Psi_N(t)\right\|_1 \le \tau^{-\frac{1}{2}}\left(\int_t^{t+\tau}\left\|\partial_s\Psi_N(s)\right\|_1^2 ds\right)^{\frac{1}{2}} \le \tau^{-\frac{1}{2}}\left(\int_t^{t+\tau}\left\|\partial_s\Psi_N(s)\right\|_2^2 ds\right)^{\frac{1}{2}}$$

and $\left\|\partial_t\Psi_N(t)\right\|_2 \le \left\|\partial_t\Psi(t)\right\|_2$, we see that

$$|G_{7,3}(t)| \le D\left\|\Delta\tilde{\Psi}(t)\right\|^2 + c_{7,4}\left\|\Psi(t) - \Psi_N(t)\right\|_1^2 + c_{7,5}\tau\int_t^{t+\tau}\left\|\partial_s\Psi(s)\right\|_2^2 ds$$

where $c_{7,5} = 2\delta^2 c_{7,4}$. Using Lemma 5.9, we deduce that

$$
\begin{aligned}
|G_{7,4}(t)| \le\ & 2\delta c\tau \left\|\Delta\tilde{\Psi}(t)\right\| \|\Delta(\Psi(t)-\Psi_N(t))\| \|D_\tau(\Psi(t)-\Psi_N(t))\|_1^{\frac{1}{2}} \\
& \cdot \|D_\tau(\Psi(t)-\Psi_N(t))\|_2^{\frac{1}{2}} \\
\le\ & D\left\|\Delta\tilde{\Psi}(t)\right\|^2 + c_{7,6}\tau^2 \|\Delta(\Psi(t)-\Psi_N(t))\|^2 \|D_\tau(\Psi(t)-\Psi_N(t))\|_1 \\
& \cdot \|D_\tau(\Psi(t)-\Psi_N(t))\|_2 \\
\le\ & D\left\|\Delta\tilde{\Psi}(t)\right\|^2 + c_{7,6}\tau \|\Delta(\Psi(t)-\Psi_N(t))\|^2 \left(\int_t^{t+\tau} \|\partial_s(\Psi(s)-\Psi_N(s))\|_1^2\, ds\right)^{\frac{1}{2}} \\
& \cdot \left(\int_t^{t+\tau} \|\partial_s(\Psi(s)-\Psi_N(s))\|_2^2\, ds\right)^{\frac{1}{2}} \\
\le\ & D\left\|\Delta\tilde{\Psi}(t)\right\|^2 + c_{7,6}\int_t^{t+\tau} \|\partial_s(\Psi(s)-\Psi_N(s))\|_1^2\, ds \\
& + \frac{1}{4}c_{7,6}\tau^2 \|\Delta(\Psi(t)-\Psi_N(t))\|^4 \int_t^{t+\tau} \|\partial_s(\Psi(t)-\Psi_N(t))\|_2^2\, ds
\end{aligned}
$$

where $c_{7,6} = \dfrac{c\delta^2}{D}$. The above statements lead to

$$
\begin{aligned}
|G_7(t)| \le\ & 4D\left\|\Delta\tilde{\Psi}(t)\right\|^2 + c_7\|\Psi(t)-\Psi_N(t)\|_1^2 + c_7\int_t^{t+\tau} \|\partial_s(\Psi(s)-\Psi_N(s))\|_1^2\, ds \\
& + c_7\tau\int_t^{t+\tau} \|\partial_s\Psi(s)\|_2^2\, ds + c_7\tau^2\|\Delta(\Psi(t)-\Psi_N(t))\|^4 \int_t^{t+\tau} \|\partial_s(\Psi(s)-\Psi_N(s))\|_2^2\, ds
\end{aligned}
$$

where $c_7 = \max(c_{7,3}+2c_{7,4}, c_{7,5}, c_{7,6})$. Similarly,

$$
\begin{aligned}
|G_8(t)| \le\ & 4\varepsilon_0\mu\tau^2\left\|D_\tau\Delta\tilde{\Psi}(t)\right\|^2 + c_8\|\Psi(t)-\Psi_N(t)\|_1^2 + c_8\int_t^{t+\tau} \|\partial_s(\Psi(s)-\Psi_N(s))\|_1^2\, ds \\
& + c_8\tau\int_t^{t+\tau} \|\partial_s\Psi(s)\|_2^2\, ds + c_8\tau^2\|\Delta(\Psi(t)-\Psi_N(t))\|^4 \int_t^{t+\tau} \|\partial_s(\Psi(s)-\Psi_N(s))\|_2^2\, ds
\end{aligned}
$$

where $c_8 = \max(c_{7,7}+2c_{7,8}, c_{7,9}, c_{7,10})$ and

$$
c_{7,7} = \frac{4\xi^2}{\varepsilon_0\mu}\|\Psi\|^2_{L^\infty(0,T;H^2)}, \qquad c_{7,8} = \frac{c\xi^2}{4\varepsilon_0\mu}\|\Psi\|^2_{L^\infty(0,T;H^r)},
$$

$$
c_{7,9} = 2\delta^2 c_{7,8}, \qquad c_{7,10} = \frac{c\delta^2\xi^2}{4\varepsilon_0\mu}.
$$

Now, by taking $D = \dfrac{\mu}{16}$ and using an argument as in the derivation of (5.84), we obtain from (5.89) that

$$
\begin{aligned}
& D_\tau\left\|\nabla\tilde{\Psi}(t)\right\|^2 + \tau(\xi-1-4\varepsilon_0)\left\|D_\tau\nabla\tilde{\Psi}(t)\right\|^2 + \tfrac{\mu}{2}\left\|\Delta\tilde{\Psi}(t)\right\|^2 + \mu\tau\left(\sigma+\tfrac{\xi}{2}\right)D_\tau\left\|\Delta\tilde{\Psi}(t)\right\|^2 \\
& + \mu\tau^2\left(\xi\sigma-\sigma-\tfrac{\xi}{2}-8\varepsilon_0\right)\left\|D_\tau\Delta\tilde{\Psi}(t)\right\|^2 \le A_1\left\|\nabla\tilde{\Psi}(t)\right\|^2 + A_2(t)\left\|\Delta\tilde{\Psi}(t)\right\|^2 + A_3(t),
\end{aligned}
\tag{5.90}
$$

where

$$
\begin{aligned}
A_1 &= c_3 + c_4 + \frac{1}{2\varepsilon_0 \mu}, \\
A_2(t) &= -\mu + \tau(c_2 + c_4) + c_1 \tau N^4 \left\| \nabla \tilde{\Psi}(t) \right\|^2, \\
A_3(t) &= \frac{c_5}{\tau} \int_t^{t+\tau} \|\partial_s(\Psi(s) - \Psi_N(s))\|^2 ds + (c_5 + c_7 + c_8) \int_t^{t+\tau} \|\partial_s(\Psi(s) - \Psi_N(s))\|_1^2 ds \\
&\quad + \tau(c_6 + c_7 + c_8) \int_t^{t+\tau} \|\partial_s \Psi(s)\|_2^2 \, ds + c_5 \tau \int_t^{t+\tau} \left\| \partial_s^2 \Psi(s) \right\|^2 ds \\
&\quad + c_5 \tau^2 \int_t^{t+\tau} \left\| \partial_s^2 \Psi(s) \right\|_1^2 ds \\
&\quad + \tau^2 (c_7 + c_8) \left\| \Delta \left(\Psi(t) - \Psi_N(t) \right) \right\|^4 \int_t^{t+\tau} \left\| \partial_s \left(\Psi(s) - \Psi_N(s) \right) \right\|_2^2 ds \\
&\quad + (c_7 + c_8) \left\| \Psi(t) - \Psi_N(t) \right\|_1^2 .
\end{aligned}
$$

Let $q_0 \geq 0$ and choose ξ as follows. For $\sigma > \frac{1}{2}$, we take

$$
\xi \geq \xi_1 = \max\left(1 + 4\varepsilon_0 + q_0, \frac{2\sigma + 16\varepsilon_0}{2\sigma - 1}\right).
$$

Then $\xi\sigma - \sigma - \frac{1}{2}\xi - 8\varepsilon_0 \geq 0$ and so

$$
\tau(1 - \xi - 4\varepsilon_0) \left\| D_\tau \nabla \tilde{\Psi}(t) \right\|^2 + \mu\tau^2 \left(\xi\sigma - \sigma - \frac{1}{2}\xi - 8\varepsilon_0 \right) \left\| D_\tau \Delta \tilde{\Psi}(t) \right\|^2 \geq q_0 \tau \left\| D_\tau \nabla \tilde{\Psi}(t) \right\|^2 . \tag{5.91}
$$

If $\sigma = \frac{1}{2}$ and $\tau = \mathcal{O}\left(N^{-4}\right)$, then we take

$$
\xi \geq \xi_2 = 1 + 4\varepsilon_0 + q_0 + \mu c_p^2 \tau N^4 \left(\frac{1}{2} + 8\varepsilon_0\right)
$$

and so (5.91) also holds. Finally, for $\sigma < \frac{1}{2}$ and $\tau N^4 < \frac{2}{\mu c_p^2 (1 - 2\sigma)}$, we put

$$
\xi \geq \xi_3 = \left(1 + 4\varepsilon_0 + q_0 + \mu c_p^2 \tau N^4 (\sigma + 8\varepsilon_0)\right) \left(1 + \mu c_p^2 \tau N^4 \left(\sigma - \frac{1}{2}\right)\right)^{-1},
$$

and then (5.91) also follows. Now, by the same procedure as in the proof of Theorem 5.7, we obtain from (5.90) an estimate similar to (5.86). In particular, we have

$$
\rho(t) = \tau \sum_{s \in \bar{R}_\tau(t-\tau)} A_3(s) = \mathcal{O}\left(\tau^2 + N^{2-2r_0} + N^{-2r_1} + \tau N^{2-2r_1}\right).
$$

If $\sigma \le \frac{1}{2}$ and $\tau = \mathcal{O}(N^{-4})$, then for all $t \in \bar{R}_\tau(T), \rho(t) \le \frac{\mu}{2c_1 \tau N^4}$. On the other hand, for $\sigma > \frac{1}{2}$, we require that $A_2(t) \le 0$, i.e.,

$$\tau \le b_4, \quad \tau N^4 \left(\tau^2 + N^{2-2r_0} + N^{-2r_1} + \tau N^{2-2r_1}\right) \le b_4,$$

b_4 being a positive constant independent of N and τ. The proof is complete. ∎

We now turn to numerical experiments. For describing the errors, let $x_q^{(j_q)}$ and $\omega^{(j_q)}$ be the Legendre-Gauss-Lobatto interpolation points and weights. The error $\tilde{E}(\psi_N, t)$ is defined as

$$\tilde{E}(\psi_N, t) = \left(\frac{\sum\limits_{j_1, j_2 = 0}^{N} \left| \Psi\left(x_1^{(j_1)}, x_2^{(j_2)}, t\right) - \psi_N\left(x_1^{(j_1)}, x_2^{(j_2)}, t\right)\right|^2 \omega^{(j_1)}\omega^{(j_2)}}{\sum\limits_{j_1, j_2 = 0}^{N} \left| \Psi\left(x_1^{(j_1)}, x_2^{(j_2)}, t\right)\right|^2 \omega^{(j_1)}\omega^{(j_2)}} \right)^{\frac{1}{2}}.$$

We take the test function

$$\Psi(x, t) = \frac{1}{10} e^{\frac{t}{10}} \left(1 + \cos \pi x_1\right)\left(1 + \cos \pi x_2\right). \tag{5.92}$$

The numerical results of scheme (5.76) are presented in Tables 5.1–5.3. The first table shows the high accuracy of (5.76) even for small N, and the good stability in long time calculation even for large τ. Table 5.2 indicates that when $N \to \infty$ and $\tau \to 0$ proportionally, the numerical solution converges to the exact solution. Table 5.3 shows that the computations are also stable for very small μ, even if $\mu = 0$.

Table 5.1. The errors $\tilde{E}(\psi_N, t), \mu = 0.5, \sigma = 0.5, \delta = 0, N = 10$.

	$\tau = 0.5$	$\tau = 0.1$	$\tau = 0.05$	$\tau = 0.01$	$\tau = 0.005$	$\tau = 0.001$
$t = 10$	$2.497E-2$	$4.995E-3$	$2.497E-3$	$4.995E-4$	$2.497E-4$	$5.000E-5$
$t = 20$	$2.485E-2$	$4.969E-3$	$2.484E-3$	$4.970E-4$	$2.485E-4$	$4.983E-5$
$t = 30$	$2.438E-2$	$4.880E-3$	$2.440E-3$	$4.879E-4$	$2.439E-4$	$4.876E-5$

In the previous part, we considered the two-level spectral scheme (5.76). Since we adopt partially implicit approximation for the nonlinear term, the convergence rate in time t is of first order. It limits the advantages of spectral approximations in the space. A modified version is the following

$$\begin{cases} (D_\tau \nabla \psi_N(t), \nabla \phi) + \frac{1}{4} J\left(\psi_N(t) + \psi_N(t+\tau), \psi_N(t) + \psi_N(t+\tau), \phi\right) + \frac{\mu}{2}\left(\Delta \psi_N(t)\right. \\ \qquad \left. + \Delta \psi_N(t+\tau), \Delta \phi\right) + \frac{1}{2}\left(g(t) + g(t+\tau), \phi\right) = 0, \qquad \forall\, \phi \in \mathbb{P}_N^{2,0},\ t \in \bar{R}_\tau(T), \\ \psi_N(0) = P_N^{2,0} \Psi_0, \qquad x \in \bar{\Omega}. \end{cases}$$

Table 5.2. The errors $\tilde{E}(\psi_N, 30), \mu = 0.5, \sigma = 0.5, \delta = 0$.

	$N = 6$	$N = 8$	$N = 10$
$\tau = 0.1$	$6.151E-3$	$4.920E-3$	$4.880E-3$
$\tau = 0.01$	$3.179E-3$	$5.367E-4$	$4.879E-4$
$\tau = 0.001$	$3.091E-3$	$1.301E-4$	$4.486E-5$

Table 5.3. The errors $\tilde{E}(\psi_N, 20), \sigma = 0.5, \delta = 0, N = 10$.

	$\mu = 10^{-3}$	$\mu = 10^{-4}$	$\mu = 0$
$\tau = 0.01$	$4.325E-4$	$4.418E-4$	$4.452E-4$
$\tau = 0.005$	$2.153E-4$	$2.224E-4$	$2.296E-4$
$\tau = 0.001$	$4.214E-5$	$4.248E-5$	$4.272E-5$

This scheme is of second order in time. But we have to solve a nonlinear equation at each time step. It costs a lot of computational time and brings some difficulties in theoretical analysis. It is well known that if we use certain reasonable prediction-correction schemes for nonlinear problems, we could get the accuracy of second order in time and evaluate the values of unknown functions explicitly, e.g., see Guo (1988a). Following this idea, we construct a Legendre prediction-correction procedure for (5.74). Let $\psi_N^p(x,t)$ be the predicted value of $\psi_N(x,t)$. The prediction-correction scheme is as follows

$$\begin{cases} \frac{1}{\tau}\left(\nabla\left(\psi_N^{(p)}(t+\tau)-\psi_N(t)\right),\nabla\phi\right)+J\left(\psi_N(t),\psi_N(t),\phi\right) \\ \quad +\frac{\mu}{2}\left(\Delta\left(\psi_N(t)+\psi_N^{(p)}(t+\tau)\right),\Delta\phi\right)+\left(g(t),\phi\right)=0, \quad \forall\,\phi\in\mathbb{P}_N^{2,0}, \\ \frac{1}{\tau}(\nabla(\psi_N(t+\tau)-\psi_N(t)),\nabla\phi)+\frac{1}{2}J(\psi_N(t),\psi_N(t),\phi)+\frac{1}{2}J(\psi_N^{(p)}(t+\tau),\psi_N^{(p)}(t+\tau),\phi) \\ \quad +\frac{\mu}{2}\left(\Delta\left(\psi_N(t)+\psi_N(t+\tau)\right),\Delta\phi\right)+\frac{1}{2}(g(t)+g(t+\tau),\phi)=0, \quad \forall\,\phi\in\mathbb{P}_N^{2,0}, \\ \psi_N(0)=P_N^{2,0}\Psi_0, \qquad x\in\bar{\Omega}. \end{cases} \tag{5.93}$$

Let $E(v,t)$ be the same as in (5.78), and $\tau = \mathcal{O}(N^{-4})$. It can be proved that if $\Psi \in L^2(0,T;H^{r_0}\cap H_0^2)\cap H^1(0,T;H^{r_1})\cap H^2(0,T;H^2)\cap H^3(0,T;H^1)$, $\Psi_0 \in H_0^2(\Omega)$ and $g \in H^2(0,T;L^2)$, then there exists a positive constant b_5 depending only on μ, Ω and the norms of Ψ and g in the mentioned spaces, such that for all $t \in \bar{R}_\tau(T)$,

$$E(\Psi-\psi_N,t) \le b_5\left(\tau^4+N^{2-2r_0}+N^{-2r_1}+\tau N^{2-2r_1}\right).$$

We now present some numerical results of scheme (5.93). The definition of error

$\tilde{E}(\psi_N, t)$ is the same as in the previous paragraph. We first take the test function (5.92) and list the errors in Tables 5.4–5.6. The first table shows the high accuracy even for small N, and the good stability in long time calculation even for large τ. Table 5.5 indicates the convergence of scheme (5.93). By comparing Table 5.2 and Table 5.5, we find that (5.93) gives more accurate numerical solutions than (5.76). Table 5.6 indicates that (5.93) is stable for small μ, even if $\mu = 0$.

Table 5.4. The errors $\tilde{E}(\psi_N, t), \mu = 0.5, N = 14$.

	$t = 1$	$t = 2$	$t = 3$	$t = 4$	$t = 5$
$\tau = 0.1$	$3.919E-6$	$4.331E-6$	$4.782E-6$	$5.279E-6$	$5.825E-6$
$\tau = 0.01$	$5.905E-8$	$6.524E-8$	$7.208E-8$	$7.964E-8$	$8.798E-8$
$\tau = 0.001$	$6.576E-10$	$7.189E-10$	$7.873E-10$	$8.635E-10$	$9.482E-10$

Table 5.5. The errors $\tilde{E}(\psi_N, 30), \mu = 0.5$.

	$N = 8$	$N = 10$	$N = 12$
$\tau = 0.5$	$3.612E-4$	$3.933E-4$	$4.070E-4$
$\tau = 0.1$	$9.155E-5$	$3.675E-5$	$3.712E-5$
$\tau = 0.01$	$9.672E-5$	$1.749E-6$	$8.469E-7$

Table 5.6. The errors $\tilde{E}(\psi_N, 20), N = 10$.

	$\mu = 10^{-3}$	$\mu = 10^{-4}$	$\mu = 0$
$\tau = 0.04$	$3.487E-6$	$2.103E-5$	$2.114E-5$
$\tau = 0.025$	$2.861E-6$	$1.658E-6$	$1.469E-6$

We next take the test function

$$\Psi(x,t) = \frac{(1-x_1^2)^2(1-x_2^2)^2}{a + bt^2 + x_1^2 + x_2^2}.$$

We use (5.76) and (5.93) to solve (5.74) with $\mu = 0.05, a = 1, b = 0.01$ and $N = 12$, and list the errors in Table 5.7. It is shown again that scheme (5.93) gives much better numerical results. In particular, scheme (5.93) is very stable in long time computation, even for small N and large τ.

Table 5.7. The errors $\tilde{E}(\psi_N, t)$.

	scheme (5.93)		scheme (5.76)
	$\tau = 0.1$	$\tau = 0.4$	$\tau = 0.01$
$t = 20$	$2.694E-6$	$3.309E-5$	$3.853E-4$
$t = 60$	$1.625E-7$	$2.588E-6$	$1.613E-4$
$t = 100$	$3.559E-8$	$5.712E-7$	$9.883E-5$

Bernardi, Coppoletta and Maday (1992a, 1992b) studied the Legendre spectral methods for a two-dimensional steady fourth order problem. Shen (1994) developed an efficient direct solver for one-dimensional steady fourth order problems.

5.4. Spectral Domain Decomposition Methods

In this section, we consider spectral domain decomposition methods. The spatial computational domain is divided into several adjoining, nonintersecting subdomains, within each of them we look for a polynomial solution. Some fulfillments are required on the common boundaries of adjoining subdomains. This approach stems from its capability of covering problems in complex geometry. Orszag (1980), and McCrory and Orszag (1980) studied the patchings of complex domains. Moreover, this approach also allows local refinement to resolve internal layers or discontinuities, maintaining however the spectral accuracy enjoyed by the standard spectral methods. In the previous sections, we already used the Fourier spectral and Fourier pseudospectral methods for an elliptic system, and the Legendre spectral method for a high order parabolic equation. In this section, we take a quasi-linear hyperbolic system as an example to describe the Chebyshev pseudospectral domain decomposition method, and prove the convergence for a special case. The numerical results coincide with the theoretical analysis.

Let Ω be an open bounded domain in $\mathbb{R}^n$. U and f are two vector functions with the components $U^{(j)}$ and $f^{(j)}, 1 \leq j \leq p$. $A_q(U)$ are $p \times p$ matrices, $1 \leq q \leq n$. Consider the problem

$$\begin{cases} \partial_t U + \sum_{q=1}^{n} A_q(U) \partial_q U = f, & x \in \Omega,\ 0 < t \leq T, \\ U(x,0) = U_0(x), & x \in \bar{\Omega} \end{cases} \tag{5.94}$$

with suitable boundary conditions on $\partial\Omega \times (0,T]$. For any $\xi \in \mathbb{R}^n$ such that $|\xi|_2 = 1$, define the characteristic matrix for the direction ξ as

$$A(\xi) = \sum_{q=1}^{n} A_q(U)\xi_q.$$

(5.94) is assumed to be hyperbolic in time. It means that for any such $\xi, A(\xi)$ has p real eigenvalues and moreover it is diagonalizable. Let λ_k be the eigenvalues of $A(\xi)$ and V_k be the corresponding left eigenvectors, $1 \le k \le p$. Then

$$V_k A(\xi) = \lambda_k V_k, \quad 1 \le k \le p. \tag{5.95}$$

Suppose that $\lambda_k > 0$ for $1 \le k \le p_0$ and $\lambda_k < 0$ for $p_0 + 1 \le k \le p$. By taking the inner product of V_k with (5.94), we obtain

$$V_k \cdot \left(\partial_t U + \sum_{q=1}^{n} A_q(U)\partial_q U \right) = V_k \cdot f, \quad 1 \le k \le p. \tag{5.96}$$

Denote by $(\tau_1, \ldots, \tau_{n-1})$ the system of Cartesian coordinates of the hyperplane orthogonal to the direction ξ. Then for each $h = 1, \ldots, n-1, \tau_h = (\tau_{h_1}, \ldots, \tau_{h_n}) \in \mathbb{R}^n, \tau_h \cdot \xi = 0$ and $|\tau_h| = 1$. Owing to the identity

$$\partial_q U = \xi_q \partial_\xi U + \sum_{h=1}^{n-1} \tau_{h_q} \partial_{\tau_h} U,$$

it follows from (5.96) that

$$V_k \cdot (\partial_t U + \lambda_k \partial_\xi U) = V_k \cdot \left(f - \sum_{q=1}^{n} \left(A_q(U) \sum_{h=1}^{n-1} \tau_{h_q} \partial_{\tau_h} U \right) \right), \quad 1 \le k \le p. \tag{5.97}$$

We refer to (5.97) as the compatibility equations for (5.94).

Now, let Γ be an $(n-1)$-dimensional manifold, and ν be the unit normal direction to Γ. Taking $\xi = \nu$, (5.97) restricted to Γ becomes

$$V_k \cdot (\partial_t U + \lambda_k \partial_\nu U) = V_k \cdot \left(f - \sum_{q=1}^{n} \left(A_q(U) \sum_{h=1}^{n-1} \tau_{h_q} \partial_{\tau_h} U \right) \right), \quad 1 \le k \le p. \tag{5.98}$$

The right-hand side of (5.98) depends on the tangential derivatives of U on Γ, while the left-hand side yields a combination of transport equations along a direction that is normal to Γ. If Γ is the boundary $\partial\Omega$ and ν is oriented outward Ω (see Figure

5.1), then for $1 \le k \le p_0$, (5.98) yields p_0 transport equations according to which information are propagated from the inside to the outside of Ω. These p_0 equations are called the compatibility equations for Ω. On the other hand, if Γ is the common boundary of the adjoining subdomains Ω_1 and Ω_2, taking as ν the normal direction to Γ, oriented from Ω_1 to Ω_2 (see Figure 5.2), then the first p_0 equations of (5.98) are the compatibility equations for Ω_1. Obviously, for $p_0+1 \le k \le p$, (5.98) provides the compatibility equations for Ω_2.

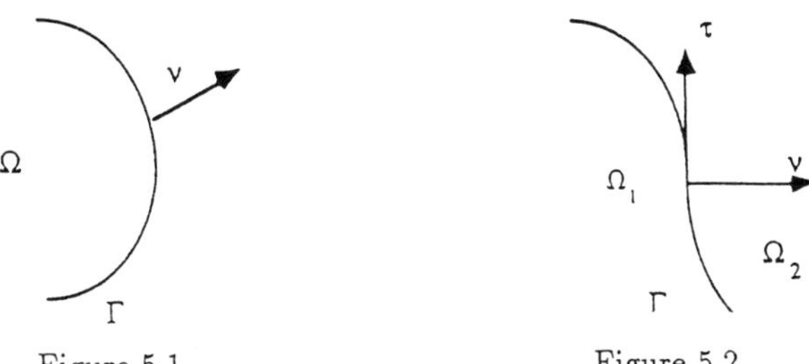

Figure 5.1. Figure 5.2.

The use of compatibility equations for single domain pseudospectral methods was advocated in Gottlieb, Gunzburger and Turkel (1982), and Canuto and Quarteroni (1987). Gottlieb, Lustman and Tadmor (1987a, 1987b) developed the analysis of the stability and convergence for the case of dissipative boundary conditions. The issue of the compatibility equations at subdomain interfaces was addressed in Cambier, Ghazzi, Veuillot and Viviand (1982).

Now, let Ω be an n-dimensional hypercube, $\bar{\Omega} = \bar{\Omega}_1 \cup \bar{\Omega}_2, \bar{\Omega}_1 \cap \bar{\Omega}_2 = \Gamma, \Omega_1 \cap \Omega_2 = \Phi$ where Γ is orthogonal to one Cartesian direction. We also assume for the moment that the solution U is continuous across Γ. Let N be any positive integer, and Σ_N be the set of Chebyshev-Gauss-Lobatto interpolation points of the reference hypercube $[-1,1]^n$. Next, let $\bar{\Omega}_{m,N}$ be the sets of the corresponding points in $\bar{\Omega}_m$, and $\Omega_{m,N} = \bar{\Omega}_{m,N} \cap \Omega_m, m = 1,2$. Moreover, $\Gamma_N = \bar{\Omega}_{1,N} \cap \Gamma = \bar{\Omega}_{2,N} \cap \Gamma$, since we use the same number of interpolation points in $\bar{\Omega}_1$ and $\bar{\Omega}_2$.

Quarteroni (1990) served a Chebyshev pseudospectral domain decomposition scheme for (5.94). It is to look for $u_{1,N}(x,t) \in \mathbb{P}_N$ for $x \in \Omega_1$ and $u_{2,N}(x,t) \in \mathbb{P}_N$ for $x \in \Omega_2$, satisfying a set of equations. Firstly, they fulfill that

$$\partial_t u_{1,N} + \sum_{q=1}^{n} A_q(u_{1,N})\, \partial_q u_{1,N} = f, \quad x \in \Omega_{1,N}, \tag{5.99}$$

$$\partial_t u_{2,N} + \sum_{q=1}^{n} A_q(u_{2,N})\,\partial_q u_{2,N} = f, \quad x \in \Omega_{2,N}, \tag{5.100}$$

$$u_{1,N} = u_{2,N}, \qquad x \in \Gamma_N. \tag{5.101}$$

In addition, we require that

$$V_k \cdot (\partial_t u_{1,N} + \lambda_k \partial_\nu u_{1,N}) = V_k \cdot \left(f - \sum_{q=1}^{n} \left(A_q(u_{1,N}) \sum_{h=1}^{n-1} \tau_{h_q} \partial_{\tau_h} u_{1,N} \right) \right), \quad 1 \le k \le p_0, \tag{5.102}$$

$$V_k \cdot (\partial_t u_{2,N} + \lambda_k \partial_\nu u_{2,N}) = V_k \cdot \left(f - \sum_{q=1}^{n} \left(A_q(u_{2,N}) \sum_{h=1}^{n-1} \tau_{h_q} \partial_{\tau_h} u_{2,N} \right) \right), \quad p_0 + 1 \le k \le p, \tag{5.103}$$

where ν is the unit outward normal direction to Ω_1 on $\Gamma, \lambda^{(k)}$ and $V^{(k)}$ are the eigenvalues and the left eigenvectors of the matrix $A(\nu)$, and τ is the tangential direction on Γ. Indeed, (5.102) are the p_0 compatibility equations for Ω_1 on Γ, while (5.103) are the $p-p_0$ ones for Ω_2 on Γ. Furthermore at each point of $\bar{\Omega}_{1,N}$ belonging to a "face" Ψ of $\partial\Omega_1/\Gamma$, we set

$$V_k \cdot (\partial_t u_{1,N} + \lambda_k \partial_\nu u_{1,N}) = V_k \cdot \left(f - \sum_{q=1}^{n} \left(A_q(u_{1,N}) \sum_{h=1}^{n-1} \tau_{h_q} \partial_{\tau_h} u_{1,N} \right) \right), \quad 1 \le k \le p_0, \tag{5.104}$$

where ν is now the outward normal direction to Ω_1 on Ψ, τ is the tangential direction on Ψ, while λ_k and V_k are the eigenvalues and the left eigenvectors of the matrix $A(\nu)$. But here p_0 is not necessarily the same as in (5.102). The remaining $p - p_0$ equations that we need at each interpolation point of Ψ must be provided by the physical boundary conditions that supplement (5.94). Similarly, at each point of $\bar{\Omega}_{2,N}$ belonging to a "face" Ψ of $\partial\Omega_2/\Gamma$, we enforce that

$$V_k \cdot (\partial_t u_{2,N} + \lambda_k \partial_\nu u_{2,N}) = V_k \cdot \left(f - \sum_{q=1}^{n} \left(A_q(u_{2,N}) \sum_{h=1}^{n-1} \tau_{h_q} \partial_{\tau_h} u_{2,N} \right) \right), \quad 1 \le k \le p_0, \tag{5.105}$$

where ν is the outward normal direction to Ω_2 on Ψ, τ is the tangential direction on Ψ, while λ_k and V_k are the eigenvalues and the left eigenvectors of the matrix $A(\nu)$. But here p_0 is not necessarily the same as before. The remaining $p - p_0$ equations are prescribed as boundary conditions.

It is noted that if the matrix A_q is not constant, then the eigenvalues as well as the eigenvectors in (5.102)–(5.105) can be generally expressed in close form in terms of ν and the components of the physical variables.

We now turn to the case in which Γ is no longer an arbitrary surface, rather the front of a shock propagating throughout Ω. Let Ω_1 and Ω_2 be the adjoining subdomains separated by Γ, and w be the speed with which Γ propagates in the direction ν, ν being the outward normal vector to Ω_1 on Γ as before. Assume that there exists a vector function F such that $A_q(z) = \partial_z F_q(z)$. Then (5.94) can be written in conservation form as

$$\partial_t U + \sum_{q=1}^{n} \partial_q F(U) = f, \quad x \in \Omega,\ 0 < t \leq T. \tag{5.106}$$

The weak solutions to (5.106) are allowed to be discontinuous across the interface Γ, satisfying the Rankine-Hugoniot condition

$$w[U] - \sum_{q=1}^{n} \nu_q \left[F_q(U)\right] = 0$$

where $[z]$ denotes the jump of the values of z on the two sides of Γ. In the context of the above Chebyshev pseudospectral approximation, the above equation reads as

$$w\left(u_{2,N} - u_{1,N}\right) - \sum_{q=1}^{n} \nu_q \left(F_q\left(u_{2,N}\right) - F_q\left(u_{1,N}\right)\right) = 0. \tag{5.107}$$

At each interpolation point $x_\Gamma \in \Gamma_N$, (5.107) yields p equations for the $2p+1$ unknowns which are $u_{1,N}^{(j)}, u_{2,N}^{(j)}, 1 \leq j \leq p$ and w. Thus we need $p+1$ further independent conditions provided by the compatibility, which, in the current situations, can be determined as follows. Let us define the characteristic matrix for Ω_1 at the point on Γ,

$$A\left(\nu^{(1)}\right) = \sum_{q=1}^{n} \nu_q A_q\left(u_{1,N}\right)$$

and denote by $\lambda_k^{(1)}$ and $V_k^{(1)}$ the eigenvalues and the left eigenvectors respectively. Similarly denote by $\lambda_k^{(2)}$ and $V_k^{(2)}$ the eigenvalues and the left eigenvectors of the matrix

$$A\left(\nu^{(2)}\right) = -\sum_{q=1}^{n} \nu_q A\left(u_{2,N}\right).$$

In general, $u_{1,N} \neq u_{2,N}$ on Γ and so $\lambda_k^{(1)} \neq -\lambda_k^{(2)}, V_k^{(1)} \neq V_k^{(2)}$. Assume that the eigenvalues are ordered as

$$\lambda_k^{(1)} < \lambda_{k+1}^{(1)}, \quad \lambda_k^{(2)} < \lambda_{k+1}^{(2)}, \quad 1 \leq k \leq p-1$$

and that Γ is the surface of propagation of a k_1-shock. So there exists an integer $1 \le k_1 \le p$ such that (see Smoller (1983))

$$\lambda^{(1)}_{k_1-1} < w < \lambda^{(1)}_{k_1}, \quad \lambda^{(2)}_{k_1} < w < \lambda^{(2)}_{k_1+1}. \tag{5.108}$$

Figure 5.3 illustrates an example of a k_1-shock. For fixed time t, we draw in bold the straight line with slope w, the speed of the shock front. On its left, we label with k the straight line whose slope equals the characteristic speed $\lambda^{(1)}_k, 1 \le k \le p$. Similarly, we label on its right the straight lines with slope $\lambda^{(2)}_k, 1 \le k \le p$.

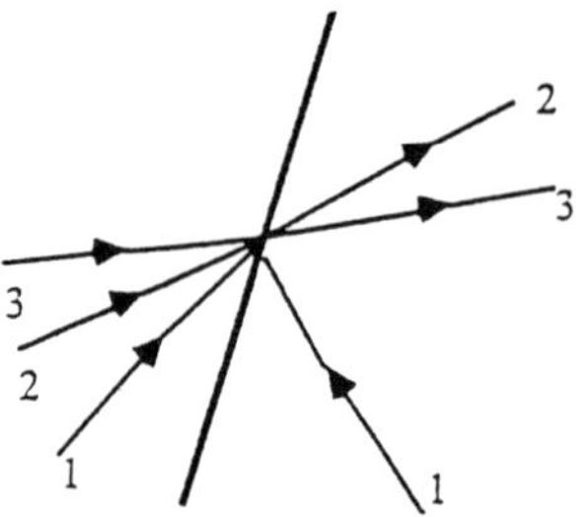

Figure 5.3. $p = 3, k_1 = 1$.

The compatibility equations for Ω_1 are given by the following $p - k_1 + 1$ equations

$$\begin{aligned} &V^{(1)}_k \cdot \left(\partial_t u_{1,N} + \lambda^{(1)}_k \partial_{\nu^{(1)}} u_{1,N}\right) \\ &\quad = V^{(1)}_k \cdot \left(f - \sum_{q=1}^{n}\left(A_q(u_{1,N}) \sum_{h=1}^{n-1} \tau_{h_q} \partial_{\tau_h} u_{1,N}\right)\right), \quad k_1 \le k \le p, \end{aligned}$$

while those for Ω_2 are given by the k_1 equations

$$\begin{aligned} &V^{(2)}_k \cdot \left(\partial_t u_{2,N} + \lambda^{(2)}_k \partial_{\nu^{(2)}} u_{2,N}\right) \\ &\quad = V^{(2)}_k \cdot \left(f - \sum_{q=1}^{n}\left(A_q(u_{2,N}) \sum_{h=1}^{n-1} \tau_{h_q} \partial_{\tau_h} u_{2,N}\right)\right), \quad p - k_1 + 1 \le k \le p. \end{aligned}$$

As usual, $\{\tau_h\}$ is the set of vectors tangential to Γ at the point x_Γ under consideration.

Note that (5.108) can be rewritten in the form

$$\lambda^{(1)}_{k_1-1} < w < \lambda^{(2)}_{k_1+1}, \quad \lambda^{(2)}_{k_1} < w < \lambda^{(1)}_{k_1}.$$

We now consider the iteration by subdomain algorithm for the solution of the domain decomposition problem. For simplicity, assume that the solutions are continuous across the interface Γ. Let $D(\nu)$ and $E(\nu)$ denote the matrix of eigenvalues of $A(\nu)$ and the matrix of the corresponding left eigenvectors, so that

$$E(\nu)A(\nu) = D(\nu)E(\nu). \tag{5.109}$$

Let $u_{1,N,m}$ and $u_{2,N,m}$ be the m-th iterated values of $u_{1,N}$ and $u_{2,N}$ respectively. At each point $x \in \Gamma_N$, set

$$\chi_{1,m} = u_{1,N,m} \qquad \chi_{2,m} = u_{2,N,m}.$$

Moreover, we denote by $E_{p_0}(\nu)$ the $p_0 \times p$ matrix given by the first p_0 rows of $E(\nu)$, and by $E_{p-p_0}(\nu)$ the $(p-p_0)\times p$ matrix given by the remaining $p-p_0$ rows of $E(\nu)$. Then $u_{1,N,m+1}$ satisfies the interior equations (5.99), the interface equation (5.102) together with

$$E_{p-p_0}(\nu)u_{1,N,m+1} = E_{p-p_0}(\nu)\chi_{2,m}, \quad x \in \Gamma_N,$$

and finally the boundary equations given by (5.104). Similarly, $u_{2,N,m+1}$ satisfies the interior equations (5.100), the interface equation (5.103) together with

$$E_{p_0}(\nu)u_{2,N,m+1} = E_{p_0}(\nu)\chi_{1,m}, \quad x \in \Gamma_N,$$

and finally the boundary equations given by (5.105). It is noted that the limit solutions $u_{1,N} = \lim\limits_{m\to\infty} u_{1,N,m}$ and $u_{2,N} = \lim\limits_{m\to\infty} u_{2,N,m}$ fulfill the conditions

$$\begin{aligned} E_{p_0}(\nu)u_{2,N} &= E_{p_0}(\nu)u_{1,N}, & x \in \Gamma_N,\\ E_{p-p_0}(\nu)u_{1,N} &= E_{p-p_0}(\nu)u_{2,N}, & x \in \Gamma_N, \end{aligned}$$

which in turn ensure the fulfillment of the continuity requirement (5.101).

We next consider a special case in which $\Omega = (-1,1), \Omega_1 = (-1,\alpha), \Omega_2 = (\alpha,1), |\alpha| < 1$. Clearly $\Gamma = \{\alpha\}$. The system (5.94) is reduced to

$$\partial_t U + A(U)\partial_x U = f, \quad x \in \Omega,\, 0 < t \le T, \tag{5.110}$$

while the compatibility equations (5.97) become

$$V_k \cdot (\partial_t U + \lambda_k \partial_x U) = V_k \cdot f, \quad 1 \le k \le p. \tag{5.111}$$

Furthermore by (5.109),

$$E\partial_t U + DE\partial_x U = Ef.$$

Obviously, D and E depend on U, if A does so.

Suppose that at the interface point $x = \alpha$, the first p_0 eigenvalues of A are positive, and the others are negative. Let D_{p_0} and E_{p_0} denote the $p_0 \times p$ matrices obtained suppressing the first p_0 rows of D and E respectively. Then the following p_0 equations

$$E_{p_0}\partial_t U + D_{p_0} E \partial_x U = E_{p_0} f, \quad x = \alpha, \tag{5.112}$$

provide the p_0 compatibility equations for Ω_1. Similarly, denoting by D_{p-p_0} and E_{p-p_0} the lower part of the matrices D and E, the equations

$$E_{p-p_0}\partial_t U + D_{p-p_0} E \partial_x U = E_{p-p_0} f, \quad x = \alpha, \tag{5.113}$$

provide the $p - p_0$ compatibility equations for Ω_2. The compatibility equations for Ω_2 at $x = 1$ are derived in a similar way, and take the form (5.112). At $x = -1$, the compatibility equations for Ω_1 take the form (5.113).

The system (5.110) is completed by the initial condition

$$U(x,0) = U_0(x), \quad x \in \bar{\Omega}, \tag{5.114}$$

and by a set of boundary conditions at $x = \pm 1$. For instance, B_1 is a $p_0 \times p$ matrix and B_2 is a $(p - p_0) \times p$ matrix, g_1 and g_2 are two given vector functions, and

$$\begin{cases} B_1 U = g_1, & \text{for } x = -1, \\ B_2 U = g_2, & \text{for } x = 1. \end{cases} \tag{5.115}$$

We now apply the Chebyshev pseudospectral domain decomposition method to (5.110). Let $y^{(j)} = \cos \frac{\pi j}{N}$, the Chebyshev-Gauss-Lobatto interpolation points in the reference interval $[-1,1]$, and

$$x_1^{(j)} = -1 + \frac{\alpha+1}{2}\left(y^{(j)}+1\right), \quad x_2^{(j)} = \alpha + \frac{1-\alpha}{2}\left(y^{(j)}+1\right), \quad 1 \le j \le N.$$

At each time t, we look for $u_{1,N} \in \mathbb{P}_N$ for $x \in \Omega_1$ and $u_{2,N} \in \mathbb{P}_N$ for $x \in \Omega_2$. They satisfy a set of equations as follows

$$
\begin{aligned}
&\partial_t u_{1,N} + A\left(u_{1,N}\right)\partial_x u_{1,N} = f, && x = x_1^{(j)},\ 1 \le j \le N-1,\\
&\partial_t u_{2,N} + A\left(u_{2,N}\right)\partial_x u_{2,N} = f, && x = x_2^{(j)},\ 1 \le j \le N-1,\\
&E_{p_0}\partial_t u_{1,N} + D_{p_0} E\partial_x u_{1,N} = E_{p_0} f, && x = \alpha,\\
&E_{p-p_0}\partial_t u_{2,N} + D_{p-p_0} E\partial_x u_{2,N} = E_{p-p_0} f, && x = \alpha,\\
&u_{1,N} = u_{2,N}, && x = \alpha,\\
&E_{p-p_0}\partial_t u_{1,N} + D_{p-p_0} E\partial_x u_{1,N} = E_{p-p_0} f, && x = -1,\\
&E_{p_0}\partial_t u_{2,N} + D_{p_0} E\partial_x u_{2,N} = E_{p_0} f, && x = 1,\\
&B_1 u_{1,N} = g_1, && x = -1,\\
&B_2 u_{2,N} = g_2, && x = 1.
\end{aligned}
$$

Obviously, the first two equations in the above set are the internal equations. The next three equations are the interface equations. The remaining ones are the boundary equations.

We now write the iteration. Let $\chi_{1,m} = u_{1,N,m}$ and $\chi_{2,m} = u_{2,N,m}$ at $x = \alpha$. Then we look for $u_{1,N,m+1} \in \mathbb{P}_N$ in Ω_1, such that

$$
\begin{aligned}
&\partial_t u_{1,N,m+1} + A\partial_x u_{1,N,m+1} = f, && x = x_1^{(j)},\ 1 \le j \le N-1,\\
&E_{p_0}\partial_t u_{1,N,m+1} + D_{p_0} E\partial_x u_{1,N,m+1} = E_{p_0} f, && x = \alpha,\\
&E_{p-p_0} u_{1,N,m+1} = E_{p-p_0}\chi_{2,m}, && x = \alpha,\\
&E_{p-p_0}\partial_t u_{1,N,m+1} + D_{p-p_0} E\partial_x u_{1,N,m+1} = E_{p-p_0} f, && x = -1,\\
&B_1 u_{1,N,m+1} = g_1, && x = -1,
\end{aligned}
$$

where A, D and E depend on $u_{1,N,m+1}$. Similarly, we look for $u_{2,N,m+1} \in \mathbb{P}_N$ in Ω_2, such that

$$
\begin{aligned}
&\partial_t u_{2,N,m+1} + A\partial_x u_{2,N,m+1} = f, && x = x_2^{(j)},\ 1 \le j \le N-1,\\
&E_{p-p_0}\partial_t u_{2,N,m+1} + D_{p-p_0} E\partial_x u_{2,N,m+1} = E_{p-p_0} f, && x = \alpha,\\
&E_{p_0} u_{2,N,m+1} = E_{p_0}\chi_{1,m}, && x = \alpha,\\
&E_{p_0}\partial_t u_{2,N,m+1} + D_{p_0} E\partial_x u_{2,N,m+1} = E_{p_0} f, && x = 1,\\
&B_2 u_{2,N,m+1} = g_2, && x = 1,
\end{aligned}
$$

where A, D and E depend on $u_{2,N,m+1}$.

We shall consider the convergence of the above iteration for problem (5.110), (5.114) and (5.115). For simplicity, let $p = 2, a$ be a constant, $|a| < 1$ and

$$
A = \begin{pmatrix} a & 1 \\ 1 & a \end{pmatrix}.
$$

Then the eigenvalues $\lambda_1 = a+1 > 0, \lambda_2 = a-1 < 0$. Let $D = \mathrm{Diag}\{\lambda_1, \lambda_2\}$ and

$$E = \frac{1}{\sqrt{2}}\begin{pmatrix} 1 & 1 \\ 1 & -1 \end{pmatrix}, \quad E_1 = \frac{1}{\sqrt{2}}(1,\ 1), \quad E_2 = \frac{1}{\sqrt{2}}(1,\ -1).$$

Clearly $A = EDE$. We require the initial value $U_0(x)$ to be a continuous vector function with a first component $U_0^{(1)}(x)$ vanishing at $x = \pm 1$. The boundary condition is that $U^{(1)}(-1,t) = U^{(1)}(1,t) = 0$ for $0 < t \le T$. Let τ be the mesh size in time $t, \beta = \frac{1}{\tau}, \lambda = \lambda_1$ and $\mu = -\lambda_2 > 0$. We use the Euler method for the temporal discretization. When we advance from the time t to the time $t+\tau$, we denote the values $u_{1,N}(x,t+\tau), u_{2,N}(x,t+\tau), \beta u_{1,N}(x,t) + f(x,t)$ and $\beta u_{2,N}(x,t) + f(x,t)$ by $u_{1,N}, u_{2,N}, f_{1,p}$ and $f_{2,p}$ for simplicity. Then we obtain that

$$\begin{cases} \beta u_{1,N} + A\partial_x u_{1,N} = f_{1,p}, & x = x_1^{(j)},\ 1 \le j \le N-1, \\ \beta u_{2,N} + A\partial_x u_{2,N} = f_{2,p}, & x = x_2^{(j)},\ 1 \le j \le N-1, \\ \beta E_1 u_{1,N} + \lambda E_1 \partial_x u_{1,N} = E_1 f_{1,p}, & x = \alpha, \\ \beta E_2 u_{2,N} - \mu E_2 \partial_x u_{2,N} = E_2 f_{2,p}, & x = \alpha, \end{cases} \tag{5.116}$$

and

$$\begin{cases} E_2 u_{1,N} = E_2 u_{2,N}, & x = \alpha, \\ E_1 u_{2,N} = E_1 u_{1,N}, & x = \alpha. \end{cases} \tag{5.117}$$

On the boundary $\partial\Omega$,

$$\begin{cases} \beta E_2 u_{1,N} - \mu E_2 \partial_x u_{1,N} = E_2 f_{1,p}, & x = -1, \\ \beta E_1 u_{2,N} + \lambda E_1 \partial_x u_{2,N} = E_1 f_{2,p}, & x = 1, \\ u_{1,N}^{(1)} = 0, & x = -1, \\ u_{2,N}^{(1)} = 0, & x = 1. \end{cases} \tag{5.118}$$

At the $(m+1)$th step of the iteration, the solutions $u_{1,N,m+1}$ and $u_{2,N,m+1}$ satisfy (5.116) and (5.118). The matching interface condition (5.117) is dealt with as follows

$$\begin{cases} E_2 u_{1,N,m+1} = E_2 u_{2,N,m}, & x = \alpha, \\ E_1 u_{2,N,m+1} = E_1 u_{1,N,m}, & x = \alpha. \end{cases} \tag{5.119}$$

The proof of the convergence of the above iteration is rather involved. It is convenient to carry it out using the characteristic form of the equations. To this end, define the error functions $e_{\gamma,m} = E\left(u_{\gamma,N,m} - u_{\gamma,N}\right), \gamma = 1,2$. Clearly,

$$e_{\gamma,m}^{(1)} = E_1\left(u_{\gamma,N,m} - u_{\gamma,N}\right), \quad e_{\gamma,m}^{(2)} = E_2\left(u_{\gamma,N,m} - u_{\gamma,N}\right), \quad \gamma = 1,2.$$

Since $DE = EA$, (5.116), (5.118) and (5.119) imply that

$$\begin{cases} \beta e_{1,m+1} + D\partial_x e_{1,m+1} = 0, & x = x_1^{(j)},\ 1 \le j \le N-1, \\ e_{1,m+1}^{(1)} + e_{1,m+1}^{(2)} = 0, & x = -1, \\ \beta e_{1,m+1}^{(2)} - \mu\partial_x e_{1,m+1}^{(2)} = 0, & x = -1, \\ \beta e_{1,m+1}^{(1)} + \lambda\partial_x e_{1,m+1}^{(1)} = 0, & x = \alpha, \\ e_{1,m+1}^{(2)} = e_{2,m}^{(2)}, & x = \alpha. \end{cases} \tag{5.120}$$

Similarly, we have that

$$\begin{cases} \beta e_{2,m+1} + D\partial_x e_{2,m+1} = 0, & x = x_2^{(j)},\ 1 \le j \le N-1, \\ e_{2,m+1}^{(1)} + e_{2,m+1}^{(2)} = 0, & x = 1, \\ \beta e_{2,m+1}^{(1)} + \lambda\partial_x e_{2,m+1}^{(1)} = 0, & x = 1, \\ \beta e_{2,m+1}^{(2)} - \mu\partial_x e_{2,m+1}^{(2)} = 0, & x = \alpha, \\ e_{2,m+1}^{(1)} = e_{1,m}^{(1)}, & x = \alpha. \end{cases} \tag{5.121}$$

We start by proving two important lemmas.

Lemma 5.13. *Let $\eta > 0$ and ε be two given real numbers, and $y^{(j)}$ be the Chebyshev-Gauss-Lobatto interpolation points in the reference interval $[-1,1]$. The function $v \in \mathbb{P}_N$ satisfies*

$$\begin{cases} v - \eta\partial_y v = 0, & y = y^{(j)},\ 1 \le j \le N, \\ v = \varepsilon, & y = y^{(0)} = 1. \end{cases} \tag{5.122}$$

Then there exists a constant $\sigma_N(\eta)$ such that $|\sigma_N(\eta)| < 1$ and $v(-1) = \sigma_N(\eta)\varepsilon$.

Proof. Let $T_l(y)$ be the Chebyshev polynomial of degree l, $\Phi_{N-1}(y) = \partial_y T_N(y)$ and $\Phi_N(y) = y\partial_y T_N(y)$. Since $y^{(j)}, 1 \le j \le N-1$ are the roots of $\partial_y T_N(y)$, we deduce from the first formula of (5.122) that

$$v - \eta\partial_y v = \xi\left(\Phi_{N-1} + \Phi_N\right), \quad |y| \le 1 \tag{5.123}$$

where ξ is a constant that can be determined by the inflow boundary condition, i.e., the second formula of (5.122). Let $\hat{\Phi}_{\gamma,l}$ be the Chebyshev coefficients of $\Phi_\gamma, \gamma = N-1, N$. Following Canuto (1988), we know that $\hat{\Phi}_{\gamma,l} \ge 0$ for $0 \le l \le \gamma, \gamma = N-1, N$. Next, let $\phi_\eta \in \mathbb{P}_N$ and $\psi_\eta \in \mathbb{P}_N$ be the unique solutions of equations

$$\phi_\eta - \eta\partial_y\phi_\eta = \Phi_{N-1} \tag{5.124}$$

and

$$\psi_\eta - \eta \partial_y \psi_\eta = \Phi_N. \tag{5.125}$$

Owing to (5.123), we state that

$$v = d\left(\phi_\eta + \psi_\eta\right). \tag{5.126}$$

Taking into account (2.72), we obtain from (5.124) and (5.125) that

$$\hat{\phi}_{\eta,l} = \hat{\Phi}_{N-1,l} + \frac{2\eta}{c_l} \sum_{\substack{p=l+1 \\ p+l \text{ odd}}} p\hat{\phi}_{\eta,p}, \quad 0 \le l \le N$$

and

$$\hat{\psi}_{\eta,l} = \hat{\Phi}_{N,l} + \frac{2\eta}{c_l} \sum_{\substack{p=l+1 \\ p+l \text{ odd}}} p\hat{\psi}_{\eta,p}, \quad 0 \le l \le N.$$

Clearly for $\eta > 0$,

$$\hat{\phi}_{\eta,l} \ge 0, \quad \hat{\psi}_{\eta,l} \ge 0, \quad 0 \le l \le N. \tag{5.127}$$

Moreover, since $T_l(1) = 1$ for any l, (5.126) leads to

$$d = \varepsilon \left(\sum_{l=0}^{N} \left(\hat{\phi}_{\eta,l} + \hat{\psi}_{\eta,l} \right) \right)^{-1}$$

and so $\mathrm{sign}(d) = \mathrm{sign}(\varepsilon)$. Furthermore, the solution of (5.123) has the following Chebyshev coefficients

$$\hat{v}_l = \varepsilon \left(\hat{\phi}_{\eta,l} + \hat{\psi}_{\eta,l} \right) \left(\sum_{p=0}^{N} \left(\hat{\phi}_{\eta,p} + \hat{\psi}_{\eta,p} \right) \right)^{-1} \tag{5.128}$$

whence

$$\mathrm{sign}\left(\hat{v}_l\right) = \mathrm{sign}(\varepsilon), \quad 0 \le l \le N.$$

Put

$$\sigma_N(\eta) = \sum_{l=0}^{N} (-1)^l \left(\hat{\phi}_{\eta,l} + \hat{\psi}_{\eta,l} \right) \left(\sum_{p=0}^{N} \left(\hat{\phi}_{\eta,p} + \hat{\psi}_{\eta,p} \right) \right)^{-1}.$$

Since $T_l(-1) = (-1)^l$, we obtain from (5.128) that $v(-1) = \sigma_N(\eta)\varepsilon$. Finally, (5.127) ensures that $|\sigma_N(\eta)| < 1$ for $\eta > 0$. ∎

Lemma 5.14. *Let $\eta, \varepsilon, y^{(j)}$ and $\sigma_N(\eta)$ be the same as in Lemma 5.13. The function $v \in \mathbb{P}_N$ satisfies*

$$\begin{cases} v + \eta \partial_y v = 0, & y = y^{(j)}, \, 0 \le j \le N-1, \\ v = \varepsilon, & y = y^{(N)} = -1. \end{cases} \tag{5.129}$$

Then $v(1) = \sigma_N(\eta)\varepsilon$.

Proof. We first note that

$$v + \eta \partial_y v = \rho \left(\Phi_{N-1} - \Phi_N \right), \quad \rho = (-1)^{N-1} \frac{\varepsilon + \eta \partial_y v(-1)}{2N^2} \tag{5.130}$$

where the constant ρ has been determined by the second formula of (5.129) and the fact that $\partial_y T_l(1) = l^2, \partial_y T(-1) = (-1)^{l+1} l^2$. Consider now the following auxiliary problem

$$\begin{cases} w - \eta \partial_y w = 0, & y = y^{(j)}, \, 1 \le j \le N, \\ w = \varepsilon, & y = y^{(0)} = 1. \end{cases}$$

We want to show that

$$v(-y) = w(y), \quad |y| \le 1. \tag{5.131}$$

As a matter of fact, $w(y)$ satisfies

$$w - \eta \partial_y w = \rho^* \left(\Phi_{N-1} + \Phi_N \right), \quad \rho^* = \frac{\varepsilon - \eta \partial_y w(1)}{2N^2},$$

and thus

$$w(-y) - \eta \partial_y w(-y) = \rho^*(1-y) \partial_y T_N(-y), \quad |y| \le 1.$$

Clearly $\partial_y T_l(-y) = (-1)^{l+1} \partial_y T_l(y)$. Also by (5.131), $\partial_y w(-y) = -\partial_y v(y)$. Thus we get that

$$v(y) + \eta \partial_y (y) = (-1)^{N-1} \rho^* (1-y) \partial_y T_N(y).$$

Since $(-1)^{N-1}\rho^* = \rho$ and $(1-y)\partial_y T_N(y) = \Phi_{N-1}(y) - \Phi_N(y)$, we conclude that v satisfies (5.130), whence v satisfies the first formula of (5.129). In view of Lemma 5.13 and (5.131), we know that $v(1) = w(-1) = \sigma_N(\eta)\varepsilon$. Furthermore, (5.128) and (5.131) lead to

$$\hat{v}_l = (-1)^l \varepsilon \left(\hat{\phi}_{\eta,l} + \hat{\psi}_{\eta,l} \right) \left(\sum_{p=0}^{N} \left(\hat{\phi}_{\eta,p} + \hat{\psi}_{\eta,p} \right) \right)^{-1}.$$

■

We are now in a position to state some results. Let

$$\lambda' = \frac{2\lambda}{\beta(1+\alpha)}, \qquad \mu' = \frac{2\mu}{\beta(1+\alpha)},$$
$$\lambda'' = \frac{2\lambda}{\beta(1-\alpha)}, \qquad \mu'' = \frac{2\mu}{\beta(1-\alpha)}.$$

Lemma 5.15. *The solution of (5.120) satisfies*

$$e^{(1)}_{1,m+1}(\alpha) = -\sigma_N(\lambda')\sigma_N(\mu')e^{(2)}_{2,m}(\alpha).$$

Proof. Let $y(x) = 2(x+1)(\alpha+1)^{-1} - 1$ be the affine transformation from $\bar{\Omega}_1$ to the reference interval $[-1,1]$. Let us define $v \in \mathbb{P}_N$ such that

$$v(y(x)) = e^{(2)}_{1,m+1}(x), \quad x \in \bar{\Omega}_1.$$

It is readily seen from the first formula, the third formula and the fifth formula of (5.120) that $v(y)$ is the solution of (5.122), provided that $\eta = \mu'$ and $\varepsilon = e^{(2)}_{2,m}(\alpha)$. Thus by Lemma 5.13,

$$e^{(2)}_{1,m+1}(-1) = v(-1) = \sigma_N(\mu')e^{(2)}_{2,m}(\alpha). \tag{5.132}$$

Next, define $w \in \mathbb{P}_N$ such that

$$w(y(x)) = e^{(1)}_{1,m+1}(x), \quad x \in \bar{\Omega}_1.$$

By virtue of the first formula, the second formula and the fourth formula of (5.120), $w(y)$ is the solution of (5.129), provided that $\eta = \lambda'$ and $\varepsilon = -e^{(2)}_{1,m+1}(-1)$. So by Lemma 5.14 and (5.132),

$$e^{(1)}_{1,m+1}(\alpha) = w(1) = \sigma_N(\lambda')\varepsilon = -\sigma_N(\lambda')\sigma_N(\mu')e^{(2)}_{2,m}(\alpha).$$

■

Lemma 5.16. *The solution of (5.121) satisfies*

$$e^{(2)}_{2,m+1}(\alpha) = -\sigma_N(\lambda'')\sigma_N(\mu'')e^{(1)}_{1,m}(\alpha).$$

Proof. Let $y(x) = 2(x-\alpha)(1-\alpha)^{-1} - 1$ be the affine mapping from $\bar{\Omega}_2$ to the interval $[-1,1]$, and define $v \in \mathbb{P}_N$ such that

$$v(y(x)) = e^{(1)}_{2,m+1}(x), \quad x \in \bar{\Omega}_2.$$

Owing to (5.121), $v(y)$ is the solution of (5.129) with $\eta = \lambda''$ and $\varepsilon = e_{1,m}^{(1)}(\alpha)$. Thus by Lemma 5.14,

$$e_{2,m+1}^{(1)}(1) = \sigma_N(\lambda'')e_{1,m}^{(1)}(\alpha). \tag{5.133}$$

On the other hand, let $w \in \mathbb{P}_N$ such that

$$w(y(x)) = e_{2,m+1}^{(2)}(x), \quad x \in \bar{\Omega}_2.$$

According to (5.121), $w(y)$ is the solution of (5.122) with $\eta = \mu''$ and $\varepsilon = -e_{2,m+1}^{(1)}(1)$. Thus by Lemma 5.13,

$$e_{2,m+1}^{(2)}(\alpha) = -\sigma_N(\mu'')e_{2,m+1}^{(1)}(1)$$

from which and (5.133), the desired result comes. ∎

Now define the interface error

$$e_m = \left(e_{1,m}^{(1)}(\alpha)\right)^2 + \left(e_{1,m}^{(2)}(\alpha)\right)^2 + \left(e_{2,m}^{(1)}(\alpha)\right)^2 + \left(e_{2,m}^{(2)}(\alpha)\right)^2, \quad m \geq 1,$$

and set

$$\sigma_N^*(\lambda, \mu, \alpha) = \max\left(\sigma_N^2(\lambda')\sigma_N^2(\mu'), \sigma_N^2(\lambda'')\sigma_N^2(\mu'')\right).$$

Obviously, $\sigma_N^*(\lambda, \mu, \alpha) < 1$.

Lemma 5.17. *For $m \geq 2$,*

$$e_{m+1} \leq \sigma_N^*(\lambda, \mu, \alpha)e_m.$$

Proof. By the last formula of (5.120), the last formula of (5.121), Lemmas 5.15 and 5.16, we find that

$$e_{2,m+1}^{(1)}(\alpha) = e_{1,m}^{(1)}(\alpha) = -\sigma_N(\lambda')\sigma_N(\mu')e_{2,m-1}^{(2)}(\alpha) = -\sigma_N(\lambda')\sigma_N(\mu')e_{1,m}^{(2)}(\alpha)$$

and

$$e_{1,m+1}^{(2)}(\alpha) = e_{2,m}^{(2)}(\alpha) = -\sigma_N(\lambda'')\sigma_N(\mu'')e_{1,m-1}^{(1)}(\alpha) = -\sigma_N(\lambda'')\sigma_N(\mu'')e_{2,m}^{(1)}(\alpha).$$

Therefore

$$\begin{aligned} e_{m+1} &= \sigma_N^2(\lambda')\sigma_N^2(\mu')\left(\left(e_{1,m}^{(2)}(\alpha)\right)^2 + \left(e_{2,m}^{(2)}(\alpha)\right)^2\right) \\ &\quad + \sigma_N^2(\lambda'')\sigma_N^2(\mu'')\left(\left(e_{1,m}^{(1)}(\alpha)\right)^2 + \left(e_{2,m}^{(1)}(\alpha)\right)^2\right) \\ &\leq \sigma_N^*(\lambda, \mu, \alpha)e_m. \end{aligned}$$

∎

Lemma 5.18. *For $m \geq 1$,*

$$\max_{-1\leq x\leq\alpha}\left(\left(e_{1,m}^{(1)}(x)\right)^2+\left(e_{1,m}^{(2)}(x)\right)^2\right)+\max_{\alpha\leq x\leq 1}\left(\left(e_{2,m}^{(1)}(x)\right)^2+\left(e_{2,m}^{(2)}(x)\right)^2\right)\leq 2e_m.$$

Proof. From the proof of Lemma 5.15, we deduce that for $x \in [-1, \alpha]$,

$$\begin{aligned}
e_{1,m+1}^{(2)}(x)=\sum_{l=0}^{N}\hat{v}_l T_l(y(x)) &= e_{2,m}^{(2)}(\alpha)\left(\sum_{l=0}^{N}\left(\hat{\phi}_{\mu',l}+\hat{\psi}_{\mu',l}\right)\right)^{-1}\sum_{l=0}^{N}\left(\hat{\phi}_{\mu',l}+\hat{\psi}_{\mu',l}\right)T_l(y(x)),\\
e_{1,m+1}^{(1)}(x)=\sum_{l=0}^{N}\hat{w}_l T_l(y(x)) &= -\sigma_N(\mu')e_{2,m}^{(2)}(\alpha)\left(\sum_{l=0}^{N}\left(\hat{\phi}_{\lambda',l}+\hat{\psi}_{\lambda',l}\right)\right)^{-1}\\
&\quad\cdot\sum_{l=0}^{N}(-1)^l\left(\hat{\phi}_{\lambda',l}+\hat{\psi}_{\lambda',l}\right)T_l(y(x)).
\end{aligned}$$

Since $|T_l(y)| \leq 1$ for $|y| \leq 1$, we know from the above two equalities and the last formula of (5.120) that for $x \in [-1, \alpha]$,

$$\left|e_{1,m+1}^{(2)}(x)\right|\leq\left|e_{1,m+1}^{(2)}(\alpha)\right|,\quad \left|e_{1,m+1}^{(1)}(x)\right|\leq|\sigma_N(\mu')|\left|e_{1,m+1}^{(2)}(\alpha)\right|.$$

In a similar way, we see from the proof of Lemma 5.16 that for $x \in [\alpha, 1]$,

$$\left|e_{2,m+1}^{(1)}(x)\right|\leq\left|e_{2,m+1}^{(1)}(\alpha)\right|,\quad \left|e_{2,m+1}^{(2)}(x)\right|\leq|\sigma_N(\lambda'')|\left|e_{2,m+1}^{(1)}(\alpha)\right|.$$

The above estimations lead to the conclusion. ■

Theorem 5.9. *The iteration applied to (5.116)–(5.118) converges, as $m \to \infty$.*

Proof. The result follows from Lemmas 5.17 and 5.18. Actually, since E is nonsingular, the convergence of $e_{j,m}$ implies that of $u_{\gamma,N,m} - u_{\gamma,N}, \gamma = 1, 2$. ■

In the final part of this section, we present some numerical results. The time discretization considered is the second-order Crank-Nicolson scheme with the mesh size $\tau = 0.01$. The test function is

$$U^{(1)}(x,t)=e^t\text{arctg}\left(ax+\frac{a}{2}\right),\quad U^{(2)}(x,t)=e^t\text{arctg}\left(ax-\frac{a}{2}\right).$$

We use the Chebyshev pseudospectral domain decomposition method to solve this problem. For comparison, we also use single domain method for this problem. In Tables 5.8 and 5.9, we list the maximum norms of the errors between the exact solution and the corresponding numerical ones at $t = 1$, with different values of a and

N. It can be seen that the domain decomposition procedure provides better numerical results, in particular, for the solutions possessing oscillations. Furthermore, if the matrices associated with the pseudospectral method are full, then the use of several subdomains allows the reduction of the overall complexity. It is noted that similar results are also valid for multi-subdomain method and the Legendre pseudospectral domain decomposition method.

Table 5.8. The errors of single domain method.

	$a = 1$	$a = 4$	$a = 10$
$N = 4$	$4.767E-4$	$1.288E-2$	$2.812E-2$
$N = 8$	$1.094E-4$	$5.079E-3$	$2.345E-2$
$N = 20$	$2.572E-7$	$2.860E-4$	$1.005E-2$

Table 5.9. The errors of two domains method.

	$a = 1$	$a = 4$	$a = 10$
$N = 4$	$1.937E-4$	$6.119E-3$	$3.668E-2$
$N = 8$	$7.377E-7$	$1.024E-3$	$1.893E-2$
$N = 20$	$2.577E-7$	$1.943E-4$	$9.231E-3$

Another early work of spectral domain decomposition methods for hyperbolic systems was due to Kopriva (1986, 1987). We also refer to Quarteroni (1987b), Macaraeg and Streett (1988), and Macaraeg, Streett and Hussaini (1989) for this technique and its applications. Carlenzoli, Quarteroni and Valli (1991) also used it for compressible fluid flows. Recently Quarteroni, Pasquarelli and Valli (1991) developed heterogeneous domain decomposition method arising in the approximation of certain physical phenomena governed by different kinds of equations in several disjoint subregions. Various applications of domain decomposition methods can be found in Chan, Glowinski, Periaux and Widlund (1988), and Canuto, Hussaini, Quarteroni and Zang (1988).

For the applications of Chebyshev pseudospectral methods to other nonlinear problems, we also refer to Haidvogel, Robinson and Schulman (1980), and Maday and Métivet (1986).

5.5. Spectral Multigrid Methods

Multigrid methods have been used for spectral approximations since the beginning of the last decade. Zang, Wong and Hussaini (1982, 1984) described this approach and its applications to periodic problems and non-periodic problems. It makes the spectral methods more efficient. In this section, we present some basic results in this topic.

We first let $\Lambda = (0, 2\pi)$ and consider the model problem

$$\begin{cases} -\partial_x^2 U = f, & -\infty < x < \infty, \\ U(x) = U(x+2\pi), & -\infty < x < \infty. \end{cases} \tag{5.134}$$

Let N be an even positive integer. Set $x^{(j)} = \frac{2\pi j}{N}$ and $f_j = f\left(x^{(j)}\right)$. Let u_j be the approximation to the value $U\left(x^{(j)}\right), 0 \le j \le N-1$. Then

$$u_j = \sum_{l=-\frac{N}{2}}^{\frac{N}{2}-1} \tilde{u}_l \exp\left(\frac{2\pi i j l}{N}\right), \tag{5.135}$$

$\tilde{u}_l$ being the discrete Fourier coefficients,

$$\tilde{u}_l = \frac{1}{N}\sum_{j=0}^{N-1} u_j \exp\left(-\frac{2\pi i j l}{N}\right), \quad l = -\frac{N}{2}, -\frac{N}{2}+1, \dots, \frac{N}{2}-1. \tag{5.136}$$

A sensible approximation to the left-hand side of (5.134) at $x^{(j)}$ is

$$\sum_{l=-\frac{N}{2}}^{\frac{N}{2}-1} l^2 \tilde{u}_l \exp\left(\frac{2\pi i j l}{N}\right).$$

Let $V = (u_0, \dots, u_{N-1})$ and $F = (f_0, \dots, f_{N-1})$. Moreover, we denote by D a diagonal matrix with the elements $D_{jk} = j^2\delta_{j,k}$, and by C a matrix with the elements

$$C_{jk} = \frac{1}{N}\exp\left(\frac{-2\pi i j k}{N}\right).$$

Clearly

$$\left(C^{-1}\right)_{jk} = \exp\left(\frac{2\pi i j k}{N}\right).$$

Set $L = C^{-1}DC$. The pseudospectral approximation may be represented by

$$LV = F. \tag{5.137}$$

The original problem becomes the calculation for unknown vectors, as in McKerrel, Phillips and Delves (1981).

Orszag (1980), and Haidvogel and Zang (1979) proposed various iterations to solve (5.137). Let ω be a relaxation parameter and V_m be the m-th iterated value of V. The Richardson iteration for (5.137) is

$$V_{m+1} = V_m + \omega\,(F - LV_m)\,. \tag{5.138}$$

It can be checked that L has the eigenvalues $\lambda_l = l^2$ and the corresponding eigenvectors w_l with the components $w_l^{(j)} = \exp\left(\dfrac{2\pi ijl}{N}\right)$. Let $\tilde{V}_m = V_m - V$. It can be resolved into an expansion in the eigenvectors of L. Each iteration reduces the l-th error component by a factor $\nu(\lambda_l) = 1 - \omega\lambda_l$. Clearly the iteration is convergent, if for all $l, |1 - \omega\lambda_l| < 1$. In the current context, $\lambda_{\min} = 1$ and $\lambda_{\max} = \frac{1}{4}N^2$. So (5.138) is convergent for $0 < \omega < \frac{8}{N^2}$. The best choice of ω results from minimizing $|\nu(\lambda_l)|$ for $\lambda \in [\lambda_{\min}, \lambda_{\max}]$. The optimal relaxation parameter for this single-grid procedure is

$$\omega_{\text{SG}} = \frac{2}{\lambda_{\max} + \lambda_{\min}}. \tag{5.139}$$

It produces the spectral radius

$$\rho_{\text{SG}} = \frac{\lambda_{\max} - \lambda_{\min}}{\lambda_{\max} + \lambda_{\min}}. \tag{5.140}$$

Unfortunately, $\rho_{\text{SG}} \approx 1 - \frac{8}{N^2}$ which imples that $\mathcal{O}(N^2)$ iterations are required to achieve the convergence. This slow convergence is the outcome of balancing the damping of the lowest frequency eigenfunctions with that of the highest frequency one in the minimax problem described in the above.

The multigrid approach takes advantage of the fact that the low-frequency modes $\left(|l| < \frac{N}{4}\right)$ can be represented just as well on a coarser grid. So it settles for balancing the middle-frequency eigenfunction $\left(|l| = \frac{N}{4}\right)$ with the highest frequency one $\left(|l| = \frac{N}{2}\right)$. Namely, it damps effectively only those modes which cannot be resolved on coarser grids. To do this, we replace $\lambda_{\min}$ in (5.139) by λ_{mid}, and obtain that

$$\omega_{\text{MG}} = \frac{2}{\lambda_{\max} + \lambda_{\text{mid}}}. \tag{5.141}$$

The multigrid smoothing factor

$$\mu_{\text{MG}} = \frac{\lambda_{\max} - \lambda_{\text{mid}}}{\lambda_{\max} + \lambda_{\text{mid}}} \tag{5.142}$$

measures the damping rate of the high frequency modes. In this example, $\lambda_{\text{mid}} = \dfrac{N^2}{16}$ and so $\mu_{\text{MG}} = 0.6$, independent of N.

We present the multigrid process by considering the interplay between two grids. The fine grid problem with N_f interpolation points is written in the form

$$L_f V_f = F_f.$$

The decision to switch to the coarser grid is made after the fine grid approximation W_f to V_f has been sufficiently smoothed by the relaxation process, i.e., after the high frequency content of the error $V_f - W_f$ has been sufficiently reduced. Let L_c be a coarse grid operator and V_c be the correction of W_f on the coarse grid with N_c interpolation points. Furthermore denote by R a restriction operator interpolating functions from the fine grid to the coarse grid, and

$$F_c = R\left(F_f - L_f V_f\right).$$

The auxiliary coarse grid problem is

$$L_c V_c = F_c.$$

After an adequate approximation W_c to V_c has been obtained, the fine grid approximation is updated by

$$W_f \longleftarrow W_f + P W_c$$

where the prolongation operator P interpolates functions from the coarse grid to the fine grid. We can repeat the above procedure with a nest of grids and a control structure to reach a desired result.

We now construct the restriction operator R and the prolongation operator P. Consider the problem

$$\partial_x \left(a(x)\partial_x U(x)\right) = f(x), \quad x \in \Lambda \tag{5.143}$$

where $\Lambda = (0, 2\pi)$ in the periodic case or $\Lambda = (-1, 1)$ in the Dirichlet case.

We first consider the periodic case. On the coarse grid, the discrete Fourier coefficients of the corrections u_j, at the coarse grid interpolation points $x^{(j)}$ are computed using (5.136) with $N = N_c$. The fine grid approximation is then updated by

$$u_j \longleftarrow u_j + \sum_{l=-\frac{N_c}{2}}^{\frac{N_c}{2}-1} \tilde{u}_l \exp\left(il\bar{x}^{(j)}\right), \quad 0 \le j \le N_f - 1,$$

where $\bar{x}^{(j)}$ are the fine grid interpolation points. The restriction operator is constructed in a similar fashion. It turns out that except for a factor of $\frac{N_f}{N_c}$, P and R are adjoint. An explicit representation of the prolongation operator is

$$P_{jk} = \frac{1}{N_c} \sum_{l=-\frac{N_c}{2}+1}^{\frac{N_c}{2}-1} \exp\left(2\pi i l \left(\frac{j}{N_f} - \frac{k}{N_c}\right)\right).$$

It sums to yield

$$P_{jk} = \frac{1}{N_c} S\left(\frac{j}{N_f} - \frac{k}{N_c}\right)$$

where

$$S(r) = \begin{cases} N_c - 1, & r \text{ is an integer,} \\ \sin(\pi r N_c)\cot(\pi r) - \cos(\pi r N_c), & \text{otherwise.} \end{cases}$$

The corresponding restriction operator is

$$R_{jk} = \frac{1}{N_f} S\left(\frac{j}{N_c} - \frac{k}{N_f}\right).$$

We next consider the Dirichlet case, and use Chebyshev approximation. The prolongation operator is

$$P_{jk} = \frac{2}{\bar{c}_k N_c} \sum_{l=0}^{N_c} \bar{c}_l^{-1} \cos\frac{\pi l j}{N_f} \cos\frac{\pi l k}{N_c}$$

where $\bar{c}_k = 2$ for $k = 0, N_c$ and $\bar{c}_k = 1$ for $1 \le k \le N_c - 1$. This sums to

$$P_{jk} = \frac{2}{\bar{c}_k N_c}\left(Q\left(\frac{j}{N_f} - \frac{k}{N_c}\right) + Q\left(\frac{j}{N_f} + \frac{k}{N_c}\right)\right)$$

where

$$Q(r) = \begin{cases} \frac{N_c}{2}, & r \text{ is an integer,} \\ \frac{1}{4} - \frac{1}{4}\cos(\pi r N_c) & \\ +\frac{1}{2}\cos\left(\frac{\pi r}{2}(N_c+1)\right)\sin\left(\frac{\pi r N_c}{2}\right)\csc\left(\frac{\pi r}{2}\right), & \text{otherwise.} \end{cases}$$

One kind of restriction operator is defined as

$$R_{jk} = \frac{2}{\tilde{c}_k N_f}\left(\bar{Q}\left(\frac{j}{N_c} - \frac{k}{N_f}\right) + \bar{Q}\left(\frac{j}{N_c} + \frac{k}{N_f}\right)\right)$$

where $\tilde{c}_k = 2$ for $k = 0, N_f$ and $\tilde{c}_k = 1$ otherwise, and

$$\bar{Q}(r) = \begin{cases} \dfrac{1}{4} + \dfrac{N_c}{2}, & r \text{ is an integer,} \\ \dfrac{1}{4} + \dfrac{1}{2} \cos\left(\dfrac{\pi r}{2}(N_c+1)\right) \sin\left(\dfrac{\pi r N_c}{2}\right) \csc\left(\dfrac{\pi r}{2}\right), & \text{otherwise.} \end{cases}$$

The other is defined by the adjoint requirement, i.e.,

$$R_{jk} = \frac{1}{\bar{c}_k N_c}\left(Q\left(\frac{j}{N_c} - \frac{k}{N_f}\right) + Q\left(\frac{j}{N_c} + \frac{k}{N_f}\right)\right).$$

For the problem (5.143), the discrete operator on the fine grid is

$$L_f = D_{N_f} A D_{N_f} \tag{5.144}$$

where A is the diagonal matrix with the elements $A_{jk} = a\left(x^{(j)}\right)\delta_{j,k}$, D_{N_f} is the matrix related to the differentiation. In the periodic case,

$$\left(D_{N_f}\right)_{jk} = \begin{cases} \dfrac{1}{2}(-1)^{j+k} \cot\left(\dfrac{\pi}{N_f}(j-k)\right), & j \neq k, \\ 0, & j = k. \end{cases} \tag{5.145}$$

In the Dirichlet case,

$$\left(D_{N_f}\right)_{jk} = \begin{cases} \dfrac{\tilde{c}_j(-1)^{j+k}}{\tilde{c}_k\left(x^{(j)} - x^{(k)}\right)}, & j \neq k, \\ \dfrac{-x^{(k)}}{2\left(1 - \left(x^{(k)}\right)^2\right)}, & 1 \leq j = k \leq N_f - 1, \\ \dfrac{2N_f^2 + 1}{6}, & j = k = 1, \\ \dfrac{-\left(2N_f^2 + 1\right)}{6}, & j = k = N_f. \end{cases} \tag{5.146}$$

Many multigrid investigators have advocated choosing the coarse grid operator so that

$$L_c = R L_f P. \tag{5.147}$$

Both the Fourier and the Chebyshev first-derivative operators defined by (5.145) and (5.146), satisfy

$$D_{N_c} = R D_{N_f} P. \tag{5.148}$$

However, (5.147) itself is not satisfied, if the coarse grid analog of (5.144) is used to define L_c, except the case in which $a(x)$ is a constant. On the other hand, much of the efficiency of the pseudospectral method is lost if (5.147) is used to define L_c. Some compromises were suggested by Zang, Wong and Hussaini (1984). The most satisfying one seems to be using (5.144), but with the restricted values of $a(x)$ in place of the pointwise values. The Chebyshev restrictions should be performed with R.

We now turn to two-dimensional problems. Let $\Omega = (-1,1)^2$ and consider the self-adjoint elliptic equation

$$\partial_1 \left(a_1(x)\partial_1 U(x)\right) + \partial_2 \left(a_2(x)\partial_2 U(x)\right) = f(x), \quad x \in (-1,1)^2 \tag{5.149}$$

with the Dirichlet boundary condition. We use pseudospectral method with $(N+1)^2$ Chebyshev-Gauss-Lobatto interpolation points. It leads to a discrete set of equations like (5.137). The convergence rate of Richardson iteration on a single grid is governed by the ratio of the largest eigenvalue to the smallest eigenvalue of L, referred to as the condition number. The multigrid condition number is the ratio of the largest eigenvalue to the smallest high frequency eigenvalue. In our current context, $\lambda_{\max} = \mathcal{O}(N^4)$, $\lambda_{\text{mid}} = \mathcal{O}(N^2)$ and $\lambda_{\min} \approx \frac{\pi^2}{4}$. An effective preconditioning is essential for multigrid as well as for single-grid iterative schemes.

The preconditioned Richardson iteration can be expressed as

$$V \longleftarrow V + \omega H^{-1}(F - LV)$$

where H is a preconditioning matrix. An obvious choice of H is a finite difference approximation H_{FD} to the differential operator in (5.149). But for multi-dimensional problems, these finite difference approximations are costly to invert. An attractive alternative is to use an approximate LU-decomposition. In this case, H is taken as the product of a lower triangular matrix $\mathcal{L}$ and an upper triangular matrix $\mathcal{U}$. Buleev (1960) and Oliphant (1961) proposed one such type of preconditioning by taking $\mathcal{L}$ as the lower triangular portion of H_{FD} and choosing $\mathcal{U}$ such that the two super diagonals of $\mathcal{LU}$ agree with those of H_{FD}, see Zang, Wong and Hussaini (1984). Zang, Wong and Hussaini (1982) also gave a similar decomposition denoted by H_{RS}, in which the diagonal elements of $\mathcal{L}$ are altered from those of H_{FD} to ensure that the row sums of H_{RS} and H_{FD} are identical. Both types of preconditioning can be computed by a simple recursion. Let $u_{j,k}$ be the values of u at the interior grid point $\left(x_1^{(j)}, x_2^{(k)}\right)$.

Suppose that a five-point finite difference approximation to (5.149) at this point is given by

$$e_{j,k}u_{j,k} + d_{j,k}u_{j-1,k} + b_{j,k}u_{j,k-1} + g_{j,k}u_{j+1,k} + h_{j,k}u_{j,k+1} = f_{j,k},$$

and that a five-diagonal incomplete LU factorization is given by

$$\begin{aligned}(\mathcal{L}V)_{j,k} &= \nu_{j,k}u_{j,k} + \beta_{j,k}u_{j-1,k} + \gamma_{j,k}u_{j,k-1},\\ (\mathcal{U}V)_{j,k} &= u_{j,k} + \lambda_{j,k}u_{j+1,k} + \sigma_{j,k}u_{j,k+1}.\end{aligned}$$

Set

$$\begin{aligned}\beta_{j,k} &= d_{j,k},\\ \gamma_{j,k} &= b_{j,k},\\ \nu_{j,k} &= e_{j,k} - \beta_{j,k}\sigma_{j,k-1} - \gamma_{j,k}\lambda_{j-1,k} - \alpha\left(\beta_{j,k}\lambda_{j,k-1} + \gamma_{j,k}\sigma_{j-1,k}\right),\\ \lambda_{j,k} &= \frac{g_{j,k}}{\nu_{j,k}},\\ \sigma_{j,k} &= \frac{h_{j,k}}{\nu_{j,k}}.\end{aligned}$$

Then $\alpha = 0$ gives H_{LU5} and H_{RS5} uses $\alpha = 1$. Straightforward modifications are made near the boundaries. A more accurate factorization can be achieved by including one extra nonzero diagonal in $\mathcal{L}$ and $\mathcal{U}$. Suppose that

$$\begin{aligned}(\mathcal{L}V)_{j,k} &= \nu_{j,k}u_{j,k} + \beta_{j,k}u_{j-1,k} + \gamma_{j,k}u_{j,k-1} + \theta_{j,k}u_{i+1,j-1},\\ (\mathcal{U}V)_{j,k} &= u_{j,k} + \lambda_{j,k}u_{j+1,k} + \sigma_{j,k}u_{j,k+1} + \chi_{j,k}u_{i-1,j+1}.\end{aligned}$$

Put

$$\begin{aligned}\gamma_{j,k} &= b_{j,k},\\ \theta_{j,k} &= -\gamma_{j,k}\lambda_{j,k-1},\\ \beta_{j,k} &= d_{j,k} - \gamma_{j,k}\chi_{j,k-1},\\ \nu_{j,k} &= e_{j,k} - \gamma_{j,k}\sigma_{j,k-1} - \theta_{j,k}\chi_{j+1,k-1} - \beta_{j,k}\lambda_{j-1,k} - \alpha\left(\lambda_{j+1,k-1}\theta_{j,k} - \chi_{j-1,k}\beta_{j,k}\right),\\ \lambda_{j,k} &= \frac{1}{\nu_{j,k}}\left(g_{j,k} - \sigma_{j+1,k-1}\theta_{j,k}\right),\\ \chi_{j,k} &= \frac{-\beta_{j,k}\sigma_{j-1,k}}{\nu_{j,k}},\\ \sigma_{j,k} &= \frac{h_{j,k}}{\nu_{j,k}}.\end{aligned}$$

Once again, $\alpha = 0$ gives the $H_{\rm LU7}$ version and $\alpha = 1$ gives the $H_{\rm RS7}$ version. Tables 5.10 and 5.11 list the multigrid condition numbers and the multigrid smoothing rates for the preconditioned procedure related to (5.149) with $a_1(x) \equiv a_2(x) \equiv 1$.

Table 5.10. Multigrid condition numbers.

N	$H_{\rm FD}^{-1}L$	$H_{\rm LU5}^{-1}L$	$H_{\rm LU7}^{-1}L$	$H_{\rm RS5}^{-1}L$	$H_{\rm RS7}^{-1}L$
8	1.733	1.863	1.756	1.984	1.796
16	1.874	2.134	1.932	2.925	2.172
32	1.941	2.324	2.054	4.425	2.934
64	1.974	2.525	2.191	8.675	5.607
128	1.990	2.763	2.307		

Table 5.11. Multigrid smoothing rates.

N	$H_{\rm FD}^{-1}L$	$H_{\rm LU5}^{-1}L$	$H_{\rm LU7}^{-1}L$
8	0.268	0.301	0.274
16	0.304	0.362	0.318
32	0.320	0.398	0.345
64	0.328	0.433	0.373
128	0.331	0.469	0.395

Finally, we present some numerical results. Let $a_1(x) \equiv a_2(x) \equiv 1$ in (5.149) and take the test function

$$U(x) = \sin\left(\alpha\pi x_1 + \frac{\pi}{4}\right)\sin\left(\alpha\pi x_2 + \frac{\pi}{4}\right).$$

We use spectral multigrid method (**SMG**) and finite difference multigrid method (**FDMG**) to solve it. The errors and the CPU times in seconds (on a CDC Cyber 175) for these methods are shown in Tables 5.12 and 5.13. Clearly the former method provides more accurate numerical solution for the same CPU time.

Table 5.12. The errors and CPU times, $\alpha = 1$.

	Errors		CPU Times	
N	FDMG	SMG	FDMG	SMG
8	$2.37E-2$	$4.42E-5$	0.08	0.35
16	$5.73E-3$	$8.53E-13$	0.24	2.32
32	$1.42E-3$	$1.25E-14$	0.76	5.62
64	$3.55E-4$		2.59	
128	$8.88E-5$		9.71	

Table 5.13. The errors and CPU times, $\alpha = 5$.

	Errors		CPU Times	
N	FDMG	SMG	FDMG	SMG
8	$1.68E+0$	$1.73E+1$	0.07	0.33
16	$1.92E-1$	$2.56E-2$	0.19	1.19
32	$4.11E-2$	$5.25E-10$	0.53	7.84
64	$9.87E-3$		2.26	
128	$2.44E-3$		9.52	

Spectral multigrid methods have been used for nonlinear problems, such as in Streett, Zang and Hussaini (1985), and Zang and Hussaini (1986).

CHAPTER 6

MIXED SPECTRAL METHODS

In studying boundary layer, channel flow, flow past a suddenly heated vertical plate and some other problems, we encounter semi-periodic problems. It is reasonable to deal with them using Fourier approximations in the periodic directions. Sometimes, the orthogonality of Fourier system reduces the resulting schemes to a set of separate equations for the Fourier coefficients of unknown functions at each time step, and thus it is easier to solve. We can use finite difference approximations or finite element approximations in the non-periodic directions. But in these cases, the accuracy in space is still limited due to these approximations, even if the genuine solution of an original problem is infinitely smooth. This deficiency can be removed by mixed Fourier-Legendre approximations or mixed Fourier-Chebyshev approximations. The purpose of this chapter is to present these two mixed spectral methods with their applications to semi-periodic unsteady incompressible fluid flow.

6.1. Mixed Fourier-Legendre Approximations

In this section, we introduce the mixed Fourier-Legendre approximations. Let $x = (x_1, x_2), \Lambda_1 = \{x_1 \,|\, -1 < x_1 < 1\}, \Lambda_2 = \{x_2 \,|\, 0 < x_2 < 2\pi\}$ and $\Omega = \Lambda_1 \times \Lambda_2$. To describe the approximation results using different orthogonal systems in different directions, we define some non-isotropic spaces as follows,

$$
\begin{aligned}
H^{r,s}(\Omega) &= L^2(\Lambda_2, H^r(\Lambda_1)) \bigcap H^s(\Lambda_2, L^2(\Lambda_1)), && r, s \geq 0,\\
M^{r,s}(\Omega) &= H^{r,s}(\Omega) \bigcap H^1(\Lambda_2, H^{r-1}(\Lambda_1)) \bigcap H^{s-1}(\Lambda_2, H^1(\Lambda_1)), && r, s \geq 1,\\
X^{r,s}(\Omega) &= H^s(\Lambda_2, H^{r+1}(\Lambda_1)) \bigcap H^{s+1}(\Lambda_2, H^r(\Lambda_1)), && r, s \geq 0
\end{aligned}
$$

equipped with the norms

$$\|v\|_{H^{r,s}(\Omega)} = \left(\|v\|^2_{L^2(\Lambda_2,H^r(\Lambda_1))} + \|v\|^2_{H^s(\Lambda_2,L^2(\Lambda_1))}\right)^{\frac{1}{2}},$$
$$\|v\|_{M^{r,s}(\Omega)} = \left(\|v\|^2_{H^{r,s}(\Omega)} + \|v\|^2_{H^1(\Lambda_2,H^{r-1}(\Lambda_1))} + \|v\|^2_{H^{s-1}(\Lambda_2,H^1(\Lambda_1))}\right)^{\frac{1}{2}},$$
$$\|v\|_{X^{r,s}(\Omega)} = \left(\|v\|^2_{H^s(\Lambda_2,H^{r+1}(\Lambda_1))} + \|v\|^2_{H^{s+1}(\Lambda_2,H^r(\Lambda_1))}\right)^{\frac{1}{2}}.$$

Denote by $\tilde{\mathcal{D}}_p(\Omega)$ the set of all infinitely differentiable functions with compact supports in Λ_1 and the period 2π for the variable x_2. The closures of $\tilde{\mathcal{D}}_p(\Omega)$ in $H^{r,s}(\Omega)$, $M^{r,s}(\Omega)$ and $X^{r,s}(\Omega)$ are denoted by $H^{r,s}_{0,p}(\Omega)$, $M^{r,s}_{0,p}(\Omega)$ and $X^{r,s}_{0,p}(\Omega)$, respectively. For the sake of simplicity, we denote the norms $\|\cdot\|_{H^{r,s}(\Omega)}$, $\|\cdot\|_{M^{r,s}(\Omega)}$ and $\|\cdot\|_{X^{r,s}(\Omega)}$ also by $\|\cdot\|_{H^{r,s}}$, $\|\cdot\|_{M^{r,s}}$ and $\|\cdot\|_{X^{r,s}}$, respectively. Furthermore, $|\cdot|_{H^{r,s}}$ and $|\cdot|_{M^{r,s}}$ stand for the semi-norms, associated with the norm $\|\cdot\|_{H^{r,s}}$ and $\|\cdot\|_{M^{r,s}}$, respectively. Besides, $H^{r,s}_p(\Omega)$ denotes the closure in $H^{r,s}(\Omega)$ of the set involving all infinitely differentiable functions with the period 2π for the variable x_2, etc.

Let M and N be any positive integers. Denote by $\mathbb{P}_{M,1}$ the set of all polynomials of degree at most M on the interval Λ_1 and $\mathbb{P}^0_{M,1} = \mathbb{P}_{M,1} \cap H^1_0(\Lambda_1)$. Denote by $\tilde{V}_{N,2}$ the set of all trigonometric polynomials of degree at most N on the interval Λ_2, and by $V_{N,2}$ the subset of $\tilde{V}_{N,2}$ containing all real-valued functions. Set $V_{M,N} = \mathbb{P}_{M,1} \otimes V_{N,2}$ and $V^0_{M,N} = \mathbb{P}^0_{M,1} \otimes V_{N,2}$. In the space $V_{M,N}$, several inverse inequalities are valid.

Theorem 6.1. *For any $\phi \in V_{M,N}$ and $2 \le p \le \infty$,*

$$\|\phi\|_{L^p} \le c\left(M^2 N\right)^{\frac{1}{2}-\frac{1}{p}} \|\phi\|.$$

Proof. Set

$$\phi_l(x_1) = \frac{1}{2\pi}\int_{\Lambda_2} \phi(x_1, x_2)\, e^{-ilx_2}\, dx_2. \tag{6.1}$$

Then $\phi_l \in \mathbb{P}_{M,1}$ and by Theorem 2.7,

$$\begin{aligned}\|\phi\|_{L^\infty} &\le \sum_{|l|\le N} \|\phi_l\|_{L^\infty(\Lambda_1)} \le cM \sum_{|l|\le N} \|\phi_l\|_{L^2(\Lambda_1)} \le cMN^{\frac{1}{2}} \left(\sum_{|l|\le N} \|\phi\|^2_{L^2(\Lambda_1)}\right)^{\frac{1}{2}} \\ &\le cMN^{\frac{1}{2}}\|\phi\|.\end{aligned}$$

Consequently,

$$\|\phi\|^p_{L^p} \le \|\phi\|^{p-2}_{L^\infty}\|\phi\|^2 \le M^{p-2}N^{\frac{p}{2}-1}\|\phi\|^p$$

which leads to the conclusion. ∎

As a consequence of Theorem 6.1, we have that for any $\phi, \psi \in V_{M,N}$,

$$\|\phi\psi\|^2 \leq \|\phi\|_{L^4}^2 \|\psi\|_{L^4}^2 \leq cM^2 N \|\phi\|^2 \|\psi\|^2. \tag{6.2}$$

Theorem 6.2. *For any* $\phi \in V_{M,N}$,

$$|\phi|_1^2 \leq \left(\frac{9}{4}M^4 + N^2\right) \|\phi\|^2.$$

Proof. Let $\phi_l(x_1)$ be the same as in (6.1). We know from the proof of Theorem 2.8 that

$$\|\partial_1 \phi\|^2 = 2\pi \sum_{|l| \leq N} \|\partial_1 \phi_l\|_{L^2(\Lambda_1)}^2 \leq \frac{18\pi}{4} M^4 \sum_{|l| \leq N} \|\phi_l\|_{L^2(\Lambda_1)}^2 = \frac{9}{4} M^4 \|\phi\|^2.$$

On the other hand,

$$\|\partial_2 \phi\|^2 = 2\pi \sum_{|l| \leq N} l^2 \|\phi\|_{L^2(\Lambda_1)}^2 \leq N^2 \|\phi\|^2.$$

Then the desired result follows. ■

The mixed Fourier-Legendre expansion of a function $v \in L^2(\Omega)$ is

$$v(x) = \sum_{m=0}^{\infty} \sum_{|l|=0}^{\infty} \hat{v}_{m,l} L_m(x_1) e^{ilx_2}$$

where $\hat{v}_{m,l}$ is the Fourier-Legendre coefficient,

$$\hat{v}_{m,l} = \frac{1}{2\pi}\left(m + \frac{1}{2}\right) \int_\Omega v(x) L_m(x_1) e^{-ilx_2}\, dx.$$

The $L^2(\Omega)$-orthogonal projection $P_{M,N} : L_p^2(\Omega) \to V_{M,N}$ is such a mapping that for any $v \in L_p^2(\Omega)$,

$$(v - P_{M,N} v, \phi) = 0, \quad \forall\, \phi \in V_{M,N},$$

or equivalently,

$$P_{M,N} v(x) = \sum_{m=0}^{M} \sum_{|l| \leq N} \hat{v}_{m,l} L_m(x_1) e^{ilx_2}.$$

Theorem 6.3. *For any* $v \in H_p^{r,s}(\Omega)$ *and* $r, s \geq 0$,

$$\|v - P_{M,N} v\| \leq c\left(M^{-r} + N^{-s}\right) \|v\|_{H^{r,s}}.$$

Proof. Let

$$v_l(x_1) = \frac{1}{2\pi}\int_{\Lambda_2} v(x_1,x_2)e^{-ilx_2}\,dx_2, \tag{6.3}$$

and $P_M : L^2(\Lambda_1) \to \mathbb{P}_{M,1}$ be the $L^2(\Lambda_1)$-orthogonal projection. Then

$$P_{M,N}v(x) = \sum_{|l|\le N} P_M v_l(x_1)e^{ilx_2}.$$

By virtue of Lemma 2.2,

$$\begin{aligned}\|v - P_{M,N}v\|^2 &= 2\pi\sum_{|l|\le N}\|v_l - P_M v_l\|^2_{L^2(\Lambda_1)} + 2\pi\sum_{|l|>N}\|v_l\|^2_{L^2(\Lambda_1)}\\ &\le cM^{-2r}\sum_{|l|\le N}\|v_l\|^2_{H^r(\Lambda_1)} + cN^{-2s}\sum_{|l|>N}|l|^{2s}\|v_l\|^2_{L^2(\Lambda_1)}\\ &\le cM^{-2r}\|v\|^2_{L^2(\Lambda_2,H^r(\Lambda_1))} + cN^{-2s}|v|^2_{H^s(\Lambda_2,L^2(\Lambda_1))}.\end{aligned}$$

■

It is noted that for any $v \in H^{r,s}_{0,p}(\Omega)$ and $r,s\ge 0$,

$$\|v - P_{M,N}v\| \le c\left(M^{-r} + N^{-s}\right)|v|_{H^{r,s}}. \tag{6.4}$$

When we use the mixed Fourier-Legendre approximations for semi-periodic problems of partial differential equations, we need different kinds of orthogonal projections to obtain the optimal error estimates. For instance, the $H^1(\Omega)$-orthogonal projection $P^1_{M,N} : H^1_p(\Omega) \to V_{M,N}$ is such a mapping that for any $v \in H^1_p(\Omega)$,

$$\left(\nabla(v - P^1_{M,N}v), \nabla\phi\right) + (v,\phi) = 0, \quad \forall\, \phi \in V_{M,N}.$$

While the $H^1_0(\Omega)$-orthogonal projection $P^{1,0}_{M,N}(\Omega) \to V^0_{M,N}$ is such a mapping that for any $v \in H^1_{0,p}(\Omega)$,

$$\left(\nabla(v - P^{1,0}_{M,N}v), \nabla\phi\right) = 0, \quad \forall\, \phi \in V^0_{M,N}.$$

Theorem 6.4. *If $v \in M^{r,s}_p(\Omega)$ with $r,s\ge 1$, then for $\mu = 0,1$,*

$$\|v - P^1_{M,N}v\|_\mu \le c\left(M^{1-r} + N^{1-s}\right)\left(M^{\mu-1} + N^{\mu-1}\right)\|v\|_{M^{r,s}}.$$

If, in addition, for certain positive constants c_1 and c_2,

$$c_1 N \le M \le c_2 N, \tag{6.5}$$

then

$$\|v - P^1_{M,N}v\| \le c\left(M^{-r} + N^{-s}\right)\|v\|_{M^{r,s}}.$$

Proof. Let $v_l(x_1)$ be as before, and P_M^1 be the $H^1(\Lambda_1)$-orthogonal projection. Set

$$v_*(x) = \sum_{|l|\leq N} \left(P_M^1 v_l(x_1)\right) e^{ilx_2}.$$

By Theorem 2.10,

$$\begin{aligned}
\|v - P_{M,N}^1 v\|_1^2 &\leq \inf_{w \in V_{M,N}} \|v - w\|_1^2 \leq \|v - v_*\|_1^2 \\
&\leq 2\pi \sum_{|l|\leq N} \left(\|v_l - P_M^1 v_l\|_{H^1(\Lambda_1)}^2 + l^2 \|v_l - P_M^1 v_l\|_{L^2(\Lambda_1)}^2\right) \\
&\quad + 2\pi \sum_{|l|>N} \left(\|v_l\|_{H^1(\Lambda_1)}^2 + l^2 \|v_l\|_{L^2(\Lambda_1)}^2\right) \\
&\leq cM^{2-2r} \sum_{|l|\leq N} \left(\|v_l\|_{H^r(\Lambda_1)}^2 + l^2 \|v_l\|_{H^{r-1}(\Lambda_1)}^2\right) \\
&\quad + cN^{2-2s} \sum_{|l|>N} l^{2s-2} \left(\|v_l\|_{H^1(\Lambda_1)}^2 + l^2 \|v_l\|_{L^2(\Lambda_1)}^2\right) \\
&\leq c\left(M^{1-r} + N^{1-s}\right)^2 \|v\|_{M^{r,s}}^2.
\end{aligned}$$

Then the first conclusion with $\mu = 1$ follows. By means of the duality, we deduce that

$$\|v - P_{M,N}^1 v\| \leq c\left(M^{1-r} + N^{1-s}\right)\left(M^{-1} + N^{-1}\right) \|v\|_{M^{r,s}}.$$

The last result comes from the above inequality and (6.5). ∎

Theorem 6.5. *If $v \in H_{0,p}^1(\Omega) \cap M_p^{r,s}(\Omega)$ with $r, s \geq 1$, then for $\mu = 0, 1$,*

$$\|v - P_{M,N}^{1,0} v\|_\mu \leq c\left(M^{1-r} + N^{1-s}\right)\left(M^{\mu-1} + N^{\mu-1}\right) \|v\|_{M^{r,s}}.$$

If, in addition, (6.5) holds, then

$$\|v - P_{M,N}^{1,0} v\| \leq c\left(M^{-r} + N^{-s}\right) \|v\|_{M^{r,s}}.$$

Proof. Let $v_l(x_1)$ be as before, and $P_M^{1,0}$ be the $H_0^1(\Lambda_1)$-orthogonal projection. Put

$$v_*(x) = \sum_{|l|\leq N} \left(P_M^{1,0} v_l(x_1)\right) e^{ilx_2}.$$

By virtue of Theorem 2.11,

$$\|v - P_{M,N}^{1,0} v\|_1^2 \leq c \inf_{w \in V_{M,N}^0} \|v - w\|_1^2 \leq c\|v - v_*\|_1^2$$

$$\begin{aligned}
&\leq c\sum_{|l|\leq N}\left(|v_l - P_M^{1,0}v_l|^2_{H^1(\Lambda_1)} + l^2\|v - P_{M,N}^{1,0}v_l\|^2_{L^2(\Lambda_1)}\right)\\
&\quad + c\sum_{|l|>N}\left(|v_l|^2_{H^1(\Lambda_1)} + l^2\|v_l\|^2_{L^2(\Lambda_1)}\right)\\
&\leq cM^{2-2r}\sum_{|l|\leq N}\left(\|v_l\|^2_{H^r(\Lambda_1)} + l^2\|v_l\|^2_{H^{r-1}(\Lambda_1)}\right)\\
&\quad + cN^{2s-2}\sum_{|l|>N} l^{2s-2}\left(|v_l|^2_{H^1(\Lambda_1)} + l^2\|v_l\|^2_{L^2(\Lambda_1)}\right)\\
&\leq c\left(M^{-r} + N^{-s}\right)^2\|v\|^2_{M^{r,s}}.
\end{aligned}$$

The remaining results come from an argument as in the last part of the proof of Theorem 6.4. ∎

In numerial analysis of mixed Fourier-Legendre spectral schemes for nonlinear problems, we also need to bound the norm $\|P_{M,N}^{1,0}v\|_{W^{r,p}}$. One of them is stated in the following theorem.

Theorem 6.6. *Let (6.5) hold and $s > \frac{1}{2}$. If $v \in H^1_{0,p}(\Omega) \cap H^s_p(\Lambda_2, H^1(\Lambda_1))$, then*

$$\|P_{M,N}^{1,0}v\|_{L^\infty} \leq c\|v\|_{H^s(\Lambda_2,H^1(\Lambda_1))}.$$

If, in addition, $v \in X_p^{r,s}(\Omega)$ with $r \geq 1$, then

$$\|P_{M,N}^{1,0}v\|_{W^{1,\infty}} \leq c\|v\|_{X^{r,s}}.$$

Proof. We first let $v_l(x_1)$ be as before, and let

$$P_{M,N}^{1,0}v(x) = \sum_{|l|\leq N} v_l^*(x_1)e^{ilx_2}.$$

Obviously $v_l \in H_0^1(\Lambda_1)$ and $v_l^* \in \mathbb{P}^0_{M,1}$. Define

$$a_l(w,z) = (\partial_1 w, \partial_1 z)_{L^2(\Lambda_1)} + l^2(w,z)_{L^2(\Lambda_1)}.$$

We assert from the orthogonality of the Fourier system and the property of $P_{M,N}^{1,0}$ that

$$a_l\left(v_l - v_l^*, \phi\right) = 0, \quad \forall\, \phi \in \mathbb{P}^0_{M,1}.$$

Obviously

$$|v_l - v_l^*|^2_{H^1(\Lambda_1)} + l^2\|v_l - v_l^*\|^2_{L^2(\Lambda_1)} = a_l\left(v_l - v_l^*, v_l - v_l^*\right) = a_l\left(v_l - v_l^*, v_l - \phi\right), \quad \forall\, \phi \in \mathbb{P}^0_{M,1}.$$

By Schwarz inequality and Poincaré inequality, we deduce that

$$\|v_l-v_l^*\|^2_{H^1(\Lambda_1)}+l^2\|v_l-v_l^*\|^2_{L^2(\Lambda_1)} \le c\left(\|v_l-\phi\|^2_{H^1(\Lambda_1)}+l^2\|v_l-\phi\|^2_{L^2(\Lambda_1)}\right), \quad \forall\,\phi\in \mathbb{P}^0_{M,1}.$$

Let $P^{1,0}_M$ be the same as in the proof of Theorem 6.5 and take $\phi = P^{1,0}_{M,1}v_l$. Using Theorem 2.11, we derive from the above inequality and (6.5) that

$$\|v_l-v_l^*\|^2_{H^1(\Lambda_1)}+l^2\|v_l-v_l^*\|^2_{L^2(\Lambda_1)} \le cM^{2-2r}\|v_l\|^2_{H^r(\Lambda_1)}.$$

By means of duality,

$$\|v_l-v_l^*\|_{H^\mu(\Lambda_1)} \le cM^{\mu-r}\|v_l\|_{H^r(\Lambda_1)}, \quad \mu=0,1. \tag{6.6}$$

We now estimate $\|P^{1,0}_{M,N}v\|_{W^{\mu,\infty}}, \mu=0,1$. Clearly,

$$\|P^{1,0}_{M,N}v\|_{W^{\mu,\infty}} \le \sum_{|l|\le N}\left(\|\partial_1^\mu v_l^*\|_{L^\infty(\Lambda_1)}+(1+|l|^\mu)\,\|v_l^*\|_{L^\infty(\Lambda_1)}\right), \quad \mu=0,1. \tag{6.7}$$

We can bound the terms on the right-hand side of (6.7). Let I_M be the Legendre-Gauss-Lobatto interpolation on the interval Λ_1. Then

$$\|\partial_1 v_l^*\|_{L^\infty(\Lambda_1)} \le \|\partial_1 v_l^*-I_M\partial_1 v_l\|_{L^\infty(\Lambda_1)}+\|\partial_1 v_l-I_M\partial_1 v_l\|_{L^\infty(\Lambda_1)}+\|\partial_1 v_l\|_{L^\infty(\Lambda_1)}. \tag{6.8}$$

According to Theorem 2.7 and (2.63), we know that for $r>\frac{1}{2}$,

$$\begin{aligned}\|\partial_1 v_l^*-I_M\partial_1 v_l\|_{L^\infty(\Lambda_1)} &\le cM\|\partial_1 v_l^*-I_M\partial_1 v_l\|_{L^2(\Lambda_1)}\\ &\le cM\left(|v_l-v_l^*|_{H^1(\Lambda_1)}+\|\partial_1 v_l-I_M\partial_1 v_l\|_{L^2(\Lambda_1)}\right)\\ &\le cM|v_l-v_l^*|_{H^1(\Lambda_1)}+cM^{1-r}\|v_l\|_{H^{r+1}(\Lambda_1)}.\end{aligned} \tag{6.9}$$

Let $0<\delta\le\frac{1}{2}$. By imbedding theory and (2.63), we have that for $r>\frac{1}{2}$,

$$\|\partial_1 v_l-I_M\partial_1 v_l\|_{L^\infty(\Lambda_1)} \le c\|\partial_1 v_l-I_M\partial_1 v_l\|_{H^{\frac{1}{2}+\delta}(\Lambda_1)} \le cM^{1-r}\|v_l\|_{H^{r+1}(\Lambda_1)}. \tag{6.10}$$

Moreover, the imbedding theory implies that for $r>\frac{1}{2}$,

$$|\partial_1 v_l|_{L^\infty(\Lambda_1)} \le c\|v_l\|_{H^{r+1}(\Lambda_1)}. \tag{6.11}$$

The combination of (6.6) and (6.8)–(6.11) leads to that for $r\ge 1$ and $s>\frac{1}{2}$,

$$\begin{aligned}\sum_{|l|\le N}\|\partial_1 v_l^*\|_{L^\infty(\Lambda_1)} &\le cM\sum_{|l|\le N}|v_l-v_l^*|_{H^1(\Lambda_1)}+c\sum_{|l|\le N}\|v_l\|_{H^{r+1}(\Lambda_1)}\\ &\le c\left(M^{1-r}+1\right)\sum_{|l|\le N}\|v_l\|_{H^{r+1}(\Lambda_1)}\\ &\le c\left(\sum_{|l|\le N}\left(1+l^2\right)^{-s}\right)^{\frac{1}{2}}\left(\sum_{|l|\le N}\left(1+l^2\right)^{s}\|v_l\|_{H^{r+1}(\Lambda_1)}\right)\\ &\le c\|v\|_{H^s(\Lambda_2,H^{r+1}(\Lambda_1))}.\end{aligned}$$

Similarly, we have that

$$\sum_{|l|\le N} \|v_l^*\|_{L^\infty(\Lambda_1)} \le c\|v\|_{H^s(\Lambda_2,H^1(\Lambda_1))}, \quad \sum_{|l|\le N} |l|\|v_l^*\|_{L^\infty(\Lambda_1)} \le c\|v\|_{H^{s+1}(\Lambda_2,H^r(\Lambda_1))}.$$

The proof is complete. ■

Finally we consider the mixed Fourier-Legendre interpolation. Let I_M be the same as in the previous paragraphs and I_N be the Fourier interpolation on the interval Λ_2 respectively. Set $I_{M,N} = I_M I_N = I_N I_M$.

Theorem 6.7. *If* $v \in H_p^\beta(\Lambda_2, H^r(\Lambda_1)) \cap H_p^s(\Lambda_2, H^\alpha(\Lambda_1)) \cap H_p^\gamma\left(\Lambda_2, H^\lambda(\Lambda_1)\right), 0 \le \alpha \le \min(r,\lambda), 0 \le \beta \le \min(s,\gamma)$ *and* $r, s, \lambda, \gamma > \frac{1}{2}$, *then*

$$\begin{aligned}\|v - I_{M,N}v\|_{H^\beta(\Lambda_2,H^\alpha(\Lambda_1))} \le\ & cM^{2\alpha-r+\frac{1}{2}}\|v\|_{H^\beta(\Lambda_2,H^r(\Lambda_1))} + cN^{\beta-s}\|v\|_{H^s(\Lambda_2,H^\alpha(\Lambda_1))} \\ & + cq(\beta)M^{2\alpha-\lambda+\frac{1}{2}}N^{\beta-\gamma}\|v\|_{H^\gamma\left(\Lambda_2,H^\lambda(\Lambda_1)\right)}\end{aligned}$$

where $q(\beta) = 0$ *for* $\beta > \frac{1}{2}$ *and* $q(\beta) = 1$ *for* $\beta \le \frac{1}{2}$.

Proof. Let I be the identity operator. We have that

$$\|v - I_{M,N}v\|_{H^\beta(\Lambda_2,H^\alpha(\Lambda_1))} \le D_1 + D_2$$

where

$$\begin{aligned} D_1 &= \|v - I_M v\|_{H^\beta(\Lambda_2,H^\alpha(\Lambda_1))} + \|v - I_N v\|_{H^\beta(\Lambda_2,H^\alpha(\Lambda_1))}, \\ D_2 &= \|(I_M - I)(I_N - I)v\|_{H^\beta(\Lambda_2,H^\alpha(\Lambda_1))}. \end{aligned}$$

From Theorems 2.4 and 2.12,

$$D_1 \le cM^{2\alpha-r+\frac{1}{2}}\|v\|_{H^\beta(\Lambda_2,H^r(\Lambda_1))} + cN^{\beta-s}\|v\|_{H^s(\Lambda_2,H^\alpha(\Lambda_1))}.$$

If $\beta > \frac{1}{2}$, then by Theorems 2.4 and 2.12,

$$D_2 \le c\|v - I_M v\|_{H^\beta(\Lambda_2,H^\alpha(\Lambda_1))} \le cM^{2\alpha-r+\frac{1}{2}}\|v\|_{H^\beta(\Lambda_2,H^r(\Lambda_1))}.$$

If $\beta \le \frac{1}{2}$, then

$$D_2 \le cN^{\beta-\gamma}\|v - I_M v\|_{H^\gamma(\Lambda_2,H^\alpha(\Lambda_1))} \le cM^{2\alpha-\lambda+\frac{1}{2}}N^{\beta-\gamma}\|v\|_{H^\gamma\left(\Lambda_2,H^\lambda(\Lambda_1)\right)}.$$

The proof is complete. ■

It is noted that if $0 \le \alpha \le 1$ and $\alpha < \min(2r-1, 2\lambda-1)$, then we can improve Theorem 6.7 by using (2.63). In this case, we have that

$$\begin{aligned}\|v - I_{M,N}v\|_{H^\beta(\Lambda_2,H^\alpha(\Lambda_1))} \le\ & cM^{\alpha-r}\|v\|_{H^\beta(\Lambda_2,H^r(\Lambda_1))} + cN^{\beta-s}\|v\|_{H^s(\Lambda_2,H^\alpha(\Lambda_1))} \\ & + cq(\beta)M^{\alpha-\lambda}N^{\beta-\gamma}\|v\|_{H^\gamma(\Lambda_2,H^\lambda(\Lambda_1))},\end{aligned}$$

$q(\beta)$ being the same as in Theorem 6.7.

Theorems 6.3 and 6.5 can be found in Bernardi, Maday and Metivet (1987).

6.2. Mixed Fourier-Chebyshev Approximations

This section is devoted to the mixed Fourier-Chebyshev approximations. Let Λ_1, Λ_2 and Ω be the same as in the previous section. Set $\omega(x) = \omega(x_1) = (1 - x_1^2)^{-\frac{1}{2}}$, and

$$(v, w)_\omega = \int_\Omega v(x)w(x)\omega(x)\,dx, \quad \|v\|_\omega = (v, v)_\omega^{\frac{1}{2}}.$$

Define

$$L^2_\omega(\Omega) = \{v \mid v \text{ is measurable}, \|v\|_\omega < \infty\}.$$

We define the spaces $L^p_\omega(\Omega), H^r_\omega(\Omega), W^{r,p}_\omega(\Omega)$ and their norms in a similar way. We also introduce some non-isotropic spaces as follows

$$\begin{aligned}H^{r,s}_\omega(\Omega) &= L^2\left(\Lambda_2, H^r_\omega(\Lambda_1)\right) \cap H^s\left(\Lambda_2, L^2_\omega(\Lambda_1)\right), \quad r, s \ge 0, \\ M^{r,s}_\omega(\Omega) &= H^{r,s}_\omega(\Omega) \cap H^1\left(\Lambda_2, H^{r-1}_\omega(\Lambda_1)\right) \cap H^{s-1}\left(\Lambda_2, H^1_\omega(\Lambda_1)\right), \quad r, s \ge 1, \\ X^{r,s}_\omega(\Omega) &= H^s\left(\Lambda_2, H^{r+1}_\omega(\Lambda_1)\right) \cap H^{s+1}\left(\Lambda_2, H^r_\omega(\Lambda_1)\right), \quad r, s \ge 0\end{aligned}$$

equipped with the norms

$$\begin{aligned}\|v\|_{H^{r,s}_\omega(\Omega)} &= \left(\|v\|^2_{L^2(\Lambda_2,H^r_\omega(\Lambda_1))} + \|v\|^2_{H^s(\Lambda_2,L^2_\omega(\Lambda_1))}\right)^{\frac{1}{2}}, \\ \|v\|_{M^{r,s}_\omega(\Omega)} &= \left(\|v\|^2_{H^{r,s}_\omega(\Omega)} + \|v\|^2_{H^1(\Lambda_2,H^{r-1}_\omega(\Lambda_1))} + \|v\|^2_{H^{s-1}(\Lambda_2,H^1_\omega(\Lambda_1))}\right)^{\frac{1}{2}}, \\ \|v\|_{X^{r,s}_\omega(\Omega)} &= \left(\|v\|^2_{H^s(\Lambda_2,H^{r+1}_\omega(\Lambda_1))} + \|v\|^2_{H^{s+1}(\Lambda_2,H^r_\omega(\Lambda_1))}\right)^{\frac{1}{2}}.\end{aligned}$$

The meaning of $\tilde{\mathcal{D}}_p(\Omega)$ is the same as in the previous section. The closures of $\tilde{\mathcal{D}}_p(\Omega)$ in $H^{r,s}_\omega(\Omega), M^{r,s}_\omega(\Omega), X^{r,s}_\omega(\Omega)$ are denoted by $H^{r,s}_{0,p,\omega}(\Omega), M^{r,s}_{0,p,\omega}(\Omega), X^{r,s}_{0,p,\omega}(\Omega)$ respectively. Their norms are denoted also by $\|\cdot\|_{H^{r,s}_\omega}, \|\cdot\|_{M^{r,s}_\omega}, \|\cdot\|_{X^{r,s}_\omega}$ for simplicity. The corresponding semi-norms are denoted by $|\cdot|_{H^{r,s}_\omega}$ and $|\cdot|_{M^{r,s}_\omega}$. The space $H^{r,s}_{p,\omega}(\Omega)$ is the closure in $H_\omega^{r,s}(\Omega)$ of the set containing all infinitely differentiable functions with the period 2π in the x_2-direction, etc.

Let M and N be any positive integers. The subspaces $\mathbb{P}_{M,1}, \mathbb{P}^0_{M,1}, V_{N,2}, \tilde{V}_{N,2}, V_{M,N}$ and $V^0_{M,N}$ have the same meanings as in Section 6.1. We first give some inverse inequalities.

Theorem 6.8. *For any $\phi \in V_{M,N}$ and $2 \le p \le \infty$,*

$$\|\phi\|_{L^p_\omega} \le c(MN)^{\frac{1}{2}-\frac{1}{p}}\|\phi\|_\omega.$$

Proof. Let $\phi_l(x_1)$ be the same as in (6.1). According to Theorem 2.13, we deduce that

$$\begin{aligned}\|\phi\|_{L^\infty} &\le \sum_{|l|\le N}\|\phi_l\|_{L^\infty(\Lambda_1)} \le cM^{\frac{1}{2}}\sum_{|l|\le N}\|\phi_l\|_{L^2_\omega(\Lambda_1)}\\ &\le c(MN)^{\frac{1}{2}}\left(\sum_{|l|\le N}\|\phi_l\|^2_{L^2_\omega(\Lambda_1)}\right)^{\frac{1}{2}} = c(MN)^{\frac{1}{2}}\|\phi\|_\omega\end{aligned}$$

whence

$$\|\phi\|^p_{L^p_\omega} \le \|\phi\|^{p-2}_{L^\infty(\Lambda_1)}\|\phi\|^2_\omega \le c(MN)^{\frac{p}{2}-1}\|\phi\|^p_\omega.$$

This completes the proof. ∎

As a consequence of Theorem 6.8, we have that for any $\phi, \psi \in V_{M,N}$,

$$\|\phi\psi\|^2_\omega \le cMN\|\phi\|^2_\omega\|\psi\|^2_\omega. \tag{6.12}$$

Theorem 6.9. *For any $\phi \in V_{M,N}$,*

$$|\phi|^2_{1,\omega} \le \left(4M^4 + N^2\right)\|\phi\|^2_\omega.$$

Proof. Let ϕ_l be the same as in (6.1). We see from the proof of Theorem 2.14 that

$$\|\partial_1\phi\|^2_\omega = 2\pi\sum_{|l|\le N}\|\partial_1\phi_l\|^2_{L^2_\omega(\Lambda_1)} \le 8\pi M^4\sum_{|l|\le N}\|\phi_l\|^2_{L^2_\omega(\Lambda_1)} = 4M^4\|\phi\|^2_\omega.$$

The rest of the proof is clear. ∎

The mixed Fourier-Chebyshev expansion of a function $v \in L^2_\omega(\Omega)$ is

$$v(x) = \sum_{m=0}^\infty\sum_{|l|=0}^\infty \hat{v}_{m,l}T_m(x_1)e^{ilx_2}$$

with the Fourier-Chebyshev coefficients

$$\hat{v}_{m,l} = \frac{1}{\pi^2 c_l} \int_\Omega v(x) T_m(x_1) e^{-ilx_2}\, dx.$$

The $L^2(\Omega)$-orthogonal projection $P_{M,N} : L^2_{p,\omega}(\Omega) \to V_{M,N}$ is such a mapping that for any $v \in L^2_{p,\omega}(\Omega)$,

$$(v - P_{M,N}v, \phi)_\omega = 0, \quad \forall\, \phi \in V_{M,N},$$

or equivalently,

$$P_{M,N}v(x) = \sum_{m=0}^{M} \sum_{|l|\le N} \hat{v}_{m,l} T_m(x_1) e^{ilx_2}.$$

Theorem 6.10. *For any $v \in H^{r,s}_{p,\omega}(\Omega)$ and $r, s \ge 0$,*

$$\|v - P_{M,N}v\|_\omega \le c\left(M^{-r} + N^{-s}\right) \|v\|_{H^{r,s}_\omega}.$$

Proof. Let $v_l(x_1)$ be the same as in (6.3), and $P_M : L^2_\omega(\Lambda_1) \to \mathbb{P}_{M,1}$ be the $L^2_\omega(\Lambda_1)$-orthogonal projection. Then

$$P_{M,N}v(x) = \sum_{|l|\le N} P_M v_l(x_1) e^{ilx_2}.$$

Using Lemma 2.5, we find that

$$\begin{aligned}
\|v - P_{M,N}v\| &= 2\pi \sum_{|l|\le N} \|v_l - P_M v_l\|^2_{L^2_\omega(\Lambda_1)} + 2\pi \sum_{|l|>N} \|v_l\|^2_{L^2_\omega(\Lambda_1)} \\
&\le cM^{-2r} \sum_{|l|\le N} \|v_l\|^2_{H^r_\omega(\Lambda_1)} + cN^{-2s} \sum_{|l|>N} |l|^{2s} \|v_l\|^2_{L^2_\omega(\Lambda_1)} \\
&\le cM^{-2r} \|v\|_{L^2(\Lambda_2, H^r_\omega(\Lambda_1))} + CN^{-2s} \|v\|^2_{H^s(\Lambda_2, L^2_\omega(\Lambda_1))}.
\end{aligned}$$

■

It is noted that for any $v \in H^{r,s}_{0,p,\omega}(\Omega)$ and $r, s \ge 0$,

$$\|v - P_{M,N}v\|_\omega \le c\left(M^{-r} + N^{-s}\right) |v|_{H^{r,s}_\omega}. \tag{6.13}$$

When we apply the mixed Fourier-Chebyshev approximations to different semi-periodic problems of partial differential equations, we adopt different orthogonal projections to obtain the optimal error estimations. For instance, the $H^1_\omega(\Omega)$-orthogonal projection $P^1_{M,N} : H^1_{p,\omega}(\Omega) \to V_{M,N}$ is such a mapping that for any $v \in H^1_{p,\omega}(\Omega)$,

$$\left(\nabla(v - P^1_{M,N}v), \nabla\phi\right)_\omega + (v, \phi)_\omega = 0, \quad \forall\, \phi \in V_{M,N}.$$

While the $H^1_{0,\omega}(\Omega)$-orthogonal projection $P^{1,0}_{M,N} : H^1_{0,p,\omega}(\Omega) \to V^0_{M,N}$ is such a mapping that for any $v \in H^1_{0,p,\omega}(\Omega)$,

$$\left(\nabla(v - P^{1,0}_{M,N}v), \nabla(\phi\omega)\right) = 0, \quad \forall\, \phi \in V^0_{M,N}.$$

Theorem 6.11. *If $v \in M^{r,s}_{p,\omega}(\Omega)$ with $r, s \geq 1$, then for $\mu = 0, 1$,*

$$\|v - P^1_{M,N}v\|_{\mu,\omega} \leq c\left(M^{1-r} + N^{1-s}\right)\left(M^{\mu-1} + N^{\mu-1}\right)\|v\|_{M^{r,s}_\omega}.$$

If in addition (6.5) holds, then

$$\|v - P^1_{M,N}v\|_{\omega} \leq c\left(M^{-r} + N^{-s}\right)\|v\|_{M^{r,s}_\omega}.$$

Proof. Let $v_l(x_1)$ be as before, and P^1_M be the $H^1_\omega(\Lambda_1)$-orthogonal projection. Put

$$v_*(x) = \sum_{|l|\leq N}\left(P^1_M v_l(x_1)\right)e^{ilx_2}.$$

By Theorem 2.16,

$$\begin{aligned}
\|v - P^1_{M,N}v\|^2_{1,\omega} &\leq \inf_{w\in V_{M,N}}\|v - w\|^2_{1,\omega} \leq \|v - v_*\|^2_{1,\omega} \\
&\leq 2\pi\sum_{|l|\leq N}\left(\|v_l - P^1_M v_l\|^2_{H^1_\omega(\Lambda_1)} + l^2\|v_l - P^1_M v_l\|^2_{L^2_\omega(\Lambda_1)}\right) \\
&\quad + 2\pi\sum_{|l|>N}\left(\|v\|^2_{H^1_\omega(\Lambda_1)} + l^2\|v_l\|^2_{L^2_\omega(\Lambda_1)}\right) \\
&\leq cM^{2-2r}\sum_{|l|\leq N}\left(\|v_l\|^2_{H^r_\omega(\Lambda_1)} + l^2\|v\|^2_{H^{r-1}_\omega(\Lambda_1)}\right) \\
&\quad + cN^{2-2s}\sum_{|l|>N} l^{2s-2}\left(\|v_l\|^2_{H^1_\omega(\Lambda_1)} + l^2\|v_l\|^2_{L^2_\omega(\Lambda_1)}\right) \\
&\leq c\left(M^{1-r} + N^{1-s}\right)^2\|v\|^2_{M^{r,s}_\omega}.
\end{aligned}$$

So the first result is true for $\mu = 1$. The rest of the proof is clear. ∎

Theorem 6.12. *If $v \in H^1_{0,p,\omega}(\Omega) \cap M^{r,s}_{p,\omega}(\Omega)$ with $r, s \geq 1$, then for $\mu = 0, 1$,*

$$\|v - P^{1,0}_{M,N}v\|_{\mu,\omega} \leq c\left(M^{1-r} + N^{1-s}\right)\left(M^{\mu-1} + N^{\mu-1}\right)\|v\|_{M^{r,s}_\omega}.$$

If in addition (6.5) holds, then

$$\|v - P^{1,0}_{M,N}v\|_{\omega} \leq c\left(M^{-r} + N^{-s}\right)\|v\|_{M^{r,s}_\omega}.$$

Proof. Let $v_l(x_1)$ be as before, and $P_M^{1,0}$ be the $H^1_{0,\omega}(\Lambda_1)$-orthogonal projection. Put

$$v_*(x) = \sum_{|l|\le N} \left(P^{1,0}_{M,1} v_l(x_1)\right) e^{ilx_2}.$$

By Theorem 2.18,

$$\begin{aligned}
\|v - P^{1,0}_{M,N} v\|^2_{1,\omega} &\le c \inf_{\omega\in V^0_{M,N}} \|v-w\|_{1,\omega} \le c\|v-v_*\|_{1,\omega} \\
&\le c\sum_{|l|\le N}\left(|v_l - P^{1,0}_M v_l|^2_{H^1_\omega(\Lambda_1)} + l^2\|v_l - P^{1,0}_M v_l\|^2_{L^2_\omega(\Lambda_1)}\right) \\
&\quad + c\sum_{|l|>N}\left(|v_l|^2_{H^1_\omega(\Lambda_1)} + l^2\|v_l\|^2_{L^2_\omega(\Lambda_1)}\right) \\
&\le cM^{2-2r}\sum_{|l|\le N}\left(\|v_l\|^2_{H^r_\omega(\Lambda_1)} + l^2\|v_l\|^2_{H^{r-1}_\omega(\Lambda_1)}\right) \\
&\quad + cN^{2-2s}\sum_{|l|>N} l^{2s-2}\left(|v_l|^2_{H^1_\omega(\Lambda_1)} + l^2\|v_l\|^2_{L^2_\omega(\Lambda_1)}\right) \\
&\le c\left(M^{1-r}+N^{1-s}\right)^2 \|v\|^2_{M^{r,s}_\omega}.
\end{aligned}$$

The rest of the proof is clear. ∎

In numerical analysis of mixed Fourier-Chebyshev spectral schemes for nonlinear problems, we have to estimate the norm $\|P^{1,0}_{M,N}v\|_{W^{r,p}_\omega}$. One of them is stated in the following theorem.

Theorem 6.13. *Let (6.5) hold and $r, s > \frac{1}{2}$. If $v \in H^1_{0,p,\omega}(\Omega)\cap H^s_p(\Lambda_2, H^r_\omega(\Lambda_1))$, then*

$$\|P^{1,0}_{M,N} v\|_{L^\infty} \le c\|v\|_{H^s(\Lambda_2, H^r_\omega(\Lambda_1))}.$$

If in addition $v \in X^{r,s}_{p,\omega}(\Omega)$, then

$$\|P^{1,0}_{M,N} v\|_{W^{1,\infty}} \le c\|v\|_{X^{r,s}_\omega}.$$

Proof. Firstly let $v_l(x_1)$ be as before, and set

$$P^{1,0}_{M,N} v(x) = \sum_{|l|\le N} v^*_l(x_1) e^{ilx_2}.$$

Clearly $v_l \in H^1_{0,\omega}(\Lambda_1)$ and $v^*_l \in \mathbb{P}^0_{M,1}$. Define

$$a_{l,\omega}(w,z) = (\partial_1 w, \partial_1(z\omega))_{L^2(\Lambda_1)} + l^2(w,z)_{L^2_\omega(\Lambda_1)}.$$

If $w \in H^1_{0,\omega}(\Lambda_1)$, then by Lemma 2.8,

$$a_{l,\omega}(w,w) \geq \frac{1}{4}\|w\|^2_{H^1_\omega(\Lambda_1)} + l^2\|w\|^2_{L^2_\omega(\Lambda_1)}.$$

Moreover, for any $w, z \in H^1_{0,\omega}(\Lambda_1)$,

$$|a_{l,\omega}(w,z)| \leq c\left(\|w\|_{H^1_\omega(\Lambda_1)} + |l|\|w\|_{L^2_\omega(\Lambda_1)}\right)\left(\|z\|_{H^1_\omega(\Lambda_1)} + |l|\|z\|_{L^2_\omega(\Lambda_1)}\right).$$

Since $a_{l,\omega}(v_l - v_l^*, \phi) = 0$ for all $\phi \in \mathbb{P}^0_{M,1}$, we see that

$$\begin{aligned}\frac{1}{4}\|v_l - v_l^*\|^2_{H^1_\omega(\Lambda_1)} + l^2\|v_l - v_l^*\|^2_{L^2_\omega(\Lambda_1)} &\leq a_{l,\omega}(v_l - v_l^*, v_l - v_l^*)\\ &= a_{l,\omega}(v_l - v_l^*, v_l - \phi), \quad \forall\, \phi \in \mathbb{P}^0_{M,1}.\end{aligned}$$

Consequently,

$$\frac{1}{4}\|v_l - v_l^*\|^2_{H^1_\omega(\Lambda_1)} + l^2\|v_l - v_l^*\|^2_{L^2_\omega(\Lambda_1)} \leq \inf_{\phi \in \mathbb{P}^0_{M,1}}\left(\frac{1}{4}\|v_l - \phi\|^2_{H^1_\omega(\Lambda_1)} + l^2\|v_l - \phi\|^2_{L^2_\omega(\Lambda_1)}\right).$$

Now, let $P^{1,0}_M$ be the same as in the proof of Theorem 6.12. By taking $\phi = P^{1,0}_M v_l$ and using Theorem 2.18, we deduce that

$$\frac{1}{4}\|v_l - v_l^*\|^2_{H^1_\omega(\Lambda_1)} + l^2\|v_l - v_l^*\|^2_{L^2_\omega(\Lambda_1)} \leq cM^{2-2r}\|v_l\|^2_{H^r_\omega(\Lambda_1)}.$$

By means of the duality,

$$\|v_l - v_l^*\|_{H^\mu_\omega(\Lambda_1)} \leq cM^{\mu-r}\|v_l\|_{H^r_\omega(\Lambda_1)}, \quad \mu = 0, 1. \tag{6.14}$$

We now estimate $\|P^{1,0}_{M,N}v\|_{W^{\mu,\infty}}, \mu = 0,1$. Obviously,

$$\|P^{1,0}_{M,N}v\|_{W^{\mu,\infty}} \leq \sum_{|l|\leq N}\left(\|\partial_1^\mu v_l^*\|_{L^\infty(\Lambda_1)} + (1 + |\mu|^\mu)\|v_l^*\|_{L^\infty(\Lambda_1)}\right), \quad \mu = 0, 1. \tag{6.15}$$

We can bound the terms on the right-hand side of (6.15). Let I_M be the Chebyshev-Gauss-Lobatto interpolation on the interval Λ_1. Then

$$\|\partial_1 v_l^*\|_{L^\infty(\Lambda_1)} \leq \|\partial_1 v_l^* - I_M\partial_1 v_l\|_{L^\infty(\Lambda_1)} + \|\partial_1 v_l - I_M\partial_1 v_l\|_{L^\infty(\Lambda_1)} + \|\partial_1 v_l\|_{L^\infty(\Lambda_1)}. \tag{6.16}$$

Using Theorems 2.13 and 2.19, we obtain that for $r > \frac{1}{2}$,

$$\begin{aligned}\|\partial_1 v_l^* - I_M\partial_1 v_l\|_{L^\infty(\Lambda_1)} &\leq cM^{\frac{1}{2}}\|\partial_1 v_l^* - I_M\partial_1 v_l\|_{L^2_\omega(\Lambda_1)}\\ &\leq cM^{\frac{1}{2}}\left(|v_l^* - v_l|_{H^1_\omega(\Lambda_1)} + \|\partial_1 v_l - I_M\partial_1 v_l\|_{L^2_\omega(\Lambda_1)}\right)\\ &\leq cM^{\frac{1}{2}}|v_l^* - v_l|_{H^1_\omega(\Lambda_1)} + cM^{\frac{1}{2}-r}\|v_l\|_{H^{r+1}_\omega(\Lambda_1)}.\end{aligned} \tag{6.17}$$

Now, let $w \in H^r(\Lambda_1)$ and $r > \frac{1}{2}$. Set $x_1 = \cos y, \tilde{\Lambda} = (0,\pi)$ and $\tilde{w}(y) = w(x_1)$. Let $\tilde{I}_M$ be the corresponding Fourier interpolation. Then $\widetilde{I_{M,1}w} = \tilde{I}_M\tilde{w}$. Let $0 < \delta < \frac{1}{2}$. We have

$$\|\tilde{I}_M\tilde{w} - \tilde{w}\|_{L^\infty(\tilde{\Lambda})} \le c\|\tilde{I}_M\tilde{w} - \tilde{w}\|^{\frac{1}{2}}_{H^{\frac{1}{2}-\delta}(\tilde{\Lambda})}\|\tilde{I}_M\tilde{w} - \tilde{w}\|^{\frac{1}{2}}_{H^{\frac{1}{2}+\delta}(\tilde{\Lambda})} \le cM^{\frac{1}{2}-r}\|\tilde{w}\|_{H^r(\tilde{\Lambda})}.$$

By the continuity of the mapping $w \to \tilde{w}$, $\|I_M w - w\|_{L^\infty(\Lambda_1)} \le cM^{\frac{1}{2}-r}\|w\|_{H^r_\omega(\Lambda_1)}$. Thus

$$\|\partial_1 v_l - I_M\partial_1 v_l\|_{L^\infty(\Lambda_1)} \le cM^{\frac{1}{2}-r}\|v_l\|_{H^{r+1}_\omega(\Lambda_1)}. \tag{6.18}$$

Clearly,

$$\|\partial_1 v_l\|_{L^\infty(\Lambda_1)} \le c\|v_l\|_{H^{r+1}(\Lambda_1)} \le c\|v_l\|_{H^{r+1}_\omega(\Lambda_1)}. \tag{6.19}$$

The combination of (6.14) and (6.16)–(6.19) yields that for $r, s > \frac{1}{2}$,

$$\begin{aligned}
\sum_{|l|\le N}\|\partial_1 v_l^*\|_{L^\infty(\Lambda_1)} &\le cM^{\frac{1}{2}}\sum_{|l|\le N}|v_l - v_l^*|_{H^1_\omega(\Lambda_1)} + c\sum_{|l|\le N}\|v_l\|_{H^{r+1}_\omega(\Lambda_1)} \\
&\le c\left(M^{\frac{1}{2}-r}+1\right)\sum_{|l|\le N}\|v_l\|_{H^{r+1}_\omega(\Lambda_1)} \\
&\le c\left(\sum_{|l|\le N}\left(1+l^2\right)^{-s}\right)^{\frac{1}{2}}\left(\sum_{|l|\le N}\left(1+l^2\right)^{s}\|v_l\|^2_{H^{r+1}_\omega(\Lambda_1)}\right)^{\frac{1}{2}} \\
&\le c\|v\|_{H^s\left(\Lambda_2, H^{r+1}_\omega(\Lambda_1)\right)}.
\end{aligned}$$

Similarly

$$\sum_{|l|\le N}\|v_l\|_{L^\infty(\Lambda_1)} \le c\|v\|_{H^s(\Lambda_2, H^r_\omega(\Lambda_1))},$$

$$\sum_{|l|\le N}|l|\|v_l\|_{L^\infty(\Lambda_1)} \le c\|v\|_{H^{s+1}(\Lambda_2, H^r_\omega(\Lambda_1))}.$$

The proof is complete. ∎

Finally, we consider the mixed Fourier-Chebyshev interpolation. Let I_M be the same as in the previous paragraphs and I_N be the Fourier interpolation on the interval Λ_2. Set $I_{M,N} = I_M I_N = I_N I_M$.

Theorem 6.14. *If* $v \in H^\beta_p(\Lambda_2, H^r_\omega(\Lambda_1)) \cap H^s_p(\Lambda_2, H^\alpha_\omega(\Lambda_1)) \cap H^\gamma_p\left(\Lambda_2, H^\lambda_\omega(\Lambda_1)\right), 0 \le \alpha \le \min(r,\lambda), 0 \le \beta \le \min(s,\gamma)$ *and* $r, s, \lambda, \gamma > \frac{1}{2}$, *then*

$$\begin{aligned}
\|v - I_{M,N}v\|_{H^\beta(\Lambda_2, H^\alpha_\omega(\Lambda_1))} \le\ & cM^{2\alpha-r}\|v\|_{H^\beta(\Lambda_2, H^r_\omega(\Lambda_1))} + cN^{\beta-s}\|v\|_{H^s(\Lambda_2, H^\alpha_\omega(\Lambda_1))} \\
&+ cq(\beta)M^{2\alpha-\lambda}N^{\beta-\gamma}\|v\|_{H^\gamma\left(\Lambda_2, H^\lambda_\omega(\Lambda_1)\right)}
\end{aligned}$$

where $q(\beta) = 0$ *for* $\beta > \frac{1}{2}$ *and* $q(\beta) = 1$ *for* $\beta \le \frac{1}{2}$.

Proof. We have

$$\|v - I_{M,N}v\|_{H^\beta(\Lambda_2, H^\alpha_\omega(\Lambda_1))} \le D_1 + D_2$$

with

$$\begin{aligned} D_1 &= \|v - I_M v\|_{H^\beta(\Lambda_2, H^\alpha_\omega(\Lambda_1))} + \|v - I_N v\|_{H^\beta(\Lambda_2, H^\alpha_\omega(\Lambda_1))}, \\ D_2 &= \|(I_M - I)(I_N - I)v\|_{H^\beta(\Lambda_2, H^\alpha_\omega(\Lambda_1))}. \end{aligned}$$

From Theorems 2.4 and 2.19,

$$D_1 \le cM^{2\alpha - r}\|v\|_{H^\beta(\Lambda_2, H^r_\omega(\Lambda_1))} + cN^{\beta - s}\|v\|_{H^s(\Lambda_2, H^\alpha_\omega(\Lambda_1))}.$$

For $\beta > \frac{1}{2}$, we have

$$D_2 \le c\|v - I_M v\|_{H^\beta(\Lambda_2, H^\alpha_\omega(\Lambda_1))} \le cM^{2\alpha - r}\|v\|_{H^\beta(\Lambda_2, H^r_\omega(\Lambda_1))},$$

while for $\beta \le \frac{1}{2}$,

$$D_2 \le cN^{\beta - \gamma}\|v - I_M v\|_{H^\gamma(\Lambda_2, H^\alpha(\Lambda_1))} \le cM^{2\alpha - \lambda}N^{\beta - \gamma}\|v\|_{H^\gamma\left(\Lambda_2, H^\lambda(\Lambda_1)\right)}.$$

The proof is complete. ∎

Theorems 6.10, 6.12 and 6.14 can be found in Guo, Ma, Cao and Huang (1992), and Guo and Li (1995a). Quarteroni (1987a) also considered this kind of mixed approximation.

The filterings are also available for the mixed spectral methods. For example, let

$$\phi(x) = \sum_{m=0}^{M} \sum_{|l|=0}^{N} \hat{\phi}_{m,l} T_m(x_1) e^{ilx_2}.$$

For $\alpha_1, \alpha_2, \beta_1, \beta_2 \ge 1$, we define the filtering $R_{M,N}(\alpha_1, \alpha_2, \beta_1, \beta_2)$ as

$$R_{M,N}(\alpha_1, \alpha_2, \beta_1, \beta_2)\,\phi(x) = \sum_{m=0}^{M} \left(1 - \left|\frac{m}{M}\right|^{\alpha_1}\right)^{\beta_1} \left(1 - \left|\frac{n}{N}\right|^{\alpha_2}\right)^{\beta_2} \hat{\phi}_{m,l} T_m(x_1) e^{ilx_2}.$$

Following the same line as in the proofs of Theorems 2.26 and 2.27, we can prove that if $0 \le r \le \alpha_1$ and $0 \le s \le \alpha_2$, then

$$\|R_{M,N}(\alpha_1, \alpha_2, \beta_1, \beta_2)\,\phi - \phi\|_\omega \le c\left(\beta_1 M^{-r} + \beta_2 N^{-s}\right)\|\phi\|_{H^{r,s}_\omega}.$$

In order to keep the spectral accuracy, we can take $\alpha_j = \alpha_j(N)$ and $\alpha_j(N) \to \infty$ as $N \to \infty, j = 1, 2$, see Guo and Li (1995b).

6.3. Applications

In the final section of this chapter, we take two-dimensional unsteady Navier-Stokes equation as an example to describe the mixed spectral methods. For simplicity, we focus on the mixed Fourier-Legendre spectral schemes. Let Λ_1, Λ_2 and Ω be the same as in the previous sections, and $\partial\Omega = \{x \mid x_1 = -1, 1, 0 \le x_2 \le 2\pi\}$. $U(x,t), P(x,t)$ and $\mu > 0$ are the speed, the pressure and the kinetic viscosity, respectively, and $U = \left(U^{(1)}, U^{(2)}\right)$. $U_0(x), P_0(x)$ and $f(x,t)$ are given functions with the period 2π for the variable x_2. We look for the solution (U, P) with the period 2π for the variable x_2, such that

$$\begin{cases} \partial_t U + (U\cdot\nabla)U - \mu\Delta U + \nabla P = f, & x\in\Omega,\ 0<t\le T, \\ \nabla\cdot U = 0, & x\in\Omega,\ 0\le t\le T, \\ U = 0, & x\in\partial\Omega,\ 0<t\le T, \\ U(x,0) = U_0(x), & x\in\bar{\Omega}, \\ P(x,0) = P_0(x), & x\in\bar{\Omega}. \end{cases} \tag{6.20}$$

Let

$$A(v) = \int_\Omega v(x)\,dx$$

and $L^q_{0,p}(\Omega) = \{v \mid v\in L^q_p(\Omega), A(v) = 0\}$. For fixing the values of P, we require that $P(x,t)\in L^2_{0,p}(\Omega)$ for all $0\le t\le T$. Let $\mathcal{V}_p = \{v \mid v\in\tilde{\mathcal{D}}_p(\Omega), \nabla\cdot v = 0\}$. Denote by H_p and V_p the closures of $\mathcal{V}_p$ in $L^2(\Omega)$ and $H^1(\Omega)$. Set

$$a(v,w) = (\nabla v, \nabla w), \qquad b(v,w) = (\nabla\cdot v, w). \tag{6.21}$$

The weak form of (6.20) is

$$\begin{cases} (\partial_t U(t), v) + ((U(t)\cdot\nabla)U(t), v) + \mu a(U(t), v) = (f, v), & \forall\, v\in V_p,\ 0<t\le T, \\ U(0) = U_0, & x\in\bar{\Omega}. \end{cases} \tag{6.22}$$

By an argument as in Lions (1969), it can be shown that if $U_0\in H_p$ and $f\in L^2(0,T;V_p')$, then (6.22) possesses a unique solution in $L^2(0,T;V_p)\cap L^\infty(0,T;H_p)$. The solution satisfies the conservation which is exactly the same as (5.67). We can approximate (6.22) directly with trial function space belonging to the space V_p. To avoid this, we recall some artificial compression methods. The first was due to Chorin (1967). It means that instead of the continuity equation, we consider the approximate one as

$$\beta\partial_t P + \nabla\cdot U = 0, \quad \beta > 0. \tag{6.23}$$

The solution of this auxiliary problem tends to the genuine solution of (6.22) as $\beta \to 0$, under some conditions, see Lions (1970). Guo (1981b) modified this approach, in which the continuity equation is approximated by

$$\beta\partial_t P + \nabla \cdot U - \alpha\beta\Delta P = 0, \quad \beta > 0,$$

α being certain positive constant. It was used for finite difference schemes of Navier-Stokes equation with strict error estimates. Recently Shen (1992) proposed a new artificial compression as

$$\nabla \cdot U - \beta\Delta P = 0, \quad \beta > 0.$$

We shall use (6.23) in this section. On the other hand, the convective term $U \cdot \nabla U$ can be expressed as $d(U, U)$ where

$$d(v, w) = \frac{1}{2}w^{(1)}\partial_1 v + \frac{1}{2}w^{(2)}\partial_2 v + \frac{1}{2}\,\partial_1\left(w^{(1)}v\right) + \frac{1}{2}\partial_2\left(w^{(2)}v\right). \tag{6.24}$$

Clearly

$$(d(v, w), z) + (d(z, w), v) = 0, \quad \forall\, v, z \in H^1(\Omega),\ w \in H_0^1(\Omega). \tag{6.25}$$

Let M and N be any positive integers, and τ be the mesh size in time. Set

$$V_{M,N} = \mathbb{P}_{M,1}^0 \otimes V_{N,2}, \quad W_{M,N} = \{w \in \mathbb{P}_{M,1} \otimes V_{N,2} \mid A(v) = 0\}.$$

The approximations of U and P are denoted by $u_{M,N}$ and $p_{M,N}$, respectively. A mixed Fourier-Legendre spectral scheme with the artificial compression for (6.20) is to find $u_{M,N}(x, t) \in V_{M,N}$ and $p_{M,N}(x, t) \in W_{M,N}$ for all $t \in \bar{R}_\tau(T)$ such that

$$\begin{cases} (D_\tau u_{M,N}(t) + d(u_{M,N}(t) + \delta\tau D_\tau u_{M,N}(t), u_{M,N}(t)), v) + \mu a(u_{M,N}(t) + \sigma\tau D_\tau u_{M,N}(t), v) \\ \quad -b\left(v, p_{M,N}(t) + \theta\tau D_\tau p_{M,N}(t)\right) = (f(t), v), \quad \forall\, v \in V_{M,N},\ t \in \bar{R}_\tau(T), \\ \beta\left(D_\tau p_{M,N}(t), w\right) + b\left(u_{M,N}(t) + \theta\tau D_\tau u_{M,N}(t), w\right) = 0, \quad \forall\, w \in W_{M,N},\ t \in \bar{R}_\tau(T), \\ u_{M,N}(0) = P_{M,N}^1 U_0, \\ p_{M,N}(0) = P_{M,N} P_0, \end{cases} \tag{6.26}$$

where δ, σ and θ are parameters and $0 \leq \delta, \sigma, \theta \leq 1$. If $\delta = \sigma = \theta = 0$, then (6.26) is an explicit scheme. Otherwise, we need an iteration to evaluate the values of $u_{M,N}$ and $p_{M,N}$ at each time step. In particular, if $\delta = \sigma = \theta = \frac{1}{2}$, then by taking $v = u_{M,N}(t) + u_{M,N}(t + \tau)$ and $w = p_{M,N}(t) + p_{M,N}(t + \tau)$ in (6.26), we derive from

(6.25) that

$$\|u_{M,N}(t)\|^2 + \beta\,\|p_{M,N}(t)\|^2 + \frac{1}{2}\mu\tau \sum_{s\in\bar{R}_\tau(t-\tau)} |u_{M,N}(s) + u_{M,N}(s+\tau)|_1^2$$
$$= \|u_{M,N}(0)\|^2 + \beta\,\|p_{M,N}(0)\|^2 + \tau \sum_{s\in\bar{R}_\tau(t-\tau)} (f(s), u_{M,N}(s) + u_{M,N}(s+\tau)).\tag{6.27}$$

This simulates the conservation (5.67) suitably.

We now analyse the generalized stability of (6.26). For simplicity, we only consider the case with $\delta = \sigma = 0, \theta > \frac{1}{2}$. Assume that $u_{M,N}(0), p_{M,N}(0), f$ and the right-hand side of the second formula of (6.26) have the errors $\tilde{u}, \tilde{p}, \tilde{f}$ and $\tilde{g}$ respectively. They induce the errors of $u_{M,N}$ and $p_{M,N}$, denoted by $\tilde{u}$ and $\tilde{p}$ respectively. Then we get from (6.26) that

$$\begin{cases} (D_\tau\tilde{u}(t) + d\,(\tilde{u}(t), u_{M,N}(t) + \tilde{u}(t)) + d\,(u_{M,N}(t), \tilde{u}(t))\,, v) + \mu a\,(\tilde{u}(t), v) \\ \quad -b\,(v, \tilde{p}(t) + \theta\tau D_\tau\tilde{p}(t)) = \left(\tilde{f}, v\right), \quad \forall\, v \in V_{M,N},\ t \in \bar{R}_\tau(T), \\ \beta\,(D_\tau\tilde{p}(t), w) + b\,(\tilde{u}(t) + \theta\tau D_\tau\tilde{u}(t), w) = (\tilde{g}, w)\,, \quad \forall\, w \in W_{M,N},\ t \in \bar{R}_\tau(T), \\ \tilde{u}(0) = \tilde{u}_0, \quad x \in \bar{\Omega}, \\ \tilde{p}(0) = \tilde{p}_0, \quad x \in \bar{\Omega}. \end{cases}\tag{6.28}$$

Let $\varepsilon > 0$. Taking $v = \tilde{u}(t) + \theta\tau D_\tau\tilde{u}(t)$ in the first formula of (6.28), we obtain from (1.22) and (6.25) that

$$\begin{aligned} & D_\tau\|\tilde{u}(t)\|^2 + \tau(2\theta - 1 - \varepsilon)\|D_\tau\tilde{u}(t)\|^2 + 2\mu|\tilde{u}(t)|_1^2 + \theta\mu\tau D_\tau|\tilde{u}(t)|_1^2 \\ & \quad - \theta\mu\tau^2|D_\tau\tilde{u}(t)|_1^2 - 2b\,(\tilde{u}(t) + \theta\tau D_\tau\tilde{u}(t), \tilde{p}(t) + \theta\tau D_\tau\tilde{p}(t)) + \sum_{j=1}^{3} F_j(t) \\ \le\ & \|\tilde{u}(t)\|^2 + \left(1 + \frac{\theta^2\tau}{\varepsilon}\right)\|\tilde{f}(t)\|^2 \end{aligned}\tag{6.29}$$

where

$$F_1(t) = 2\,(d\,(u_{M,N}(t), \tilde{u}(t))\,, \tilde{u}(t))\,,$$
$$F_2(t) = 2\theta\tau\,(d\,(\tilde{u}(t), u_{M,N}(t) + \tilde{u}(t))\,, D_\tau\tilde{u}(t))\,,$$
$$F_3(t) = 2\theta\tau\,(d\,(u_{M,N}(t), \tilde{u}(t))\,, D_\tau\tilde{u}(t))\,.$$

Next taking $w = \tilde{p}(t) + \theta\tau D_\tau\tilde{p}(t)$ in the second formula of (6.28), we get that

$$\begin{aligned} & \beta D_\tau\|\tilde{p}(t)\|^2 + \beta\tau(2\theta - 1 - \varepsilon)\|D_\tau\tilde{p}(t)\|^2 + 2b\,(\tilde{u}(t) + \theta\tau D_\tau\tilde{u}(t), \tilde{p}(t) + \theta\tau D_\tau\tilde{p}(t)) \\ & \quad \le \beta\|\tilde{p}(t)\|^2 + \left(\tfrac{1}{\beta} + \tfrac{\theta^2\tau}{\beta\varepsilon}\right)\|\tilde{g}(t)\|^2. \end{aligned}\tag{6.30}$$

Putting (6.29) and (6.30) together, it follows that

$$\begin{aligned}
&D_\tau\left(\|\tilde{u}(t)\|^2+\beta\|\tilde{p}(t)\|^2\right)+\tau(2\theta-1-\varepsilon)\left(\|D_\tau\tilde{u}(t)\|^2+\beta\|D_\tau\tilde{p}(t)\|^2\right)\\
&\quad+2\mu|\tilde{u}(t)|_1^2+\theta\mu\tau D_\tau|\tilde{u}(t)|_1^2-\theta\mu\tau^2|D_\tau\tilde{u}(t)|_1^2+\sum_{j=1}^{3}F_j(t)\\
&\quad\le\|\tilde{u}(t)\|^2+\beta\|\tilde{p}(t)\|^2+\left(1+\frac{\theta^2\tau}{\varepsilon}\right)\left(\|\tilde{f}(t)\|^2+\frac{1}{\beta}\|\tilde{g}(t)\|^2\right). \qquad (6.31)
\end{aligned}$$

We now estimate $|F_j(t)|$. To do this, we need some bounds for $\|d(v,w)\|$. It is easy to see that

$$\|d(v,w)\|\le c\|v\|_{W^{1,\infty}}\|w\|_1,$$
$$\|d(v,w)\|\le c\|v\|_1\|w\|_{W^{1,\infty}}.$$

By virtue of Theorem 6.1, we can also see that for any $v,w\in V_{M,N}$,

$$\|d(v,w)\|\le cMN^{\frac{1}{2}}\left(\|v\||w|_1+\|w\||v|_1\right).$$

Using the above three estimates, we obtain

$$\begin{aligned}
&|F_1(t)|\le\frac{1}{2}\mu|\tilde{u}(t)|_1^2+c\left(1+\frac{1}{\mu}\|u_{M,N}(t)\|^2_{W^{1,\infty}}\right)\|\tilde{u}(t)\|^2,\\
&|F_2(t)|\le\varepsilon\tau\|D_\tau\tilde{u}(t)\|^2+\frac{c\theta^2\tau}{\varepsilon}\|u_{M,N}(t)\|^2_{W^{1,\infty}}\|\tilde{u}(t)\|_1^2+\frac{c\theta^2\tau M^2N}{\varepsilon}\|\tilde{u}(t)\|^2|\tilde{u}(t)|_1^2,\\
&|F_3|\le\varepsilon\tau\|D_\tau\tilde{u}(t)\|^2+\frac{c\theta^2\tau}{\varepsilon}\|u_{M,N}(t)\|^2_{W^{1,\infty}}\|\tilde{u}(t)\|_1^2.
\end{aligned}$$

Substituting the above estimates into (6.31) yields that

$$\begin{aligned}
&D_\tau\left(\|\tilde{u}(t)\|^2+\beta\|\tilde{p}(t)\|^2\right)+\tau(2\theta-1-3\varepsilon)\left(\|D_\tau\tilde{u}(t)\|^2+\beta\|D_\tau\tilde{p}(t)\|^2\right)\\
&\quad+\mu|\tilde{u}(t)|_1^2+\theta\mu\tau D_\tau|\tilde{u}(t)|_1^2-\theta\mu\tau^2|D_\tau\tilde{u}(t)|_1^2\\
&\quad\le A_1\left(\|\tilde{u}(t)\|^2+\beta\|\tilde{p}(t)\|^2\right)+A_2(t)|\tilde{u}(t)|_1^2+A_3(t) \qquad (6.32)
\end{aligned}$$

where

$$\begin{aligned}
&A_1(t)=c+c\left(\frac{1}{\mu}+\frac{\theta^2\tau}{\varepsilon}\right)\|u_{M,N}\|^2_{C(0,T;W^{1,\infty})},\\
&A_2(t)=-\frac{\mu}{2}+\frac{c\theta^2\tau}{\varepsilon}\left(\|u_{M,N}\|^2_{C(0,T;W^{1,\infty})}+M^2N\|\tilde{u}(t)\|^2\right),\\
&A_3(t)=c\left(1+\frac{\theta^2\tau}{\varepsilon}\right)\left(\|\tilde{f}(t)\|^2+\frac{1}{\beta}\|\tilde{g}(t)\|^2\right).
\end{aligned}$$

Assume that ε and q_0 are suitably small positive constants such that

$$2\theta - 1 - 3\varepsilon - q_0 \geq \theta\mu\tau\left(\frac{9}{4}M^4 + N^2\right). \tag{6.33}$$

Then by Theorem 6.2,

$$\tau(2\theta - 1 - 3\varepsilon)\|D_\tau\tilde{u}(t)\|^2 - \theta\mu\tau^2|D_\tau\tilde{u}(t)|_1^2 \geq q_0\tau\|D_\tau\tilde{u}(t)\|^2. \tag{6.34}$$

Let

$$E(v,w,t) = \|v(t)\|^2 + \beta\|w(t)\|^2 + \tau\sum_{s\in\bar{R}_\tau(t-\tau)}\left(q_0\tau\|D_\tau v(s)\|^2 + q_0\beta\tau\|D_\tau w(s)\|^2 + |v(s)|_1^2\right) \tag{6.35}$$

and

$$\rho(t) = \|\tilde{u}_0\|^2 + \beta\|\tilde{p}_0\|^2 + \theta\mu\tau|\tilde{u}_0|_1^2 + \tau\sum_{s\in\bar{R}_\tau(t-\tau)} A_3(s). \tag{6.36}$$

By summing (6.32) for $t \in \bar{R}_\tau(t-\tau)$, we derive from (6.34) that

$$E(\tilde{u},\tilde{p},t) \leq \rho(t) + \tau\sum_{s\in\bar{R}_\tau(t-\tau)}\left(A_1E(\tilde{u},\tilde{p},s) + A_2(s)|\tilde{u}(s)|_1^2\right).$$

By applying Lemma 3.2 to the above inequality, we obtain the following result.

Theorem 6.15. *Let $\delta = \sigma = 0, \theta > \frac{1}{2}, \tau \leq b_1\mu$ in (6.26), and (6.33) hold. If $\rho(t_1)e^{b_2t_1} \leq \dfrac{b_3}{\tau M^2N}$ for some $t_1 \in R_\tau(T)$, then for all $t \in \bar{R}_\tau(t_1)$,*

$$E(\tilde{u},\tilde{p},t) \leq \rho(t)e^{b_2t}$$

where b_1, b_2 and b_3 are some positive constants depending on μ and $\|u_{M,N}\|_{C(0,T;W^{1,\infty})}$.

We next deal with the convergence of (6.26) with $\delta = \sigma = 0$ and $\theta > \frac{1}{2}$. Let $U_{M,N} = P_{M,N}^{1,0}U, \tilde{U} = u_{M,N} - U_{M,N}$ and $\tilde{P} = p_{M,N} - P_{M,N}P$. We get from (6.20) and (6.26) that

$$\begin{cases} (D_\tau\tilde{U}(t) + d(\tilde{U}(t), U_{M,N}(t) + \tilde{U}(t)) + d(U_{M,N}(t),\tilde{U}(t)), v) + \mu a(\tilde{U}(t), v) - b(v, \tilde{P}(t) \\ \quad +\theta\tau D_\tau\tilde{P}(t)) = (G_1(t) + G_2(t) + G_3(t), v) - b(v, G_4(t)), \quad \forall\, v \in V_{M,N}, t \in \bar{R}_\tau(T), \\ \beta(D_\tau\tilde{P}(t), w) + b(\tilde{U}(t) + \theta\tau D_\tau\tilde{U}(t), w) = -\beta(D_\tau P(t), w) + (G_5(t), w), \\ \qquad\qquad\qquad\qquad\qquad\qquad\qquad\qquad \forall\, w \in W_{M,N}, t \in \bar{R}_\tau(T), \\ \tilde{U}(0) = \tilde{p}(0) = 0 \end{cases}$$

where

$$G_1(t) = \partial_t U(t) - D_\tau U_{M,N}(t),$$
$$G_2(t) = (U(t)\cdot\nabla)\,U(t) - (U_{M,N}(t)\cdot\nabla)\,U_{M,N}(t),$$
$$G_3(t) = -\frac{1}{2}(\nabla\cdot U_{M,N}(t))\,U_{M,N}(t),$$
$$G_4(t) = P(t) - P_{M,N}P(t) - \theta\tau D_\tau P_{M,N}P(t),$$
$$G_5(t) = \nabla\cdot(U(t) - U_{M,N}(t) - \theta\tau D_\tau U_{M,N}(t)).$$

By Theorem 6.5 and (4.24),

$$\|G_1(t)\|^2 \le c\tau\|U\|^2_{H^2(t,t+\tau;L^2)} + \frac{c}{\tau}\left(M^{-2r}+N^{-2s}\right)\|U\|^2_{H^1(t,t+\tau;M^{r,s})}.$$

By Theorems 6.5 and 6.6, we deduce that for $r,s>1$,

$$\begin{aligned}\|G_2(t)\|^2 &\le 2\|\,(U(t)\cdot\nabla)\,U(t) - (U_{M,N}(t)\cdot\nabla)\,U(t)\|^2\\
&\quad + \|\,(U_{M,N}(t)\cdot\nabla)\,U(t) - (U_{M,N}(t)\cdot\nabla)\,U_{M,N}(t)\|^2\\
&\le c\left(\|U(t)\|^2_{W^{1,\infty}}\|U(t)-U_{M,N}(t)\|^2 + \|U_{M,N}(t)\|^2_{L^\infty}\|U(t)-U_{M,N}(t)\|_1^2\right)\\
&\le c\left(M^{-2r}+N^{-2s}\right)\left(\|U(t)\|^4_{M^{r+1,s+1}} + \|U(t)\|^4_{W^{1,\infty}}\right).\end{aligned}$$

Since $\nabla\cdot U = 0$, we derive in a similar manner that

$$\|G_3(t)\|^2 = \left\|G_3(t) - \frac{1}{2}(\nabla\cdot U)U\right\|^2 \le c\left(M^{-2r}+N^{-2s}\right)\left(\|U(t)\|^4_{M^{r+1,s+1}} + \|U(t)\|^4_{W^{1,\infty}}\right).$$

Obviously,

$$\|G_4(t)\|^2 \le c\left(M^{-2r}+N^{-2s}\right)\|P(t)\|^2_{H^{r,s}} + c\theta^2\tau\|P\|^2_{H^1(t,t+\tau;L^2)},$$
$$\|G_5(t)\|^2 \le c\left(M^{-2r}+N^{-2s}\right)\|U(t)\|^4_{M^{r+1,s+1}} + c\theta^2\tau\|U\|^2_{H^1(t,t+\tau;M^{1,1})}.$$

Finally, by an argument as in the proof of Theorem 6.15, we get the following result.

Theorem 6.16. *Let $\delta=\sigma=0, \theta>\frac{1}{2}, \tau\le b_4\mu$ in (6.26), and (6.33) hold. If for $r,s\ge 1$,*

$$U\in C\left(0,T;H^1_{0,p}\cap M_p^{r+1,s+1}\cap W^{1,\infty}\right)\cap H^1\left(0,T;M_p^{r,s}\right)\cap H^2\left(0,T;L_p^2\right),$$
$$P\in C\left(0,T;H_p^{r,s}\right)\cap H^1\left(0,T;L_p^2\right),$$

then for all $t\in\bar{R}_\tau(T)$,

$$E\left(u_{M,N}-U_{M,N},p_{M,N}-P_{M,N}P,t\right)\le b_5\left(\frac{1}{\beta}\left(\tau^2+M^{-2r}+N^{-2s}\right)+\beta\right)$$

where b_4 and b_5 are some positive constants depending on μ and the norms of U and P in the spaces mentioned above.

We find from Theorem 6.16 that the best choice is $\beta = \mathcal{O}(\tau) = \mathcal{O}\left(M^{-2r} + N^{-2s}\right)$, and the corresponding accuracy is of the order $\beta^{\frac{1}{2}}$. It seems that the artificial compression brings the convenience in calculations, but lowers the accuracy. Indeed, if we compare the numerical solution to other projection of the exact solution, called the Stokes projection, we can remove this problem, and get a better error estimation for any $\beta \geq 0$. We shall introduce this technique in the next chapter.

Obviously, since we use the approximation (6.24) for the nonlinear term $(U \cdot \nabla)U$, a reasonable simulation of conservation (6.27) follows, even the trial functions do not satisfy the incompressibility. Also because of this reason, the effects of the main nonlinear error terms are cancelled, i.e.,

$$\left(d\left(\tilde{u}(t), \tilde{u}(t)\right), \tilde{u}(t)\right) = \left(d\left(\tilde{U}(t), \tilde{U}(t)\right), \tilde{U}(t)\right) = 0.$$

Therefore the nice error estimates follow. Conversely, if the trial function space $V_{M,N}$ fulfills the incompressibility, then it is better to approximate the convective term by $\tilde{d}(u_{M,N}, u_{M,N})$,

$$\tilde{d}(v,w) = \partial_1\left(w^{(1)}v\right) + \partial_2\left(w^{(2)}v\right).$$

In this case, the conservation can also be simulated properly. For the error analysis, we have

$$\left(\tilde{d}\left(\tilde{u}(t), \tilde{u}(t)\right), \tilde{u}(t)\right) = \frac{1}{2}\left(|\tilde{u}(t)|^2, \nabla \cdot \tilde{u}(t)\right).$$

As $\nabla \cdot \tilde{u}(t) \neq 0$ usually, this term cannot be cancelled. This fact shows that a good approximation to the nonlinear convective term plays an important role.

Bernardi, Maday and Metivet (1987) applied the mixed Fourier-Legendre spectral methods first to semi-periodic problems of steady Navier-Stokes equation.

The mixed Fourier-Chebyshev spectral methods have also been used for semi-periodic problems of various nonlinear differential equations, such as the vorticity equation in Guo, Ma, Cao and Huang (1992), and the Navier-Stokes equation in Guo and Li (1996). Recently, the Fourier-Chebyshev pseudospectral methods have also been developed, e.g., see Guo and Li (1993, 1995a).

CHAPTER 7

COMBINED SPECTRAL METHODS

In many practical problems, the domains are not rectangular. We might use some mappings to change them to rectangular and then use pure spectral methods to solve the resulting problems numerically. But it is not easy to do so. On the other hand, such procedures usually bring in the complexity and errors. Thus it is better to adopt the finite element methods in those cases. However, sometimes the intersections of the domains may be rectangular, such as a cylindrical container. So it is reasonable to use the combined spectral-finite element methods or the combined pseudospectral-finite element methods. Since finite element methods are based on Galerkin method with interpolations of functions, the latter seem more attractive. In this chapter, we present the combined Fourier spectral-finite element methods for semi-periodic problems, and the combined Legendre spectral-finite element methods and the combined Chebyshev spectral-finite element methods for non-periodic problems. Finally, we take Navier-Stokes equation as an example to show the applications of these combined methods.

7.1. Some Basic Results in Finite Element Methods

In this section, we recall some basic results of finite element methods in one dimension, as a preparation of the forthcoming discussions. Let $\Lambda = (-1,1)$ and $x \in \bar{\Lambda}$. Decompose Λ into $2M$ subintervals $\Delta_j, -M \le j \le M-1$. h_j represents the length of Δ_j. Let $h = \max\limits_{-M \le j \le M-1} h_j$. Assume that there exists a positive constant ρ independent of the partition such that $\max\limits_{-M \le j \le M-1} \dfrac{h}{h_j} \le \rho$. Let k be non-negative integer and $\sqcap_{h,k}$ be the Lagrange interpolation of degree k over each Δ_j. Define

$$\tilde{S}_{h,k} = \{v \mid v \text{ is a polynomial of degree} \le k \text{ on } \Delta_j, -M \le j \le M-1\}.$$

Moreover $S_{h,k} = \tilde{S}_{h,k} \cap H^1(\Lambda)$ and $S^0_{h,k} = \tilde{S}_{h,k} \cap H^1_0(\Lambda)$. In the space $\tilde{S}_{h,k}$, several inverse inequalities exist.

Lemma 7.1. *For any $\phi \in \tilde{S}_{h,k}$ and $1 \le p \le q \le \infty$,*

$$\|\phi\|_{L^q} \le c \left(\frac{h}{k^2}\right)^{\frac{1}{q}-\frac{1}{p}} \|\phi\|_{L^p}.$$

Proof. ϕ is a polynomial of degree at most k on each Δ_j with the left extreme point $x^{(j)}$. Let $y = \frac{2}{h_j}\left(x - x^{(j)}\right) - 1, \tilde{\Lambda} = \{y \mid |y| < 1\}$ and $\tilde{\phi}(y) = \phi(x)$. By virtue of Theorem 2.7,

$$\|\phi\|_{L^q(\Delta_j)} = \left(\frac{h_j}{2}\right)^{\frac{1}{q}} \|\tilde{\phi}\|_{L^q(\tilde{\Lambda})} \le c \left(\frac{h_j}{2}\right)^{\frac{1}{q}} k^{\frac{2}{p}-\frac{2}{q}} \|\tilde{\phi}\|_{L^p(\tilde{\Lambda})}.$$

On the other hand,

$$\|\tilde{\phi}\|_{L^p(\tilde{\Lambda})} = \left(\frac{2}{h_j}\right)^{\frac{1}{p}} \|\phi\|_{L^p(\Delta_j)}.$$

Consequently,

$$\|\phi\|_{L^q(\Delta_j)} \le c \left(\frac{h_j}{k^2}\right)^{\frac{1}{q}-\frac{1}{p}} \|\phi\|_{L^p(\Delta_j)}$$

from which the desired result follows. ∎

Lemma 7.2. *For any $\phi \in \tilde{S}_{h,k}$ and $0 \le r \le \mu$,*

$$\|\phi\|_{\mu} \le c \left(\frac{h}{k^2}\right)^{r-\mu} \|\phi\|_{r}.$$

Proof. By Theorem 2.8 and an argument as in the proof of the previous theorem, we deduce that

$$\|\phi\|_{H^\mu(\Delta_j)} \le c \left(\frac{h_j}{k^2}\right)^{r-\mu} \|\phi\|_{H^r(\Delta_j)}, \quad \mu = 0, 1.$$

Hence

$$\|\phi\|_{\mu} \le c \left(\frac{h}{k^2}\right)^{r-\mu} \|\phi\|_{r}, \quad \mu = 0, 1.$$

The general result comes from induction and space interpolation. ∎

The $L^2(\Lambda)$-orthogonal projection $P_{h,k} : L^2(\Lambda) \to \tilde{S}_{h,k}$ is such a mapping that for any $v \in L^2(\Lambda)$,

$$(v - P_{h,k}v, \phi) = 0, \quad \forall\, \phi \in \tilde{S}_{h,k}. \tag{7.1}$$

For simplicity, let $c(k)$ be a generic positive constant depending on k. We have from Ciarlet (1978) that for any $v \in H^r(\Lambda), \frac{1}{2} < r \leq k+1$ and $0 \leq \mu \leq \min(r,1)$,

$$\|v - \Pi_{h,k}v\|_\mu \leq c(k)h^{r-\mu}|v|_r, \tag{7.2}$$

$$\|v - \Pi_{h,k}v\|_{W^{\mu,\infty}} \leq c(k)h^{r-\mu}|v|_{W^{r,\infty}}. \tag{7.3}$$

Lemma 7.3. *For any* $v \in H^r(\Lambda), \frac{1}{2} < r \leq k+1$ *and* $0 \leq \mu \leq \min(r,1)$,

$$\|v - P_{h,k}v\|_\mu \leq c(k)h^{r-\mu}\|v\|_r.$$

Proof. We first assume that $r \geq 1$. Since $\Pi_{h,k}v \in \tilde{S}_{h,k}$, it follows from Lemma 7.2 and (7.2) that

$$\begin{aligned}\|v - P_{h,k}v\|_\mu &\leq \|v - \Pi_{h,k}\|_\mu + \|\Pi_{h,k}v - P_{h,k}v\|_\mu \\ &\leq c(k)h^{r-\mu}|v|_r + ch^{-\mu}\|\Pi_{h,k}v - P_{h,k}v\|.\end{aligned}$$

Since

$$\|\Pi_{h,k}v - P_{h,k}v\|^2 = (\Pi_{h,k}v - P_{h,k}v, \Pi_{h,k}v - v) \leq \|\Pi_{h,k}v - v\|\,\|\Pi_{h,k}v - P_{h,k}v\|,$$

we obtain from (7.2) that

$$\|v - P_{h,k}v\|_\mu \leq c(k)h^{r-\mu}\|v\|_r. \tag{7.4}$$

Next let $0 \leq r < 1$. By (7.4) and projection theory,

$$\|v - P_{h,k}v\| \leq c(k)h\|v\|_1,$$

$$\|v - P_{h,k}v\| \leq c(k)\|v\|.$$

Thanks to the space interpolation, we assert that for any $0 \leq \beta \leq 1$,

$$\|v - P_{h,k}v\| \leq c(k)h^\beta\|v\|_\beta.$$

On the other hand, by (7.4),

$$\|v - P_{h,k}v\|_1 \leq c(k)\|v\|_1.$$

Using the space interpolation again, we obtain that for $0 \leq \beta \leq 1$,

$$\|v - P_{h,k}v\|_\mu \leq c(k)h^{\beta(1-\mu)}\|v\|_{H^{\mu+(1-\mu)\beta}}.$$

Letting $\beta = \frac{r-\mu}{1-\mu}$, we complete the proof. ■

The $H^1(\Lambda)$-orthogonal projection: $H^1(\Lambda) \to S_{h,k}$ is such a mapping that for any $v \in H^1(\Lambda)$,

$$\left(\partial_x \left(v - P^1_{h,k}v\right), \partial_x \phi\right) + (v, \phi) = 0, \quad \forall\, \phi \in S_{h,k}. \tag{7.5}$$

The $H^1_0(\Lambda)$-orthogonal projection: $H^1_0(\Lambda) \to S^0_{h,k}$ is such a mapping that for any $v \in H^1_0(\Lambda)$,

$$\left(\partial_x \left(v - P^{1,0}_{h,k}v\right), \partial_x \phi\right) = 0, \quad \forall\, \phi \in S^0_{h,k}. \tag{7.6}$$

Lemma 7.4. *If $v \in H^r(\Lambda), 1 \le r \le k+1$ and $1-k \le \mu \le 1$, then*

$$\left\|v - P^1_{h,k}v\right\|_\mu \le c(k)h^{r-\mu}\|v\|_r.$$

Lemma 7.5. *If $v \in H^r(\Lambda) \cap H^1_0(\Lambda), 1 \le r \le k+1$ and $1-k \le \mu \le 1$, then*

$$\left\|v - P^{1,0}_{h,k}v\right\|_\mu \le c(k)h^{r-\mu}|v|_r.$$

Lemma 7.5 comes from Douglas and Dupont (1974). Lemma 7.4 can be proved in the same manner. Guo and Ma (1991) gave a result similar to Lemma 7.3.

Finally, we give a special Poincaré inequality which will be used in the forthcoming sections. Let $\tilde{\Lambda} = \{y \mid a \le y \le b\}, v \in H^1(\tilde{\Lambda})$ and $|v(y^*)| = \inf\limits_{y\in\tilde{\Lambda}} |v(y)|$. Then

$$\begin{aligned} v^2(y) &= 2\int_{y^*}^{y} |v(z)||\partial_z v(z)|\,dz + |v(y^*)|^2 \\ &\le 2\|v\|_{L^2(\tilde{\Lambda})}|v|_{H^1(\tilde{\Lambda})} + \frac{1}{b-a}\|v\|^2_{L^2(\tilde{\Lambda})}, \end{aligned}$$

and so

$$\|v\|^2_{L^\infty(\tilde{\Lambda})} \le \left(4 + \frac{2}{(b-a)^2}\right)^{\frac{1}{2}} \|v\|_{L^2(\tilde{\Lambda})}\|v\|_{H^1(\tilde{\Lambda})}. \tag{7.7}$$

If, in addition, the mean of $v(y)$ over $\tilde{\Lambda}$ equals zero, then by the continuity, $\inf\limits_{y\in\tilde{\Lambda}} |v(y)| = 0$. Thus

$$\|v\|^2_{L^\infty(\tilde{\Lambda})} \le 2\|v\|_{L^2(\tilde{\Lambda})}|v|_{H^1(\tilde{\Lambda})}.$$

7.2. Combined Fourier-Finite Element Approximations

We discuss the combined Fourier-finite element approximations in this section. Let $\Lambda_1 = \{x_1 \mid |x_1| < 1\}, \Lambda_2 = \{x_2 \mid 0 < x_2 < 2\pi\}$ and $\Omega = \Lambda_1 \times \Lambda_2$. We use the notations in Section 6.1, such as $H_p^{r,s}(\Omega), H_{0,p}^{r,s}(\Omega), M_p^{r,s}(\Omega)$ and $X_p^{r,s}(\Omega)$. k and N are integers, $k \geq 0, N > 0$. $\tilde{S}_{h,k,1}$ and $S_{h,k,1}$ are the sets of functions on Λ_1, as defined in the above section. $S_{h,k,1}^0 = S_{h,k,1} \cap H_0^1(\Lambda_1)$. $V_{N,2}$ is the set of trigonometric functions of degree $\leq N$, on Λ_2. Set $\tilde{V}_{h,k,N} = \tilde{S}_{h,k,1} \otimes V_{N,2}, V_{h,k,N} = S_{h,k,1} \otimes V_{N,2}$ and $V_{h,k,N}^0 = S_{h,k,1}^0 \otimes V_{N,2}$. Several inverse inequalities are valid in the space $\tilde{V}_{h,k,N}$.

Theorem 7.1. *For any $\phi \in \tilde{V}_{h,k,N}$ and $2 \leq p \leq \infty$,*

$$\|\phi\|_{L^p} \leq c \left(\frac{h}{k^2 N} \right)^{\frac{1}{p} - \frac{1}{2}} \|\phi\|.$$

Proof. Let

$$\phi_l(x_1) = \frac{1}{2\pi} \int_{\Lambda_2} \phi(x) e^{-ilx_2} \, dx_2.$$

Then $\phi_l \in \tilde{S}_{h,k,1}$ and so by Lemma 7.1,

$$\|\phi\|_{L^\infty} \leq \sum_{|l| \leq N} \|\phi_l\|_{L^\infty(\Lambda_1)} \leq c \left(\frac{h}{k^2} \right)^{-\frac{1}{2}} \sum_{|l| \leq N} \|\phi_l\|_{L^2(\Lambda_1)} \leq c \left(\frac{h}{k^2 N} \right)^{-\frac{1}{2}} \|\phi\|.$$

Therefore

$$\|\phi\|_{L^p}^p \leq \|\phi\|_{L^\infty}^{p-2} \|\phi\|^2 \leq c \left(\frac{h}{k^2 N} \right)^{1 - \frac{p}{2}} \|\phi\|^p$$

which gives the desired result. ∎

Theorem 7.2. *For any $\phi \in \tilde{V}_{h,k,N}$ and $0 \leq r \leq \mu$,*

$$\|\phi\|_\mu \leq c \left(\frac{h}{k^2} + \frac{1}{N} \right)^{r - \mu} \|\phi\|_r.$$

Proof. Let ϕ_l be as before. By Lemma 7.2,

$$\|\phi_l\|_{H^\mu(\Lambda_1)} \leq c \left(\frac{h}{k^2} \right)^{r - \mu} \|\phi_l\|_{H^r(\Lambda_1)}, \quad \mu = 0, 1. \tag{7.9}$$

Therefore

$$\|\partial_1^\mu \phi\|^2 \leq c \left(\frac{h}{k^2} \right)^{2r - 2\mu} \sum_{|l| \leq N} \|\phi_l\|_{H^r(\Lambda_1)}^2 \leq c \left(\frac{h}{k^2} \right)^{2r - 2\mu} \|\phi\|_r^2, \quad \mu = 0, 1.$$

It is easy to verify that

$$\|\partial_2^\mu \phi\|^2 \le c\left(\frac{1}{N}\right)^{2r-2\mu} \sum_{|l|\le N} l^{2r}\|\phi_l\|^2_{L^2(\Lambda_1)} \le c\left(\frac{1}{N}\right)^{2r-2\mu} \|\phi\|_r^2, \quad \mu = 0,1.$$

Finally the desired result comes from induction and space interpolation. ■

The $L^2(\Omega)$-orthogonal projection $P_{h,k,N} : L^2_p(\Omega) \to \tilde{V}_{h,k,N}$ is such a mapping that for any $v \in L^2_p(\Omega)$,

$$(v - P_{h,k,N}v, \phi) = 0, \quad \forall\, \phi \in \tilde{V}_{h,k,N}.$$

Set $\bar{r} = \min(r, k+1)$. For simplicity, $P_{h,k}$ denotes the $L^2(\Lambda_1)$-orthogonal projection on $\tilde{S}_{h,k,1}$, and P_N denotes the $L^2(\Lambda_2)$-orthogonal projection on $V_{N,2}$. Similarly, $P^1_{h,k}$ stands for the $H^1(\Lambda_1)$-orthogonal projection on $S_{h,k,1}$ and $P^{1,0}_{h,k}$ stands for the $H^1_0(\Lambda_1)$-orthogonal projection on $S^0_{h,k,1}$.

Theorem 7.3. *If $v \in H^{r,s}_p(\Omega), r > \frac{1}{2}$ and $s \ge 0$, then*

$$\|v - P_{h,k,N}v\| \le c(k)\left(h^{\bar{r}} + N^{-s}\right)\|v\|_{H^{r,s}}.$$

Proof. Let

$$v_l(x_1) = \frac{1}{2\pi}\int_{\Lambda_2} v(x)e^{-ilx_2}\,dx_2.$$

Then

$$P_{h,k,N}v(x) = \sum_{|l|\le N} P_{h,k}v_l(x_1)e^{ilx_2}.$$

Using Lemma 7.3, we obtain that

$$\begin{aligned}
\|v - P_{h,k,N}v\|^2 &= \sum_{|l|\le N}\|v_l - P_{h,k}v_l\|^2_{L^2(\Lambda_1)} + \sum_{|l|>N}\|v_l\|^2_{L^2(\Lambda_1)} \\
&\le c(k)h^{2\bar{r}}\sum_{|l|\le N}\|v_l\|^2_{H^r(\Lambda_1)} + cN^{-2s}\sum_{|l|>N}|l|^{2s}\|v_l\|^2_{L^2(\Lambda_1)} \\
&\le c(k)h^{2\bar{r}}\|v\|^2_{L^2(\Lambda_2,H^r(\Lambda_1))} + cN^{-2s}|v|^2_{H^s(\Lambda_2,L^2(\Lambda_1))}.
\end{aligned}$$

■

In particular, for any $v \in H^{r,s}_{0,p}(\Omega), r > \frac{1}{2}$ and $s \ge 0$,

$$\|v - P_{h,k,N}v\| \le c(k)\left(h^{\bar{r}} + N^{-s}\right)|v|_{H^{r,s}}.$$

The $H^1(\Omega)$-orthogonal projection $P^1_{h,k,N} : H^1_p(\Omega) \to V_{h,k,N}$ is such a mapping that for any $v \in H^1_p(\Omega)$,

$$\left(\nabla\left(v - P^1_{h,k,N}v\right), \nabla\phi\right) + (v,\phi) = 0, \quad \forall\, \phi \in V_{h,k,N}.$$

The $H_0^1(\Omega)$-orthogonal projection $P_{h,k,N}^{1,0} : H_{0,p}^1(\Omega) \to V_{h,k,N}^0$ is such a mapping that for any $v \in H_{0,p}^1(\Omega)$,

$$\left(\nabla\left(v - P_{h,k,N}^{1,0} v\right), \nabla\phi\right) = 0, \quad \forall\, \phi \in V_{h,k,N}^0.$$

Theorem 7.4. *If $v \in M_p^{r,s}(\Omega)$ and $r, s \geq 1$, then*

$$\left\|v - P_{h,k,N}^1 v\right\|_\mu \leq c(k)\left(h^{\bar{r}-1} + N^{1-s}\right)\left(h^{1-\mu} + N^{\mu-1}\right)\|v\|_{M^{r,s}}, \quad \mu = 0, 1.$$

If, in addition, for certain positive constants c_1 and c_2,

$$c_1 N \leq \frac{1}{h} \leq c_2 N, \tag{7.10}$$

then

$$\left\|v - P_{h,k,N}^1 v\right\| \leq c(k)\left(h^{\bar{r}} + N^{-s}\right)\|v\|_{M^{r,s}}.$$

Proof. Let $v_l(x_1)$ be as before, and

$$v_*(x) = \sum_{|l| \leq N} \left(P_{h,k}^1 v_l(x_1)\right) e^{ilx_2}.$$

By Lemma 7.4,

$$\begin{aligned}
\left\|v - P_{h,k,N}^1 v\right\|_1^2 &\leq \|v - v_*\|_1^2 \\
&\leq \sum_{|l| \leq N}\left(\left\|v_l - P_{h,k}^1 v_l\right\|_{H^1(\Lambda_1)}^2 + l^2\left\|v_l - P_{h,k}^1 v_l\right\|_{L^2(\Lambda_1)}^2\right) \\
&\quad + \sum_{|l| > N}\left(\|v_l\|_{H^1(\Lambda_1)}^2 + l^2\|v_l\|_{L^2(\Lambda_1)}^2\right) \\
&\leq c(k) h^{2\bar{r}-2} \sum_{|l| \leq N}\left(\|v_l\|_{H^r(\Lambda_1)}^2 + l^2\|v_l\|_{H^{r-1}(\Lambda_1)}^2\right) \\
&\quad + cN^{2-2s} \sum_{|l| > N} l^{2s-2}\left(\|v_l\|_{H^1(\Lambda_1)}^2 + l^2\|v_l\|_{L^2(\Lambda_1)}^2\right) \\
&\leq c(k)\left(h^{\bar{r}-1} + N^{1-s}\right)^2\|v\|_{M^{r,s}}^2.
\end{aligned} \tag{7.11}$$

So the first conclusion is true for $\mu = 1$. The rest of the proof is clear. ∎

Theorem 7.5. *If $v \in M_p^{r,s}(\Omega) \cap H_{0,p}^1(\Omega)$ and $r, s \geq 1$, then*

$$\left\|v - P_{h,k,N}^{1,0} v\right\|_\mu \leq c(k)\left(h^{\bar{r}-1} + N^{1-s}\right)\left(h^{1-\mu} + N^{\mu-1}\right)|v|_{M^{r,s}}, \quad \mu = 0, 1.$$

If, in addition, (7.10) holds, then

$$\left\|v - P_{h,k,N}^{1,0} v\right\| \leq c(k)\left(h^{\bar{r}} + N^{-s}\right)|v|_{M^{r,s}}.$$

Proof. Let $v_l(x_1)$ be as before and

$$v_*(x) = \sum_{|l|\leq N} \left(P^{1,0}_{h,k} v_l(x_1)\right) e^{ilx_2}.$$

Then

$$\left\| v - P^{1,0}_{h,k,N} v \right\|_1 \leq c \inf_{w\in V^0_{h,k,N}} |v-w|_1 \leq c\, |v-v_*|_1 .$$

By an argument as in the derivation of (7.11), we prove the first result for $\mu = 1$. Finally by means of the duality and (7.10), we complete the proof. ∎

We can also bound the norm $\left\| P^{1,0}_{h,k,N} v\right\|_{W^{r,p}}$.

Theorem 7.6. *Let (7.10) hold, $k \geq 1$ and $r, s > \frac{1}{2}$. For any $v \in H^1_{0,p}(\Omega) \cap H^s_p(\Lambda_2, H^r(\Lambda_1))$,*

$$\left\| P^{1,0}_{h,k,N} v\right\|_{L^\infty} \leq c(k) \|v\|_{H^s(\Lambda_2,H^r(\Lambda_1))}.$$

If, in addition, $v \in X^{r,s}_p(\Omega)$, then

$$\left\| P^{1,0}_{h,k,N} v\right\|_{W^{1,\infty}} \leq c(k)\|v\|_{X^{r,s}}.$$

Proof. Let $v_l(x_1)$ be as before and

$$P^1_{h,k,N} v(x) = \sum_{|l|\leq N} v^*_l(x_1) e^{ilx_2}.$$

Obviously $v^*_l \in S^0_{h,k,1}$. By an argument as in the proof of Theorem 6.6, we know that for any $\phi \in S^0_{h,k,1}$,

$$\|v_l - v^*_l\|^2_{H^1(\Lambda_1)} + l^2 \|v_l - v^*_l\|^2_{L^2(\Lambda_1)} \leq c(\|v_l - \phi\|^2_{H^1(\Lambda_1)} + l^2\|v_l-\phi\|^2_{L^2(\Lambda_1)}).$$

By taking $\phi = P^{1,0}_{h,k} v_l$, we obtain from Lemma 7.5 that

$$\|v_l - v^*_l\|_{H^1(\Lambda_1)} \leq c(k) h^{\bar{r}-1} \|v\|_{H^r(\Lambda_1)}.$$

By means of the duality,

$$\|v_l - v^*_l\|_{H^\mu(\Lambda_1)} \leq c(k) h^{\bar{r}-\mu} \|v\|_{H^r(\Lambda_1)}, \quad \mu = 0, 1. \tag{7.12}$$

It is clear that for $\mu = 0, 1$,

$$\left\| P^{1,0}_{h,k,N} v\right\|_{W^{\mu,\infty}} \leq \sum_{|l|\leq N} \left(\|\partial^\mu_1 v^*_l\|_{L^\infty(\Lambda_1)} + (1 + |l|^\mu)\, \|v^*_l\|_{L^\infty(\Lambda_1)}\right). \tag{7.13}$$

We estimate the terms on the right-hand side of (7.13). Firstly, we have

$$\|\partial_1 v_l^*\|_{L^\infty(\Lambda_1)} \le \|\partial_1 v_l^* - \Pi_{h,k}\partial_1 v_l\|_{L^\infty(\Lambda_1)} + \|\partial_1 v_l - \Pi_{h,k}\partial_1 v_l\|_{L^\infty(\Lambda_1)} + \|\partial_1 v_l\|_{L^\infty(\Lambda_1)}. \tag{7.14}$$

From Lemma 7.1, (7.2) and (7.12),

$$\begin{aligned}\|\partial_1 v_l^* - \Pi_{h,k}\partial_1 v_l\|_{L^\infty(\Lambda_1)} &\le c(k)h^{-\frac{1}{2}}\left(\|v_l^* - v_l\|_{H^1(\Lambda_1)} + \|\partial_1 v_l - \Pi_{h,k}\partial_1 v_l\|_{L^2(\Lambda_1)}\right)\\ &\le c(k)\left(\|v_l\|_{H^{\frac{3}{2}}(\Lambda_1)} + h^{r-\frac{1}{2}}\|v_l\|_{H^{r+1}(\Lambda_1)}\right).\end{aligned}$$

By (7.3),

$$\|\partial_1 v_l - \Pi_{h,k}\partial_1 v_l\|_{L^\infty(\Lambda_1)} \le c(k)\,\|v_l\|_{H^{r+1}(\Lambda_1)}.$$

By imbedding theory,

$$\|\partial_1 v_l\|_{L^\infty(\Lambda_1)} \le \|v_l\|_{H^{r+1}(\Lambda_1)}.$$

The above statements yield that

$$\begin{aligned}\sum_{|l|\le N}\|\partial_1 v_l^*\|_{L^\infty(\Lambda_1)} &\le c(k)\sum_{|l|\le N}\|v_l\|_{H^{r+1}(\Lambda_1)}\\ &\le c(k)\left(\sum_{|l|\le N}\left(1+l^2\right)^{-s}\right)^{\frac{1}{2}}\left(\sum_{|l|\le N}\left(1+l^2\right)^{s}\|v_l\|^2_{H^{r+1}(\Lambda_1)}\right)^{\frac{1}{2}}\\ &\le c(k)\|v\|_{H^s(\Lambda_2,H^{r+1}(\Lambda_1))}.\end{aligned} \tag{7.15}$$

Similarly, it can be shown that

$$\sum_{|l|\le N}\|v_l^*\|_{L^\infty(\Lambda_1)} \le c(k)\|v\|_{H^s(\Lambda_2,H^r(\Lambda_1))}, \tag{7.16}$$

$$\sum_{|l|\le N}|l|\,\|v_l^*\|_{L^\infty(\Lambda_1)} \le c(k)\|v\|_{H^{s+1}(\Lambda_2,H^r(\Lambda_1))}. \tag{7.17}$$

The combination of (7.13) and (7.15)–(7.17) completes the proof. ∎

Finally we consider the combined Fourier pseudospectral-finite element approximation. Let I_N be the Fourier interpolation on the interval Λ_2, associated with $\Lambda_{2,N}$, the set of interpolation points. Define $I_{h,k,N} = P_{h,k}I_N = I_N P_{h,k}$. Set

$$(v,w)_N = \frac{2\pi}{2N+1}\sum_{x_2\in\Lambda_{2,N}}\int_{\Lambda_1} v(x)w(x)\,dx_1, \quad \|v\|_N = (v,v)_N^{\frac{1}{2}}.$$

Then for any $v \in C\left(\Lambda_2, L^2(\Lambda_1)\right)$,

$$(v - I_{h,k,N}v, \phi)_N = 0, \quad \forall\, \phi \in \tilde{V}_{h,k,N}.$$

Theorem 7.7. *If* $v \in H_p^{r,s}(\Omega)$ *and* $r, s > \frac{1}{2}$, *then*

$$\|v - I_{h,k,N}v\| \le c(k)\left(h^{\bar{r}} + N^{-s}\right)\|v\|_{H^{r,s}}.$$

Proof. Let $E_{h,k,N} = P_{h,k,N}v - I_{h,k,N}v$. Then

$$\|E_{h,k,N}\|^2 = \int_{\Lambda_1} \|P_{h,k,N}v - I_{h,k,N}v\|^2_{L^2(\Lambda_2)}\, dx_1.$$

By using Theorem 7.3 and an argument as in the proof of Theorem 2.4, we achieve the aim. ∎

Canuto, Maday and Quarteroni (1982) first proved Theorem 7.3. Guo and Ma (1991) obtained some results similar to Theorem 7.6. Here we prove them in a different way and improve the corresponding results.

7.3. Combined Legendre-Finite Element Approximations

In this section and the next section, we consider non-periodic problems with rectangular intersection of the domains. In this case, we can use the combined Legendre-finite element approximations. For the sake of simplicity, we only consider two-dimensional problems. Clearly it is not necessary to use such kind of combined approximations for two-dimensional problems. But it is easy to generalize the corresponding results to the three-dimensional problems.

Let $\Lambda_1 = \{x_1 \mid |x_1| < 1\}, \Lambda_2 = \{x_2 \mid |x_2| < 1\}$ and $\Omega = \Lambda_1 \times \Lambda_2$. The integers $k \ge 0$ and $N > 0$. The notations $h, \tilde{S}_{h,k,1}, S_{h,k,1}, S^0_{h,k,1}, P_{h,k}, P^1_{h,k}$ and $P^{1,0}_{h,k}$ have the same meanings as in the previous section. Let $\mathbb{P}_{N,2}$ be the set of all polynomials of degree at most N defined on Λ_2, and $\mathbb{P}^0_{N,2} = \mathbb{P}_{N,2} \cap H^1_0(\Lambda_2)$. For simplicity, let P_N, P^1_N and $P^{1,0}_N$ denote the $L^2(\Lambda_2)$-orthogonal projection on $\mathbb{P}_{N,2}$, the $H^1(\Lambda_2)$-orthogonal projection on $\mathbb{P}_{N,2}$ and the $H^1_0(\Lambda_2)$-orthogonal projection on $\mathbb{P}^0_{N,2}$, respectively. Set $\tilde{V}_{h,k,N} = \tilde{S}_{h,k,1} \otimes \mathbb{P}_{N,2}, V_{h,k,N} = S_{h,k,1} \otimes \mathbb{P}_{N,2}$ and $V^0_{h,k,N} = V_{h,k,N} \cap H^1_0(\Omega)$. In the space $\tilde{V}_{h,k,N}$, several inequalities are valid.

Theorem 7.8. *For any* $\phi \in V_{h,k,N}$ *and* $2 \le p \le \infty$,

$$\|\phi\|_{L^p} \le c\left(\frac{h}{k^2N^2}\right)^{\frac{1}{p}-\frac{1}{2}} \|\phi\|.$$

Proof. For each fixed $x_2 \in \Lambda_2, \phi(x_1,\cdot) \in \tilde{S}_{h,k,1}$ and for each fixed $x_1 \in \Lambda_1, \phi(\cdot,x_2) \in \mathbb{P}_{N,2}$. By Lemma 7.1 and Theorem 2.7,

$$\begin{aligned}
\|\phi\|_{L^\infty} &\le \sup_{x_2\in\Lambda_2} \|\phi\|_{L^\infty(\Lambda_1)} \le \frac{ck}{\sqrt{h}} \sup_{x\in\Lambda_2} \|\phi\|_{L^2(\Lambda_1)} \\
&\le \frac{ck}{\sqrt{h}} \left(\int_{\Lambda_1} \sup_{x_2\in\Lambda_2} \phi^2(x)\, dx_1 \right)^{\frac{1}{2}} \le \frac{ckN}{\sqrt{h}} \|\phi\|.
\end{aligned}$$

Thus

$$\|\phi\|_{L^p}^p \le \|\phi\|_{L^\infty}^{p-2} \|\phi\|^2 \le c \left(\frac{k^2 N^2}{h} \right)^{\frac{p}{2}-1} \|\phi\|^p$$

and the conclusion follows. ∎

Theorem 7.9. *For any $\phi \in \tilde{V}_{h,k,N}$ and $0 \le r \le \mu$,*

$$\|\phi\|_\mu \le c \left(\frac{h}{k^2} + \frac{1}{N^2} \right)^{r-\mu} \|\phi\|_r.$$

Proof. Let

$$\phi(x) = \sum_{l=0}^{N} \phi_l(x_1) L_l(x_2).$$

Then $\phi_l(x_1) \in \tilde{S}_{h,k,1}$. As in the proof of Theorem 7.2, we have

$$\|\partial_1^\mu \phi\|^2 \le c \left(\frac{h}{k^2} \right)^{2r-2\mu} \|\phi\|_r^2. \tag{7.18}$$

On the other hand,

$$\partial_2 \phi(x) = \sum_{l=0}^{N} \phi_l(x_1) \partial_2 L_l(x_2).$$

By (2.49) and the same argument as in the proof of Theorem 2.8,

$$\|\partial_2 \phi(x_1,\cdot)\|_{L^2(\Lambda_2)} \le \sum_{l=0}^{N} \sqrt{l(l+1)}\, |\phi_l(x_1)| \le cN^2 \|\phi(x_1,\cdot)\|_{L^2(\Lambda_2)}.$$

Therefore

$$\|\partial_2 \phi\|^2 \le cN^4 \int_{\Lambda_1} \|\phi(x_1,\cdot)\|_{L^2(\Lambda_2)}^2\, dx_1 = cN^4 \|\phi\|^2$$

which together with (7.18) imply the result for $\mu = 1$ and $r = 0$. The proof is complete by induction and space interpolation. ∎

The $L^2(\Omega)$-orthogonal projection $P_{h,k,N} : L^2(\Omega) \to \tilde{V}_{h,k,N}$ is such a mapping that for any $v \in L^2(\Omega)$,

$$(v - P_{h,k,N} v, \phi) = 0, \quad \forall\, \phi \in \tilde{V}_{h,k,N}.$$

Theorem 7.10. *If $v \in H^{r,s}(\Omega), r > \frac{1}{2}$ and $s \geq 0$, then*

$$\|v - P_{h,k,N}v\| \leq c(k)\left(h^{\bar{r}} + N^{-s}\right)\|v\|_{H^{r,s}}.$$

Proof. Using Theorem 2.9 and Lemma 7.3, we find that

$$\begin{aligned}\|v - P_{h,k,N}v\| &\leq \|v - P_N P_{h,k}v\| \leq \|P_N(v - P_{h,k})\| + \|v - P_N v\| \\ &\leq c(k)\left(h^{\bar{r}}\|v\|_{L^2(\Lambda_2, H^r(\Lambda_1))} + N^{-s}\|v\|_{H^s(\Lambda_2, L^2(\Lambda_1))}\right) \\ &\leq c(k)\left(h^{\bar{r}} + N^{-s}\right)\|v\|_{H^{r,s}}.\end{aligned}$$

■

In particular, for $v \in H_0^{r,s}(\Omega)$,

$$\|v - P_{h,k,N}v\| \leq c(k)\left(h^{\bar{r}} + N^{-s}\right)|v|_{H^{r,s}}.$$

The $H^1(\Omega)$-orthogonal projection $P^1_{h,k,N} : H^1(\Omega) \to V_{h,k,N}$ is such a mapping that for any $v \in H^1(\Omega)$,

$$\left(\nabla\left(v - P^1_{h,k,N}v\right), \nabla\phi\right) + (v, \phi) = 0, \quad \forall\, \phi \in V_{h,k,N}.$$

The $H_0^1(\Omega)$-orthogonal projection $P^{1,0}_{h,k,N} : H_0^1(\Omega) \to V^0_{h,k,N}$ is such a mapping that for any $v \in H_0^1(\Omega)$,

$$\left(\nabla\left(v - P^{1,0}_{h,k,N}v\right), \nabla\phi\right) = 0, \quad \forall\, \phi \in V^0_{h,k,N}.$$

Theorem 7.11. *If $v \in M^{r,s}(\Omega)$ and $r, s \geq 1$, then for $\mu = 0, 1$,*

$$\left\|v - P^1_{h,k,N}v\right\|_\mu \leq c(k)\left(h^{\bar{r}-1} + h^{-1}N^{-s} + N^{1-s}\right)\left(h^{1-\mu} + N^{\mu-1}\right)\|v\|_{M^{r,s}}.$$

If in addition (7.10) holds, then

$$\left\|v - P^1_{h,k,N}v\right\| \leq c(k)\left(h^{\bar{r}} + N^{-s}\right)\|v\|_{M^{r,s}}.$$

Proof. Let $v_* = P_{h,k}P^1_N v \in V_{h,k,N}$. It is easy to see that

$$\left\|v - P^1_{h,k,N}v\right\|_1 \leq \|v - v_*\|_1 \leq \|v - P_{h,k}v\|_1 + \left\|P_{h,k}\left(v - P^1_N v\right)\right\|_1. \tag{7.19}$$

Using Lemma 7.3, we obtain

$$\begin{aligned}\|v - P_{h,k}v\|_1^2 &= \int_{\Lambda_2}\|v - P_{h,k}v\|^2_{H^1(\Lambda_1)}\,dx_2 + \int_{\Lambda_2}\|\partial_2 v - P_{h,k}\partial_2 v\|_{L^2(\Lambda_1)}\,dx_2 \\ &\leq c(k)h^{2\bar{r}-2}\left(\|v\|^2_{L^2(\Lambda_2, H^r(\Lambda_1))} + \|v\|^2_{H^1(\Lambda_2, H^{r-1}(\Lambda_1))}\right).\end{aligned} \tag{7.20}$$

By Lemma 7.2 and Theorem 2.10,

$$\begin{aligned}\left\|P_{h,k}\left(v-P_N^1 v\right)\right\|_1^2 &= \int_{\Lambda_2}\left\|P_{h,k}\left(v-P_N^1 v\right)\right\|_{H^1(\Lambda_1)}^2 dx_2 \\ &\quad + \int_{\Lambda_2}\left\|P_{h,k}\partial_2\left(v-P_N^1 v\right)\right\|_{L^2(\Lambda_1)}^2 dx_2 \\ &\le c(k)h^{-2}N^{-2s}\|v\|_{H^s(\Lambda_2,L^2(\Lambda_1))} + c(k)N^{2-2s}\|v\|^2_{H^s(\Lambda_2,L^2(\Lambda_1))}.\end{aligned}$$

Therefore the first result holds for $\mu=1$. The remaining proof is clear. ∎

Theorem 7.12. *If $v\in M^{r,s}(\Omega)\cap H_0^1(\Omega)$ and $r,s\ge 1$, then for $\mu=0,1$,*

$$\left\|v-P_{h,k,N}^{1,0}v\right\|_\mu \le c(k)\left(h^{\bar r-1}+h^{-1}N^{-s}+N^{1-s}\right)\left(h^{1-\mu}+N^{\mu-1}\right)\|v\|_{M^{r,s}}.$$

If, in addition, (7.10) holds, then

$$\left\|v-P_{h,k,N}^{1,0}v\right\| \le c(k)\left(h^{\bar r}+N^{-s}\right)|v|_{M^{r,s}}.$$

Proof. By letting $v_*=P_{h,k}P_N^{1,0}v$ and following the same line as in the proof of Theorem 7.11, we can prove this theorem easily. ∎

We now estimate $\left\|P_{h,k,N}^{1,0}v\right\|_{W^{r,\infty}}$.

Theorem 7.13. *Let (7.10) hold, $k\ge 1$ and $r>\frac{1}{2}$. For any $v\in M^{\frac{3}{2},\frac{3}{2}}(\Omega)\cap H_0^1(\Omega)\cap H^{\frac{3}{4}}(\Lambda_2,H^r(\Lambda_1))$,*

$$\left\|P_{h,k,N}^{1,0}v\right\|_{L^\infty} \le c(k)\|v\|_{M^{\frac{3}{2},\frac{3}{2}}(\Omega)\cap H^{\frac{3}{4}}(\Lambda_2,H^r(\Lambda_1))}.$$

If, in addition, $k\ge 2$ and $v\in M^{\frac{5}{2},\frac{5}{2}}(\Omega)\cap X^{\frac{1}{2},\frac{1}{2}}(\Omega)\cap X^{r,\frac{3}{4}}(\Omega)$, then

$$\left\|P_{h,k,N}^{1,0}v\right\|_{W^{1,\infty}} \le c(k)\|v\|_{M^{\frac{5}{2},\frac{5}{2}}\cap X^{\frac{1}{2},\frac{1}{2}}\cap X^{r,\frac{3}{4}}}.$$

Proof. We first have that

$$\left\|P_{h,k,N}^{1,0}v\right\|_{L^\infty} \le \left\|P_{h,k,N}^{1,0}v-\Pi_{h,k}P_N v\right\|_{L^\infty} + \|\Pi_{h,k}P_N v\|_{L^\infty}. \tag{7.21}$$

Further, using Theorems 2.9, 7.8 and 7.12, and (7.2) yields that

$$\begin{aligned}\left\|P_{h,k,N}^{1,0}v-\Pi_{h,k}P_N v\right\|_{L^\infty} &\le \frac{c(k)N}{\sqrt{h}}\left\|P_{h,k,N}^{1,0}v-\Pi_{h,k}P_N v\right\| \\ &\le \frac{c(k)N}{\sqrt{h}}\left(\left\|P_{h,k,N}^{1,0}v-v\right\|+\|v-\Pi_{h,k}P_N v\|\right) \\ &\le c(k)\|v\|_{M^{\frac{3}{2},\frac{3}{2}}}.\end{aligned} \tag{7.22}$$

On the other hand, we use an $L^\infty(\Lambda_2)$-estimate in Section 2.4 to obtain that

$$\begin{aligned}\|P_N v\|_{L^\infty(\Lambda_2)} &\le \|P_N v - v\|_{L^\infty(\Lambda_2)} + \|v\|_{L^\infty(\Lambda_2)}\\ &\le cN^{\frac{3}{4}-\mu}\|v\|_{H^\mu(\Lambda_2)} + c\|v\|_{H^{\frac{3}{4}}(\Lambda_2)} \le c\|v\|_{H^{\frac{3}{4}}(\Lambda_2)}.\end{aligned} \tag{7.23}$$

By imbedding theory, (7.2) and (7.23),

$$\|\Pi_{h,k}P_N v\|_{L^\infty} \le c\,\|\Pi_{h,k}P_N v\|_{L^\infty(\Lambda_2,H^r(\Lambda_1))} \le c\|v\|_{H^{\frac{3}{4}}(\Lambda_2,H^r(\Lambda_1))}. \tag{7.24}$$

Then the first conclusion comes from (7.21), (7.22) and (7.24).

We now turn to prove the second result. Let $v_* = P^{1,0}_{h,k,N}v - P^{1,0}_{h,k}P^{1,0}_N v \in V^0_{h,k,N}$. By the definitions of $P^{1,0}_{h,k}$ and $P^{1,0}_{h,k,N}$, we deduce that

$$\begin{aligned}|v_*|_1^2 &= \left(\nabla\left(P^{1,0}_{h,k,N}v - P^{1,0}_{h,k}P^{1,0}_N v\right), \nabla v_*\right) = \left(\nabla\left(v - P^{1,0}_{h,k}P^{1,0}_N v\right), \nabla v_*\right)\\ &= \left(\nabla\left(v - P^{1,0}_N v\right), \nabla v_*\right) + \left(\nabla P^{1,0}_N\left(v - P^{1,0}_{h,k}v\right), \nabla v_*\right)\\ &= \left(\partial_1\left(v - P^{1,0}_N v\right), \partial_1 v_*\right) + \left(\partial_2\left(v - P^{1,0}_{h,k}v\right), \partial_2 v_*\right)\\ &\le |v_*|_1\left(\left\|\partial_1\left(v - P^{1,0}_N v\right)\right\| + \left\|\partial_2\left(v - P^{1,0}_{h,k}v\right)\right\|\right).\end{aligned}$$

Hence by Theorem 2.11 and Lemma 7.5,

$$|v_*|_1 \le c(k)\left(h + \frac{1}{N}\right)^{\frac{3}{2}}\|v\|_{X^{\frac{1}{2},\frac{1}{2}}}. \tag{7.25}$$

The combination of Theorem 7.8 and (7.25) yields

$$\|v_*\|_{W^{1,\infty}} \le \frac{c(k)N}{\sqrt{h}}\|v_*\|_1 \le c(k)\|v\|_{X^{\frac{1}{2},\frac{1}{2}}}. \tag{7.26}$$

On the other hand, we have that

$$\left\|\partial_1 P^{1,0}_{h,k}P^{1,0}_N v\right\|_{L^\infty} \le A_1 + A_2 + A_3 \tag{7.27}$$

where

$$\begin{aligned}A_1 &= \left\|\partial_1 P^{1,0}_{h,k}P^{1,0}_N v - \partial_1 P^{1,0}_{h,k}P_N v\right\|_{L^\infty},\\ A_2 &= \left\|\partial_1 P^{1,0}_{h,k}P_N v - \Pi_{h,k}P_N\partial_1 v\right\|_{L^\infty},\\ A_3 &= \|\Pi_{h,k}P_N\partial_1 v\|_{L^\infty}.\end{aligned}$$

By Theorems 2.9, 2.11 and 7.8, and Lemma 7.5,

$$\begin{aligned}A_1 &\le \frac{c(k)N}{\sqrt{h}}\left\|\partial_1 P^{1,0}_{h,k}\left(P^{1,0}_N v - P_N v\right)\right\|\\ &\le \frac{c(k)N}{\sqrt{h}}\left\|P^{1,0}_N v - P_N v\right\|_{L^2(\Lambda_2,H^1(\Lambda_1))}\\ &\le c(k)\|v\|_{H^{\frac{3}{2}}(\Lambda_2,H^1(\Lambda_1))}.\end{aligned}$$

By Theorems 2.9 and 7.8, Lemma 7.5 and (7.2),

$$A_2 \le \frac{c(k)N}{\sqrt{h}}\left(\left\|\partial_1 P_{h,k}^{1,0} P_N v - \partial_1 P_N v\right\| + \|\partial_1 P_N v - \sqcap_{h,k}\partial_1 P_N v\|\right) \le c(k)\|v\|_{L^2(\Lambda_2, H^{\frac{5}{2}}(\Lambda_1))}.$$

By imbedding theory, (7.2) and (7.23),

$$A_3 \le c(k)\|v\|_{H^{\frac{3}{4}}(\Lambda_2, H^{\frac{3}{2}+\delta}(\Lambda_1))}, \quad \delta > 0.$$

Similarly, we have that

$$\left\|\partial_2 P_{h,k}^{1,0} P_N^{1,0} v\right\|_{L^\infty} \le B_1 + B_2 + B_3 \tag{7.28}$$

where

$$\begin{aligned} B_1 &= \left\|P_{h,k}^{1,0}\partial_2 P_N^{1,0} v - \sqcap_{h,k}\partial_2 P_N^{1,0} v\right\|_{L^\infty}, \\ B_2 &= \left\|\sqcap_{h,k}\left(\partial_2 P_N^{1,0} v - P_N\partial_2 v\right)\right\|_{L^\infty}, \\ B_3 &= \|\sqcap_{h,k} P_N \partial_2 v\|_{L^\infty}. \end{aligned}$$

It can be verified that

$$\begin{aligned} B_1 &\le \frac{c(k)N}{\sqrt{h}}\left\|\left(P_{h,k}^{1,0} - \sqcap_{h,k}\right)\left(\partial_2 P_N^{1,0} v\right)\right\| \le c(k)\|v\|_{H^1(\Lambda_2, H^{\frac{3}{2}}(\Lambda_1))}, \\ B_2 &\le \frac{c(k)N}{\sqrt{h}}\left(\left\|\partial_2 P_N^{1,0} v - \partial_2 v\right\| + \|P_N\partial_2 v - \partial_2 v\|\right) \le c(k)\|v\|_{H^{\frac{5}{2}}(\Lambda_2, L^2(\Lambda_1))} \end{aligned}$$

and

$$B_3 \le c(k)\|v\|_{H^{\frac{7}{4}}(\Lambda_2, H^{\frac{1}{2}+\delta}(\Lambda_1))}, \quad \delta > 0.$$

By substituting the above estimates into (7.27) and (7.28), we assert that

$$\left|P_{h,k}^{1,0} P_N^{1,0} v\right|_{W^{1,\infty}} \le c(k)\|v\|_{M^{\frac{5}{2},\frac{5}{2}} \cap X^{\frac{1}{2},\frac{1}{2}} \cap X^{r,\frac{3}{4}}}. \tag{7.29}$$

Finally we complete the proof using (7.26) and (7.29). ■

We now deal with the combined Legendre pseudospectral-finite element approximation. Let I_N be the Legendre-Gauss-Lobatto interpolation on the interval Λ_2, associated with the interpolation points $x_2^{(j)}$ and the weights $\omega^{(j)}, 0 \le j \le N$. Define $I_{h,k,N} = P_{h,k} I_N = I_N P_{h,k}$. Set

$$(v, w)_N = \sum_{j=0}^{N}\int_{\Lambda_1} v\left(x_1, x_2^{(j)}\right) w\left(x_1, x_2^{(j)}\right)\omega^{(j)}\,dx_1, \quad \|v\|_N = (v, v)_N^{\frac{1}{2}}.$$

Then for any $v \in C\left(\Lambda_2, L^2(\Lambda_1)\right)$,

$$(v - I_{h,k,N} v, \phi)_N = 0, \quad \forall\, \phi \in \tilde{V}_{h,k,N}.$$

Theorem 7.14. *If $v \in H^{r,s}(\Omega)$ and $r, s > \frac{1}{2}$, then*

$$\|v - I_{h,k,N}v\| \le c(k)\left(h^{\bar{r}} + N^{-s}\right)\|v\|_{H^{r,s}}.$$

Proof. The conclusion follows from Lemma 7.3 and (2.63). ∎

7.4. Combined Chebyshev-Finite Element Approximations

This section is devoted to the combined Chebyshev-finite element approximations. Let Λ_1, Λ_2 and Ω be the same as in Section 7.3. Also, the notations $\tilde{S}_{h,k,1}, S_{h,k,1}, S^0_{h,k,1}$, $\mathbb{P}_{N,2}, \mathbb{P}^0_{N,2}, \tilde{V}_{h,k,N}, V_{h,k,N}, V^0_{h,k,N}$ and $P_{h,k}, P^1_{h,k}, P^{1,0}_{h,k}$ have the same meanings as in that section. Let $\omega(x) = \omega(x_2) = (1 - x_2^2)^{-\frac{1}{2}}$. Define

$$(v, w)_\omega = \int_\Omega v(x)w(x)\omega(x_2)\,dx, \quad \|v\|_\omega = (v, v)^{\frac{1}{2}}_\omega.$$

We still use the notations $L^p_\omega(\Omega), H^r_\omega(\Omega), H^r_{0,\omega}(\Omega), W^{r,p}_\omega(\Omega), H^{r,s}_\omega(\Omega), H^{r,s}_{0,\omega}(\Omega), M^{r,s}_\omega(\Omega)$ and $X^{r,s}_\omega(\Omega)$ as before. Moreover, P_N, P^1_N and $P^{1,0}_N$ stand for the $L^2_\omega(\Lambda_2)$-orthogonal projection on $\mathbb{P}_{N,2}$, the $H^1_\omega(\Lambda_2)$-orthogonal projection on $\mathbb{P}_{N,2}$ and the $H^1_{0,\omega}(\Lambda_2)$-orthogonal projection on $\mathbb{P}^0_{N,2}$ respectively. We first give some inverse inequalities.

Theorem 7.15. *For any $\phi \in \tilde{V}_{h,k,N}$ and $2 \le p \le \infty$,*

$$\|\phi\|_{L^p_\omega} \le c\left(\frac{h}{k^2 N}\right)^{\frac{1}{p}-\frac{1}{2}}\|\phi\|_\omega.$$

Proof. By Lemma 7.1 and Theorem 2.13, we have that

$$\begin{aligned}\|\phi\|_{L^\infty} &\le \sup_{x_2\in\Lambda_2}\|\phi\|_{L^\infty(\Lambda_1)} \le \frac{ck}{\sqrt{h}}\sup_{x_2\in\Lambda_2}\|\phi\|_{L^2(\Lambda_1)} \le \frac{ck}{\sqrt{h}}\left(\int_{\Lambda_1}\sup_{x_2\in\Lambda_2}\phi^2(x)\,dx_1\right)^{\frac{1}{2}}\\ &\le ck\sqrt{\frac{N}{h}}\left(\int_{\Lambda_1}\|\phi(x_1,\cdot)\|^2_{L^2_\omega(\Lambda_2)}\,dx_1\right)^{\frac{1}{2}} = ck\sqrt{\frac{N}{h}}\|\phi\|_\omega.\end{aligned}$$

Therefore

$$\|\phi\|^p_{L^p_\omega} \le \|\phi\|^{p-2}_{L^\infty}\|\phi\|^2_\omega \le c\left(\frac{k^2N}{h}\right)^{\frac{p}{2}-1}\|\phi\|^p_\omega$$

and the desired result follows. ∎

Theorem 7.16. *For any $\phi \in \tilde{V}_{h,k,N}$ and $0 \le r \le \mu$,*

$$\|\phi\|_{\mu,\omega} \le c\left(\frac{h}{k^2} + \frac{1}{N^2}\right)^{r-\mu}\|\phi\|_{r,\omega}.$$

Proof. Let $T_l(x_2)$ be the Chebyshev polynomial of degree l on Λ_2, and

$$\phi(x) = \sum_{l=0}^{N} \phi_l(x_1) T_l(x_2).$$

Then $\phi_l(x_1) \in \tilde{S}_{h,k,1}$. By (7.9),

$$\begin{aligned} \|\partial_1^\mu \phi\|_\omega^2 &= \frac{\pi}{2} \sum_{l=0}^{N} c_l \|\partial_1^\mu \phi_l\|_{L^2(\Lambda_1)}^2 \le c\left(\frac{h}{k^2}\right)^{2r-2\mu} \sum_{l=0}^{N} c_l \|\phi_l\|_{H^r(\Lambda_1)}^2 \\ &\le c\left(\frac{h}{k^2}\right)^{2r-2\mu} \|\phi\|_{r,\omega}^2, \quad \mu = 0, 1. \end{aligned}$$

On the other hand,

$$\partial_2 \phi(x) = \sum_{l=1}^{N} \phi_l(x_1) \partial_2 T_l(x_2).$$

By (2.72) and an argument as in the proof of Theorem 2.14,

$$\|\partial_2 \phi(x_1, \cdot)\|_{L_\omega^2(\Lambda_2)} \le c N^2 \|\phi(x_1, \cdot)\|_{L_\omega^2(\Lambda_2)}.$$

Therefore

$$\|\partial_2 \phi\|_\omega^2 \le c N^4 \int_{\Lambda_1} \|\phi(x_1, \cdot)\|_{L_\omega^2(\Lambda_2)}^2 \, dx_1 \le c N^4 \|\phi\|_\omega^2.$$

The previous statements imply the result for $\mu = 1$ and $r = 0$. The proof is complete by induction and space interpolation. ∎

The $L_\omega^2(\Omega)$-orthogonal projection $P_{h,k,N} : L_\omega^2(\Omega) \to \tilde{V}_{h,k,N}$ is such a mapping that for any $v \in L_\omega^2(\Omega)$,

$$(v - P_{h,k,N} v, \phi)_\omega = 0, \quad \forall \phi \in \tilde{V}_{h,k,N}.$$

Theorem 7.17. *If $v \in H_\omega^{r,s}(\Omega), r > \frac{1}{2}$ and $s \ge 0$, then*

$$\|v - P_{h,k,N} v\|_\omega \le c(k) \left(h^{\bar{r}} + N^{-s}\right) \|v\|_{H_\omega^{r,s}}.$$

Proof. Using Theorem 2.15 and Lemma 7.3, we find that

$$\begin{aligned} \|v - P_{h,k,N} v\|_\omega &\le \|v - P_N P_{h,k} v\|_\omega \le \|P_N (v - P_{h,k})\|_\omega + \|v - P_N v\|_\omega \\ &\le c(k) \left(h^{\bar{r}} + N^{-s}\right) \|v\|_{H_\omega^{r,s}}. \end{aligned}$$

∎

In particular, for $v \in H^{r,s}_{0,\omega}(\Omega)$,

$$\|v - P_{h,k,N}v\|_\omega \le c(k)\left(h^{\bar{r}} + N^{-s}\right)|v|_{H^{\bar{r},s}_\omega}.$$

The $H^1_\omega(\Omega)$-orthogonal projection $P^1_{h,k,N} : H^1_\omega(\Omega) \to V_{h,k,N}$ is such a mapping that for any $v \in H^1_\omega(\Omega)$,

$$\left(\nabla\left(v - P^1_{h,k,N}v\right), \nabla\phi\right)_\omega + (v,\phi)_\omega = 0, \quad \forall\, \phi \in V_{h,k,N}.$$

The $H^1_{0,\omega}(\Omega)$-orthogonal projection $P^{1,0}_{h,k,N} : H^1_{0,\omega}(\Omega) \to V^0_{h,k,N}$ is such a mapping that for any $v \in H^1_{0,\omega}(\Omega)$,

$$\left(\nabla\left(v - P^{1,0}_{h,k,N}v\right), \nabla(\phi\omega)\right) = 0, \quad \forall\, \phi \in V^0_{h,k,N}.$$

Theorem 7.18. *If $v \in M^{r,s}_\omega(\Omega)$ and $r, s \ge 1$, then*

$$\left\|v - P^1_{h,k,N}v\right\|_{\mu,\omega} \le c(k)\left(h^{\bar{r}-1} + h^{-1}N^{-s} + N^{1-s}\right)\left(h^{1-\mu} + N^{\mu-1}\right)\|v\|_{M^{\bar{r},s}_\omega}, \quad \mu = 0, 1.$$

If, in addition, (7.10) holds, then

$$\left\|v - P^1_{h,k,N}\right\|_\omega \le c(k)\left(h^{\bar{r}} + N^{-s}\right)\|v\|_{M^{\bar{r},s}_\omega}.$$

Proof. Let $v_* = P_{h,k}P^1_N v \in V_{h,k,N}$. Then

$$\left\|v - P^1_{h,k,N}v\right\|_{1,\omega} \le \|v - v_*\|_{1,\omega} \le \|v - P_{h,k}v\|_{1,\omega} + \left\|P_{h,k}\left(v - P^1_N v\right)\right\|_{1,\omega}.$$

By Lemma 7.3,

$$\begin{aligned}\|v - P_{h,k}v\|^2_{1,\omega} &= \int_{\Lambda_2}\left(\|v - P_{h,k}v\|^2_{H^1(\Lambda_1)} + \|\partial_2 v - P_{h,k}\partial_2 v\|^2_{L^2(\Lambda_1)}\right)\omega(x_2)\,dx_2 \\ &\le c(k)h^{2\bar{r}-2}\left(\|v\|^2_{L^2_\omega(\Lambda_2,H^{\bar{r}}(\Lambda_1))} + \|v\|^2_{H^1_\omega(\Lambda_2,H^{\bar{r}-1}(\Lambda_1))}\right). \qquad (7.30)\end{aligned}$$

By Lemma 7.2, Theorem 2.16 and projection theorem,

$$\begin{aligned}\left\|P_{h,k}\left(v - P^1_N v\right)\right\|^2_{1,\omega} &= \int_{\Lambda_2}\left\|P_{h,k}\left(v - P^1_N v\right)\right\|^2_{H^1(\Lambda_1)}\omega(x_2)\,dx_2 \\ &\quad + \int_{\Lambda_2}\left\|P_{h,k}\partial_2\left(v - P^1_N v\right)\right\|^2_{L^2(\Lambda_1)}\omega(x_2)\,dx_2 \\ &\le c(k)h^{-2}N^{-2s}\|v\|^2_{H^s_\omega(\Lambda_2,L^2(\Lambda_1))} \\ &\quad + c(k)N^{2-2s}\|v\|^2_{H^s_\omega(\Lambda_2,L^2(\Lambda_1))}. \qquad (7.31)\end{aligned}$$

Therefore the first result is true for $\mu = 1$. The rest of the proof is clear. ∎

Theorem 7.19. *If $v \in M_\omega^{r,s}(\Omega) \cap H^1_{0,\omega}(\Omega)$ and $r, s \geq 1$, then for $\mu = 0, 1$,*

$$\left\| v - P^{1,0}_{h,k,N} v \right\|_{\mu,\omega} \leq c(k) \left(h^{\bar{r}-1} + h^{-1} N^{-s} + N^{1-s} \right) \left(h^{1-\mu} + N^{\mu-1} \right) |v|_{M_\omega^{r,s}}.$$

If, in addition, (7.10) holds, then

$$\left\| v - P^{1,0}_{h,k,N} v \right\|_{\omega} \leq c(k) \left(h^{\bar{r}} + N^{-s} \right) |v|_{M_\omega^{r,s}}.$$

Proof. We have from Lemma 2.8 that

$$\left\| v - P^{1,0}_{h,k,N} v \right\|^2_{1,\omega} \leq c \inf_{w \in V^0_{h,k,N}} \|v - w\|^2_{1,\omega}.$$

Take $w = v_* = P_{h,k} P_N^{1,0} v \in V^0_{h,k,N}$. Then by Lemmas 7.2 and 7.3, and Theorem 2.18, we can follow the same line as in the derivations of (7.30) and (7.31) to get the conclusion. ∎

We now estimate the norm $\left\| P^{1,0}_{h,k,N} v \right\|_{W_\omega^{r,p}}$.

Theorem 7.20. *Let (7.10) hold, $k \geq 1$ and $r, s > \frac{1}{2}$. For any $v \in H^1_0(\Omega) \cap H^s_\omega(\Lambda_2, H^r(\Lambda_1))$,*

$$\left\| P^{1,0}_{h,k,N} v \right\|_{L^\infty} \leq c(k) \|v\|_{H^1_\omega \cap H^s_\omega(\Lambda_2, H^r(\Lambda_1))}.$$

If, in addition, $v \in M_\omega^{2,2}(\Omega) \cap X_\omega^{r,s}(\Omega)$, then

$$\left\| P^{1,0}_{h,k,N} v \right\|_{W^{1,\infty}} \leq c(k) \|v\|_{M_\omega^{2,2} \cap X_\omega^{r,s}}.$$

Proof. We have

$$\left\| P^{1,0}_{h,k,N} v \right\|_{L^\infty} \leq \left\| P^{1,0}_{h,k,N} v - \sqcap_{h,k} P_N v \right\|_{L^\infty} + \|\sqcap_{h,k} P_N v\|_{L^\infty}.$$

Moreover, using Theorems 2.15, 7.15 and 7.19, and (7.2), we get that

$$\begin{aligned} \left\| P^{1,0}_{h,k,N} v - \sqcap_{h,k} P_N v \right\|_{L^\infty} &\leq c(k) \sqrt{\frac{N}{h}} \left\| P^{1,0}_{h,k,N} v - \sqcap_{h,k} P_N v \right\|_{\omega} \\ &\leq c(k) \sqrt{\frac{N}{h}} \left(\left\| P^{1,0}_{h,k,N} - v \right\|_{\omega} + \|v - \sqcap_{h,k} P_N v\|_{\omega} \right) \leq c(k) \|v\|_{1,\omega}. \end{aligned}$$

On the other hand, we use an $L^\infty(\Lambda_2)$-estimate in Section 2.5 to obtain that for any $\mu > 0$,

$$\begin{aligned} \|P_N v\|_{L^\infty(\Lambda_2)} &\leq c \, \|P_N v - v\|_{L^\infty(\Lambda_2)} + \|v\|_{L^\infty(\Lambda_2)} \\ &\leq c(1 + \ln N) N^{-\mu} \|v\|_{W_\omega^{\mu,\infty}(\Lambda_2)} + \|v\|_{L^\infty(\Lambda_2)}. \end{aligned}$$

Taking $\mu < s - \frac{1}{2}$, we get

$$\|P_N v\|_{L^\infty(\Lambda_2)} \le \|v\|_{H^s_\omega(\Lambda_2)}. \tag{7.32}$$

Therefore, we have from imbedding theory that

$$\|\sqcap_{h,k} P_N v\|_{L^\infty} \le c(k)\,\|P_N v\|_{L^\infty} \le \|v\|_{H^s_\omega(\Lambda_2, H^r(\Lambda_1))}.$$

The above statements lead to the first conclusion.

We next prove the second result. Let $v_* = P^{1,0}_{h,k,N} v - P^{1,0}_{h,k} P^{1,0}_N v \in V^0_{h,k,N}$. Then by Lemma 2.8,

$$\begin{aligned}
\frac{1}{4}\|v_*\|^2_{1,\omega} &\le \left(\nabla\left(P^{1,0}_{h,k,N} v - P^{1,0}_{h,k} P^{1,0}_N v\right), \nabla(v_*\omega)\right) = \left(\nabla\left(v - P^{1,0}_{h,k} P^{1,0}_N v\right), \nabla(v_*\omega)\right) \\
&\le \left(\nabla\left(v - P^{1,0}_N v\right), \nabla(v_*\omega)\right) + \left(\nabla P^{1,0}_N \left(v - P^{1,0}_{h,k} v\right), \nabla(v_*\omega)\right) \\
&= \left(\partial_1\left(v - P^{1,0}_N v\right), \partial_1 v_*\right)_\omega + \left(\partial_2\left(v - P^{1,0}_{h,k} v\right), \partial_2(v_*\omega)\right) \\
&\le |v_*|_{1,\omega}\left(\left\|\partial_1\left(v - P^{1,0}_N v\right)\right\|_\omega + \left\|\partial_2\left(v - P^{1,0}_{h,k} v\right)\right\|_\omega\right).
\end{aligned}$$

Thus by Theorem 2.18 and Lemma 7.5,

$$|v_*|_{1,\omega} \le c(k)\left(N^{-1} + h\right)\|v\|_{H^1_\omega(\Lambda_2, H^1(\Lambda_1))}$$

from which and Theorem 7.15, we assert that

$$\|v_*\|_{W^{1,\infty}} \le c(k)\sqrt{\frac{N}{h}}\,\|v_*\|_{1,\omega} \le c(k)\|v\|_{H^1_\omega(\Lambda_2, H^1(\Lambda_1))}. \tag{7.33}$$

On the other hand,

$$\begin{aligned}
\left\|\partial_1 P^{1,0}_{h,k} P^{1,0}_N v\right\|_{L^\infty} &\le A_1 + A_2 + A_3, \\
\left\|\partial_2 P^{1,0}_{h,k} P^{1,0}_N v\right\|_{L^\infty} &\le B_1 + B_2 + B_3
\end{aligned}$$

where A_j and B_j are of the same forms as in (7.27) and (7.28). The only difference is that $P_N, P^{1,0}_N$ are now the orthogonal projections associated with the norms $\|\cdot\|_{L^2_\omega(\Lambda_2)}$ and $\|\cdot\|_{H^1_\omega(\Lambda_2)}$, respectively. Therefore by Theorems 2.15, 2.18 and 7.15, and Lemma 7.5,

$$A_1 \le c(k)\sqrt{\frac{N}{h}}\left\|\partial_1 P^{1,0}_{h,k}\left(P^{1,0}_N v - P_N v\right)\right\|_\omega \le c(k)\|v\|_{H^1_\omega(\Lambda_2, H^1(\Lambda_1))}.$$

By Theorems 2.15 and 7.15, Lemma 7.5 and (7.2),

$$A_2 \le c(k)\sqrt{\frac{N}{h}}\left\|\partial_1 P^{1,0}_{h,k} P_N v - \sqcap_{h,k}\partial_1 P_N v\right\|_\omega \le c(k)\|v\|_{L^2_\omega(\Lambda_2, H^2(\Lambda_1))}.$$

By imbedding theory, (7.2) and (7.32),

$$A_3 \le c(k)\|v\|_{H_\omega^{\frac{1}{2}+\delta}(\Lambda_2, H^{\frac{3}{2}+\delta}(\Lambda_1))}, \qquad \delta > 0.$$

In the same manner, we verify that

$$\begin{aligned} B_1 &\le c(k)\|v\|_{H_\omega^1(\Lambda_2, H^1(\Lambda_1))}, \\ B_2 &\le c(k)\|v\|_{H_\omega^2(\Lambda_2, L^2(\Lambda_1))}, \\ B_3 &\le c(k)\|v\|_{H_\omega^{\frac{3}{2}+\delta}(\Lambda_2, H^{\frac{1}{2}+\delta}(\Lambda_1))}, \qquad \delta > 0. \end{aligned}$$

The above six estimations together with (7.33) provide the bound of $\left\|P^{1,0}_{h,k,N}v\right\|_{W^{1,\infty}}$.
■

Finally we consider the combined Chebyshev pseudospectral-finite element approximation. Let I_N be the Chebyshev-Gauss-Lobatto interpolation on the interval Λ_2, associated with the interpolation points $x_2^{(j)}$ and the weights $\omega^{(j)}, 0 \le j \le N$. Define $I_{h,k,N} = P_{h,k}I_N = I_N P_{h,k}$. Set

$$(v, w)_N = \sum_{j=0}^{N} \int_{\Lambda_1} v\left(x_1, x_2^{(j)}\right) w\left(x_1, x_2^{(j)}\right) \omega^{(j)}\, dx_1, \qquad \|v\|_N = (v, v)_N^{\frac{1}{2}}.$$

Then for any $v \in C\left(\Lambda_2, L^2(\Lambda_1)\right)$,

$$\left(v - I_{h,k,N}v, \phi\right)_N = 0, \quad \forall\, \phi \in \tilde{V}_{h,k,N}.$$

Theorem 7.21. *If* $v \in H_\omega^{r,s}(\Omega)$ *and* $r, s > \frac{1}{2}$, *then*

$$\|v - I_{h,k,N}v\|_\omega \le c(k)\left(h^{\bar{r}} + N^{-s}\right)\|v\|_{H_\omega^{r,s}}.$$

Proof. The result follows from Lemma 7.3 and Theorem 2.19. ■

Theorems 7.17, 7.19 and 7.20 can be found in Guo, He and Ma (1995).

7.5. Applications

In this section, we again take the two-dimensional unsteady Navier-Stokes equation as an example to describe the combined spectral-finite element methods. We focus on the combined Fourier pseudospectral-finite element schemes. But the main idea and the technique used can be generalized to non-periodic problems defined on a three-dimensional complex domain, and to other kinds of combined spectral-finite element methods and combined pseudospectral-finite element methods presented in the previous sections.

Let U, P and $\mu > 0$ be the speed vector, the pressure and the kinetic viscosity. We consider the semi-periodic problem (6.22) and require $P(x,t)$ to be in the space $L^2_{0,p}(\Omega)$, i.e., P has the period 2π for x_2 and $A(P(x,t)) = 0$ for $0 \le t \le T$. The bilinear forms $a(v,w)$ and $b(v,w)$ are given by (6.21). We adopt the artificial compression (6.23) with the parameter $\beta \ge 0$. When $\beta = 0$, it is reduced to the original problem and so requires the incompressibility automatically. As explained in Section 6.3, a reasonable approximation of non-linear convective term plays an important role in preserving the global property of the genuine solution and in increasing the computational stability and the total accuracy of numerical solution. But it is not easy to do so for pseudospectral methods, since the interpolation and the differentiation do not commute. Besides, for the improvement of stability, a filtering will be used.

Now let $k \ge 0$ and $N > 0$ be integers. We use all the notations of Section 7.2. For simplicity, let $V_{h,N}$ be the set of vector functions and $W_{h,N}$ be the set of scalar functions, defined by

$$V_{h,N} = \left(S^0_{h,k+1,1} \otimes V_{N,2}\right)^2, \qquad W_{h,N} = \left\{w \mid w \in \tilde{S}_{h,k,1} \otimes V_{N,2}, A(w) = 0\right\}.$$

The interpolation $\sqcap_{h,k+1}$ is denoted by $\sqcap_{k+1}$ for simplicity. Furthermore, for

$$v(x) = \sum_{l=0}^{N} v_l(x_1)e^{ilx_2},$$

the filtering with single parameter $\alpha \ge 1$ is given by

$$R_N v = R_N(\alpha)v = \sum_{l=0}^{N} \left(1 - \left|\frac{l}{N}\right|^{\alpha}\right) v_l(x_1)e^{ilx_2}.$$

Let

$$D(v,w,z) = \frac{1}{2}((w \cdot \nabla)v, z) - \frac{1}{2}((w \cdot \nabla)z, v).$$

If $\nabla \cdot w = 0$, then $D(v,w,z) = ((w \cdot \nabla)v, z)$. Thus we define the following trilinear form

$$d(v,w,z) = \frac{1}{2}\left(I_N((w \cdot \nabla)v), z\right) - \frac{1}{2}\left(I_N((w \cdot \nabla)z), v\right).$$

We shall approximate the nonlinear term by $d(v,v,z)$. Clearly

$$d(v,w,z) + d(z,w,v) = 0. \tag{7.34}$$

Guo and Ma (1993) proposed a scheme for (6.22). Let τ be the mesh size in time. $u_{h,N}$ and $p_{h,N}$ are the approximations to U and P, respectively. δ, σ and θ are parameters, $0 \le \delta, \sigma, \theta \le 1$. The combined Fourier pseudospectral-finite element scheme for (6.22) is to find $u_{h,N}(x,t) \in V_{h,N}, p_{h,N}(x,t) \in W_{h,N}$ for all $t \in \bar{R}_\tau(T)$ such that

$$\begin{cases} (D_\tau u_{h,N}(t), v) + d\left(R_N u_{h,N}(t) + \delta\tau R_N D_\tau u_{h,N}(t), R_N u_{h,N}(t), R_N v\right) \\ +\mu a\left(u_{h,N}(t) + \sigma\tau D_\tau u_{h,N}(t), v\right) - b\left(v, p_{h,N}(t) + \theta\tau D_\tau p_{h,N}(t)\right) \\ = \left(I_N \sqcap_{k+1} f(t), v\right), \qquad v \in V_{h,N},\ t \in \bar{R}_\tau(T), \\ \beta\left(D_\tau p_{h,N}(t), w\right) + b\left(u_{h,N}(t) + \theta\tau D_\tau u_{h,N}(t), w\right) = 0, \quad \forall\, w \in W_{h,N},\ t \in \bar{R}_\tau(T), \\ u_{h,N}(0) = I_N \sqcap_{k+1} U_0, \\ p_{h,N}(0) = I_N \sqcap_{k+1} P_0 \end{cases} \tag{7.35}$$

where P_0 is determined by

$$\Delta P_0 = \nabla \cdot (f - (U_0 \cdot \nabla) U_0).$$

If $\delta = \sigma = \theta = 0$, then (7.35) is an explicit scheme. Otherwise a linear iteration is needed to evaluate the values of unknown functions at the interpolation points. In particular, if $\delta = \sigma = \theta = \frac{1}{2}$, then we get from (7.34) and (7.35) that

$$\begin{aligned} &\|u_{h,N}(t)\|^2 + \beta\|p_{h,N}(t)\|^2 + \frac{1}{2}\mu\tau \sum_{s \in \bar{R}_\tau(t-\tau)} |u_{h,N}(s) + u_{h,N}(s+\tau)|_1^2 \\ &\quad = \|u_{h,N}(0)\|^2 + \beta\|p_{h,N}(0)\|^2 + \tau \sum_{s \in \bar{R}_\tau(t-\tau)} \left(I_N \sqcap_{k+1} f(s), u_{h,N}(s) + u_{h,N}(s+\tau)\right). \end{aligned}$$

Clearly it simulates the conservation (5.67) properly.

We analyze the generalized stability of scheme (7.35). Let $c_I = c_I(k)$ be a positive constant such that for any $v \in V_{h,N}$,

$$|v|_1^2 \le c_I\left(h^{-2} + N^2\right)\|v\|^2. \tag{7.36}$$

Assume that

$$\sigma > \frac{1}{2}, \quad \text{or} \quad \lambda = \tau\left(h^{-2} + N^2\right) < \frac{2}{\mu c_I(1 - 2\sigma)}. \tag{7.37}$$

Let ε be a suitably small positive constant and $q_0 > 0$. We consider the following two cases,

$$\text{(i)}\ \ 2\theta\sigma \ge \theta + \sigma + 2\varepsilon \text{ and } 2\theta \ge 1 + 2\varepsilon + q_0, \tag{7.38}$$

(ii) $2\theta\sigma \le \theta + \sigma + 2\varepsilon$ and

$$2\theta \ge (1 + 2\varepsilon + q_0 + \mu\lambda c_I(\sigma + 2\varepsilon))\left(1 - \mu\lambda c_I\left(\frac{1}{2} - \sigma\right)\right)^{-1}. \tag{7.39}$$

Suppose that $u_{h,N}(0), p_{h,N}(0), I_N \sqcap_{k+1} f$ and the right-hand side of the second formula of (7.35) have the errors $\tilde{u}_0, \tilde{p}, \tilde{f}$ and $\tilde{g}$, respectively, which induce the errors of $u_{h,N}$ and $p_{h,N}$, denoted by $\tilde{u}$ and $\tilde{p}$ for simplicity. Then from (7.35), it follows that

$$\begin{cases} (D_\tau \tilde{u}(t), v) + d\left(R_N \tilde{u}(t) + \delta\tau R_N D_\tau \tilde{u}(t), R_N u_{h,N}(t) + R_N \tilde{u}(t), R_N v\right) \\ \quad +d\left(R_N u_{h,N}(t) + \delta\tau R_N D_\tau u_{h,N}(t), R_N \tilde{u}(t), R_N v\right) + \mu a\left(\tilde{u}(t)\right. \\ \quad \left.+\sigma\tau D_\tau \tilde{u}(t), v\right) - b\left(v, \tilde{p}(t) + \theta\tau D_\tau \tilde{p}(t)\right) = \left(\tilde{f}(t), v\right), \quad v \in V_{h,N},\ t \in \bar{R}_\tau(T), \\ \beta\left(D_\tau \tilde{p}(t), w\right) + b\left(\tilde{u}(t) + \theta\tau D_\tau \tilde{u}(t), w\right) = \left(\tilde{g}(t), w\right), \quad \forall\, w \in W_{h,N},\ t \in \bar{R}_\tau(T). \end{cases} \tag{7.40}$$

Take $v = \tilde{u} + \theta\tau D_\tau \tilde{u}$ in the first formula of (7.40). Since

$$|2\theta\tau\left(D_\tau \tilde{u}(t), f(t)\right)| \le \varepsilon\mu\tau^2 \left|D_\tau \tilde{u}(t)\right|_1^2 + \frac{c\theta^2}{\varepsilon\mu}\left\|\tilde{f}(t)\right\|_{-1}^2,$$

we get from (1.22) and (7.34) that

$$\begin{aligned} &D_\tau \|\tilde{u}(t)\|^2 + \tau(2\theta-1)\|D_\tau \tilde{u}(t)\|^2 + \mu(2-\varepsilon)|\tilde{u}(t)|_1^2 + \mu\tau(\sigma+\theta)D_\tau|\tilde{u}(t)|_1^2 \\ &\quad +\mu\tau^2(2\theta\sigma - \sigma - \theta - \varepsilon)|D_\tau \tilde{u}(t)|_1^2 - 2b\left(\tilde{u}(t) + \theta\tau D_\tau \tilde{u}(t), \tilde{p} + \theta\tau D_\tau \tilde{p}(t)\right) \\ &\quad +\sum_{j=1}^{3} F_j(t) \le \frac{c(1+\theta^2)}{\varepsilon\mu}\left\|\tilde{f}(t)\right\|_{-1}^2 \end{aligned} \tag{7.41}$$

where

$$\begin{aligned} F_1(t) &= 2d\left(R_N u_{h,N}(t) + \delta\tau R_N D_\tau u_{h,N}(t), R_N \tilde{u}(t), R_N \tilde{u}(t)\right), \\ F_2(t) &= 2\theta\tau d\left(R_N u_{h,N}(t) + \delta\tau R_N D_\tau u_{h,N}(t), R_N \tilde{u}(t), R_N D_\tau \tilde{u}(t)\right) \\ &\quad + 2\tau(\theta-\delta)d\left(R_N \tilde{u}(t), R_N u_{h,N}(t), R_N D_\tau \tilde{u}(t)\right), \\ F_3(t) &= 2\tau(\theta-\delta)d\left(R_N \tilde{u}(t), R_N \tilde{u}(t), R_N D_\tau \tilde{u}(t)\right). \end{aligned}$$

Next take $w = \tilde{p} + \theta\tau D_\tau \tilde{p}$ in the second formula of (7.40). We obtain that

$$\begin{aligned} &\beta D_\tau \|\tilde{p}(t)\|^2 + \beta\tau(2\theta - 1 - \varepsilon)\|D_\tau \tilde{p}(t)\|^2 \\ &\quad +2b\left(\tilde{u}(t) + \theta\tau D_\tau \tilde{u}(t), \tilde{p}(t) + \theta\tau D_\tau \tilde{p}(t)\right) \le \beta\|\tilde{p}(t)\|^2 + \left(\frac{1}{\beta} + \frac{\tau\theta^2}{\varepsilon\beta}\right)\|\tilde{g}(t)\|^2. \end{aligned} \tag{7.42}$$

Putting (7.41) and (7.42) together, we derive that

$$\begin{aligned} &D_\tau\left(\|\tilde{u}(t)\|^2 + \beta\|\tilde{p}(t)\|^2\right) + \tau(2\theta-1-\varepsilon)\left(\|D_\tau \tilde{u}(t)\|^2 + \beta\|D_\tau \tilde{p}(t)\|^2\right) \\ &\quad +\mu(2-\varepsilon)|\tilde{u}(t)|_1^2 + \mu\tau(\sigma+\theta)D_\tau|\tilde{u}(t)|_1^2 + \mu\tau^2(2\theta\sigma - \sigma - \theta - \varepsilon)|D_\tau \tilde{u}(t)|_1^2 \\ &\quad +\sum_{j=1}^{3} F_j(t) \le \beta\|\tilde{p}(t)\|^2 + \frac{c}{\varepsilon\mu}\left\|\tilde{f}(t)\right\|_{-1}^2 + \left(\frac{1}{\beta} + \frac{\tau\theta^2}{\varepsilon\beta}\right)\|\tilde{g}(t)\|^2. \end{aligned} \tag{7.43}$$

We are going to estimate $|F_j(t)|$. Let $v, w, z \in V_{h,N}$ and

$$B_q(v,w,z) = |(I_N(w\partial_q v), z)| + |(I_N(w\partial_q z), v)|, \quad q = 1,2.$$

By virtue of (2.23), Theorems 2.1, and Theorem 2.4,

$$\begin{aligned} B_q(v,w,z) &\leq |(w\partial_q v, z)_N| + |(w\partial_q z, v)_N| \\ &\leq c(k)\|w\|(\|I_N(z\partial_q v)\| + \|v\|_{L^\infty}|z|_1) \\ &\leq c(k)\|w\|\left(\|z\partial_q v\| + \|v\|_{L^\infty}|z|_1\right). \end{aligned}$$

Moreover, for any $0 < \gamma < \frac{1}{2}$,

$$\begin{aligned} \|z\partial_q v\| &\leq \left(\int_{\Lambda_1} \|\partial_q v\|^2_{L^{\frac{2}{1-2\gamma}}(\Lambda_2)} \|z\|^2_{L^{\frac{1}{\gamma}}(\Lambda_2)} dx_1\right)^{\frac{1}{2}} \\ &\leq c(k)\left(\int_{\Lambda_1} \|\partial_q v\|^2_{H^\gamma(\Lambda_2)} \|z\|^2_{H^{\frac{1}{2}-\gamma}(\Lambda_2)} dx_1\right)^{\frac{1}{2}} \leq c(k)\,\|\partial_q v\|_{H^\gamma(\Lambda_2,L^2(\Lambda_1))}\,|z|_1. \end{aligned}$$

Also by imbedding theory, $H^{r,s}(\Omega) \hookrightarrow C(\bar{\Omega})$ for $\dfrac{1}{r} + \dfrac{1}{s} < 2$. Therefore

$$B_q(v,w,z) \leq c(k)\|v\|_{M^{1,\gamma+1}}\|w\||z|_1. \tag{7.44}$$

Similarly

$$B_q(v,w,z) \leq c(k)\|v\|\|w\|_{M^{1,\gamma+1}}|z|_1. \tag{7.45}$$

By Theorems 2.1 and 2.4, Lemma 4.1,(2.23) and integration by parts, we can prove that

$$\begin{aligned} B_q(v,w,z) &= |(w\partial_q v, z)_N| + |(\partial_q I_N(vw), z)| \leq c(k)(|v|_1\|w\|_{L^\infty} + |vw|_1)\|z\| \\ &\leq c(k)\left(\|v\|_{L^\infty}|w|_1 + |v|_1\|w\|_{L^\infty}\right)\|z\|. \end{aligned} \tag{7.46}$$

Using Theorem 2.26, (7.44) and (7.45), it is not difficult to see that

$$\begin{aligned} |F_1(t)| &\leq \varepsilon\mu\,|\tilde{u}(t)|_1^2 + \frac{c(k)}{\varepsilon\mu}\,\|u_{h,N}\|^2_{C(0,T;M^{1,\gamma+1})}\,\|\tilde{u}(t)\|^2, \\ |F_2(t)| &\leq \varepsilon\mu\tau^2\,|D_\tau\tilde{u}(t)|_1^2 + \frac{c(k)}{\varepsilon\mu}\left(\theta^2 + (\theta-\delta)^2\right)\|u_{h,N}\|^2_{C(0,T;M^{1,\gamma+1})}\,\|\tilde{u}(t)\|^2. \end{aligned}$$

Furthermore, we use Theorem 2.26, Theorem 7.1 and (7.46) to get that

$$|F_3(t)| \leq \varepsilon\tau\,\|D_\tau\tilde{u}(t)\|^2 + \frac{c(k)\tau N(\theta-\delta)^2}{\varepsilon h}\,\|\tilde{u}(t)\|^2\,|\tilde{u}(t)|_1^2.$$

By substituting the above estimates into (7.43), we obtain that

$$\begin{aligned} &D_\tau\left(\|\tilde{u}(t)\|^2+\beta\|\tilde{p}(t)\|^2\right)+\tau(2\theta-1-2\varepsilon)\left(\|D_\tau\tilde{u}(t)\|^2+\beta\|D_\tau\tilde{p}(t)\|^2\right)\\ &\quad+\mu|\tilde{u}(t)|_1^2+\mu\tau(\sigma+\theta)D_\tau|\tilde{u}(t)|_1^2+\mu\tau^2(2\theta\sigma-\sigma-\theta-2\varepsilon)|D_\tau\tilde{u}(t)|_1^2\\ &\qquad\le A_1\left(\|\tilde{u}(t)\|^2+\beta\|\tilde{u}(t)\|^2\right)+A_2(t)|\tilde{u}(t)|_1^2+A_3(t)\end{aligned}$$

where

$$\begin{aligned} A_1&=\frac{c(k)}{\varepsilon\mu}\left(1+\theta^2+(\theta-\delta)^2\right)\|u_{h,N}\|^2_{C(0,T;M^{1,\gamma+1})}+1,\\ A_2(t)&=-\mu(1-2\varepsilon)+\frac{c(k)\tau N}{\varepsilon h}(\theta-\delta)^2\|\tilde{u}(t)\|^2,\\ A_3(t)&=\frac{c(k)}{\varepsilon\mu}\left\|\tilde{f}(t)\right\|_{-1}^2+\left(\frac{1}{\beta}+\frac{\tau\theta^2}{\varepsilon\beta}\right)\|\tilde{g}(t)\|^2.\end{aligned}$$

In the case (7.38), the previous inequality reads

$$\begin{aligned} &D_\tau\left(\|\tilde{u}(t)\|^2+\beta\|\tilde{p}(t)\|^2\right)+q_0\tau\left(\|D_\tau\tilde{u}(t)\|^2+\beta\|D_\tau\tilde{p}(t)\|^2\right)+\mu|\tilde{u}(t)|_1^2\\ &\quad+\mu\tau(\sigma+\theta)D_\tau|\tilde{u}(t)|_1^2\le A_1\left(\|\tilde{u}(t)\|^2+\beta\|\tilde{p}(t)\|^2\right)+A_2(t)|\tilde{u}(t)|_1^2+A_3(t).\end{aligned}\tag{7.47}$$

In the case (7.39), we get from (7.36) and (7.37) that

$$\begin{aligned} &\tau(2\theta-1-2\varepsilon)\left(\|D_\tau\tilde{u}(t)\|^2+\beta\|D_\tau\tilde{p}(t)\|^2\right)\\ &\quad+\mu\tau^2(2\theta\sigma-\sigma-\theta-2\varepsilon)|D_\tau\tilde{u}(t)|_1^2\ge q_0\tau\left(\|D_\tau\tilde{u}(t)\|^2+\beta\|D_\tau\tilde{p}(t)\|^2\right)\end{aligned}$$

and thus (7.47) is still valid. Let $E(v,w,t)$ and $\rho(t)$ be the same as in (6.35) and (6.36). Then the summation of (7.47) for $t\in\bar{R}_\tau(T)$ yields

$$E(\tilde{u},\tilde{p},t)\le\rho(t)+\tau\sum_{s\in\bar{R}_\tau(t-\tau)}\left(A_1E(\tilde{u},\tilde{p},s)+A_2(s)|\tilde{u}(s)|_1^2\right).$$

Finally, we get from Lemma 3.2 the following result.

Theorem 7.22. *Assume that*

(i) (7.37) holds, and (7.38) or (7.39) is valid;

(ii) for certain $t_1\in R_\tau(T),\rho(t_1)e^{b_1t_1}\le\dfrac{\mu hc(k)}{\tau N(\theta-\delta)^2}$.

Then for all $t \in \bar{R}_\tau(t_1)$,

$$E(\tilde{u}, \tilde{p}, t) \le c\rho(t)e^{b_1 t},$$

b_1 *being some positive constant depending on* k, β, μ *and the norm* $\|u_{h,N}\|_{C(0,T;M^{1,\gamma+1})}$, $\gamma > 0$.

We know from Theorem 7.22 that if $\sigma > \frac{1}{2}$, then the restriction on λ disappears and so we can use larger τ. If $\theta = \delta > \frac{\sigma}{2\sigma-1}$ for $\sigma > \frac{1}{2}$, then the restriction on $\rho(t)$ also disappears and so the index of generalized stability, $s = -\infty$. This improves the stability essentially. Indeed, it is the best result. On the other hand, since we adopt the operator $d(v, w, z)$, the effect of the main nonlinear term is cancelled, i.e.,

$$d(\tilde{u}(t), \tilde{u}(t), \tilde{u}(t)) = 0.$$

This fact again shows the importance of a reasonable treatment with the leading nonlinear terms in the approximation of nonlinear problems.

We are going to analyze the convergence of scheme (7.35). In order to obtain the uniform optimal error estimation for $\beta \ge 0$, we first consider the corresponding Stokes problem, which is itself very important in fluid dynamics. Let V' be the dual space of $V = \left(H^1_{0,p}(\Omega)\right)^2$ and $W = L^2_{0,p}(\Omega)$. $\langle \cdot, \cdot \rangle$ stands for the duality between V' and V. Let $\eta \in V'$ and $\xi \in W' = W$. The Stokes problem is to find $u \in V$ and $p \in W$ such that

$$\begin{cases} \mu a(u, v) - b(v, p) = \langle \eta, v \rangle, & \forall\, v \in V, \\ b(u, w) = (\xi, w), & \forall\, w \in W. \end{cases} \tag{7.48}$$

Let $V^*_{h,N}$ be the trial function space of the speed vector u. We approximate u and p by $u^{h,N} \in V^*_{h,N}$ and $p^{h,N} \in W_{h,N}$ satisfying

$$\begin{cases} \mu a\left(u^{h,N}, v\right) - b\left(v, p^{h,N}\right) = \langle \eta, v \rangle, & \forall\, v \in V^*_{h,N}, \\ b\left(u^{h,N}, w\right) = (\xi, w), & \forall\, w \in W_{h,N}. \end{cases} \tag{7.49}$$

(7.50) is a combined Fourier pseudospectral-finite element approximation to (7.49). Following the same line as in Girault and Raviart (1979), it can be shown that (7.50) has a unique solution provided that the following BB condition (the inf-sup condition) is fulfilled,

$$\sup_{v \in V^*_{h,N}} \frac{b(v, w)}{\|v\|_1} \ge c\|w\|, \quad \forall\, w \in W_{h,N}. \tag{7.50}$$

This condition is fulfilled for $V^*_{h,N} = \left(S^0_{h,k+2,1} \otimes V_{N,2}\right)^2$, see Canuto, Fujii and Quarteroni (1983). If the components of trial functions belong to the different spaces $A_{h,N}$ and $B_{h,N}$, then it is denoted by $V^*_{h,N} = A_{h,N} \times B_{h,N}$. It was pointed out that (7.51) holds for $V^*_{h,N} = \left(S^0_{h,k+2,1} \otimes V_{N,2}\right) \times \left(S^0_{h,k+1,1} \otimes V_{N,2}\right)$. Guo and Ma (1993) improved this result essentially, stated as in the following lemma.

Lemma 7.6. *If* $V^*_{h,N} = \left(S^0_{h,k+1,1} \otimes V_{N,2}\right) \times \left(S^0_{h,1,1} \otimes V_{N,2}\right)$ *and (7.10) holds, then (7.51) is fulfilled.*

We define the linear operator $P_* : V \times W \to V^*_{h,N} \times W_{h,N}$ by $P_*(u,p) = (P_*u, P_*p) = \left(u^{h,N}, p^{h,N}\right)$ where $\left(u^{h,N}, p^{h,N}\right)$ is the solution of (7.50). Let $\tilde{r} = \min(r, k+2)$. Using Lemma 7.6 and an abstract approximation result in Girault and Raviart (1979), we obtain the following result.

Lemma 7.7. *Let* $V^*_{h,N} = V_{h,N}$. *(7.50) has a unique solution. Moreover if* $u \in M^{r,s}_p(\Omega) \cap H^1_{0,p}(\Omega)$, $p \in H^{r-1,s-1}_p(\Omega) \cap L^2_{0,p}(\Omega)$ *and* $r, s \geq 1$, *then*

$$\|u - P_*u\|_1 + \|p - P_*p\| \leq c(k)\left(h^{\tilde{r}-1} + N^{1-s}\right)\left(\|u\|_{M^{\tilde{r},s}} + \|p\|_{H^{\tilde{r}-1,s-1}}\right). \tag{7.51}$$

If $u \in H^{r,s}_p(\Omega) \cap H^1_{0,p}(\Omega), p \in H^{r-1,s-1}_p(\Omega) \cap L^2_{0,p}(\Omega)$ *and* $r, s \geq 1$, *then*

$$\|u - P_*u\| \leq c(k)\left(h^{\tilde{r}} + N^{-s}\right)\left(\|u\|_{H^{\tilde{r},s}} + \|p\|_{H^{\tilde{r}-1,s-1}}\right). \tag{7.52}$$

By (7.52) and Theorem 2.26, it can be shown that if $u \in M^{1,s}_p(\Omega) \cap H^1_0(\Omega), p \in H^{0,s-1}_p(\Omega) \cap L^2_{0,p}(\Omega)$ and $1 \leq s \leq \alpha$, then

$$\|R_N P_*u\|_{M^{1,s}} + \|R_N P_*p\|_{H^{s-1}(\Lambda_2,L^2(\Lambda_1))} \leq c(k)\left(\|u\|_{M^{1,s}} + \|p\|_{H^{s-1}(\Lambda_2,L^2(\Lambda_1))}\right). \tag{7.53}$$

We now deal with the convergence. Let $U^* = P_*U, P^* = P_*P$. We deduce from (6.22) that

$$\begin{cases} (D_\tau U^*(t), v) + d\left(R_N U^*(t) + \delta\tau R_N D_\tau U^*(t), R_N U^*(t), R_N v\right) \\ \quad +\mu a\left(U^*(t) + \sigma\tau D_\tau U^*(t), v\right) - b\left(v, P^*(t) + \theta\tau D_\tau P^*(t)\right) \\ \quad = \left(I_N \sqcap_{k+1} f(t), v\right) + \sum\limits_{j=1}^{7} G_j(v,t), \quad \forall\, v \in V_{h,N},\ t \in \bar{R}_\tau(T), \\ \beta\left(D_\tau P^*(t), w\right) + b\left(U^*(t) + \theta\tau D_\tau U^*(t), w\right) = \left(G_8(t), w\right), \quad \forall\, w \in W_{h,N},\ t \in \bar{R}_\tau(T), \end{cases} \tag{7.54}$$

where

$$\begin{aligned}
G_1(v,t) &= (D_\tau U(t) - \partial_t U(t), v), \\
G_2(v,t) &= (D_\tau U^*(t) - D_\tau U(t), v), \\
G_3(v,t) &= d(R_N U^*(t), R_N U^*(t), R_N v) - D(U(t), U(t), v), \\
G_4(v,t) &= \delta\tau d(R_N D_\tau U^*(t), R_N U^*(t), R_N v), \\
G_5(v,t) &= \sigma\mu\tau a(D_\tau U^*(t), v), \\
G_6(v,t) &= -\theta\tau b(v, D_\tau P^*(t)), \\
G_7(v,t) &= (f(t) - I_N \sqcap_{k+1} f(t), v), \\
G_8(v,t) &= \beta D_\tau P^*(t).
\end{aligned}$$

Let $\tilde{U} = u_{h,N} - U^*$ and $\tilde{P} = p_{h,N} - P^*$. Then by (7.35) and (7.55), we obtain that

$$\begin{cases}
\left(D_\tau \tilde{U}(t), v\right) + d\left(R_N \tilde{U}(t) + \delta\tau R_N D_\tau \tilde{U}(t), R_N U^*(t) + R_N \tilde{U}(t), Rv\right) \\
\quad + d\left(R_N U^*(t) + \delta\tau R_N D_\tau U^*(t), R_N \tilde{U}(t), R_N v\right) + \mu a\left(\tilde{U}(t) + \sigma\tau D_\tau \tilde{U}(t), v\right) \\
\quad - b\left(v, \tilde{P}(t) + \theta\tau D_\tau \tilde{P}(t)\right) = -\sum\limits_{j=1}^{7} G_j(v,t), \quad \forall\, v \in V_{h,N},\ t \in \bar{R}_\tau(T), \\
\beta\left(D_\tau \tilde{P}(t), w\right) + b\left(\tilde{U}(t) + \theta\tau D_\tau \tilde{U}(t), w\right) = -\left(G_8(t), w\right), \quad \forall\, w \in W_{h,N},\ t \in \bar{R}_\tau(T), \\
\tilde{U}(0) = I_N \sqcap_{k+1} U_0 - U^*(0), \\
\tilde{P}(0) = I_N \sqcap_{k+1} P_0 - P^*(0).
\end{cases} \tag{7.55}$$

We now estimate the terms on the right-hand side of (7.56). Firstly by (4.24) and (7.53),

$$\begin{aligned}
&\tau \sum_{s \in \bar{R}_\tau(t-\tau)} \|D_\tau U(s) - \partial_t U(s)\|_{-1}^2 \le c(k)\tau^2 \|U\|_{H^2(0,t;H^{-1})}^2, \\
&\tau \sum_{s \in \bar{R}_\tau(t-\tau)} \|D_\tau U^*(s) - D_\tau U(s)\|^2 \\
&\quad \le c(k)\left(h^{2\tilde{r}} + N^{-2s}\right)\left(\|U\|_{H^1(0,t;H^{\tilde{r},s})}^2 + \|P\|_{H^1(0,t;H^{\tilde{r}-1,s-1})}^2\right).
\end{aligned}$$

To estimate the nonlinear terms, we first let $v, w, z \in V_{h,N}$, $s \ge \gamma + \frac{3}{2}$ and $\gamma > 0$. Set

$$\begin{aligned}
A_q &= \left|\left((R_N I_N - I)(R_N w \partial_q R_N v), z\right)\right|, \qquad & q = 1,2, \\
B_q &= \left|\left(I_N(R_N w \partial_q R_N z) - R_N w \partial_q z, R_N v\right)\right|, \qquad & q = 1,2.
\end{aligned}$$

By Theorem 2.26 and an argument as in the proof of Lemma 10 of Guo and Ma (1991), we know that if $\alpha \ge \max(s-\mu, 1)$, $\mu = 0, -1$ and $s > \frac{1}{2}$, then for any

$v, w \in V_{N,2}$,

$$\begin{aligned}&\|R_N I_N (R_N v R_N w) - R_N v R_N w\|_{H^\mu(\Lambda_2)} \\ &\quad \le c(k) N^{\mu - s} \left(|v|_{H^s(\Lambda_2)} \|w\|_{H^{\gamma+\frac{1}{2}}(\Lambda_2)} + |w|_{H^s(\Lambda_2)} \|v\|_{H^{\gamma+\frac{1}{2}}(\Lambda_2)} \right). \end{aligned} \tag{7.56}$$

By virtue of imbedding theory and (7.57), we deduce that

$$\begin{aligned} A_q \;&\le\; \|(R_N I_N - I)(R_N w \partial_q R_N v)\|_{-1} \|z\|_1 \\ &\le\; c(k) N^{-s} \left(\int_{\Lambda_1} \left(|\partial_q v|^2_{H^{s-1}(\Lambda_2)} \|w\|^2_{H^{\gamma+\frac{1}{2}}(\Lambda_2)} \right.\right. \\ &\qquad \left.\left. + |w|^2_{H^{s-1}(\Lambda_2)} \|\partial_q v\|^2_{H^{\gamma+\frac{1}{2}}(\Lambda_2)} \right) dx_1 \right)^{\frac{1}{2}} \|z\|_1 \\ &\le\; c(k) N^{-s} \left(\|\partial_q v\|_{H^{s-1}(\Lambda_2, L^2(\Lambda_1))} \|w\|_{H^{\gamma+\frac{1}{2}}(\Lambda_2, H^{\gamma+\frac{1}{2}}(\Lambda_1))} \right. \\ &\qquad \left. + \|\partial_q v\|_{H^{\gamma+\frac{1}{2}}(\Lambda_2, L^2(\Lambda_1))} \|w\|_{H^{s-1}(\Lambda_2, H^{\gamma+\frac{1}{2}}(\Lambda_1))} \right) \|z\|_1. \end{aligned}$$

Similarly,

$$\begin{aligned} B_q \;&\le\; |(R_N I_N (R_N v R_N w) - R_N v R_N w, \partial_q z)| \\ &\le\; c(k) N^{-s} \left(\int_{\Lambda_1} \left(|v|^2_{H^s(\Lambda_2)} \|w\|^2_{H^{\gamma+\frac{1}{2}}(\Lambda_2)} + |w|^2_{H^s(\Lambda_2)} \|v\|^2_{H^{\gamma+\frac{1}{2}}(\Lambda_2)} \right) dx_1 \right)^{\frac{1}{2}} \|z\|_1 \\ &\le\; c(k) N^{-s} \left(\|v\|_{H^s(\Lambda_2, L^2(\Lambda_1))} \|w\|_{H^{\gamma+\frac{1}{2}}(\Lambda_2, H^{\gamma+\frac{1}{2}}(\Lambda_1))} \right. \\ &\qquad \left. + \|w\|_{H^s(\Lambda_2, L^2(\Lambda_1))} \|v\|_{H^{\gamma+\frac{1}{2}}(\Lambda_2, H^{\gamma+\frac{1}{2}}(\Lambda_1))} \right) \|z\|_1. \end{aligned}$$

The above arguments imply that

$$A_q + B_q \le c(k) N^{-s} \|v\|_{M^{1,s}} \|w\|_{M^{1,s}} \|z\|_1$$

from which and (7.54), it follows that

$$\begin{aligned}&|d(R_N U^*(t), R_N U^*(t), R_N v) - D(R_N U^*(t), R_N U^*(t), v)| \\ &\quad \le c(k) N^{-s} \left(\|U(t)\|_{M^{1,s}} + \|P(t)\|_{H^{s-1}(\Lambda_2, L^2(\Lambda_1))} \right)^2 \|v\|_1. \end{aligned} \tag{7.57}$$

Similar to (7.44) and (7.45), we have from Theorem 2.26 and (7.53) that for $r \ge 1$,

$$\begin{aligned}&|D(R_N U^*(t), R_N U^*(t), v) - D(U(t), U(t), v)| \\ &\quad \le c(k) \left(\|U(t)\|_{M^{1,\gamma+1}} + \|U^*(t)\|_{M^{1,\gamma+1}} \right) \|R_N U^*(t) - U(t)\| \, \|v\|_1 \\ &\quad \le c(k) \left(h^{\tilde{r}} + N^{-s} \right) \left(\|U(t)\|_{M^{\tilde{r},s}} + \|P(t)\|_{H^{\tilde{r}-1,s-1}} \right)^2 \|v\|_1. \end{aligned} \tag{7.58}$$

The estimates (7.58) and (7.59) lead to the bound for $\|G_3(v,t)\|_{-1}$. Furthermore we have from Theorem 2.26 and (7.45) that

$$|G_4(v,t)| \le c(k)\tau \, \|U^*(t)\|_{M^{1,\gamma+1}} \, \|D_\tau U^*(t)\| \, \|v\|_1.$$

Moreover, by (7.53) and (7.54),

$$\begin{aligned}
&\tau^3 \sum_{s\in \bar{R}_\tau(t-\tau)} \|U^*(s)\|^2_{M^{1,\gamma+1}} \, \|D_\tau U^*(s)\|^2 \\
&\quad \le c(k)\tau^2 \left(\|U\|^2_{C(0,t;M^{1,\gamma+1})} + \|P\|^2_{C(0,t;H^\gamma(\Lambda_2,L^2(\Lambda_1)))}\right)\left(\|U\|^2_{H^1(0,t;H^1)} + \|P\|^2_{H^1(0,t;L^2)}\right).
\end{aligned}$$

We can estimate $|G_5(v,t)|$ and $|G_6(v,t)|$ in the same manner. Also by Theorem 2.4 and (7.2),

$$\|G_7(v,t)\| \le c(k)\left(h^{\tilde{r}} + N^{-s}\right) \|f(t)\|_{H^{\tilde{r},s}} \|v\|.$$

In addition,

$$\begin{aligned}
\left\|\tilde{U}(0)\right\| &\le c(k)\left(h^{\tilde{r}} + N^{-s}\right)\left(\|U_0\|_{H^{\tilde{r},s}} + \|P_0\|_{H^{\tilde{r}-1,s-1}}\right), \\
\left\|\tilde{P}(0)\right\| &\le c(k)\left(\|U_0\|_1 + \|P_0\|_{H^{\frac{1}{2}+\delta}(\Lambda_2, H^{\frac{1}{2}+\delta}(\Lambda_1))}\right), \quad \delta > 0.
\end{aligned}$$

Finally we apply Theorem 7.22 to (7.56), and the convergence follows.

Theorem 7.23. *Assume that*

(i) (7.10) and condition (i) of Theorem 7.22 are fulfilled;

(ii) for certain constants $r \ge 1$ and $s > \frac{3}{2}$,

$$\begin{aligned}
&U \in C\left(0,T; M_p^{r,s} \cap H^1_{0,p}\right) \cap H^1\left(0,T; H^{r,s}\right) \cap H^2\left(0,T; H^{-1}\right), \\
&P \in C\left(0,T; L^2_{0,p}\right) \cap H^1\left(0,T; H_p^{r-1,s-1}\right), \\
&f \in C\left(0,T; H_p^{r,s}\right);
\end{aligned}$$

(iii) for certain $t_1 \in R_\tau(T)$ and certain positive constant b_2,

$$e^{b_2 t_1}\left(\tau^2 + h^{2\tilde{r}} + N^{-2s} + \beta\right) \le \frac{\mu h c(k)}{\tau N(\delta-\theta)^2}.$$

Then for all $t \in \bar{R}_\tau(t_1)$,

$$\|u_{h,N}(t) - U^*(t)\|^2 \le c e^{b_2 t}\left(\tau^2 + h^{2\tilde{r}} + N^{-2s} + \beta\right)$$

where b_2 is a certain positive constant depending on k, β, μ and the norms of U, P and f in the mentioned spaces.

We find that since we compare $(u_{h,N}, p_{h,N})$ to (U^*, P^*), the solution of the corresponding Stokes equation (7.50), the factor $\frac{1}{\beta}$ in the error estimates disappears. It improves the error estimates, and covers the special case $\beta = 0$. If we use this technique for the numerical method discussed in Section 6.3, we can also improve the corresponding results. On the other hand, if we take $\delta = \theta > \frac{1}{2}$, then $t_1 = T$. It means the global convergence of scheme (7.35).

Finally, we present some numerical results. We use (7.35) with $\sigma = \theta = 1, \delta = 0, k = 0$ and the uniform mesh size $h = \frac{2}{M}$ (MFPSFE). We also use the corresponding finite element scheme (FE) with the solution u_h. We compare the error $U - u_{h,N}$ and the error $U - u_h$. In particular, we compare the errors for the numerical simulations with very small viscosity $\mu = 10^{-6}$. For the description of the errors, let

$$\Omega_{M,N} = \left\{ x \,\middle|\, x_1 = mh, x_2 = \frac{l\pi}{N}, -\frac{M}{2} + 1 \le m \le \frac{M}{2} - 1, 1 \le l \le 2N \right\}$$

and

$$E^{(j)}(t) = \left(\frac{\sum\limits_{x \in \Omega_{M,N}} \left| U^{(j)}(x,t) - v^{(j)}(x,t) \right|^2}{\sum\limits_{x \in \Omega_{M,N}} \left| U^{(j)}(x,t) \right|^2} \right), \qquad v = u_{h,N} \text{ or } u_h.$$

We take the test function

$$U = (\partial_2 \psi, -\partial_1 \psi),$$
$$P = \exp\left(\cos x_1 + \cos 2x_2 + \frac{t}{5} \right)$$

where

$$\psi = x_1^2 (1 - x_1)^2 (\exp(\sin 3x_2) + \sin 6x_2)\, e^{\frac{1}{20}t}.$$

In all calculations, $M = 14, N = 8$ and $\tau = 0.002$. The numerical results with various values of the artificial compressibility parameter β and filtering parameter α are given in Tables 7.1 and 7.2. Evidently, the combined Fourier pseudospectral-finite element scheme provides better numerical results. It is indicated that a good choice of the filterings improves the stability. We also find that the errors decrease when β decreases. All of the above facts agree with the theoretical analysis.

Table 7.1. The errors $E^{(j)}(1.0), \beta = 10^{-4}$.

	α	$E^{(1)}(1.0)$	$E^{(2)}(1.0)$
MFPSFE	∞	$2.838E{-1}$	$3.426E{-1}$
	3	$9.229E{-2}$	$1.411E{-1}$
	2	$7.630E{-2}$	$1.165E{-1}$
FE		$5.797E{-1}$	$4.080E{-1}$

Table 7.2. The errors $E^{(j)}(1.0), \beta = 10^{-5}$.

	α	$E^{(1)}(1.0)$	$E^{(2)}(1.0)$
MFPSFE	∞	$1.243E{-1}$	$1.879E{-1}$
	4	$5.914E{-2}$	$4.585E{-2}$
	3	$5.931E{-2}$	$4.495E{-2}$
FE		$1.815E{-1}$	$2.948E{-1}$

The combined Fourier-finite element approximations were first used by Mercier and Raugel (1982). They have been used for various nonlinear problems, such as the steady Navier-Stokes equation in Canuto, Maday and Quarteroni (1984), the unsteady Navier-Stokes equation in Guo and Cao (1991, 1992a), Cao and Guo (1993a), Guo and Ma (1991, 1993), and Ma and Guo (1992), the compressible fluid flow in Guo and Cao (1992b). At the same time, the combined Chebyshev-finite element methods developed rapidly, with their applications to nonlinear problems, e.g., see Guo, He and Ma (1995), He and Guo (1995), and Guo, Ma and Hou (1996).

For the convenience of computations, the Fourier spectral-finite difference approximations are also used widely in practice. Some of early work was due to Murdock (1977, 1986), Ingham (1984, 1985), Beringen (1984), Milinazo and Saffman (1985). In particular, Guo (1987, 1988b) set up a framework for the numerical analysis of this approach, used for the equations describing incompressible fluid flows, compressible fluid flows and atmospheric flows, see Guo (1987,1988b), Guo and Xiong (1989), and Guo and Huang (1993). Also, the Fourier pseudospectral-finite difference approximations and their applications have been developed, e.g., see Guo and Zheng (1992), and Guo and Xiong (1994).

CHAPTER 8

SPECTRAL METHODS ON THE SPHERICAL SURFACE

In the previous chapters, we discussed various spectral methods in Descartes coordinates with their applications to numerical simulations in fluid dynamics, quantum mechanics, heat conduction and other topics. However, in meteorology, ocean science, potential magnetostatic field and some other fields, we have to consider problems defined on an elliptic plane, on a spherical surface or in a spherical gap. Several numerical algorithms have been developed for such problems, e.g., see Boyd (1978a, 1978b), Haltiner and Williams (1980), Jarraud and Baede (1985), Bramble and Pasciak (1985), and Dennis and Quartapelle (1985). Some of them are related to the spectral methods on a spherical surface or in a spherical gap. But there were no rigorous approximation results for a long period as pointed out in Canuto, Hussaini, Quarteroni and Zang (1988). Recently Guo and Zheng (1994), and Guo (1995) developed a framework for the numerical analysis of spectral methods on the spherical surface. We shall present some results in this chapter. We first discuss the spectral methods and then the pseudospectral methods. The final part of this chapter is for the applications of these methods.

8.1. Spectral Approximation on the Spherical Surface

This section is devoted to the spectral approximation on the spherical surface. Let λ and θ be the longitude and the latitude respectively. Denote by S the unit spherical surface,

$$S = \left\{ (\lambda, \theta) \middle| 0 \leq \lambda < 2\pi, -\frac{\pi}{2} \leq \theta < \frac{\pi}{2} \right\}.$$

The differentiations with respect to λ and θ are denoted by ∂_λ and ∂_θ. For a scalar function v, some commonly used differential operators are defined as

$$\nabla v = \left(\frac{1}{\cos\theta}\partial_\lambda v, \partial_\theta v\right),$$
$$J(v,w) = \frac{1}{\cos\theta}\left(\partial_\lambda v \partial_\theta w - \partial_\theta v \partial_\lambda w\right),$$
$$\Delta v = \frac{1}{\cos\theta}\partial_\theta\left(\cos\theta\partial_\theta v\right) + \frac{1}{\cos^2\theta}\partial_\lambda^2 v.$$

Let $\mathcal{D}(S)$ be the set of all infinitely differentiable functions with the regularity at $\theta = \pm\frac{\pi}{2}$, and the period 2π for the variable λ. $\mathcal{D}'(S)$ stands for the duality of $\mathcal{D}(S)$. We define the distributions in $\mathcal{D}'(S)$ and their derivatives in the usual way, see Lions and Magenes (1972). If

$$\int_S v\partial_\lambda w\, dS = -\int_S zw\, dS, \quad \forall\, w \in \mathcal{D}(S),$$

then we say that z is the derivative of v with respect to λ, denoted by $\partial_\lambda v$, etc. Furthermore, we can define the gradient, the Jacobi operator and the Laplacian in the sense of distributions. For example, if

$$\int_S v\Delta w\, dS = \int_S zw\, dS, \quad \forall\, w \in \mathcal{D}(S),$$

then we say that $z = \Delta v$, etc. Now, let

$$(v,w) = \int_S vw\, dS, \quad \|v\| = (v,v)^{\frac{1}{2}},$$

and

$$L^2(S) = \{v \in \mathcal{D}'(S) \mid \|v\| < \infty\}.$$

Furthermore, define

$$H^1(S) = \left\{v \,\middle|\, v, \frac{1}{\cos\theta}\partial_\lambda v, \partial_\theta v \in L^2(S)\right\}$$

equipped with the semi-norm and the norm as

$$|v|_1 = \left(\left\|\frac{1}{\cos\theta}\partial_\lambda v\right\|^2 + \|\partial_\theta v\|^2\right)^{\frac{1}{2}}, \quad \|v\|_1 = \left(\|v\|^2 + |v|_1^2\right)^{\frac{1}{2}}.$$

For any positive integer r, we can define the space $H^r(S)$ with the norm $\|\cdot\|_r$ by induction. In particular, the norm $\|\cdot\|_2$ is equivalent to $(\|v\|^2 + \|\Delta v\|^2)^{\frac{1}{2}}$. For any real $r \geq 0$, the space $H^r(S)$ is defined by the interpolation between the spaces $H^{[r]}$

and $H^{[r]+1}$. For $r < 0$, the space $H^r(S)$ is the duality of $H^{-r}(S)$. Particularly, $H^0(S) = L^2(S)$ and $\|v\|_0 = \|v\|$. It can be verified that for any $v, w \in H^2(S)$,

$$-(\Delta v, w) = (\nabla v, \nabla w). \tag{8.1}$$

The space $C(\Omega)$ is the set of all continuous functions on S, with the norm $\|\cdot\|_C$.

Lemma 8.1. *For any* $0 \le \mu \le r, H^r(S) \hookrightarrow H^\mu(S)$. *For any* $r > 0, H^{r+1}(S) \hookrightarrow C(S)$.

Proof. The first assertion follows immediately from their definitions. We now prove the second one. Let B be the unit ball in the three-dimensional Euclidean space, and w be a function defined on B. We denote by $\tilde{T}w$ the restriction to S of w. We can take $H^{r+1}(S)$ to be the trace space of $H^{r+\frac{3}{2}}(B)$ with the norm

$$\|v\|_{H^{r+1}(S)} = \inf_{\substack{w \in H^{r+\frac{3}{2}}(B) \\ \tilde{T}w = v}} \|w\|_{H^{r+\frac{3}{2}}(B)}.$$

By imbedding theory, $H^{r+\frac{3}{2}}(B) \hookrightarrow C(\bar{B})$, and so for any $w \in H^{r+\frac{3}{2}}(B)$,

$$\|w\|_{C(\bar{B})} \le c\|w\|_{H^{r+\frac{3}{2}}(B)}.$$

On the other hand, for any $v \in H^{r+1}(S)$, there exists $\tilde{w} \in H^{r+\frac{3}{2}}(B)$ such that $\tilde{T}\tilde{w} = v$ and

$$\|\tilde{w}\|_{H^{r+\frac{3}{2}}(B)} \le 2\|v\|_{H^{r+1}(S)}.$$

Consequently,

$$\|v\|_{C(S)} = \sup_{x \in S} |v(x)| = \sup_{x \in S} |\tilde{w}(x)| \le \sup_{x \in \bar{B}} |\tilde{w}(x)| \le c\,\|\tilde{w}(x)\|_{H^{r+\frac{3}{2}}(B)} \le 2c\|v\|_{H^{r+1}(S)}.$$

This implies the assertion. ∎

Let

$$L_0^2(S) = \left\{v \in L^2(S)\,|A(v) = 0\right\}, \quad A(v) = \int_S v\,dS.$$

Lemma 8.2. *For any* $v \in H^1(S) \cap L_0^2(S)$,

$$\|v\| \le c_0|v|_1,$$

c_0 being a positive constant independent of v.

Proof. By the Poincaré inequality, we claim that

$$\|v\|^2 \le c_0^2 \left(|v|_1^2 + A^2(v)\right)$$

and the desired result follows. ■

Lemma 8.3. *If* $v \in L^2(S), w \in H^{r+1}(S)$ *and* $r > 0$, *then*

$$\|vw\| \le c\|v\|\|w\|_{r+1}.$$

Proof. By Lemma 8.1,

$$\|vw\| \le c\|v\|^2\|w\|_C^2 \le c\|v\|^2\|w\|_{r+1}^2.$$

■

We next turn to the $L^2(S)$-orthogonal system. Let $L_l(x)$ be the Legendre polynomial of degree l. The normalized associated Legendre function is given by

$$\begin{cases} L_{l,m}(x) = \sqrt{\dfrac{(2m+1)(m-l)!}{2(m+l)!}}\,(1-x^2)^{\frac{l}{2}}\,\partial_x^l L_m(x), & \text{for } l \ge 0,\ m \ge |l|, \\ L_{l,m}(x) = L_{-l,m}(x), & \text{for } l < 0,\ m \ge |l|. \end{cases} \tag{8.2}$$

Moreover, the spherical harmonic function $Y_{l,m}(\lambda,\theta)$ is

$$Y_{l,m}(\lambda,\theta) = \frac{1}{\sqrt{2\pi}} e^{il\lambda} L_{l,m}(\sin\theta), \quad m \ge |l|. \tag{8.3}$$

The set of these functions is the $L^2(S)$-orthogonal system on S, i.e.,

$$\int_0^{2\pi}\int_{-\frac{\pi}{2}}^{\frac{\pi}{2}} Y_{l,m}(\lambda,\theta)\bar{Y}_{l',m'}(\lambda,\theta)\cos\theta\, d\theta d\lambda = \delta_{l,l'}\delta_{m,m'}.$$

The spherical harmonic expansion of a function $v \in L^2(S)$ is

$$v = \sum_{l=-\infty}^{\infty}\sum_{m\ge|l|} \hat{v}_{l,m} Y_{l,m}(\lambda,\theta)$$

where the coefficient $\hat{v}_{l,m}$ is given by

$$\hat{v}_{l,m} = \int_0^{2\pi}\int_{-\frac{\pi}{2}}^{\frac{\pi}{2}} v(\lambda,\theta)\bar{Y}_{l,m}(\lambda,\theta)\cos\theta\, d\theta d\lambda.$$

It can be checked that

$$-\Delta Y_{l,m}(\lambda,\theta) = m(m+1)Y_{l,m}(\lambda,\theta). \tag{8.4}$$

So $Y_{l,m}(\lambda,\theta)$ is the eigenfunction of the operator $-\Delta$ on S, corresponding to the eigenvalue $m(m+1)$. Thus in the space $H^r(S)$, the norm $\|v\|_r$ is equivalent to

$$\left(\sum_{l=-\infty}^{\infty}\sum_{m\geq|l|} m^r(m+1)^r\left|\hat{v}_{l,m}\right|^2\right)^{\frac{1}{2}}. \tag{8.5}$$

Now let N be any positive integer, and define the finite dimensional space $\tilde{V}_N$ as

$$\tilde{V}_N = \operatorname{span}\left\{Y_{l,m} \mid |l|\leq N, |l|\leq m\leq M(l)\right\}.$$

Clearly, $\tilde{V}_N$ depends on the choice of $M(l)$. Usually we take $M(l)=N$ or $N+|l|$. For fixing the discussion, we take $M(l)=N$. Let V_N be the subset of $\tilde{V}_N$ containing all real-valued functions. In this space, we have the following inverse inequality.

Theorem 8.1. *For any* $\phi\in V_N$ *and* $0\leq r\leq\mu$,

$$\|\phi\|_\mu \leq cN^{\mu-r}\|\phi\|_r.$$

Proof. Let

$$\phi = \sum_{l=-N}^{N}\sum_{m=|l|}^{N}\hat{\phi}_{l,m}Y_{l,m}(\lambda,\theta).$$

Then by (8.5),

$$\begin{aligned}
\|\phi\|_\mu^2 &\leq c\sum_{l=-N}^{N}\sum_{m=|l|}^{N} m^\mu(m+1)^\mu\left|\hat{\phi}_{l,m}\right|^2\\
&\leq cN^{2\mu-2r}\sum_{l=-N}^{N}\sum_{m=|l|}^{N} m^r(m+1)^r\left|\hat{\phi}_{l,m}\right|^2 \leq cN^{2\mu-2r}\|\phi\|_r^2.
\end{aligned}$$

■

The $L^2(S)$-orthogonal projection $P_N : L^2(S)\to V_N$ is such a mapping that for any $v\in L^2(S)$,

$$(v-P_Nv,\phi)=0,\quad \forall\,\phi\in V_N,$$

or equivalently,

$$P_Nv(\lambda,\theta) = \sum_{l=-N}^{N}\sum_{m=|l|}^{N}\hat{v}_{l,m}Y_{l,m}(\lambda,\theta).$$

Theorem 8.2. *For any $v \in H^r(S)$ and $0 \le \mu \le r$,*

$$\|v - P_N v\|_\mu \le cN^{\mu-r}\|v\|_r.$$

Proof. By (8.5),

$$\begin{aligned}
\|v - P_N v\|_\mu^2 &\le c\sum_{l=-N}^{N}\sum_{m=N+1}^{\infty} m^\mu(m+1)^\mu |\hat{v}_{l,m}|^2 + c\sum_{|l|>N}\sum_{m=|l|}^{\infty} m^\mu(m+1)^\mu |\hat{v}_{l,m}|^2 \\
&\le c\sum_{l=-N}^{N}\sum_{m=N+1}^{\infty} m^\mu(m+1)^\mu |\hat{v}_{l,m}|^2 + c\sum_{|l|>N}\sum_{m=N+1}^{\infty} m^\mu(m+1)^\mu |\hat{v}_{l,m}|^2 \\
&\le cN^{2\mu-2r}\sum_{l=-\infty}^{\infty}\sum_{m=N+1}^{\infty} m^r(m+1)^r |\hat{v}_{l,m}|^2 \le cN^{2\mu-2r}\|v\|_r^2.
\end{aligned}$$

■

The proof of the above two theorems is due to Guo and Zheng (1994).

8.2. Pseudospectral Approximation on the Spherical Surface

We now present another approximation on the spherical surface, based on the interpolation. Let $x^{(j)}$ and $\omega^{(j)}$ be the Legendre-Gauss interpolation points and the weights on the interval $\Lambda = (-1,1)$. Define S_N as a set of grid points,

$$S_N = \left\{ \left(\lambda^{(k)}, \theta^{(j)}\right) \middle| \lambda^{(k)} = \frac{2k\pi}{2N+1}, \theta^{(j)} = \arcsin x^{(j)}, 0 \le k \le 2N, 0 \le j \le N \right\}.$$

The discrete inner point $(\cdot,\cdot)_N$ and the discrete norm $\|\cdot\|_N$ are defined by

$$(v,w)_N = \frac{2\pi}{2N+1}\sum_{k=0}^{2N}\sum_{j=0}^{N} v\left(\lambda^{(k)},\theta^{(j)}\right)\bar{w}\left(\lambda^{(k)},\theta^{(j)}\right)\omega^{(j)}, \quad \|v\|_N = (v,v)^{\frac{1}{2}}.$$

The interpolation $I_N : C(S) \longrightarrow V_N$ is such a mapping that for any $v \in C(S)$,

$$I_N v = \sum_{l=-N}^{N}\sum_{m=|l|}^{N} \tilde{v}_{l,m} Y_{l,m}(\lambda,\theta), \tag{8.6}$$

where

$$\tilde{v}_{l,m} = \frac{2\pi}{2N+1}\sum_{k=0}^{2N}\sum_{j=0}^{N} v\left(\lambda^{(k)},\theta^{(j)}\right)\bar{Y}_{l,m}\left(\lambda^{(k)},\theta^{(j)}\right)\omega^{(j)}.$$

Clearly

$$\tilde{v}_{l,m} = (v, Y_{l,m})_N\,.$$

The degree of freedom of S_N is $(2N+1)(N+1)$, while the dimension of V_N is $\sum_{l=0}^{N}(2l+1)=(N+1)^2$. Thus, $I_N v\left(\lambda^{(k)},\theta^{(j)}\right)\neq v\left(\lambda^{(k)},\theta^{(j)}\right)$ generally. Consequently, I_N is not an interpolation in the usual sense. It is actually the projection from $C(S)$ onto V_N, corresponding to the discrete inner product $(\cdot,\cdot)_N$. This fact can be seen in the proof of the following lemma.

Lemma 8.4. *For any $v\in C(S)$ and $\phi\in V_N$, we have that*

(i) $I_N\phi=\phi$,

(ii) $(I_N v,\phi)=(I_N v,\phi)_N=(v,\phi)_N$.

Proof. For the first conclusion, it is sufficient to show that for all $|l|\leq m\leq N$,

$$I_N Y_{l,m}=Y_{l,m}.$$

In fact, for all $|l'|\leq m'\leq N$,

$$(Y_{l,m},Y_{l',m'})_N=\frac{1}{2N+1}\sum_{k=0}^{2N}\sum_{j=0}^{N}e^{i(l-l')\lambda^{(k)}}L_{l,m}\left(\sin\theta^{(j)}\right)L_{l',m'}\left(\sin\theta^{(j)}\right)\omega^{(j)}.$$

Clearly

$$\frac{1}{2N+1}\sum_{k=0}^{2N}e^{i(l-l')\lambda^{(k)}}=\delta_{l,l'}.$$

On the other hand, $L_{l,m}(x)L_{l,m'}(x)$ is an algebraic polynomial of degree $m+m'\leq 2N$. By (2.34), the Legendre-Gauss quadrature formula with $N+1$ points is exact for it. Hence we have from the orthogonality of $\{L_{l,m}(x)\}$ that

$$\sum_{j=0}^{N}L_{l,m}\left(x^{(j)}\right)L_{l,m'}\left(x^{(j)}\right)\omega^{(j)}=\int_{-1}^{1}L_{l,m}(x)L_{l,m'}(x)\,dx=\delta_{m,m'}.$$

Therefore for all $|l|\leq m\leq N$ and $|l'|\leq m\leq N$,

$$(Y_{l,m},Y_{l',m'})_N=\delta_{l,l'}\delta_{m,m'}=(Y_{l,m},Y_{l',m'})\,. \tag{8.7}$$

Consequently, $I_N Y_{l,m}=Y_{l,m}$.

Next, we examine the second conclusion. Let

$$\phi(\lambda,\phi)=\sum_{l=-N}^{N}\sum_{m=|l|}^{N}\hat{\phi}_{l,m}Y_{l,m}(\lambda,\theta). \tag{8.8}$$

By (8.6) and (8.7),

$$\begin{aligned}(I_N v, \phi) &= \sum_{l=-N}^{N} \sum_{m=|l|}^{N} \tilde{v}_{l,m} (Y_{l,m}, \phi) = \sum_{l=-N}^{N} \sum_{l'=-N}^{N} \sum_{m=|l|}^{N} \sum_{m'=|l'|}^{N} \tilde{v}_{l,m} \hat{\phi}_{l',m'} (Y_{l,m}, Y_{l',m'})_N \\ &= (I_N v, \phi)_N .\end{aligned}$$

Furthermore,

$$\begin{aligned}(v, \phi)_N &= \sum_{l=-N}^{N} \sum_{m=|l|}^{N} (v, Y_{l,m})_N \, \hat{\phi}_{l,m} = \sum_{l=-N}^{N} \sum_{m=|l|}^{N} \tilde{v}_{l,m} \hat{\phi}_{l,m} \\ &= (I_N v, \phi) = (I_N v, \phi)_N .\end{aligned}$$

■

Lemma 8.5. *For any $v \in C(S)$, we have that*

(i) $\|I_N v\| = \|I_N v\|_N \le \|v\|_N$,

(ii) $\|v - I_N v\|_N = \inf_{\phi \in V_N} \|v - \phi\|_N$.

Proof. It follows from Lemma 8.4 that

$$\|I_N v\|^2 = \|I_N v\|_N^2 = (v, I_N v)_N \le \|v\|_N \, \|I_N v\|_N .$$

Then the first conclusion follows immediately. The second one follows from the projection theorem and the last result of Lemma 8.4. ■

Lemma 8.6. *For any $\phi \in V_N$, we have that*

(i) $\left\| \frac{1}{\cos\theta} \partial_\lambda \phi \right\|_N = \left\| \frac{1}{\cos\theta} \partial_\lambda \phi \right\|$,

(ii) $\|\partial_\theta \phi\|_N = \|\partial_\theta \phi\|$.

Proof. We see from (8.8) that

$$\begin{aligned}\frac{1}{\cos\theta} \partial_\lambda \phi &= \sum_{m=0}^{N} \sum_{|l| \le m} il \hat{\phi}_{l,m} e^{il\lambda} L_{l,m}(\sin\theta)(\cos\theta)^{-1}, \\ \partial_\theta \phi &= \sum_{m=0}^{N} \sum_{|l| \le m} \hat{\phi}_{l,m} e^{il\lambda} \partial_x L_{l,m}(\sin\theta) \cos\theta .\end{aligned}$$

Let $\sum_{l,m}{}'$ denote the sum in the above formulae. Then

$$\begin{aligned}\left\|\frac{1}{\cos\theta}\partial_\lambda\phi\right\|_N^2 &= \left(\frac{1}{\cos\theta}\partial_\lambda\phi,\frac{1}{\cos\theta}\partial_\lambda\phi\right)_N\\ &= \sum_{l,m}{}'\sum_{l',m'}{}' ll'\hat{\phi}_{l,m}\hat{\phi}_{l',m'}\left(e^{il\lambda}L_{l,m}(\sin\theta)(\cos\theta)^{-1},e^{il'\lambda}L_{l',m'}(\sin\theta)(\cos\theta)^{-1}\right)_N.\end{aligned}$$

For $l\neq 0$, $L_{l,m}(x)L_{l,m'}(x)\left(1-x^2\right)^{-1}$ is an algebraic polynomial of degree $m+m'-2\leq 2N-2$, and so the Legendre-Gauss quadrature formula is exact for it. Thus we obtain that

$$\begin{aligned}\left\|\frac{1}{\cos\theta}\partial_\lambda\phi\right\|_N^2 &= \sum_{l,m}{}'\sum_{l',m'}{}' ll'\hat{\phi}_{l,m}\hat{\phi}_{l',m'}\left(e^{il\lambda}L_{l,m}(\sin\theta)(\cos\theta)^{-1},e^{il'\lambda}L_{l',m'}(\sin\theta)(\cos\theta)^{-1}\right)\\ &= \left\|\frac{1}{\cos\theta}\partial_\lambda\phi\right\|^2.\end{aligned}$$

On the other hand, $\partial_x L_{l,m}(x)\partial_x L_{l,m'}(x)\left(1-x^2\right)$ is an algebraic polynomial of degree $m+m'\leq 2N$. Therefore the second conclusion can be proved in the same manner. ∎

Theorem 8.3. *For any $v\in H^r(S), 0\leq\mu\leq r$ and $r>1$,*

$$\|v-I_Nv\|_\mu\leq cN^{\mu+1+\delta-r}\|v\|_r,$$

δ being an arbitrary positive constant.

Proof. By Lemma 8.1, I_Nv is well defined. We first let $\mu=0$. It follows from conclusion (ii) of Lemma 8.5 that

$$\|v-I_Nv\|_N\leq\|v-P_Nv\|_N\leq c\,\|v-P_Nv\|_C. \tag{8.9}$$

Since $I_Nv, P_Nv\in V_N$, we have from Lemma 8.4 and (8.9) that

$$\begin{aligned}\|v-I_Nv\| &\leq \|v-P_Nv\|+\|I_Nv-P_Nv\|=\|v-P_Nv\|+\|I_Nv-P_Nv\|_N\\ &\leq \|v-P_Nv\|+\|v-I_Nv\|_N+\|v-P_Nv\|_N\leq c\,\|v-P_Nv\|_C.\end{aligned}$$

Further, we get from Lemma 8.1 and Theorem 8.2 that for $\delta>0$,

$$\|v-I_Nv\|\leq c\,\|v-P_Nv\|_{1+\delta}\leq cN^{1+\delta-r}\|v\|_r. \tag{8.10}$$

If $\mu>0$, then we deduce from Theorem 8.1 and (8.10) that

$$\|P_Nv-I_Nv\|_\mu\leq cN^\mu\,\|P_Nv-I_Nv\|\leq cN^\mu\left(\|v-P_Nv\|+\|v-I_Nv\|\right)\leq cN^{\mu+1+\delta-r}\|v\|_r.$$

This estimate together with Theorem 8.2 imply the desired result. ∎

All results in this section come from Cao and Guo (1997).

8.3. Applications

In this section, we take the vorticity equation as an example to describe the spectral methods on the surface. Let H, Ψ and $\mu > 0$ be the vorticity, the stream function and the kinetic viscosity, respectively. The vorticity equation on the spherical surface S is of the form

$$\begin{cases} \partial_t H + J(H,\Psi) - \mu\Delta\Psi = f, & (\lambda,\theta) \in S,\ 0 < t \leq T, \\ -\Delta\Psi = H, & (\lambda,\theta) \in S,\ 0 \leq t \leq T, \\ H(0) = H_0, & (\lambda,\theta) \in S \end{cases} \tag{8.11}$$

where f and H_0 are given functions. It is natural to assume that all functions have the period 2π for the variable λ, and are regular at the pole points $\theta = \pm\frac{\pi}{2}$. For fixing Ψ, we also require that $\Psi(\lambda,\theta,t) \in L_0^2(\Omega)$ for $0 \leq t \leq T$.

The weak form of (8.11) is to find $H \in L^2(0,T;H^1(S)) \cap L^\infty(0,T;L^2(S))$ and $\Psi \in L^2(0,T;H^1(S) \cap L_0^2(S))$ such that

$$\begin{cases} (\partial_t H(t) + J(H(t),\Psi(t)),v) + \mu(\nabla H(t),\nabla v) = (f(t),v), & \forall\, v \in H^1(S),\ 0 < t \leq T, \\ (\nabla\Psi(t),\nabla v) = (H(t),v), & \forall\, v \in H^1(S),\ 0 \leq t \leq T, \\ H(0) = H_0, & (\lambda,\theta) \in S. \end{cases} \tag{8.12}$$

The existence and uniqueness of the solution of (8.12) was discussed in Zen (1979). In particular, if $H_0 \in L^\infty(S)$ and $f \in L^\infty(0,T;L^\infty(S))$, then (8.12) possesses a unique solution and $H \in L^2(0,T;H^1(S)) \cap H^1(0,T,H^{-1}(S))$. The solution is conservative. To prove it, we first show that for any $v, z \in H^{r+1}(S), w \in H^1(S)$ and $r > 0$,

$$(J(v,w),z) + (J(z,w),v) = 0. \tag{8.13}$$

Indeed, the integration by parts yields

$$\begin{aligned} (J(v,w),z) &= \int_0^{2\pi}\int_{-\frac{\pi}{2}}^{\frac{\pi}{2}} z\left(\partial_\lambda v \partial_\theta w - \partial_\lambda w \partial_\theta v\right) d\theta d\lambda \\ &= -\int_0^{2\pi}\int_{-\frac{\pi}{2}}^{\frac{\pi}{2}} v\left(\partial_\lambda z \partial_\theta w - \partial_\theta z \partial_\lambda w\right) d\theta d\lambda \\ &\quad - \int_0^{2\pi} z\left(\lambda,\frac{\pi}{2}\right) v\left(\lambda,\frac{\pi}{2}\right) \partial_\lambda w\left(\lambda,\frac{\pi}{2}\right) d\lambda \\ &\quad + \int_0^{2\pi} z\left(\lambda,-\frac{\pi}{2}\right) v\left(\lambda,-\frac{\pi}{2}\right) \partial_\lambda w\left(\lambda,-\frac{\pi}{2}\right) d\lambda. \end{aligned}$$

From Courant and Hilbert (1953), the regularity of w implies that w approaches the limits independently of λ, as $\theta \to \pm\frac{\pi}{2}$. This means that $\partial_\lambda w = 0$ at $\theta = \pm\frac{\pi}{2}$, and so

(8.13) follows. Next let $v = H$ in (8.12) and integrate the resulting equation from 0 to t. We obtain from (8.13) that

$$\|H(t)\|^2 + 2\mu \int_0^t |H(s)|_1^2\, ds = \|H_0\|^2 + 2\int_0^t (f(s), H(s))\, ds. \tag{8.14}$$

Now let $V_N^0 = V_N \cap L_0^2(S)$. H and Ψ are approximated by η_N and ψ_N. The spectral scheme for (8.12) is to find $\eta_N(\lambda,\theta,t) \in V_N, \psi_N(\lambda,\theta,t) \in V_N^0$ for all $t \in \bar{R}_\tau(T)$ such that

$$\begin{cases} (D_\tau \eta_N(t) + J(\eta_N(t) + \delta\tau D_\tau \eta_N(t), \psi_N(t)), \phi) & \\ \quad -\mu(\Delta(\eta_N(t) + \sigma\tau D_\tau\eta_N(t)), \phi) = (f(t),\phi), & \forall\, \phi \in V_N,\ t \in \bar{R}_\tau(T), \\ -(\Delta\psi_N(t), \phi) = (\eta_N(t), \phi), & \forall\, \phi \in V_N^0,\ t \in \bar{R}_\tau(T), \\ \eta_N(0) = P_N H_0, & (\lambda,\theta) \in S \end{cases} \tag{8.15}$$

where δ and σ are the parameters, $0 \le \delta, \sigma \le 1$. If both of them are zero, then (8.15) is an explicit scheme. Otherwise a linear iteration is needed at each time $t \in R_\tau(T)$.

We now consider the existence and uniqueness of the solution of (8.15) with $\delta \neq 0$ or $\sigma \neq 0$. Let

$$\begin{aligned} F(t) = {} & \eta_N(t-\tau) - \tau(1-\delta)J(\eta_N(t-\tau), \psi_N(t-\tau)) \\ & + \mu\tau(1-\sigma)\Delta\eta_N(t-\tau) + \tau f(t-\tau). \end{aligned}$$

Then for any $t \in R_\tau(T)$,

$$(\eta_N(t), \phi) + \delta\tau(J(\eta_N(t), \psi_N(t-\tau)), \phi) - \mu\sigma\tau(\Delta\eta_N(t), \phi) = (F(t), \phi), \quad \forall\, \phi \in V_N.$$

Clearly it is a linear algebraic system for the coefficients of the spherical harmonic expansion of $\eta_N(t)$. Hence we only have to verify that $\eta_N(t)$ is a trival solution when $F(t) \equiv 0$. In fact, by taking $\phi = \eta_N(t)$ and using (8.13),

$$\|\eta_N(t)\|^2 + \mu\sigma\tau\, |\eta_N(t)|_1^2 = 0.$$

It implies $\eta_N(t) \equiv 0$, whence $\eta_N(t)$ is determined uniquely at each time $t \in \bar{R}_\tau(T)$. In particular, if $\delta = \sigma = \frac{1}{2}$, then we have from (8.13) and (8.15) that

$$\begin{aligned} & \|\eta_N(t)\|^2 + \frac{\mu\tau}{2} \sum_{s \in \bar{R}_\tau(t-\tau)} |\eta_N(s) + \eta_N(s+\tau)|_1^2 \\ & \quad = \|\eta_N(0)\|^2 + \tau \sum_{s \in \bar{R}_\tau(t-\tau)} (f(s), \eta_N(s) + \eta_N(s+\tau)). \end{aligned} \tag{8.16}$$

This is a reasonable analogy of (8.14).

We now analyze the generalized stability of scheme (8.15). Suppose that $\eta_N(0), f$ and the right-hand side of (8.15) have the errors $\tilde{\eta}_0, \tilde{f}$ and $\tilde{g}$, respectively. They induce the errors of η_N and ψ_N, denoted by $\tilde{\eta}_N$ and $\tilde{\psi}_N$. Then

$$\begin{cases} \left(D_\tau\tilde{\eta}(t) + J\left(\tilde{\eta}(t) + \delta\tau D_\tau\tilde{\eta}(t), \psi_N(t) + \tilde{\psi}(t)\right) + J\left(\eta_N(t) + \delta\tau D_\tau\eta_N(t), \tilde{\psi}(t)\right), \phi\right) \\ \qquad -\mu\left(\Delta\left(\tilde{\eta}(t) + \sigma\tau D_\tau\tilde{\eta}(t)\right), \phi\right) = \left(\tilde{f}(t), \phi\right), \qquad \forall\, \phi \in V_N,\ t \in \bar{R}_\tau(T), \\ -\left(\Delta\tilde{\psi}(t), \phi\right) = \left(\tilde{\eta}(t) + \tilde{g}(t), \phi\right), \qquad \forall\, \phi \in V_N^0,\ t \in \bar{R}_\tau(T), \\ A\left(\tilde{\psi}(t)\right) = 0, \qquad t \in \bar{R}_\tau(T), \\ \tilde{\eta}(0) = \tilde{\eta}_0, \qquad (\lambda, \theta) \in S. \end{cases} \tag{8.17}$$

Let ξ be an undetermined positive constant. By taking $\phi = 2\tilde{\eta} + \xi\tau D_\tau\tilde{\eta}$ in the first formula of (8.17), we get from (1.22) and (8.13) that

$$\begin{aligned} &D_\tau \|\tilde{\eta}(t)\|^2 + \tau(\xi - 1)\|D_\tau\tilde{\eta}(t)\|^2 + 2\mu|\tilde{\eta}(t)|_1^2 + \mu\tau\left(\sigma + \frac{\xi}{2}\right)|D_\tau\tilde{\eta}(t)|_1^2 \\ &\quad + \mu\tau^2\left(\xi\sigma - \sigma - \frac{\xi}{2}\right)|D_\tau\tilde{\eta}(t)|_1^2 + \sum_{j=1}^{4} F_j(t) = \left(\tilde{f}(t), 2\tilde{\eta}(t) + \xi\tau D_\tau\tilde{\eta}(t)\right) \end{aligned} \tag{8.18}$$

where

$$\begin{aligned} F_1(t) &= 2\left(J\left(\eta_N(t) + \delta\tau D_\tau\eta_N(t), \tilde{\psi}(t)\right), \tilde{\eta}(t)\right), \\ F_2(t) &= \xi\tau\left(J\left(\eta_N(t) + \delta\tau D_\tau\eta_N(t), \tilde{\psi}(t)\right), D_\tau\tilde{\eta}(t)\right), \\ F_3(t) &= \tau(\xi - 2\delta)\left(J\left(\tilde{\eta}(t), \psi_N(t)\right), D_\tau\tilde{\eta}(t)\right), \\ F_4(t) &= \tau(\xi - 2\delta)\left(J\left(\tilde{\eta}(t), \tilde{\psi}(t)\right), D_\tau\tilde{\eta}(t)\right). \end{aligned}$$

Further, we put $\phi = \tilde{\psi}$ in the second formula of (8.17), and then obtain that

$$\left|\tilde{\psi}(t)\right|_1^2 \leq \frac{1}{2c_0^2}\left\|\tilde{\psi}(t)\right\|^2 + c\left(\|\tilde{\eta}(t)\|^2 + \|\tilde{g}(t)\|^2\right).$$

By Lemma 8.2, we find that

$$\left|\tilde{\psi}(t)\right|_1^2 \leq c\left(\|\tilde{\eta}(t)\|^2 + \|\tilde{g}(t)\|^2\right) \tag{8.19}$$

from which and Lemma 8.2,

$$\left\|\tilde{\psi}(t)\right\|_2^2 \leq c\left(\left\|\tilde{\psi}(t)\right\|^2 + \left\|\Delta\tilde{\psi}(t)\right\|^2\right) \leq c\left(\|\tilde{\eta}(t)\|^2 + \|\tilde{g}(t)\|^2\right). \tag{8.20}$$

We are going to estimate $|F_j(t)|$. Let $\varepsilon > 0$. By (8.13), (8.19) and Lemma 8.1, we find that for $\gamma > 0$,

$$\begin{aligned}|F_1(t)| &\leq 2\left|\left(J\left(\tilde{\eta}(t),\tilde{\psi}(t)\right),\eta_N(t)+\delta\tau D_\tau\eta_N(t)\right)\right| \\ &\leq \varepsilon\mu\left|\tilde{\eta}(t)\right|_1^2+\frac{c}{\varepsilon\mu}\left\|\eta_N\right\|_{C(0,T;L^\infty)}^2\left\|\tilde{\psi}(t)\right\|_1^2 \\ &\leq \varepsilon\mu\left|\tilde{\eta}(t)\right|_1^2+\frac{c}{\varepsilon\mu}\left\|\eta_N\right\|_{C(0,T;H^{\gamma+1})}^2\left(\left\|\tilde{\eta}(t)\right\|^2+\left\|\tilde{g}(t)\right\|^2\right).\end{aligned}$$

Similarly,

$$|F_2(t)| \leq \varepsilon\mu\tau^2\left|D_\tau\tilde{\eta}(t)\right|_1^2+\frac{c\xi^2}{\varepsilon\mu}\left\|\eta_N\right\|_{C(0,T;H^{\gamma+1})}^2\left(\left\|\tilde{\eta}(t)\right\|^2+\left\|\tilde{g}(t)\right\|^2\right).$$

Furthermore, the use of Lemma 8.3 yields

$$|F_3(t)| \leq \varepsilon\tau\left\|D_\tau\tilde{\eta}(t)\right\|^2+\frac{c\tau(\xi-2\delta)^2}{\varepsilon}\left\|\psi_N\right\|_{C(0,T;H^{\gamma+2})}^2\left|\tilde{\eta}(t)\right|_1^2.$$

Since

$$\left\|\psi_N\right\|_2^2 \leq c\left(\left\|\psi_N(t)\right\|^2+\left\|\Delta\psi_N(t)\right\|^2\right) \leq c\left\|\eta_N(t)\right\|^2, \tag{8.21}$$

we deduce from Theorem 8.1 that

$$|F_3(t)| \leq \varepsilon\tau\left\|D_\tau\tilde{\eta}(t)\right\|^2+\frac{c\tau N^{2\gamma}(\xi-2\delta)^2}{\varepsilon}\left\|\eta_N\right\|_{C(0,T;L^2)}^2\left|\tilde{\eta}(t)\right|_1^2.$$

Also, by Lemma 8.3, Theorem 8.1 and (8.20),

$$\begin{aligned}|F_4(t)| &\leq \varepsilon\tau\left\|D_\tau\tilde{\eta}_N\right\|^2+\frac{c\tau(\xi-2\delta)^2}{\varepsilon}\left\|\tilde{\psi}(t)\right\|_{\gamma+2}^2\left|\tilde{\eta}(t)\right|_1^2 \\ &\leq \varepsilon\tau\left\|D_\tau\tilde{\eta}(t)\right\|^2+\frac{c\tau(\xi-2\delta)^2}{\varepsilon}\left(\left\|\tilde{\eta}(t)\right\|_\gamma^2+\left\|\tilde{g}(t)\right\|_\gamma^2\right)\left|\tilde{\eta}(t)\right|_1^2 \\ &\leq \varepsilon\tau\left\|D_\tau\tilde{\eta}(t)\right\|^2+\frac{c\tau N^{2\gamma}(\xi-2\delta)^2}{\varepsilon}\left(\left\|\tilde{\eta}(t)\right\|^2+\left\|\tilde{g}(t)\right\|^2\right)\left|\tilde{\eta}(t)\right|_1^2.\end{aligned}$$

Clearly,

$$\left|\left(\tilde{f}(t),\tilde{\eta}(t)+\xi\tau D_\tau\tilde{\eta}(t)\right)\right| \leq \varepsilon\tau\left\|D_\tau\tilde{\eta}(t)\right\|^2+c\left\|\tilde{\eta}(t)\right\|^2+c\left(1+\frac{\tau\xi^2}{\varepsilon}\right)\left\|\tilde{f}(t)\right\|^2.$$

By substituting the above estimates into (8.18), we claim that

$$\begin{aligned}&D_\tau\left\|\tilde{\eta}(t)\right\|^2+\tau(\xi-1-3\varepsilon)\left\|D_\tau\tilde{\eta}(t)\right\|^2+\mu\left|\tilde{\eta}(t)\right|_1^2+\mu\tau\left(\sigma+\frac{\xi}{2}\right)D_\tau\left|\tilde{\eta}(t)\right|_1^2 \\ &+\mu\tau^2\left(\xi\sigma-\sigma-\frac{\xi}{2}-\varepsilon\right)\left|D_\tau\tilde{\eta}(t)\right|_1^2 \leq A_1\left\|\tilde{\eta}(t)\right\|^2+A_2(t)\left|\tilde{\eta}(t)\right|_1^2+A_3(t)\end{aligned} \tag{8.22}$$

where

$$A_1 = c\left(1 + \frac{1+\xi^2}{\varepsilon\mu}\|\eta_N\|^2_{C(0,T;H^{\gamma+1})}\right),$$

$$A_2(t) = -\mu + \varepsilon\mu + \frac{c\tau N^{2\gamma}}{\varepsilon}(\xi - 2\delta)^2\left(\|\eta_N\|^2_{C(0,T;L^2)} + \|\tilde{\eta}(t)\|^2 + \|\tilde{g}(t)\|^2\right),$$

$$A_3(t) = c\left(1 + \frac{\tau\xi^2}{\varepsilon}\right)\left\|\tilde{f}(t)\right\|^2 + \frac{c}{\varepsilon\mu}\left(1+\xi^2\right)\|\eta_N\|^2_{C(0,T;H^{\gamma+1})}\|\tilde{g}(t)\|^2.$$

Let $q_0 > 0$ and ε be suitably small. We choose ξ in three different cases.

(i) $\sigma > \frac{1}{2}$. We take

$$\xi \geq \xi_1 = \max\left(1 + q_0 + 3\varepsilon, \frac{2\sigma + 2\varepsilon}{2\sigma - 1}\right).$$

Then

$$\tau(\xi - 1 - 3\varepsilon)\|D_\tau\tilde{\eta}(t)\|^2 + \mu\tau^2\left(\xi\sigma - \sigma - \frac{\xi}{2} - \varepsilon\right)|D_\tau\tilde{\eta}(t)|_1^2 \geq q_0\tau\|D_\tau\tilde{\eta}(t)\|^2. \quad (8.23)$$

(ii) $\sigma = \frac{1}{2}$. According to Theorem 8.1, there exists a positive constant c_I such that $\|v\|_1^2 \leq c_I N^2\|v\|^2$ for all $v \in V_N$. For this reason, we take

$$\xi \geq \xi_2 = 1 + q_0 + 3\varepsilon + \mu c_I\tau N^2\left(\frac{1}{2} + \varepsilon\right)$$

and so (8.23) also holds.

(iii) $\sigma < \frac{1}{2}$ and $\tau N^2 < \dfrac{2}{\mu c_I(1-2\sigma)}$. Then we take

$$\xi \geq \xi_3 = \frac{1 + q_0 + 3\varepsilon + \mu c_I\tau N^2(\sigma + \varepsilon)}{1 + \mu c_I\tau N^2\left(\sigma - \frac{1}{2}\right)}.$$

whence (8.23) is still valid.

Now let

$$E(v,t) = \|v(t)\|^2 + \tau\sum_{s\in\bar{R}_\tau(t-\tau)}\left(\mu|v(s)|_1^2 + q_0\tau\|D_\tau v(s)\|^2\right),$$

$$\rho(t) = \|\tilde{\eta}(0)\|^2 + \mu\tau\left(\sigma + \frac{\xi}{2}\right)|\tilde{\eta}(0)|_1^2 + \tau\sum_{s\in\bar{R}_\tau(t-s)}A_3(s).$$

Then the summation (8.22) over $\bar{R}_\tau(t)$ yields

$$E(\tilde{\eta},t) \leq \rho(t) + \tau\sum_{s\in\bar{R}_\tau(t-\tau)}\left(A_1 E(\tilde{\eta},s) + A_2(s)E^2(\tilde{\eta},s)\right). \quad (8.24)$$

Applying Lemma 3.2 to (8.24), we claim the result as follows.

Theorem 8.4. *Assume that*

(i) $\tau N^{2\gamma}$ *is suitably small,* γ *being an arbitrary small positive constant;*

(ii) $\sigma > \dfrac{1}{2}$ *or* $\tau N^2 < \dfrac{2}{c_I\mu(1-2\sigma)}$;

(iii) $\|\tilde{g}(t)\|^2 \le \dfrac{b_1}{\tau N^{2\gamma}}$ *and* $\rho(t_1)e^{b_2 t_1} \le \dfrac{b_3}{\tau N^{\gamma}}$ *for some* $t_1 \in R_\tau(T)$.

Then for all $t \in \bar{R}_\tau(t_1)$

$$E(\tilde{\eta}, t) \le \rho(t)e^{b_2 t} \tag{8.25}$$

where $b_1 - b_3$ *are some positive constants depending on* μ *and* $\|\eta_N\|_{C(0,T;H^{\gamma+1})}$.

We now consider a special case, i.e.,

$$\begin{cases} 2\delta \ge \xi_1, & \text{for } \sigma > \frac{1}{2}, \\ 2\delta \ge \xi_2, & \text{for } \sigma = \frac{1}{2}, \\ 2\delta \ge \xi_3, & \text{for } \sigma < \frac{1}{2}. \end{cases} \tag{8.26}$$

We can take $\xi = 2\delta$ and so $A_2(t) \equiv -\mu + \varepsilon\mu$. It leads to the following result.

Theorem 8.5. *If (8.26) and condition (ii) of Theorem 8.4 are fulfilled, then for all* $\rho(t)$ *and* $t \in \bar{R}_\tau(T)$, *the estimate (8.25) is valid.*

We know again from Theorem 8.5 that a suitable implicit approximation of non-linear term can improve the stability essentially. Indeed in the case of Theorem 8.5, scheme (8.15) has the generalized stability index $s = -\infty$.

In the final part of this section, we deal with the convergence of (8.15). First of all, we derive from (8.4) that for $v \in H^2(S)$,

$$\begin{aligned} P_N \Delta v(\lambda, \theta) &= -\sum_{|l|\le N}\sum_{m\ge|l|} m(m+1)\hat{v}_{l,m} Y_{l,m}(\lambda,\theta) \\ &= \sum_{|l|\le N}\sum_{m\ge|l|} \hat{v}_{l,m}\Delta Y_{l,m}(\lambda,\theta) = \Delta P_N v(\lambda,\theta). \end{aligned} \tag{8.27}$$

Let $H_N = P_N H$ and $\Psi_N = P_N \Psi$. Then we have from (8.12) and (8.27) that

$$\begin{cases} (D_\tau H_N(t) + J(H_N(t) + \delta\tau D_\tau H_N(t), \Psi_N(t)), \phi) + \mu(\nabla(H_N(t) + \sigma\tau D_\tau H_N(t)), \nabla\phi) \\ \qquad = \left(\sum_{j=1}^{3} G_j(t) + f(t), \phi\right) + (\nabla G_4(t), \nabla\phi), \quad \forall\, \phi \in V_N,\ t \in \bar{R}_\tau(T), \\ (\nabla\Psi_N(t), \nabla\phi) = (H_N(t), \phi), \qquad \forall\, \phi \in V_N^0,\ t \in \bar{R}_\tau(T), \\ A(\Psi_N(t)) = 0, \qquad t \in \bar{R}_\tau(T), \\ H_N(0) = P_N H_0, \qquad (\lambda,\theta) \in S \end{cases} \tag{8.28}$$

where

$$\begin{aligned}
G_1(t) &= D_\tau H_N(t) - \partial_t H_N(t), \\
G_2(t) &= J\left(H_N(t), \Psi_N(t)\right) - J\left(H(t), \Psi(t)\right), \\
G_3(t) &= \delta\tau J\left(D_\tau H_N(t), \Psi_N(t)\right), \\
G_4(t) &= \mu\sigma\tau D_\tau H_N(t).
\end{aligned}$$

Next, set $\tilde{H} = \eta_N - H_N$ and $\tilde{\Psi} = \psi_N - \Psi_N$. We get from (8.15) and (8.28) that

$$\begin{cases}
\left(D_\tau \tilde{H}(t) + J\left(\tilde{H}(t) + \delta\tau D_\tau \tilde{H}(t), \Psi_N(t) + \tilde{\Psi}(t)\right)\right. \\
\quad \left. + J\left(H_N(t) + \delta\tau D_\tau H_N(t), \tilde{\Psi}(t)\right), \phi\right) - \mu\left(\Delta\left(\tilde{H}(t) + \sigma\tau D_\tau \tilde{H}(t)\right), \phi\right) \\
\quad = -\left(\sum\limits_{j=1}^{3} G_j(t), \phi\right) - \left(\nabla G_4(t), \nabla\phi\right), \qquad \forall\, \phi \in V_N,\, t \in \bar{R}_\tau(T), \\
-\left(\Delta\tilde{\Psi}(t), \phi\right) = \left(\tilde{H}(t), \phi\right), \qquad \forall\, \phi \in V_N^0,\, t \in \bar{R}_\tau(T), \\
A\left(\tilde{\Psi}(t)\right) = 0, \qquad t \in \bar{R}_\tau(T), \\
\tilde{H}(0) = 0, \qquad (\lambda, \theta) \in S.
\end{cases}$$

By an argument as in the proof of Theorem 8.4, we can deduce an estimate similar to (8.24). But η_N and $\tilde{\eta}$ are replaced by H_N and $\tilde{H}$ respectively, while $\rho(t)$ is replaced by

$$\tilde{\rho}(t) = \tau \sum_{s \in \bar{R}_\tau(t-\tau)} \left(\sum_{j=1}^{3} \|G_j(s)\|^2 + |G_4(s)|_1^2\right).$$

As before, we have

$$\tau \sum_{s \in \bar{R}_\tau(t-\tau)} \|G_1(s)\|^2 \le c\tau^2 \|H\|^2_{H^2(0,t;L^2)}.$$

By Lemma 8.1, Theorem 8.2 and a similar estimate as (8.21), we verify that for any $\gamma > 0$,

$$\begin{aligned}
\|G_2(t)\|^2 &\le c\,\|\Psi_N\|^2_{\gamma+2}\,\|H_N(t) - H(t)\|_1^2 + c\|H(t)\|_1^2\,\|\Psi_N(t) - \Psi(t)\|^2_{\gamma+2} \\
&\le cN^{-2r}\left(\|\Psi(t)\|^2_{\gamma+2}\|H(t)\|^2_{r+1} + \|\Psi(t)\|^2_{r+\gamma+2}\|H(t)\|_1^2\right) \\
&\le cN^{-2r}\left(\|H(t)\|^2_\gamma\|H(t)\|^2_{r+1} + \|H(t)\|^2_{r+\gamma}\|H(t)\|_1^2\right).
\end{aligned}$$

By virtue of Lemma 8.3 and Theorem 8.2, it is easy to see that

$$\begin{aligned}
\tau \sum_{s \in \bar{R}_\tau(t-\tau)} \|G_3(s)\|^2 &\le c\tau^2\,\|\Psi_N\|^2_{C(0,T;H^{\gamma+2})}\,\|H_N\|^2_{H^1(0,T;H^1)} \\
&\le c\tau^2\|\Psi\|^2_{C(0,T;H^{\gamma+2})}\|H\|^2_{H^1(0,T;H^1)} \\
&\le c\tau^2\|H\|^2_{C(0,T;H^\gamma)}\|H\|^2_{H^1(0,T;H^1)}.
\end{aligned}$$

Finally

$$|G_4(t)|_1^2 \le c\tau^2 \|H(t)\|^2_{H^1(t,t+\tau;H^1)}.$$

Therefore $\tilde{\rho} = \mathcal{O}(\tau^2 + N^{-2r})$ as long as the related norms of H exist. The above statements lead to the following conclusion.

Theorem 8.6. *Assume that*

(i) (8.26) or condition (i) of Theorem 8.4 is valid;

(ii) condition (ii) of Theorem 8.4 is fulfilled;

(iii) For $r > 0, H \in C(0,T;H^{r+1}) \cap H^1(0,T;H^1) \cap H^2(0,T;L^2)$.

Then for all $t \in \bar{R}_\tau(T)$,

$$E(\eta_N - H_N, t) \le b_4\left(\tau^2 + N^{-2r}\right),$$

b_4 *being a positive constant depending on* μ *and the norms of* H *in the mentioned spaces.*

The above numerical analysis comes from Guo (1995).

The spectral methods on the spherical surface have also been used for the barotropic vorticity equation in Guo and Zheng (1994), and for a system of nonlinear partial differential equations governing the fluid flows with low Mach number in Guo and Cao (1995). The applications of the pseudospectral methods on the surface can be found in Cao and Guo (1997).

References

Abe, K. and Inoue, O. (1980), Fourier expansion solution of the KdV equation, *J. Comp. Phys.*, **34**, 202–210.

Adams, R. A. (1975), *Sobolev Spaces*, Academic Press, New York.

Alpert, B. K. and Rokhlin, V. (1991), A fast algorithm for the evaluation of Legendre expansions, *SIAM J. Sci. Stat. Comput.*, **12**, 158–179.

Bateman, H. (1953), *Higher Transcendental Functions II*, McGraw Hill Book Company, New York.

Bergh, J. and Löfström, J. (1976), *Interpolation Spaces, An Introduction*, Springer-Verlag, Berlin.

Beringen, S. (1984), Final stages of transition to turbulence in plane channel flow, *J. Fluid Mech.*, **148**, 413–442.

Bernardi, C., Canuto, C. and Maday, Y. (1986), Generalized inf-sup condition for Chebyshev approximations to Navier-Stokes equations, *C. R. Acad. Sci. Paris*, **303**, *Series I*, 971–974.

Bernardi, C., Coppoletta, G. and Maday, Y. (1992a), Some spectral approximations of two-dimensional fourth-order polynomials, *SIAM J. Sci. Comput.*, **15**, 1489–1505.

Bernardi, C., Coppoletta, G. and Maday, Y. (1992b), Some spectral approximations of two-dimensional fourth-order problems, *Math. Comp.*, **59**, 63–76.

Bernardi, C. and Maday, Y. (1989), Properties of some weighted Sobolev spaces and applications to spectral approximation, *SIAM J. Numer. Anal.*, **26**, 769–829.

Bernardi, C. and Maday, Y. (1992a), *Approximations Spectrales de Problèmes aux Limites Elliptiques*, Springer-Verlag, Berlin.

Bernardi, C. and Maday, Y. (1992b), Polynomial interpolation results in Sobolev spaces, *J. Comp. Appl. Math.*, **43**, 53–80.

Bernardi, C. and Maday, Y. (1994), *Basic Results on Spectral Methods*, R94022, Univ. Pierre et Marie Curie, Paris.

Bernardi, C., Maday, Y. and Metivet, B. (1987), Spectral approximation of the periodic-nonperiodic Navier-Stokes equations, *Numer. Math.*, **51**, 655–700.

Boyd, J. P. (1978a), The choice of spectral functions on a sphere for boundary and eigenvalue problems: a comparison of Chebyshev, Fourier and associated Legendre expansion, *Mon. Weather Rev.*, **106**, 1148–1191.

Boyd, J. P. (1978b), Spectral and pseudospectral methods for eigenvalue and separable boundary value problems, *Mon. Weather Rev.*, **106**, 1192–1203.

Boyd, J. P. (1987), Spectral method using rational basis functions on an infinite interval, *J. Comp. Phys.*, **69**, 112–142.

Bramble, J. H. and Pasciak, J. E. (1985), A boundary parametric approximation to linearized scalar potential magnetostatic field problem, *Appl. Numer. Math.*, **1**, 493–514.

Brezzi, F., Rappaz, J. and Raviart, P. A. (1980), Finite dimensional approximation of nonlinear problems, Part I: Branches of nonsingular solutions, *Numer. Math.*, **36**, 1–25.

Butzer, P. L. and Nessel, R. J. (1971), *Fourier Analysis and Approximation*, Birkhäuser, Basel.

Butzer, P. L., Nessel, R. J. and Trebels, W. (1972), Summation processes of Fourier expansion in Banach spaces II, Saturation theorems, *Tōhoku Math. J.*, **24**, 551–569.

Cai, W., Gottlieb, D. and Shu, C. W. (1989), Essentially nonoscillatory spectral Fourier methods for shock wave calculations, *Math. Comp.*, **52**, 389–410.

Cai, W., Gottlieb, D. and Shu, C. W. (1992), On one-side filters for spectral Fourier approximations of discontinuous functions, *SIAM J. Numer. Anal.*, **29**, 905–916.

Cai, W. and Shu, C. W. (1991), Uniform high order spectral methods for one and two dimensional Euler equations, *ICASE Report*, **91–26**.

Cambier, L., Ghazzi, W., Veuillot, J. P. and Viviand, H. (1982), Une approche par domaines pour le Calcul d'ecoulements compressible, in *Methods in Applied Sciences and Engineering*, ed. by Glowinski, R. and Lions, J. L., North-Holland, Amsterdam, 423–446.

Canosa, J. and Gazdag, J. (1977), The Korteweg-de Vries-Burgers' equation, *J. Comp. Phys.*, **23**, 393–403.

Canuto, C. (1988), Spectral methods and a maximum principle, *Math. Comp.*, **51**, 615–630.

Canuto, C., Fujii, H. and Quarteroni, A. (1983), Approximation of symmetric breaking bifurcation for the Rayleigh convection problem, *SIAM J. Numer. Anal.*, **20**, 873–884.

Canuto, C., Hussaini, M. Y., Quarteroni, A. and Zang, T. A. (1988), *Spectral Methods in Fluid Dynamics*, Springer-Verlag, Berlin.

Canuto, C., Maday, Y. and Quarteroni, A. (1982), Analysis for the combined finite element Fourier interpolation, *Numer. Math.*, **39**, 205–220.

Canuto, C., Maday, Y. and Quarteroni, A. (1984), Combined finite element and spectral approximation of the Navier-Stokes equations, *Numer. Math.*, **44**, 201–217.

Canuto, C. and Quarteroni, A. (1982a), Approximation results for orthogonal polynomials in Sobolev spaces, *Math. Comp.*, **38**, 67–86.

Canuto, C. and Quarteroni, A. (1982b), Error estimates for spectral and pseudospectral approximations of hyperbolic equations, *SIAM J. Numer. Anal.*, **19**, 629–642.

Canuto, C. and Quarteroni, A. (1987), On the boundary treatment in spectral methods for hyperbolic systems, *J. Comp. Phys.*, **71**, 100–110.

Cao, W. M. and Guo, B. Y. (1991), A pseudospectral methods for solving Navier-Stokes equations, *J. Comp. Math.*, **9**, 278–289.

Cao, W. M. and Guo, B. Y. (1993a), Spectral-finite element method for solving three-dimensional unsteady Navier-Stokes equations, *Acta Math. Sinica*, **9**, 27–38.

Cao, W. M. and Guo, B. Y. (1993b), Fourier collocation method for solving nonlinear Klein-Gordon equation, *J. Comp. Phys.*, **108**, 296–305.

Cao, W. M. and Guo, B. Y. (1997), A pseudospectral method for solving vorticity equations on spherical surface, *Acta Math. Appl. Sinica,* **13**,176-187.

Carlenzoli, C., Quarteroni, A. and Valli, A. (1991), *Spectral domain decomposition methods for compressible Navier-Stokes equations*, AHPCRC preprint 91-46, Univ. of Minnesota.

Carleson, L. (1966), On convergence and growth of partial sums of Fourier series, *Acta Math.*, **116**, 135–157.

Chan, F. F., Glowinski, R., Periaux, J. and Widlund, O. (1988), *Domain Decomposition Methods for P.D.E.s II*, SIAM, Philadelphia.

Chen, G. Q., Du, Q. and Tadmor, E. (1993), Spectral viscosity approximations to multidimensional scalar conservation laws, *Math. Comp.*, **61**, 629–643.

Cheng, M. T. and Chen, Y. H. (1956), On the approximation of function of several variables by trigonometrical polynomials, *Proc. of Beijing Univ.*, No.4, 412–428.

Chorin (1967), The numerical solution of the Navier-Stokes equations for an incompressible fluid, *Bull. Amer. Math. Soc.*, 928–931.

Christov, C. I. (1982), A complete orthogonal system of functions in $L^2(-\infty,\infty)$ space, *SIAM J. Appl. Math.*, **42**, 1337–1344.

Ciarlet, P. G. (1978), *The Finite Element Method for Elliptic Problems*, North-Holland, Amsterdam.

Cooley, J. W. and Tukey, J. W. (1965), An algorithm for the machine calculation of complex Fourier series, *Math Comp.*, **19**, 297–301.

Cornille, P. (1982), A pseudospectral scheme for the numerical calculation of shocks, *J. Comp. Phys.*, **47**, 146–159.

Courant, R., Friedrichs, K. O. and Lewy, H. (1928), Über die partiellen differezengleichungen der mathematischen physik, *Math. Ann.*, **100**, 32–74.

Courant, R. and Hilbert, D. (1953), *Methods of Mathematical Physics, Vol. 1*, Interscience, New York.

Dennis, S. C. and Quartapelle, L. (1985), Spectral algorithms for vector elliptic equations, *J. Comp. Phys.*, **61**, 319–342.

Don, W. S. and Gottlieb, D. (1994), The Chebyshev-Legendre method: implementing Legendre methods on Chebyshev points, *SIAM J. Numer. Anal.*, **31**, 1519–1534.

Douglas, J. and Dupont, T. (1974), Galerkin approximations for the two-point boundary value problem using continuous piecewise polynomial space, *Numer. Math.*, **22**, 99–109.

Eliasen, E., Machenhauer, B. and Rasmussen, E. (1970), *On a numerical method for integration of the hydrodynamical equations with a spectral representation of the horizontal fields*, Report No. 2, Institute for Teoretisk Meteorologi, Univ. of Copenhagen.

Feit, M. D., Flack, J. A. and Steiger, A. (1982), Solution of the Schrödinger equation by a spectral method, *J. Comp. Phys.*, **47**, 412–433.

Fox, D. J. and Orszag, S. A. (1973a), Pseudo-spectral approximation to two dimensional turbulence, *J. Comp. Phys.*, **11**, 612–619.

Fox, D. J. and Orszag, S. A. (1973b), Inviscid dynamics of two dimensional turbulence, *Phys. Fluids*, **16**, 169–171.

Fronberg, C. B. (1977), A numerical study of two dimensional turbulence, *J. Comp. Phys.*, **25**, 1–31.

Fronberg, C. B. and Whitham, G. B. (1978), A numerical and theoretical study of certain nonlinear wave phenomena, *Philos. Trans. Royal Soc. London*, **289** (1361), 313–404.

Fulton, S. R. and Taylor, G. D. (1984), On the Gottlieb-Turkel time filter for Chebyshev spectral methods, *J. Comp. Phys.*, **37**, 70–92.

Funaro, D. and Gottlieb, D. (1988), A new method of imposing boundary conditions in pseudospectral approximations of hyperbolic equations, *Math. Comp.*, **51**, 599–613.

Funaro, D. and Gottlieb, D. (1991), Convergence results for pseudospectral approximations of hyperbolic systems by a penalty-type boundary treatment, *Math. Comp.*, **57**, 585–596.

Funaro, D. and Kavian, O. (1990), Approximation of some diffusion evolution equations in unbounded domains by Hermite functions, *Math Comp.*, **57**, 597–619.

Gazdag, J. (1976), Time-differencing schemes and transform methods, *J. Comp. Phys.*, **20**, 196–207.

Girault, V. and Raviart, P. A. (1979), *Finite Element Approximation of the Navier-Stokes Equations, Lecture Notes in Mathematics 794*, Springer-Verlag, Berlin.

Gottlieb, D. (1981), The stability of pseudospectral Chebyshev methods, *Math. Comp.*, **36**, 107–118.

Gottlieb, D. (1985), Spectral methods for compressible flow problems, in *Proc. 9th Int. Conf. Numerical Methods in Fluid Dynamics*, ed. by Soubbarameyer, J. P. B., Springer-Verlag, Berlin, 48–61.

Gottlieb, D., Gunzburger, H. and Turkel, E. (1982), On numerical boundary treatment for hyperbolic systems, *SIAM J. Numer. Anal.*, **19**, 671–697.

Gottlieb, D., Hussaini, M. Y. and Orszag, S. A. (1984), Theory and Applications of spectral methods, in *Spectral Methods for Partial Differential Equations*, ed. by Voigt, R. G., Gottlieb, D. and Hussaini, M. Y., SIAM-CBMS, Philadelphia, 1–54.

Gottlieb, D., Lustman, L. and Tadmor, E. (1987a), Stability analysis of spectral methods for hyperbolic initial-boundary value problem, *SIAM J. Numer. Anal.*, **24**, 241–256.

Gottlieb, D., Lustman, L. and Tadmor, E. (1987b), Convergence of spectral methods for hyperbolic initial-boundary value systems, *SIAM J. Numer. Anal.*, **24**, 532–537.

Gottlieb, D. and Orszag, S. A. (1977), *Numerical Analysis of Spectral Methods, Theory and Applications*, SIAM-CBMS, Philadelphia.

Gottlieb, D. and Orszag, S. A. and Turkel, E. (1981), Stability of pseudospectral and finite difference methods for variable coefficient problems, *Math. Comp.*, **37**, 293–305.

Gottlieb, D. and Shu, C. W. (1994), On the Gibbs phenomenon II, Resolution properties of Fourier methods for discontinuous waves, *Comp. Meth. Appl. Mech. Eng.*, **116**, 27–37.

Gottlieb, D. and Shu, C. W. (1995a), On the Gibbs phenomenon IV, Recovering exponential accuracy in a subinterval from a Gegenbauer partial sum of a piecewise analytic function, *Math. Comp.*, **64**, 1081–1095.

Gottlieb, D. and Shu, C. W. (1995b), Recovering the exponential accuracy from collocation point values of piecewise analytic functions, *Numer. Math.*, **71**, 511–526.

Gottlieb, D. and Shu, C. W. (1996), On the Gibbs phenomenon III, Recovering exponential accuracy in a sub-interval from the spectral partial sum of a piecewise analytic function, *SIAM J. Numer. Anal.*, **33**. 280–290.

Gottlieb, D., Shu, C. W., Solomonoff, A. and Vandeven, O. H. (1992), On the Gibbs phenomenon I, Recovering exponential accuracy from the Fourier partial sum of a nonperiodic analytic function, *J. Comp. Appl. Math.*, **43**, 81–98.

Gottlieb, D. and Tadmor, E. (1985), Recovering pointwise values of discontinuous data within spectral accuracy, in *Process and Supercomputing in Computational Fluid Dynamics*, ed. by Murman, E. M. and Abarbanel, S. S., Birkhäuser, Boston, 357–375.

Gottlieb, D. and Turkel, E. (1980), On time discretization for spectral methods, *Stud. Appl. Math.*, **63**, 67–86.

Gresho, P. M. and Sani, R. L. (1987), On pressure boundary conditions for the incompressible Navier-Stokes equations, *Inter. J. Numerical Methods in Fluids*, **7**, 1111–1145.

Griffiths, D. F. (1982), The stability of difference approximation to nonlinear partial differential equation, *Bull. IMA*, **18**, December, 210–215.

Guo, B. Y. (1965), A class of difference schemes of two-dimensional viscous fluid flow, Research Report of SUST, Also see *Acta Math. Sinica*, **17**, 242–258.

Guo, B. Y. (1981a), Error estimations of the spectral method for solving KdV–Burgers' equation, Lecture in Universidad Complutense de Madrid, Also see *Acta Math. Sinica*, **28**, 1–15.

Guo, B. Y. (1981b), Difference method for fluid dynamics–numerical solution of primitive equations, *Scientia Sinica*, **24**, 297–312.

Guo, B. Y. (1982), On stability of discretization, *Scientia Sinica*, **25**, 702–715.

Guo, B. Y. (1985), Spectral method for Navier-Stokes equation, *Scientia Sinica*, **28A**, 1139–1153.

Guo, B. Y. (1987), Spectral-difference method for baroclinic primitive equation and its error estimation, *Scientia Sinica*, **30A**, 696–713.

Guo, B. Y. (1988a), *Difference Methods for Partial Differential Equations*, Science Press, Beijing.

Guo, B. Y. (1988b), A spectral-difference method for solving two-dimensional vorticity equation, *J. Comp. Math.*, **6**, 238–257.

Guo, B. Y. (1989), Weak stability of discretization for operator equation, *Scientia Sinica*, **32A**, 897–907.

Guo, B. Y. (1994), Generalized stability of discretization and its applications to numerical solutions of nonlinear differential equations, *Contemporary Mathematics*, **163**, 33–54.

Guo, B. Y. (1995), A spectral method for the vorticity equation on the surface, *Math. Comp.*, **64**, 1067–1079.

Guo, B. Y. (1996), *The State of Art in Spectral Methods*, City Univ. of Hong Kong, Hong Kong.

Guo, B. Y. and Cao, W. M. (1988), The Fourier pseudo-spectral method with a restrain operator for the RLW equation, *J. Comp. Phys.*, **74**, 110–126.

Guo, B. Y. and Cao, W. M. (1991), Spectral-finite element method for solving two-dimensional vorticity equation, *Acta Math. Appl. Sinica*, **7**, 257–271.

Guo, B. Y. and Cao, W. M. (1992a), A combined spectral-finite element method for solving two-dimensional unsteady Navier-Stokes equations, *J. Comp. Phys.*, **101**, 375–385.

Guo, B. Y. and Cao, W. M. (1992b), Spectral-finite element method for compressible fluid flows, *RAIRO Math. Model. Numer. Anal.*, **26**, 469–491.

Guo, B. Y. and Cao, W. M. (1995), A spectral method for the fluid flow with low Mach number on the spherical surface, *SIAM J. Numer. Anal.*, **32**, 1764–1777.

Guo, B. Y., Cao, W. M. and Tahira, N. (1993), A Fourier spectral scheme for solving nonlinear Klein-Gordon equation, *Numerical Mathematics*, **2**, 38–56.

Guo, B. Y. and He, L. P. (1996), The fully discrete Legendre spectral approximation of two-dimensional unsteady incompressible fluid flow in stream function form, to appear in *SIAM J. Numer. Anal.*.

Guo, B. Y., He, L. P. and Mao, D. K. (1997), On two-dimensional Navier-Stokes equation in stream function form, *JMAA*, **205**, 1-31.

Guo, B. Y., He, S. N. and Ma, H. P. (1995), Chebyshev spectral-finite element method for two-dimensional unsteady Navier-Stokes equation, to appear in *HM-SCMA J.*

Guo, B. Y. and Huang, W. (1993), The spectral-difference method for compressible flow, *J. Comp. Math.*, **11**, 37–49.

Guo, B. Y. and Li, J. (1993), Fourier-Chebyshev pseudospectral method for two-dimensional vorticity equation, *Numer. Math.*, **66**, 329–346.

Guo, B. Y. and Li, J. (1995a), Fourier-Chebyshev pseudospectral methods for two-dimensional Navier-Stokes equations, *RAIRO Math. Model. Numer. Anal.*, **29**, 303–337.

Guo, B. Y. and Li, J. (1995b), Fourier-Chebyshev pseudospectral approximation with mixed filtering for two-dimensional viscous incompressible flow, *J. Appl. Sci.*, **13**, 253–271.

Guo, B. Y. and Li, J. (1996), Fourier-Chebyshev spectral method for two-dimensional Navier-Stokes equations, *SIAM J. Numer. Anal.*, **33**, 1169–1187.

Guo, B. Y., Li, X. and Vazquez, L. (1996), A Legendre spectral method for solving the nonlinear Klein-Gordon equation, *J. Comp. Appl. Math.*, **15**, 19–36.

Guo, B. Y. and Ma, H. P. (1987), Strict error estimation for a spectral method of compressible fluid flow, *CALCOLO*, **24**, 263–282.

Guo, B. Y. and Ma, H. P. (1991), A pseudospectral-finite element method for solving two-dimensional vorticity equation, *SIAM J. Numer. Anal.*, **28**, 113–132.

Guo, B. Y. and Ma, H. P. (1993), Combined finite element and pseudospectral method for the two-dimensional evolutionary Navier-Stokes equations, *SIAM J. Numer. Anal.*, **30**, 1066–1083.

Guo, B. Y., Ma, H. P., Cao, W. M. and Huang, H. (1992), The Fourier-Chebyshev spectral method for solving two-dimensional unsteady vorticity equations, *J. Comp. Phys.*, **101**, 207–217.

Guo, B. Y., Ma, H. P. and Hou, J. Y. (1996), Chebyshev pseudospectral-hybrid finite element method for two-dimensional vorticity equation, *RAIRO Math. Model. Numer. Anal.*. 873-905.

Guo, B. Y. and Manoranjan, V. S. (1985), Spectral method for solving R. L. W. equation, *IMA J. Numer. Anal.*, **5**, 307–318.

Guo, B. Y. and Huang, W. (1993), The spectral-difference method for compressible flow, *J. Comp. Math.*, **11**, 37–49.

Guo, B. Y. and Xiong, Y. S. (1989), A spectral-difference method for two-dimensional viscous flow, *J. Comp. Phys.*, **84**, 259–278.

Guo, B. Y. and Xiong, Y. S. (1994), Fourier pseudospectral-finite difference method for two-dimensional vorticity equation, *Chin. Ann. Math.*, **15B**, 469–488.

Guo, B. Y. and Zheng, J. D. (1992), Pseudospectral-difference method for baroclinic primitive equation and its error estimation, *Science in China*, **35A**, 1–13.

Guo, B. Y. and Zheng, J. D. (1994), A spectral approximation of the barotropic vorticity equation, *J. Comp. Math.*, **12**, 173–184.

Haidvogel, D. B., Robinson, A. R. and Schulman, E. E. (1980), The accuracy, efficiency and stability of three numerical models with application to open ocean problem, *J. Comp. Phys.*, **34**, 1–15.

Haidvogel, D. B. and Zang, T. A. (1979), The accurate solution of Poisson's equation by expansion in Chebyshev polynomials, *J. Comp. Phys.*, **30**, 167–180.

Hald, O. H. (1981), Convergence of Fourier methods for Navier-Stokes equations, *J. Comp. Phys.*, **40**, 305–317.

Haltiner, G. J. and Williams, R. T. (1980), *Numerical Prediction and Dynamical Meteorology*, John Wiley & Sons, New York.

Hardy, G. H., Littlewood, J. E. and Pólya, G. (1952), *Inequalities*, 2nd ed., Cambridge University Press, Cambridge.

Harten, A. (1989), ENO schemes with subcell resolutions, *J. Comp. Phys.*, **83**, 148–184.

Harten, A., Engquist, B., Osher, S. and Chakravarthy, S. (1987), Uniformly high order accurate non-oscillatory schemes III, *J. Comp. Phys.*, **71**, 231–303.

Harvey, J. E. (1981), Fourier integral treatment yielding insight into the control of Gibbs phenomena, *Amer. J. Phys.*, **49**, 747–750.

He, S. N. and Guo, B. Y. (1995), Chebyshev spectral-finite element method for three-dimensional unsteady Navier-Stokes equation, To appear in *Appl. Math. Comp.*.

Hille, E. and Phillips, R. S. (1957), *Functional Analysis and Semi-Groups*, Providence, R.I..

Hussaini, M. Y., Salas, M. D. and Zang, T. A. (1985), Spectral methods for inviscid compressible flows, in *Advances in Computational Transonics*, ed. by Habashi, W. G., Pineridge, Swansea, U. K., 875–912.

Ingham, D. B. (1984), Unsteady separation, *J. Comp. Phys.*, **53**, 90–99.

Ingham, D. B. (1985), Flow past a suddenly heated vertical plate, *Proc. Royal Soc. London*, **A402**, 109–134.

Jarraud, M. and Baede, A. P. M. (1985), The use of spectral techniques in numerical weather prediction, in *Large-Scale Computations in Fluid Mechanics, Lectures in Applied Mathematics*, **22**, 1–41.

Kantorovich, L. V. (1934), On a new method of approximate solution of partial differential equation, *Dokl. Akad. Nauk SSSR*, **2**, 532–536.

Kantorovich, L. V. (1948), Functional analysis and applied mathematics, *YHM*, **3**, 89–185.

Keller, H. B. (1975), Approximation methods for nonlinear problems with application to two-point boundary value problems, *Math. Comp.*, **29**, 464–476.

Kenneth, P. B. (1978), C^m convergence of trigonometric interpolations, *SIAM J. Numer. Anal.*, **15**, 1258–1266.

Kopriva, D. A. (1986), A spectral multidomain method for the solution of hyperbolic systems, *Appl. Numer. Math.*, **2**, 221–241.

Kopriva, D. A. (1987), *Domain decomposition with both spectral and finite difference methods for the accurate computation of flow with shocks*, Tallahassee, FL 32306-4052, Florida State Univ..

Kreiss, H. O. and Oliger, J. (1972), Comparison of accurate methods for the integration of hyperbolic equations, *Tellus*, **24**, 199–215.

Kreiss, H. O. and Oliger, J. (1979), Stability of the Fourier method, *SIAM J. Numer. Anal.*, **16**, 421–433.

Kuo, P. Y. (1983), The convergence of spectral scheme for solving two-dimensional vorticity equation, *J. Comp. Math.*, **1**, 353–362.

Lanczos, C. (1938), Trigonometric interpolation of empirical and analytical functions, *J. Math. Phys.*, **17**, 123–199.

Lax, P. D. (1972), *Hyperbolic Systems of Conservation Laws and the Mathematical Theory of Shock Waves, ABMS-NSF Regional Conference Series in Applied Mathematics*, SIAM, Philadelphia.

Lax, P. D. and Richtmyer, R. D. (1956), Survey of the stability of linear finite difference equations, *CPAM*, **9**, 267–293.

Li, X. and Guo, B. Y. (1997), A Legendre pseudospectral method for solving nonlinear Klein-Gordon equation, *J. Comp. Math.*, 105 - 126.

Lions, J. L. (1967), *Problèmes aux Limites dans Les Equations aux Derivée Patielles*, 2'édition, Les Press de L'université de Montréal.

Lions, J. L. (1969), *Quèlques Méthods de Résolution des Problèmes aux Limites Non Linéaires*, Dunod, Paris.

Lions, J. L. (1970), On the numerical approximation of some equations arising in hydrodynamics, in *Numerical Solution of Field Problems in Continuum Physics, SIAM-AMS Proc. II*, ed. by Birkhoff, G. and Varga, R. S., AMS, Providence, Rhode Island, 11–23.

Lions. J. L. and Magenes, E. (1972), *Nonhomogeneous Boundary Value Problems and Applications, Vol. 1*, Springer-Verlag, Berlin.

Ma, H. P. (1996a), Chebyshev-Legendre spectral viscosity method for nonlinear conservation laws (unpublished).

Ma, H. P. (1996b), Chebyshev-Legendre super spectral viscosity method for nonlinear conservation laws (unpublished).

Ma, H. P. and Guo, B. Y. (1986), The Fourier pseudo-spectral method with a restraint operator for the Korteweg-de Vries equation, *J. Comp. Phys.*, **65**, 120–137.

Ma, H. P. and Guo, B. Y. (1987), The Fourier pseudo-spectral method for solving two-dimensional vorticity equation, *IMA J. Numer. Anal.*, **5**, 47–60.

Ma, H. P. and Guo, B. Y. (1988), The Chebyshev spectral method for Burgers'–like equations, *J. Comp. Math.*, **6**, 48–53.

Ma, H. P. and Guo, B. Y. (1992), Combined finite element and pseudospectral method for the three-dimensional Navier-Stokes equations, *Chin. Ann. Math.*, **13B**, 96–111.

Ma, H. P. and Guo, B. Y. (1994), The Chebyshev spectral methods with a restraint operator for Burgers' equation, *Appl. Math.–JCU*, **9B**, 213–222.

Macaraeg, M. and Streett, C. L. (1988), An analysis of artificial viscosity effects on reacting flows using a spectral multi-domain technique, in *Computational Fluid Dynamics*, ed. by Davis, G. de Vahl and Fletcher, C., North-Holland, Amsterdam, 503–514.

Macaraeg, M., Streett, C. L. and Hussaini, M. Y. (1989), A spectral multi-domain technique applied to high speed chemically reacting flow, in *Domain Decomposition Methods for P.D.E.s*, ed. by Chan, T. F., Glowinski, R., Periaux, J. and Widlund, O., SIAM, Philadelphia, 361–369.

Maday, Y. (1990), Analysis of spectral projectors in one-dimensional domain, *Math. Comp.*, **55**, 537–562.

Maday, Y. (1991), Résultats d'approximation optimaux pour les opérateurs d'interpolation polynomiale, *C. R. Acad. Sci. Paris*, **312**, 705–710.

Maday, Y., Kaber, S. M. O. and Tadmor, E. (1993), Legendre pseudospectral viscosity method for nonlinear conservation laws, *SIAM J. Numer. Anal.*, **30**, 321–342.

Maday, Y. and Métivet, B. (1986), Chebyshev spectral approximation of Navier-Stokes equation in a two-dimensional domain, *RAIRO Math. Model. Numer. Anal.*, **21**, 93–123.

Maday, Y., Pernaud-Thomas, B. and Vandeven, H. (1985), Une réhabilitation des méthodes spectrales de type Laguerre, *La Recherche Aérospatiale*, **6**, 353–379.

Maday, Y. and Quarteroni, A. (1981), Legendre and Chebyshev spectral approximations of Burgers' equation, *Numer. Math.*, **37**, 321–332.

Maday, Y. and Quarteroni, A. (1982a), Approximation of Burgers' equation by pseudo-spectral methods, *RAIRO Numer. Anal.*, **16**, 375–404.

Maday, Y. and Quarteroni, A. (1982b), Spectral and pseudospectral approximations of the Navier-Stokes equations, *SIAM J. Numer. Anal.*, **19**, 761–780.

Maday, Y. and Tadmor, E. (1989), Analysis of the spectral viscosity method for periodic conservation laws, *SIAM J. Numer. Anal.*, **26**, 854–870.

Majda, A., McDonough, J. and Osher, S. (1978), The Fourier method for nonsmooth initial data, *Numer. Math.*, **32**, 1041–1081.

McCrory, R. L. and Orszag, S. A. (1980), Spectral methods for multi-dimensional diffusion problem, *J. Comp. Phys.*, **37**, 93–112.

McKerrell, A., Phillips, C. and Delves, L. M. (1981), Chebyshev expansion methods for the solution of elliptic partial differential equation, *J. Comp. Phys.*, **40**, 444–452.

Mercier, B. (1981), *Analyse Numérique des Méthodes Spectrales*, Note CEA-N-2278, Commissariat a l'Énergie Atomique Centre d'Études de Limeil, 94190, Villeneuve-Saint Georges.

Mercier, B. and Raugel, G. (1982), Resolution d'un problème aux limites dans un ouvert axisymétrique par elements finis en r, z et series de Fourier en θ, *RAIRO Anal. Numer.*, **16**, 405–461.

Milinazo, F. A. and Saffman, P. G. (1985), Finite complitude steady waves in plane viscous shear flows, *J. Fluid Mech.*, **160**, 281–295.

Moin, P. and Kim, J. (1980), On the numerical solution of time-dependent viscous incompressible fluid flows involving solid boundaries, *J. Comp. Phys.*, **35**, 381–392.

Mulholland, S. and Sloan, D. M. (1991), The effect of filtering in the pseudospectral solution of evolutionary partial differential equations, *J. Comp. Phys.*, **96**, 369–390.

Murdock, J. W. (1977), A numerical study of nonlinear effects on the boundary-layer stability, *AIAA J.*, **15**, 1167–1173.

Murdock, J. W. (1986), Three dimensional, numerical study of boundary-layer stability, *AIAA J.*, **86**–0434.

Nessel, R. J. and Wilmes, G. (1976), On Nikolskii-type inequalities for orthogonal expansion, in *Approximation Theory II*, ed. by Lorentz, G. G., Chui, C. K. and Schumaker, L. L., Academic Press, New York, 479–484.

Von Neumann, J. and Goldstine, H. H. (1947), Numerical inverting of matrices of high order, *Bull Amer. Math. Soc.*, **53**, 1021–1099.

Nevai, P. (1979), *Orthogonal Polynomials, Men. Amer. Math. Soc.*, **213**, AMS, Providence, R.I..

Nikolskii, S. M. (1951), Inequalities for entire functions of finite degree and their application to the theory of differentiable functions of several variables, *Dokl. Akad. Nauk. SSSR*, **58**, 244–278.

Orszag, S. A. (1969), Numerical methods for the simulation of turbulence, *Phys. Fluid, Suppl. II*, **12**, 250–257.

Orszag, S. A. (1970), Transform method for calculation of vector sums, Application to the spectral form of the vorticity equation, *J. Atmosph. Sci.*, **27**, 890–895.

Orszag, S. A. (1971a), On the elimination of aliasing in finite-difference schemes by filtering high wave number components, *J. Atmosph. Sci.*, **28**, 1074.

Orszag, S. A. (1971b), Numerical simulations of incompressible flows within simple boundaries: Accuracy, *J. Fluid Mech.*, **49**, 75–112.

Orszag, S. A. (1972), Comparison of pseudospectral and spectral approximations, *Stud. Appl. Math.*, **51**, 253–259.

Orszag, S. A. (1980), Spectral methods for problems in complex geometries, *J. Comp. Phys.*, **37**, 70–92.

Pasciak, J. E. (1980), Spectral and pseudospectral methods for advection equations, *Math. Comp.*, **35**, 1081–1092.

Pasciak, J. E. (1982), Spectral methods for a nonlinear initial value problem involving pseudodifferential operators, *SIAM J. Numer. Anal.*, **19**, 142–154.

Quarteroni, A. (1984), Some results of Bernstein and Jackson type for polynomial approximation in L_p-space, *Jpn. J. Appl. Math.*, **1**, 173–181.

Quarteroni, A. (1987a), Blending Fourier and Chebyshev interpolation, *J. Approx. Theor.*, **51**, 115–126.

Quarteroni, A. (1987b), Domain decomposition techniques using spectral methods, *CALCOLO*, **24**, 141–177.

Quarteroni, A. (1990), Domain decomposition methods for systems of conservation laws, spectral collocation approximations, *SIAM J. Sci. Stat. Comput.*, **11**, 1029–1052.

Quarteroni, A. (1991), An introduction to spectral methods for partial differential equations, in *Advance on Numerical Analysis, Vol. 1*, ed. by Light, W., Clarendon Press, Oxford, 96–146.

Quarteroni, A., Pasquarelli, F. and Valli, A. (1991), Heterogeneous domain decomposition: Principles, algorithms, applications, *AHPCRC preprint 91-48*, Univ. of Minnesota.

Richtmeyer, R. D. and Morton, K. W. (1967), *Finite Difference Methods for Initial-Value Problems*, 2nd edition, Interscience, New York.

Rushid, A., Cao, W. M. and Guo, B. Y. (1994), Three level Fourier spectral approximations for fluid flow with low Mach number, *Appl. Math. Comput.*, **63**, 131–150.

Schwarz, L. (1966), *Théorie des distributions*, Hermann, Paris.

Schwraz, L. (1969), *Nonlinear Functional Analysis*, Gordon and Breach Science Publishers, New York.

Shamel, H. and Elsässer, K. (1976), The application of the spectral method to nonlinear wave propagation, *J. Comp. Phys.*, **22**, 501–516.

Shen, J. (1992), On pressure stabilization method and projection method for unsteady Navier-Stokes equation, in *Advances in Computer Methods for Partial Differential Equation VII*, ed. by Vichnevetsky, R., Knight, D. and Richter, G., 658–662.

Shen, J. (1994), Efficient spectral-Galerkin method I, Direct solvers of second and fourth order equations using Legendre polynomials, *SIAM J. Sci. Comput.*, **15**, 1489–1505.

Shen, J. (1995), Efficient spectral-Galerkin method II, Direct solvers of second and fourth equations by Chebyshev polynomials, *SIAM J. Sci. Comput.*, **16**, 74–87.

Silberman, I. (1954), Planetary waves in the atmosphere, *J. Meterol.*, **11**, 27–34.

Slater, J. C. (1934), Electronic energy bands in metal, *Phys. Rev.*, **45**, 794–801.

Smoller, J. (1983), *Shock Waves and Reaction Diffusion Equations*, Springer-Verlag, Berlin.

Stetter, H. J. (1966), Stability of nonlinear discretization algorithms, in *Numerical Solutions of Partial Differential Equations*, ed. by Bramble, J., Academic Press, New York, 111–123.

Stetter, H. J. (1973), *Analysis of Discretization Method for Ordinary Differential Equations*, Springer-Verlag, Berlin.

Strang, G. (1965), Necessary and insufficient conditions for well-posed Cauchy problems, *J. of Diff. Equations*, **2**, 107–114.

Streett, C. L., Zang, T. A. and Hussaini, M. Y. (1985), Spectral multigrid methods with application to transonic potential flow, *J. Comp. Phys.*, **57**, 43–76.

Szabados, J. and Vèrtesi, P. (1992), A survey on mean convergence of interpolatory processes, *J. Comp. App. Math.*, **43**, 3–18.

Tadmor, E. (1989), Convergence of spectral viscosity methods for nonlinear conservation laws, *SIAM J. Numer. Anal.*, **26**, 30–44.

Tadmor, E. (1990), Shock capturing by the spectral viscosity method, *Comp. Methods Appl. Mech. Engrg.*, **80**, 197–208.

Tadmor, E. (1991), Local error estimates for discontinuous solutions of nonlinear hyperbolic equations, *SIAM J. Numer. Anal.*, **28**, 891–906.

Tadmor, E. (1993a), Total variation and error estimates for spectral viscosity approximation, *Math. Comp.*, **60**, 245–246.

Tadmor, E. (1993b), Super viscosity and spectral approximations of nonlinear conservation laws, in *Numerical Methods for Fluid Dynamics IV*, ed. by Baines, M. J. and Morton, K. W., Clarendo Press, Oxford, 69–82.

Tartar, L. (1975), Compensated compactness and applications to partial differential equations, in *Research Notes in Mathematics 39. Nonlinear Analysis and Mechanics, Heriot-Watt Symposium*, ed. by Knopps, J., Vol. 4, 136–211.

Timan, A. F. (1963), *Theory of Approximation of Functions of a Real Variable*, Pergamon, Oxford.

Vandeven, H. (1991), Family of spectral filters for discontinuous problems, *J. Sci. Comput.*, **6**, 159–192.

Weideman, J. A. C. (1992), The eigenvalues of Hermite and rational spectral differential matrices, *Numer. Math.*, **61**, 409–432.

Weideman, J. A. C. and Trefethen, L. N. (1988), The eigenvalues of second-order spectral differentiation matrices, *SIAM J. Numer. Anal.*, **25**, 1279–1298.

Wengle, H. and Seifeld, J. H. (1978), Pseudospectral solution of atmospheric diffusion problems, *J. Comp. Phys.*, **26**, 87–106.

Woodward, P. and Collela, P. (1984), The numerical simulation of two-dimensional fluid flows with strong shock, *J. Comp. Phys.*, **54**, 115–117.

Zang, T. A. and Hussaini, M. Y. (1986), On spectral multigrid methods for the time-dependent Navier-Stokes equations, *Appl. Math. Comput.*, **19**, 359–373.

Zang, T. A., Wong, Y. S. and Hussaini, M. Y. (1982), Spectral multigrid methods for elliptic equations, *J. Comp. Phys.*, **48**, 485–501.

Zang, T. A., Wong, Y. S. and Hussaini, M. Y. (1984), Spectral multigrid methods for elliptic equations II, *J. Comp. Phys.*, **54**, 489–507.

Zen, Q. C. (1979), *Physical-Mathematical Basis of Numerical Prediction, Vol. 1*, Science Press, Beijing.

Tadmor E. (1989): Super viscosity and spectral approximations of nonlinear con[illegible] laws, in Numerical Methods for Fluid Dynamics IV, ed. by Baines, M. J. and Morton, K. W., Clarendon Press, Oxford, 69–82.

[illegible] L. [illegible]: Compensated compactness and applications to partial differential equations, in Research Notes in Mathematics 39, Nonlinear Analysis and Mechanics, Heriot-Watt Symposium, ed. by Knops, R. J., Vol. 4, 136–211.

[illegible]